GSM and UMTS

GSM and UMTS

The Creation of Global Mobile Communication

Edited by

Friedhelm Hillebrand
Consulting Engineer, Germany

With contributions from 37 key players involved in the work for GSM and UMTS

JOHN WILEY & SONS, LTD

Other Wiley Editorial Offices

John Wiley & Sons, Inc., 605 Third Avenue,
New York, NY 10158-0012, USA

WILEY-VCH Verlag GmbH
Pappelallee 3, D-69469 Weinheim, Germany

John Wiley & Sons Australia Ltd, 33 Park Road, Milton,
Queensland 4064, Australia

John Wiley & Sons (Canada) Ltd, 22 Worcester Road
Rexdale, Ontario, M9W 1L1, Canada

John Wiley & Sons (Asia) Pte Ltd, 2 Clementi Loop #02-01,
Jin Xing Distripark, Singapore 129809

British Library Cataloguing in Publication Data

A catalogue record for this book is available from the British Library

ISBN 0470 84322 5

Typeset in Times by Deerpark Publishing Services Ltd, Shannon, Ireland.
Printed and bound in Great Britain by T. J. International Ltd, Padstow, Cornwall.

This book is printed on acid-free paper responsibly manufactured from sustainable forestry, in which at
least two trees are planted for each one used for paper production.

Contents

Chapter 1: GSM's Achievements

Section 1: Introduction

Friedhelm Hillebrand

1.1.1 Introduction to the Content of the Book

This book describes how global mobile communication was made. It is written for those who want or need to know how this was achieved e.g.:

- Young professionals who want to build their career on GSM and UMTS and need to understand the basics
- Strategic and technical planners who want to drive the future GSM and UMTS development
- Strategists who plan to repeat GSM's success in the fourth generation
- Academics, who want to understand and analyse the development of GSM and UMTS;
- Activists in other large scale international communication projects who want to use experiences gained

But the book is also written for those about two thousand colleagues who participated in the work and want to have a record of the events or to have a more comprehensive image, of what happened in the different branches of the very big network of groups.

GSM is the system which started in Europe and was accepted by the world. It provides global mobile communication to:

- anybody: 500 million users from professionals to children in May 2001
- anywhere: 168 countries in all continents in May 2001
- any media: voice, messaging, data, multimedia

UMTS is built on the GSM footprint and plans to repeat the GSM success.

This book is focused on the pre-competitive sphere, the big co-operative effort, which enabled the huge market success world-wide. The clarification of the strategies, the strategic decisions on the broad avenues in service and system design and commercial concepts are described. This book provides an insight into the process of how this was achieved and when. In selected cases it shows the complexity of the process and the antagonism of the interests of the different parties and the consensus building process in detail. The output of the process is well documented in the technical specifications produced by the different groups. They can be retrieved from the Internet.[1]

---

[1] www.ETSI.org, www.3GPP.org, www.gsmworld.org or www.umtsforum.org. The GSM phase 1 standard can be found on the attached CD ROM in folder A3, since it is not available from the Internet.

This book describes the building of the will and momentum to create a Pan-European system to end the market segmentation and barriers to growth. Principles were agreed. Advanced services' requirements including international roaming were agreed. To fulfil these requirements an advanced new digital system was developed. The system provided for a competition of several operators in a country. Advanced low cost terminals were achieved by large markets and manufacturers competition. Advanced low cost infrastructure was enabled by large markets, multivendor concepts and manufacturers competition. All major decisions were made in time, even the most difficult ones. A far-reaching system evolution – even leading from second to third generation – was implemented. Manufacturers and operators promoted GSM in Europe and beyond. The world was invited to become a partner with equal rights in this process.

The book covers intensively the two phases which lead to the long-term strategic orientation of GSM and UMTS. There were protracted and deep controversial debates, which lead to a consensus:

- The debate about the concepts and the basic parameters of the GSM standard from the end of 1986 to mid-1987 showed that the Europeans could agree on one solution and meant business with GSM. It lead to the creation of a new more open organisation by moving all work from CEPT GSM to ETSI GSM opening the doors for manufacturers to participate with equal rights and the GSM MoU Group to participate as the operators' club.
- The debate about the strategy, the concepts and the basic parameters of UMTS in 1996–1997 lead to a re-orientation of the UMTS concept and an agreement on its cornerstones within ETSI and with key players in North America and in Asia. This required also a new more open and more efficient organisation of the work in 3GPP, which allowed access with equal rights to non-European players.

In both cases a stable base and framework for the following phase of more detailed work was achieved.

The much more competitive situation in the market created by the licensing of several operators in a country did in principle not deteriorate the consensus building process, since the new players understood very quickly that a constructive co-operation in the pre-competitive sphere was the prerequisite of the success. These new players brought often more demanding requirements. This was essential for the vivid and fertile system evolution.

The founding documents of the GSM/UMTS system are the GSM/UMTS Technical Specifications and Standards and the Permanent Reference Documents (PRDs). The Technical Specifications contain the basic technical definitions: services, system architecture, selected interfaces and operation and maintenance functions, and test specifications. Some of these, which are needed for regulatory purposes, are converted into formal Standards. The Technical Specifications and Standards were elaborated by groups who varied over time: CEPT GSM, ETSI GSM and SMG, ANSI T1P1 and 3GPP. The PRDs cover commercial and operational aspects, e.g. service and commercial requirements, test specifications for roaming, security algorithms, protocols for the interchange of charging data for roamers. They were elaborated by working groups in the GSM MoU Group, later called GSM MoU Association and now GSM Association.

The book provides in the rest of Chapter 1 key milestones and success statistics. It describes in Chapters 2–9 the GSM phases and the evolution towards UMTS built on GSM. Chapters 10–20 provide more details on technical aspects and working methods.

Chapter 21 deals with operators co-operation and the elaboration of the PRDs. Chapter 22 describes the world-wide acceptance of GSM and UMTS. Chapter 23 tries to explain from our point of view, which factors enabled this success which surpassed all expectations. A CD-ROM is attached to the book, which contains all reference documents mentioned in the footnotes of the different contributions.

The success of GSM and UMTS was created by the working together of a very large number of people in a network. The catalyst of this process was the co-operation of a smaller number of people in the pre-competitive sphere using different forums. All these colleagues created the technical system specifications for GSM and UMTS and the technical and commercial documents needed for the marketing and operation of GSM and UMTS with special emphasis on international roaming. All these people worked in a network. They shared visions and strategy. It was for all of them a privilege to have the opportunity to contribute to this "inner circle".

The book is structured into contributions, which are written as named contributions by key players who played a long-term key role in the development of GSM and UMTS. The views expressed in the different contributions are those of the authors and do not necessarily reflect the views of their respective affiliation entities. There has been a lot of dialogue between all authors during the writing of the book. However, it was the intention to provide different views from different perspectives to the reader. We did not try to iron out all differences of opinion. The book shows how different personalities could work together like a big orchestra, create and play a great symphony. In selected cases the reader will find cross references by the editor in footnotes highlighting to key differences. It is a major achievement of this book that so many people – despite their loaded agenda took the time to report about their experiences and views.

It was a pleasure for me to act as editor of this book and work together with so many excellent colleagues to provide this record of events and our explanations. We are open to dialogue with our readers. Our CVs and e-mail addresses are provided in the attached CD ROM File G.

1.1.2 Practical Advice on how to use the Book

The **content list** is structured into chapters and sections containing the individual contributions written by named authors. A short CV and e-mail address of each author is provided in the attached CD ROM file G. All footnotes in a contribution are numbered locally.

The **decision making plenary meetings** changed over time from GSM (Groupe Spécial Mobil) to SMG (Special Mobile Group) to 3GPP TSG SA (Technical Specification Group Services and System Aspects). The Plenary meetings are numbered in sequence: GSM#1 to 32, SMG#1 to 32 and SA#1...

Similarly the **subgroups reporting to the plenary** changed their names. The group responsible for services was WP1 (Working Party), then GSM1, then SMG1 and finally TSG SA WG SA1. Equal developments had the radio groups using the number 2 and the network aspects group using the number 3. The data group started as IDEG and became GSM4 and SMG4.

Reference documents, which are relevant and often difficult to retrieve are provided on the attached CD-ROM. Reference documents are mentioned in footnotes in the sections. The vast majority are temporary documents of the different standardisation groups. Their number contains a serial number and a year number. Often the format is SSS/YY (e.g. 123/87 = temporary document 123 of 1987). This was changed in later years to YY-SSS. The CD-ROM uses file names of uniform format with the year followed by the serial number for an easy automatic sorting. The CD-ROM provides the quoted reference documents and all Plenary Meeting Reports of the technical standardisation groups: CEPT, ETSI GSM and SMG and 3GPP. In addition key documents of other areas are provided e.g. the GSM Memorandum of Underatanding and the UMTS Task Force report. Finally a folder F contains funny things and a video and some photographs. More details are given in the contents list of the CD-ROM in the file "Introduction".

Several overview lists and descriptions are provided:

- Milestones: Chapter 1, Section 2
- Plenary Meeting Lists in Annex 1
- Technical groups and their evolution in Annex 2
- Chairpersons lists in Annex 3
- Key abbreviations in Annex 4

The current version of the complete **GSM and UMTS Specifications** can be found at the following website: www.3gpp.org. The GSM Phase 1 specifications of 1990/1 can be found on the attached CD-ROM in folder A3.

Chapter 1: GSM's Achievements

Section 2: GSM and UMTS Milestones

Friedhelm Hillebrand

1982	CEPT allocates 900 MHz spectrum for the use by a Pan-European mobile communication system, forms the "Groupe Spécial Mobile" (GSM) and recommends the reservation of frequencies in the 900 MHz band for the future Pan-European cellular system.
1982 (December)	First meeting of the Group Spécial Mobile (GSM) in Stockholm.
1984 (August)	Decision of France and Germany to terminate the planned common 900 MHz analogue system development and to concentrate on a standardised Pan-European digital system.
1986 (June)	GSM Permanent Nucleus established in Paris.
1986	Trials of different digital radio transmission schemes and different speech codecs in several countries, comparative evaluation by CEPT GSM in Paris.
1987 (February)	**CEPT GSM#13 meeting in Madeira: decision on the basic parameters of the GSM system.**
1987 (May)	Finalisation of the decision on the basic parameters of the GSM system.
1987 (June)	The European council agrees to the issue of a Directive reserving 900 MHz frequency blocks.
1987 (7 September)	**GSM Memorandum of Understanding, an agreement to support the development of GSM and to implement it in 1991, signed in Copenhagen by 14 operators from 13 European countries.**
1988 (I Quarter)	Completion of first set of detailed GSM specifications for infrastructure tendering purposes.
1988 (29 February)	Simultaneous issue of invitation to tender for networks by ten GSM network operators.
1988 Autumn	**Ten GSM infrastructure contracts signed by ten network operators.**
1988 (October)	Public presentation of the first set of GSM specifications at a conference in Hagen (Germany), which attracts 600 participants from Europe, USA and Japan and where copies of the specifications could be bought.
1989	Standardisation work transferred from CEPT to ETSI. CEPT Groupe Spécial Mobile becomes ETSI Technical Committee GSM.
1990	**GSM Phase 1 specifications "frozen" in ETSI Technical Committee GSM.**
1991	GSM MoU Permanent Secretariat established in Dublin.
1991 (October)	Pilot GSM networks demonstrated at ITU's Telecom '91 in Geneva.
1991 (October)	ETSI Technical Committee GSM put in charge of UMTS specification activities in addition to the GSM work and renamed "SMG" (= Special Mobile Group).
1991-1992	Search for a system name leads to "Global System for Mobile Communications".
1992	**First commercial GSM networks come into service.**
1992 (June)	First true hand-portable terminals become available.
1992 (17 June)	First international roaming agreement signed between the GSM networks of Telecom Finland and the UK's Vodafone.
1993	1 million GSM users.

1993	**Australian operators are first non-European operators who decide to implement GSM and to sign the MoU.**
1993	**ETSI Technical Committee SMG agrees objectives and methodology for an open evolution of GSM beyond phase 2, to be implemented as phase 2 + .**
1993 (September)	The world's first DCS 1800 (now GSM 1800) personal communication network opened in the UK by Mercury One-2-One (now One 2 One).
1994	Data capabilities launched in GSM networks.
1995 (June)	GSM MoU Group becomes a legal body, registered as a GSM MoU Association in Switzerland.
1995 (Autumn)	**10 million GSM users in 100 GSM networks on air in (60) countries world-wide**
1995 (October)	**GSM Phase 2 standardisation frozen in ETSI Technical Committee SMG.**
1995 (November)	**The first North American PCS 1900 (now GSM 1900) network opened by American Personal Communications in Washington, DC.**
1995	Fax, data and SMS services started, video over GSM demonstrated.
1996 (March)	**UMTS Task Force Report on a UMTS strategy for Europe completed.**
1996 (Summer)	**PCS-1900 (now GSM 1900) service, provided to the US Republican National Convention which was held in San Diego, CA.**
1996 (Spring)	Creation of the UMTS Forum as a world-wide body, dealing with market, regulation and spectrum aspects of UMTS.
1996 (August)	GSM MoU Association signs co-operation agreement with ETSI.
1996	DCS 1800 renamed GSM 1800 and PCS 1900 renamed GSM1900.
1997 (February)	GSM release 96, the first release of phase 2 + , completed by ETSI Technical Committee SMG.
1997 (February)	Consensus on UMTS strategy achieved by ETSI Technical Committee SMG.
1997 (Autumn)	100 countries on air (70 million users in 200 networks).
1997 (End)	GSM release 97 completed by ETSI Technical Committee SMG.
1998 (I Quarter)	**Decision on the basic concepts of the UMTS standard including services, radio and network aspects.**
1998 (Mid)	100 million GSM users world-wide.
1998 (End)	GSM release 98 completed by ETSI Technical Committee SMG.
1998 (December)	**Creation of the Third Generation Partnership Project (3GPP), transfer of the UMTS standardisation work.**
2000 (March)	**GSM/UMTS release 99, the basis for GSM evolution and UMTS opening in 2002, completed by 3GPP and ETSI Technical Committee SMG and ANSI T1P1.**
2000 (June)	Transfer of the remaining GSM specification work to 3GPP, closing of ETSI Technical Committee SMG, creation of a new Technical Committee MSG (= Mobile Standards Group) responsible for European regulatory standards for terminals.
2001 (May)	**500 million GSM users world-wide.**

Chapter 1: GSM's Achievements

Section 3: GSM's Success Measured in Numbers

Friedhelm Hillebrand

This section presents figures achieved in the years from 1992 to the end of 2000. It looks at the values for GSM in total, i.e. GSM 900, 1800 and 1900.

1.3.1 GSM User Numbers World-wide

Figure 1.3.1 shows the explosive growth from 0.25 million in 1992 the first year of commercial operation to 450 million at the end of 2000. The never expected number of 500 million was passed in May 2001. A duplication to 1 billion users can be expected in 2005.

million

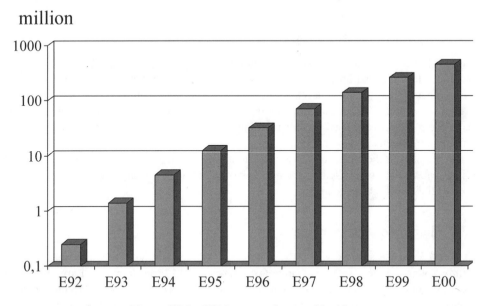

Figure 1.3.1   GSM user numbers world-wide

1.3.2 GSM Networks and Countries on Air (Figure 1.3.2)

From the initial base of 13 networks in seven European countries in 1992 these figures grew to 392 networks in 147 countries on all continents. International roaming connects all these countries. This means that GSM is present in more countries than nearly all the famous mass market brands like Coca Cola, McDonalds.

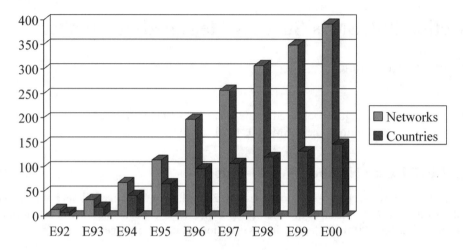

Figure 1.3.2   GSM networks and countries on air

1.3.3 International Roaming

The GSM user enjoys a truly global infrastructure.[1] All GSM operators have international roaming. The advanced operators have 100 or more roaming relations working. Every new operator has typically after 6 months of operation at least ten roaming relations in operation. In total more than 20 000 roaming agreements were concluded in Autumn 1999. There were 40 operators, who have roaming with 70 or more countries, based on 5600 roaming agreements.

The roaming traffic exceeded the expectations. In August 1999 there were more than 400 million calls/month originated by roamers in visited networks. Many networks have 10% or more of their traffic from roamers.

1.3.4 Short Message Service

The GSM Short Message Service is the first mobile data service that reached the mass market. Mainly young people love to exchange text messages and little drawings. Large European operators transmitted between 500 and 1000 million short messages in December 2000. World-wide 50 billion short messages were sent in the first quarter of 2001 after 12 billion in the first quarter of 2000 (source: GSMA)

---

[1] Japan and Korea have not admitted GSM. Japan offers interstandard roaming GSM/PDC.

1.3.5 Competition with other Standards

The competition on the world market in the first half of the 1990s was mainly between GSM and analogue systems since the other digital systems were either not available outside their home market or not yet ready. In most developed market the digital systems dominated clearly at the end of 2000. In Europe GSM serves 98% of the mobile users.

The competition between digital mobile systems took place mainly between four systems (Table 1.3.1).

The competition lead to a clear decision in the world market. GSM served 69% of all digital mobile users at the end of 2000 (Figure 1.3.3).

Table 1.3.1   Four standards were ready for the world market competition in the early 1990s

System	Radio	Network	Standard
GSM	Advanced TDMA	ISDN MAP	1991
PDC	Basic TDMA	ISDN MAP	1992
ANSI 54/136 TDMA	Basic TDMA	ANSI 41 evolution	1991
ANSI 95 CDMA	Narrow band CDMA	ANSI 41 evolution	1991

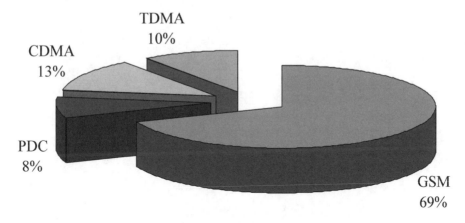

Figure 1.3.3   Competition in the world market

1.3.6 Forecasts and Reality in Europe

Compared to today's reality in Europe, the mobile and wireless communications evolution perspective of the world, at the beginning of 1985, looks retrospectively rather conservative. While the promise of an accelerated development of mobile communications was sensed as a likely possibility, due notably to the anticipated success of GSM, the most optimistic scenarios for market deployment called for a few million subscribers at the turn of the century.

A famous study by a recognised consulting company[2] forecasted at the end of 1990 a total number of 22.5...24.5 million mobile users in the year 2000 in Europe (European Union with

15 member states and Norway and Sweden). They expected that 18.5...21.3 million of them would use GSM.

In the mid 90s, there were about 13 million users of mobile communications in Europe and estimates suggested that the market potential would be of the order of 50 million at the turn of the century and 100 million by 2005.

These forecasts are to be compared with the 245 million users (only 4 million of them still on analogue systems, the rest on GSM) reached in Europe at the end of 2000.

Acknowledgements

All user figures were provided by EMC World Cellular Database. The figures on networks and countries on air were provided by the GSM Association.

---

[2] "A study on the analysis of the introduction of GSM in the European Community" by PA Consulting, September 1990, commissioned by the Commission of the European Communities

Chapter 2: The Agreement on the Concepts and the Basic Parameters of the GSM Standard (mid-1982 to mid-1987)

Section 1: The Market Fragmentation in Europe and the CEPT Initiatives in 1982

Thomas Haug[1]

2.1.1 The Situation in the Early 1980s: A Spectrum Allocation Opens the Possibility to Overcome the European Patchwork of Incompatible Systems

At the World Administrative Radio Conference in 1979 (WARC '79), a decision was taken to set aside a block of radio spectrum in the 900 MHz range for use in land mobile communication systems in Zone 1, which in the terminology of the Radio Regulations means Europe. Beyond this, little was said about how the spectrum should be used, such as the allocation to public systems versus private ones.

The European telecommunications market was for a long time badly fragmented. In an attempt to improve the situation, the organisation of 26 PTT Administrations of Western Europe and a few other countries (Conférence Européenne des Postes et Télécommunications, abbreviated (CEPT)) was since 1959 actively engaged in standardisation of telecommunications, but the progress was often hampered by differences in policy in the member countries. It is, however, of interest to note that contrary to what is often assumed by observers outside the PTTs, CEPT was not at all a purely regulatory body, but dealt extensively with technical issues in many fields and had for that purpose set up a large number of working groups. (In 1989, the technical specification work was transferred to the then newly created European Telecommunication Specification Institute (ETSI).)

Around 1980 the European situation in mobile communications was that a number of mutually incompatible mobile systems were in operation or in preparation. The lack of

---

[1] The views expressed in this sectiom are those of the author and do not necessarily reflect the views of his affiliation entity.

compatibility led to other problems, such as small market segments, small number of users in most systems, high terminal cost and high subscription fees. Consequently, mobile communications was seen as "the rich man's toy" in many countries. The idea of one standardised system, capable of being used throughout a large part of Europe, such as the CEPT member countries, had never really caught on among the authorities. Unfortunately, the entire telecommunications field had often been used as an instrument for protecting the national manufacturing and operating companies, a fact which had led to a situation where each country more or less had its own telecom standard, incompatible with the standards in other countries. The effect of this policy was no exception from the general rule that implementation of protectionist policies at the expense of standardisation leads to a loss of the economy-of-scale benefits, a loss borne by but not always directly observable by the user. In the field of mobile communications, however, (as opposed to standards in the fixed network), international standardisation would in addition to the economy-of-scale benefits offer a benefit which would be directly observable by the user, i.e. the ability to use one's own equipment when moving around in foreign networks. This feature was not implemented in any network except the Nordic NMT system, opened in 1981, which allowed users in Denmark, Finland, Iceland, Norway and Sweden (later expanded to the Netherlands and Switzerland) to use their terminals in the countries where the NMT system was installed, and in the German Network B, which was built in the Benelux countries and Austria. In general, however, the political and regulatory difficulties concerning use of radio equipment in foreign countries were very much against creating a Pan-European system which could be used when travelling abroad.

2.1.2 Vienna Meeting of CEPT Telecommunications Commission (COM-T) in 1982

The NMT network was only in the start-up phase when the Netherlands presented a proposal[2] at the meeting of the CEPT Telecommunications Commission in Vienna in June 1982. The proposal pointed out that unless a concerted action to work out plans for a common European system was started very soon, there would be a serious risk that the 900 MHz band would be taken into use for various incompatible systems of many kinds in Europe. The last chance to build a Pan-European system in the 20th Century would then be lost, since there would be no sufficiently wide spectrum available below 1 GHZ, i.e. in a part of the radio spectrum which could be considered suitable for mobile communications, given the state of the technology in 1982.

The Nordic Administrations had recently put the 450 MHz NMT system into operation and there were already indications that a large number of users would be interested in the feature of international roaming, a new feature at that time. In addition, a common standard would obviously result in considerable economic gains. To build a new European network along these lines had been discussed by the Nordic Administrations, which therefore strongly supported the Dutch point of view and proposed[3] that the total responsibility for progress and co-ordination of the specification task for a new Pan-European system should be left to a group of experts.

The CEPT Telecommunications Commission accepted this proposal and decided to start a new study question, entitled "Harmonisation of the technical and operational characteristics

[2] GSM Doc 3/82 and GSM Doc 5/82.
[3] GSM Doc 4/82.

of a public mobile communications system in the 900 MHz band". Rather than setting up a new committee, reporting directly to the Com-T according to the proposal, however, it was decided to entrust the Committee for Co-ordination of Harmonisation (CCH) with the task of setting up a Special Group for the purpose of studying the question[4] and the Swedish offer of making Thomas Haug available for the chairmanship of the group was accepted. (I was a bit surprised when I heard this, since nobody had asked me in advance.) In line with the CEPT naming rules, the new group was given the name Groupe Spécial Mobile, GSM for short.

2.1.3 The Com-T Decision and its Consequences

The decision taken by the CEPT Telecommunication Commission was very vague, and in effect left it to the new group to propose its own terms of reference. Representatives of the Netherlands and the Nordic Administrations therefore met during the summer of 1982 in order to write a proposal for an Action Plan, which was subsequently presented to CCH for approval in November of that year. That document[5] was approved and was used as the basis for the work of GSM for a long time, with a mainly editorial updating of Section 3 in 1985[6].

The decision only mentioned "harmonisation", which indicates that the compatibility aspect was the dominating factor behind the decision, and there is considerable doubt that the delegates at the CEPT Telecommunications Commission meeting believed that free circulation of radio users across international borders could be achieved within the foreseeable future, given the formidable obstacles that existed.

It was clear, anyway, that many more aspects than the mere harmonisation had to be given great attention if the goal of creating a Pan-European service, acceptable to the broad European public, was to be reached. When the above mentioned proposal for an Action Plan was presented by the newly appointed GSM Chairman to CCH in November 1982, it was therefore stressed that the goal envisaged in the decision taken by the CEPT Telecommunications Commission in Vienna, according to which the work should be completed by the end of 1996, could not be met if major new developments were to be taken into consideration. A "harmonisation" of mobile communication systems would mainly consist of specifying a system according to the techniques already established, in other words an analogue system. It was therefore proposed to the CCH that by the end of 1986, only an outline specification comprising the basic system parameters for the various parts of the system and their interfaces should be finalised. Furthermore, it was proposed that at the same time, technical specifications should be established for the system as a whole, to the point of guaranteeing compatibility. They should comprise the main building blocks of the system, which then were understood to be the switching centre, the base station, the mobile station and the man–machine interface. The proposal was accepted by the CCH and subsequently by the CEPT Telecommunications Commission.

In retrospect, I think that it was a very wise decision which was taken when the aim was changed to present just an outline specification, not a final, detailed one, by the end of 1986. Had that not been done, we would probably have been stuck with an analogue system today, based on somewhat modified versions of the technical solutions that existed in the early

---

[4] GSM Doc 1/82.
[5] Doc T/CCH (82) 21 R, later renamed GSM Doc 2/82.
[6] GSM Doc 73/85.

1980s. That system would have been unable to fulfil many of the requirements now placed on the GSM system.

In the CCH meeting in November 1982, there was a fairly long discussion on whether GSM should also be mandated to study the possibility of an interim system which would meet the needs until the new GSM system could be put into operation. Such ideas had been presented at the CEPT Telecommunications Commission meeting and several countries felt a need for such a system. In particular Dr Klaus Spindler of the Deutsche Bundespost warned against spending time and resources on an interim system, and in the end CCH decided that the resources available to GSM should be concentrated on the new system since work on an interim system would be a waste of time. Seeing afterwards how much work there was to be done on the new system, this was undoubtedly the right decision.

2.1.4 Basic Requirements for the New System

The essential contents of the Basic Requirements in the Study Plan of 1982 are given below.

- The system must be able to operate in the entire frequency bands 890–915 and 935–960 MHz.
- Co-existence with the already existing systems in the 900 MHz band must be guaranteed.
- The GSM mobile stations must be able to operate in all participating countries, preferably all CEPT countries.
- Services other than speech will be required in the new system. Since there is uncertainty about those services, however, a modular system structure allowing for a maximum of flexibility will be required. In order to reach this goal, the same philosophy as for ISDN and OSI should be applied, and standards for protocols, etc. should as far as practicable seek to obtain compatibility with such developments.
- High spectrum efficiency and state-of-the-art subscriber facilities must be achieved.
- The same facilities as those offered in the public switched telephone and data networks should be available in the mobile system.
- An identification plan, compatible with the numbering plan and routing possibilities in the PSTN and PSDN, must be worked out.
- The system must be capable of providing for handheld mobile stations.
- The demand for voice security (encryption) must be taken into account.
- No significant modification of the fixed national telephone networks must be necessary.
- The system must allow the participating countries to maintain their existing charging systems.
- An internationally standardised signalling system must be used for interconnection of the switching centres.

The influence of the NMT principles from 1971 in this set of requirements is obvious to anyone familiar with that system.

Furthermore, an action plan was prepared, listing the various other bodies in and outside CEPT with whom the GSM should stay in contact in order to avoid repeating work that was already being done.

Chapter 2: The Agreement on the Concepts and the Basic Parameters of the GSM Standard (mid-1982 to mid-1987)

Section 2: The GSM Standardisation Work 1982–1987

Thomas Haug[1]

2.2.1 The Start of the Work

The first meeting of the new Groupe Spécial Mobile (GSM) was held in Stockholm in December 1982.[2] The interest in this new activity was great, as can be seen from the fact that no less than 31 delegates from 11 CEPT Administrations participated. The starting points for the various administrations were very different. Some countries had First Generation systems in operation while other countries had none, and most of those systems were different from each other in one way or another as was mentioned earlier. A certain amount of operational experience was available from those systems, but it was agreed that much of it would not be applicable.

The meeting discussed at length the spectrum situation in the CEPT countries, and there was a great concern that it might prove impossible to avoid using the 900 MHz band for other purposes for such a long time as envisaged in the work plan for GSM. The pressure for spectrum was just too great, and there was a clear risk that allocating parts of the spectrum to other systems might become a serious threat to the future Pan-European system.

Another issue which was taken up at the first meeting was the question of free circulation of users. The idea of free international roaming was very much in line with the views of the delegates, but the existing restrictions in most European countries on the use of radio equipment by foreign visitors would clearly be a serious obstacle to the system since free circulation was one of the great advantages in the standardisation process. A French contribution[3] to the meeting pointed out that this was a purely political question which would not present very

---

[1] The views expressed in this module are those of the autor and do not necessarily reflect the views of his affiliation entity.

[2] Meeting report of GSM#1 in folder A1 of the attached CD ROM.

[3] GSM Doc 8/82.

difficult technical problems once the political issues had been cleared. As this question was already under study by the R 21 working group, there was no immediate need for GSM to take any action beyond following the studies in R 21. However, the group agreed that elimination of those obstacles as far as possible would have to be one of the key goals of the work.

2.2.2 Work Rules and Working Language(s)

The written and unwritten work rules of CEPT were quite different from those generally used in purely national bodies, but this fact caused very little trouble. It was understood that we had to work on the basis of consensus, so there would be no voting. This principle may not be the speediest way to reach a decision, but on the other hand, a consensus makes it almost certain that everyone is going to stick to the decision. Another rule which was to prove important for the working conditions of the group but which had nothing to do with the technical problems, was the issue of working language. It was specified by CEPT that committees and working groups (but not their subgroups) would have three official languages, English, French and German, with simultaneous translation. I thought from the start that it would be necessary to make English the only working language in GSM, even if that would violate the CEPT rules.

There were several reasons for this approach. Firstly, in the mobile telecommunication field (as opposed to e.g. the political field), it could safely be assumed that just about everybody would have a reasonably good command of the English language, at least good enough to communicate more efficiently than through interpreters who would have no idea about the technical field we were to discuss. Secondly, it would be far easier to arrange meetings without all the equipment and personnel required for simultaneous translation in three languages, and I expected that there would be a lot of meetings. Thirdly, there would inevitably be a large number of points in the specifications on which we would have to make precise definitions, and to carry on discussions of definitions through interpreters would be very awkward, to say the least. I did not expect this to go unchallenged (despite the fact that it was tacitly accepted by the delegates), and before meetings 1 and 2 there were official objections from some administrations which felt that simultaneous translation would be necessary in order to make things easier for all delegates. The GSM delegates themselves found that the principle of one language worked very well, however. The French and German delegates were not allowed by their administrations to agree to a permanent change in the CEPT working procedures, but they could accept an exception for any particular meeting.

After a few years the issue resurfaced because of some decisions in the CEPT to make multi-language working mandatory. A compromise solution was found with the help of some GSM delegates who informed their superiors that the work in the group went very well with just one language and that the work would actually be hampered if the group was forced to use several languages. The solution was that the chairman was to ask GSM at the end of each meeting if simultaneous translation in the next meeting was required. If nobody demanded translation, one language would suffice. This was done, and since nobody ever asked for translation the whole issue was quietly dropped after some time. I am convinced that the single language mode of operation was of great value for creating an environment which enabled us to work very informally and efficiently.

2.2.3 Political Direction

During the first few meetings of GSM, there was a great deal of uncertainty as to what course to pursue and what would be politically possible. Since the superior bodies in CEPT had given no directive as to the choice of an analogue or digital solution, it was obviously up to GSM itself to propose a solution. Clearly, an analogue system would be quite straightforward, offering no really difficult problems. A proposal for a digital system, however, would in order to convince the operators and the manufacturers, have to be based on compelling arguments showing the advantages in choosing a technology which had never been tried before in terrestrial public mobile communications. Moreover, any new analogue system would not have represented a sufficient breakthrough from the already existing systems such as NMT and AMPS to justify its adoption by the market. Thus, it was natural that the focus of the studies would be on a thorough investigation of the characteristics of the digital solution.

Regardless of whether the future system would be analogue or digital, some assumptions had to be made for testing the technical solutions, one of them would be a traffic model. Thus, a traffic model was a very important issue during the first year of the group. In retrospect, of course, those traffic figures look almost ridiculously low today.

2.2.4 Co-operation with other Groups in CEPT and other European Bodies

The goals described in GSM Doc 2/82 were, as someone said, "a tall order" which involved far more work than could be managed within GSM alone, so it was necessary to enlist the assistance of a number of other bodies. Within CEPT, there were working groups, active in fields of interest to GSM, e.g. data transmission, signalling, services and facilities and also groups dealing with commercial issues such as tariff harmonisation, etc. In addition to CEPT groups, there were bodies set up in co-operation with the EEC, such as the COST groups, which were studying wave propagation, modulation techniques, etc, and we were to liase with some industrial groups like EUCATEL. Also, it was felt necessary to follow the developments in certain fields which were not covered by other CEPT or COST groups, but which were nonetheless considered to be important to the future mobile system. An example is the field of radiation hazards, an issue which since the early 1980s had attracted great interest after some results had been published in a Soviet journal. The hazards – perceived or real – in the use of mobile telephones would no doubt be a potential threat to the entire field of mobile communications, so the results had to be studied. The GSM group obviously had no resources to pursue this matter, so a rapporteur was appointed to follow what was being published on this question. However, it was found after some years that the results were very inconclusive, so we dropped the question. As far as I can see, the issue regarding the radiation hazards to the user of mobile handsets is still (January 2001) open, but there is concern in many areas and several bodies are dealing with all aspects of this issue.

One final point must be mentioned here. The contacts with the Quadri-partite group, which have been described by Philippe Dupuis in his chapter (Chapter 2, Section 3), were of great value to us in the critical decision-making period, since they helped us develop a common strategy. Our contacts with the group were not of the formal kind, of course, but much more efficient since so many Quadri-partite delegates were also delegates to GSM meetings.

2.2.5 The Advent of Competition

During the late 1970s, a political factor became noticeable which would gradually turn out to be very important, namely the pressure in many countries (starting in the US) for the introduction of competition in the telecom field. Traditionally, the European telecom operators had always been state monopolies, but in 1982, the UK government decided to open the field of mobile communications for competition. The first system to be used this way would be TACS, a modified version of the AMPS system used in the US. Today, when there is competition in the telecom field in most countries, it is a bit strange to read the discussions of the early 1980s with predictions of the dire consequences of competition in telecommunications. Human nature is such that when a novel feature with political implications is proposed, there is usually a great deal of resistance from those who feel threatened and there is no doubt that most PTTs were very much against the competition and they did not do much to promote it. In this context, it is interesting to read the description of some of the features in the then tentatively specified TACS system (report of meeting number 2, p. 5[4]) where it is rather optimistically said that there will be "automatic handover between competing companies". That feature was considered to be of vital importance for the spectrum economy, and it was thought by the UK authorities that it would be easy to implement as indeed it probably would, technically. For many years, however, automatic handover between competing operators in the same country was not implemented anywhere, since the operators resisted the idea of accepting subscribers from a competitor. (Lately, an agreement has been worked out in Germany between the T-Mobil network and Viag Interkom for national roaming between the networks without any pressure from the regulator.) In comparison, handover agreements between companies in different countries, i.e. in a non-competitive situation, have normally caused very little difficulty, although the national and the international roaming features are technically quite similar.

2.2.6 The Creation of Working Parties and a Permanent Nucleus in GSM

Gradually, the workload increased and it became clear that it was no longer possible to deal with all issues in the plenary, so GSM decided to set up three ad-hoc working parties, namely WP 1 for services and facilities, WP 2 for radio questions and WP 3 for network aspects, mainly in order to prepare proposals for the plenary. The WPs were intended for that meeting alone, but as might be expected, they had to continue during several meetings as the workload increased. The CCH was anxious to prevent the proliferation of groups and decided that the WPs should not meet outside the ordinary meetings of GSM. However, after some time it became obvious that the workload made it necessary for the WPs to meet between the GSM meetings, and in 1985, the CCH accepted this.

In addition, it became obvious that there was a need for a body working on a permanent basis in order to take care of the co-ordination of the WPs, issuing documents and many other tasks that were difficult to handle for groups that only met at intervals of several months. The solution to this problem was to set up a Permanent Nucleus (PN) whose Terms of Reference are given in Doc 126/85, which after lengthy discussions between the CEPT Administrations concerning the location of the new body was set up in Paris and started operations in June, 1986 with Bernard Mallinder of BT as co-ordinator. Many of the countries which participated

---

[4] Meeting report of GSM#2 in folder A1 of the attached CD ROM.

in the GSM work contributed to the work of the PN. The use of a PN is not uncommon in international co-operation, of course, but this case was somewhat unusual in the sense that GSM already had its very strong WPs 1, 2 and 3, a fact which to some extent might limit the authority of the PN. Thus, one of the tasks of the PN was to assist the WPs in their work by supporting them with Program Managers. The burden of seconding skilled engineers to the PN would be totally borne by the PTTs themselves (later even the manufacturers), and the PN was to be hosted by the French Administration which (in addition to manpower seconded to the technical work of GSM) supplied the premises, secretarial assistance and telecommunication facilities. Thus, we never dealt with support in terms of money, a fact which relieved us of the tedious work with budgets, accounting. etc.

2.2.7 Co-operation with the European Commission and with Manufacturers

The European Commission had indicated their willingness to support the activity financially, but the PTTs found that it would be better for them to carry that burden themselves. By and large the contacts between GSM and the European Commission were relatively few, since the PTTs had sufficient resources to deal with the technical problems that arose. It was also my personal conviction, based on the experience from the NMT work, that despite the criticism for slowness often raised against CEPT, the PTTs were perfectly capable of carrying the project through. I was for that reason reluctant to accept EEC intervention in the work. On the political level, however, there were many contacts, and one contribution by the EEC was going to be of fundamental importance for the future of the GSM system. That was when the Frequency Directive was approved, reserving the GSM frequency spectrum for use by that system. This action stopped any attempts, not only in the EEC countries but also in many countries bordering on the EEC area, to use the spectrum for other purposes and it thereby made the success of the system possible.

On several occasions, there was pressure from the outside world to speed up the work so as to prepare the detailed specifications earlier than decided. Understandably, the manufacturing industries were concerned about a late delivery of the specifications, since they knew how much time they would need to implement the system. This question was to crop up several times during the mid-1980s and had to be discussed by GSM, but the result always was that it would not be possible to advance the starting date of the system, but preliminary specifications might be produced. After the creation of ETSI, this question became of less importance since there was then far more openness in the way GSM worked and hence, the manufacturers were well informed about the progress of the work.

2.2.8 Co-operation with non-European Bodies

The thought of spreading GSM to non-European countries was hardly in the minds of anyone in the early days. However, some ideas of co-operation with non-European bodies emerged now and then. For instance, there was a proposal from Bellcore[5] in late 1984 for co-operation, but CCH turned it down. The idea was discussed again in the Berlin meeting in October 1985, when delegates from Bellcore visited GSM for a discussion, but it was found that the difference in time plans and ideas was too great. Also, there was a discussion in early

[5] GSM Doc 72/84.

1986 on whether co-operation should be sought with a Working Party (IWP 8/13) in CCIR in order to find a way to launch GSM as a world standard. The conclusion of that discussion was that GSM must not in any circumstances allow its time schedule to be affected, and the conclusion was that although contacts with IWP 8/13 would be useful in order to spread information about the GSM system, something which could easily be achieved since many GSM delegates also took part in the CCIR meetings, the different time schedules of the two fora made it unrealistic to aim for acceptance of GSM in IWP 8/13.

2.2.9 Other Important Issues

The question of IPRs emerged early in the work of GSM. The first occurrence was probably when it was found that some areas, above all speech coding, were loaded with patents. In 1985 this question was brought to the attention of CCH, which stated that the policy of CEPT was to avoid standardising a feature or a method which was not freely available without royalties. On two points, GSM succeeded in securing royalty free agreements with the holders of two patents, i.e. the speech encoder and the Voice Activity Detector. Apart from those cases, the IPR issue was going to cause a lot of difficulties in the work on the new Pan-European system. After 1988, the question was taken over by the MoU, and the ensuing struggle is described in Stephen Temple's chapter on that body (Chapter 2, Section 4).

Among the required features of the system was the ability to protect the transmitted speech. Therefore, after discussions with the CEPT group on data transmission, it was decided to set up an expert group on security issues (SEG), which started working in May 1985, chaired by the GSM Chairman. It worked out basic protocols for authentication and for protection of speech, location and subscriber identity, which were presented in 1988 and subsequently accepted as GSM standards[6]. Among the results of the work of this group was the first specification of the subscriber identity module (SIM), a kind of smart card. The SIM was to perform a number of functions, not only those which concerned the identity of the subscriber, but also several important functions concerning the protection of the information, subscriber and network operator data, etc. The SEG also initiated a subgroup, the Algorithm Experts Group (AEG), which worked in great secrecy because of the sensitive nature of the material. The other parts of the SEG were in 1988 converted into the SIM Experts Group (SIMEG) in order to prepare specifications in co-operation with the manufacturers.

2.2.10. Choice of Speech Codec, Access Method and Modulation

The crucial issue during the first 5 years of the GSM work was the question concerning the choice of access and modulation system, a question which as described earlier was not decided by CEPT in 1982. The tests of the various proposals, eight in total, were performed in Paris by CNET assisted by the PN in late 1986 as described by Philippe Dupuis in his section about the tripartite and quadripartite activity (Chapter 2, Section 3), and the results were presented to GSM for discussion and decision in February 1987 in Funchal, Madeira. That meeting, which coincidentally happened to bear the ominous number 13 in the series of GSM meetings, turned out to be a very difficult meeting and very different from all the other GSM meetings before and after. The reason for this was that it concerned the choice of technology in an area where many countries had been very active and had invested a lot of

[6] GSM Specs. 02.09 and 03.20 (see attached CD ROM folder A3).

money and therefore had an interest in advocating and defending their preferences. Furthermore, the views differed concerning what the main use of the system would be. It was known from earlier discussions that some countries would prefer to put the emphasis on use in densely populated areas, while others wanted to consider use even in very thinly populated areas. This would necessarily have an impact on the choice of access system.

Several important issues concerning the access and modulation methods were to be decided and a number of documents were presented, perhaps the most important documents being the WP 2 contributions dealing with e.g. the Paris trials[7].

The question concerning analogue versus digital modulation was unanimously decided in favour of a digital solution, since the tests pointed clearly at the digital solution as superior to the analogue one. Thus, a question which had been on GSM's agenda for more than 4 years had finally been decided. The next question, however, turned out to be an intractable one, since most delegations were in favour of a narrowband TDMA system, but France and Germany were in favour of wideband TDMA. They had been working on several proposals for a long time, and there was substantial political pressure behind the Alcatel proposal for a wideband system, while the working level was behind the narrowband proposal.

As to the characteristics of the two proposed solutions, the view was that a system based on a bandwidth sufficiently narrow to make channel equalisation unnecessary, would result in a very low speech quality, while a wideband system (2 MHz) would require very advanced equalisation to avoid multipath fading. The system chosen was somewhere between those two extremes, i.e. an FDM channel with a bandwidth of 200 kHz, in which eight speech channels were put on a TDMA basis. The choice of TDMA had several advantages, one of them being that seamless handover between base stations was easily implemented. Also, a grouping of speech channels to facilitate HSCSD was possible.

GSM spent long nights discussing this – I recall one night when we went on until 2.30 in the morning and since we were almost dying for something to eat, somebody had the brilliant idea of raiding the refrigerators in the kitchen, adjoining the meeting hall. Unfortunately, there was nothing to eat except a large supply of sardine tins, so we had to eat a lot of sardines. If the hosts noticed the lack of sardines the following morning, they did not say anything. I hope they – when and if they read this – will forgive us in view of the hard work we put in.

Sardines or no sardines, we had arrived at an impasse. So on that point, by the end of the meeting it was still an open question whether there would be a common European system or not. In fact, those who held the key that could resolve the deadlock were not present, since it was a question on the highest political level in France and Germany, while there was no doubt that the people on the working level were quite happy to go along with the narrowband proposal.

At that moment, an innovation which would prove to be of major importance in our work, both in Funchal and later, was put forward by Stephen Temple of DTI. He proposed that we could at least agree on what he called a Working Assumption, a concept which enabled GSM to go on with the technical work on a narrowband TDMA system on a more or less temporary basis without actually deciding in favour of a system which some delegates were not authorised to accept. This concept was fruitfully used many times, and was often compared with slowly hardening concrete, i.e. it was stated that a parameter, covered by a Working Assumption, would be easily changed at an early date if there was reason to do so, but it would be more and more difficult as time went by.

[7] GSM Doc 21/87 and GSM 22/87.

This was a big step forward. The meeting could, based on the principle described above, agree on a final report[8] concerning the technical standard for a Pan-European digital cellular radio system, which describes the unanimous decision on a TDMA system and the 13 to 2 decision on a narrowband approach. Furthermore, a set of Working Assumptions for a particular characteristics of a narrowband TDMA system were agreed unanimously, in other words, it was agreed that *if* there was to be a narrowband system, that was the way to do it. In a package deal, GSM agreed to specify a narrowband TDMA system with a coder from PKI (the German division of Philips), ADPM[9] according to the WP 2 proposal, frequency hopping, possibility of a half rate coder and a number of eight time slots per carrier. This way, GSM had found a way to move forward in the technical work without formally agreeing on everything.

Thus, the meeting was a success after all, but the struggle was not over, of course, since the most important issue of narrowband versus broadband had been left open. During the following months, the activity behind the stage did not stop, and there was a lot of discussion going on at the highest levels in the major countries. Also the EEC was very concerned about the risk of failure to reach agreement and they exerted a lot of pressure on the various governments involved. Finally, in a meeting of the ministers of communication of France, Germany, Italy and the UK in May, a decision[10] was taken to accept the narrowband TDMA solution agreed in Funchal, "enhanced in the areas of modulation and coding...". This meant that the SEL delay equalisation, which was not compatible with the ADPM method chosen in Funchal, should not be excluded. The consequence of this was that there would be a change to another modulation method, i.e. GMSK.[11] Probably even those who were in favour of ADPM found that the change to GMSK was a small price to pay for European unity, and in the next meeting of GSM in Brussels in June 1987, we could therefore start working on a very sound basis to work out a tremendous amount of detail, safely convinced that in a few years we would be able to present a complete system that would be suitable for all of Europe.

---

[8] GSM Doc 46/87.
[9] Adaptive differential phase modulation.
[10] GSM Doc 68/87.
[11] Gaussian minimum shift keying.

Chapter 2: The Agreement on the Concepts and the Basic Parameters of the GSM Standard (mid-1982 to mid-1987)

Section 3: The Franco-German, tripartite and quadripartite co-operation from 1984 to 1987

Philippe Dupuis[1]

2.3.1 Introduction

As explained in Chapter 2, Section 1, the development of a common European standard for a mobile telecommunication system in the 900 MHz frequency bands designated for cellular communications by the ITU World Administrative Radio Conference (WARC) in 1979 has been proposed initially by the Nordic countries and the Netherlands. In the major European countries the main actors of the telecommunication sector, monopolistic operators and their favoured national equipment manufacturers, had not thought of a European standard before and certainly doubted that CEPT could produce one. Despite this France, Germany, Italy and the UK in the following years decided to support the GSM concept, and gave a decisive contribution to its success.

This section attempts to show how this evolution took place.

As assistant for mobile communications to the French Director General of Telecommunications from 1981 to 1988, I have been deeply involved in this process on the French side. Through the various co-operation efforts in which DGT took part I could also witness the events taking place in the other three countries.

As the four quadripartite countries are members of the European Communities it is also appropriate to review the contribution of the Commission of the European Communities. This is the subject of Section 5 of this chapter.

---

[1] The views expressed in this section are those of the author and do not necessarily reflect the views of his affiliation entity.

2.3.2 An Unsuccessful Co-operation Attempt in 1982

When Jacques Dondoux took over as French DGT in 1981 he became rapidly aware that some other European countries had done better than France in the area of mobile telephony.

Because of this he was of the opinion that DGT should not follow in this particular field its traditional approach of developing a national solution in co-operation with French manufacturers, and trying to export it to other countries later. This traditional approach had for instance been used for time division switching and videotex.

At the same time a consensus was emerging among the engineers in charge of operation and those in charge of R&D at CNET in favour of pursuing an "interim" solution which would be an adaptation of the NMT standard to the 900 MHz band. The term "interim" was already used because one started to understand that a 900 MHz digital solution would soon become feasible. A CNET research project on digital cellular technology was indeed proposed to be launched in parallel.

In October 1981 I led a French DGT delegation to Stockholm. The purpose was to receive more information about the NMT system. We met people from both Televerket and Ericsson. I have a vivid recollection of the meeting we had with Thomas Haug. We presented our plans to develop an adaptation of the NMT standard to the 900 MHz band. Thomas Haug explained that for the 900 MHz band the Nordic countries had in mind to go directly to a digital solution, but that they would nevertheless help us to develop an NMT variant at 900 MHz.[2] Regarding the digital solution he revealed that a Nordic group was even already conducting some preliminary work[3]

Very soon we discovered that the French equipment manufacturers were reluctant to produce equipment based on the NMT standard. Instead they offered to develop a new analogue cellular standard which would be better than the two existing "foreign" standards, AMPS and NMT.

At the same time British Telecom appeared to have an interest in implementing NMT at 900 MHz and started some joint activities with DGT in this area. This culminated in a meeting in London in June 1982 where BT and the DGT exposed their plans to representatives of other European countries. David Court of the DTI also made a presentation proposing that out of the 2×25 MHz set aside by CEPT for cellular systems in the 900 MHz band, 2×15 MHz would be used by the proposed interim system, while 2×10 MHz would be reserved for the later introduction of the harmonised European system that the GSM was mandated to specify, the scheme which later regulated the coexistence of the early analogue systems and of the GSM.

In the summer of 1982 the British scene was considerably modified. The Government decided to licence two competing operators. At the same time views were expressed, both in the UK and France, that an alternative solution would be to use an adaptation of the US cellular system known as AMPS. Philips and CIT-Alcatel were also actively campaigning in favour of their proposed cellular standard called MATS-E. In the UK the government decided that the two operators would have to select a compatible air interface among these different proposals.

---

[2] The tremendous success of NMT eventually forced the Nordic operators to introduce NMT 900, which was also adopted in countries like Switzerland and the Netherlands.

[3] I remember that, to make sure I had correctly understood, I asked the question: "Do you have in mind a fully digital system which could for instance work in TDMA?". Thomas confirmed that this was the case... and TDMA was indeed adopted 6 years later by the GSM group.

BT and the DGT arranged a second meeting in Paris in October 1982 to inform the participants of the June meeting of these new developments. A report of this October meeting was presented at the GSM inaugural meeting in Stockholm in December 1982.[4]

In the meantime the selection of an air interface was even brought up at the ministers level. Under the pressure of Philips and CIT-Alcatel the French minister Louis Mexandeau wrote to his UK counterpart, Kenneth Baker, to inform him that France had made the decision to use MATS-E. The reply explained that the UK operators, Cellnet and Vodafone, had decided to use an adaptation of AMPS which was then named TACS.

2.3.3 The Franco-German Co-operation on a 900 MHz Interim Analogue System

The diverging decisions announced by the French and UK ministers marked the end of the Franco-British co-operation. Soon after DGT received an offer from the German Bundespost (DBP) to select a common 900 MHz analogue interim standard. Both countries had a need for a future solution with high capacities. France needed capacity in the Paris area in the near future and then capacity for the whole country. Germany was in the process of introducing the C network in the 450 MHz band. But since there were only 2×4.4 MHz of spectrum available, this solution provided only a very limited capacity (initial target 100000 customers). Therefore DBP started an activity for a next generation system in the 900 MHz band.

An agreement was reached by French PTT and DBP and signed on 15 July 1983 in Bonn to jointly develop and introduce the same radio communication system in France and Germany in order to cover the demand for 15 years in both countries and possibly in other European countries. A work programme was agreed which consisted of jointly elaborating technical requirements for an interim system, launching co-ordinated Requests for Proposals in the two countries for the construction of the first phase of the network's implementation. Manufacturers were free to select any air interface but encouraged to produce joint, or at least co-ordinated, proposals allowing full roaming between France and Germany. A positive aspect is that preparation of the co-ordinated requests for proposals gave the French and German engineers the opportunity to know each other and to learn to work together. At that time Jean-Luc Garneau joined my team to act as project manager on the French side. My counterpart on the German side was Klaus Spindler who was responsible for all public and private radio communication services in Germany. He later became one of the most convinced supporters of GSM. The project manager on the German side was Friedhelm Hillebrand, the deputy project manager was Frieder Pernice.

The Franco-German teams developed and agreed concepts for services and for the network. A complete specification for an interim 900 MHz analogue system was developed. A request for proposals was issued on 15th December 1983. Five proposals were received on 26th March 1984. Among them four proposals were for analogue systems and one for the digital system CD900 from SEL. The evaluation and negotiations lead to MATS-E offered by Philips, Siemens and CIT of France as the preferred solution.

During the evaluation of the proposals doubts arose as to whether a Franco-German analogue 900 MHz system would be sufficiently future proof. It also became also obvious that a new analogue system could not be developed in time to meet urgent capacity require-

---

[4] Report of the meeting on an interim public land mobile system in the 900 MHz band, GSM Doc 18/82.

ments. At the same time, in the research centre of DGT, CNET, the R&D programme on digital cellular communications, named Marathon, was gathering momentum. The results of other research projects and the digital proposal by SEL also showed that the feasibility of a digital technology was not far away.

Because of this, both DGT and the Bundespost started considering abandoning the project to implement an interim 900 MHz solution and wait for the 900 MHz European digital standard elaborated by the GSM. Both had developed solutions for the limited spectrum available to them in the 400 MHz band, Radiocom 2000 in France and system C in Germany and could attempt to boost their capacity. Both realised also that the GSM group needed much more support and momentum to achieve the definition of a European standard.

The major move in this direction came from the German side. On 4th June the German evaluation team lead by F. Hillebrand and F. Pernice proposed to withdraw the request for proposals, to concentrate all efforts on a Pan-European digital system and to seek offers for different radio transmission schemes. This was discussed with the French team on 13th–14th June 84 in Paris. The discussions on this far reaching proposal continued during the summer in parallel with the continued evaluation and negotiations of the analogue system offers. In August 1984, the Bundespost director in charge of telecommunications, Waldemar Haist, brought the matter to the Bundespost Minister, Christian Schwarz-Schilling. They decided to formally propose France to re-direct the Franco-German co-operation towards digital cellular technology, with the aim of France and Germany to give a more powerful contribution to the GSM group, and possibly accelerate its work.

2.3.4 The Franco-German Co-operation on Digital Cellular Technology

To formulate this proposal Christian Schwarz-Schilling immediately tried to reach on the telephone his French opposite, Louis Mexandeau. In his absence, he spoke to the French DGT, Jacques Dondoux, then at his vacation place in the mountains of Ardèche. Jacques Dondoux was immediately in favour of the German proposal.[5]

During a high level Franco-German co-ordination meeting on 18th–19th September 1984 the new scheme was fully endorsed by both sides. The French and German engineers thus met again, but this time to elaborate the framework of a joint R&D programme on digital cellular technology. They rapidly agreed on the contents of the programme. DGT and the DBP were to fund the construction of several cellular digital technology demonstrators by the French and German manufacturers, with the purpose of demonstrating the feasibility of a digital solution and give indications for the selection of the more appropriate radio techniques. The results obtained were to be made available to the GSM group. The following time schedule was agreed:

December 1984	Request for proposals.
March 1985	Contracts awarded.
End 1985	Beginning of the field testing of the demonstrators.

At the political level the initiation of this programme was given full recognition by a joint declaration by Christian Schwarz-Schilling and Louis Mexandeau on the occasion of a Franco-German summit meeting in Bad Kreuznach on the 30th October 1984.

---

[5] Mrs Dondoux, who was born in Germany, acted as an interpreter during the conversation.

Table 2.3.1   The four Franco-German demonstrators

Name	Manufacturers	Access technique
CD 900	Alcatel/SEL/ANT/SAT	Wideband TDMA (24 channels per carrier, 1.65 MHz spacing[a]) combined with spectrum spreading
MATS-D	Philips (Tekade)	Downlink: wideband TDMA (32 channels per carrier, 1.25 MHz spacing) combined with spectrum spreading Uplink: FDMA (25 kHz spacing)
S 900-D	ANT/Bosch	Narrowband TDMA (ten channels per carrier, 250 kHz or 150 kHz spacing)
SFH 900	LCT (later Matra)	Narrowband TDMA (three channels, 150 kHz spacing) combined with slow frequency hopping

[a] Or 63 channels per carrier with 4.5 MHz spacing.

The program was also presented to GSM#6 in November 1984 in London[6] This was an important kick for the momentum of GSM, since it indicated clearly that France and Germany would not continue to develop another interim system, but that they instead intended to implement the Pan-European standardised (digital) GSM system. Furthermore it was proposed to carry out trials on radio transmission and speech coding, the two most critical new technologies. This initiative was received well. A number of other proposals for field trials were made by other parties.

The French and German engineers went on meeting each month for several days, alternatively in Paris and Bonn. The CNET engineers in charge the Marathon project became actively involved, in particular Bernard Ghillebaert and Alain Maloberti who were also taking part in the GSM work. On the German side Friedhelm Hillebrand and Frieder Pernice continued to be the leading technical representatives.

Proposals from the manufacturers were received in March 1985. After a joint evaluation the Franco-German Working Group recommended to award four contracts, two by DGT and two by the DBP, for the construction of the four demonstrators briefly described in Table 2.3.1.

DGT was to issue the CD 900 and SFH 900 contracts and the DBP the S 900-D and the MATS-D contracts.

This raised some problems inside DGT. SEL, then still an ITT company was the leader of the CD 900 project but Alcatel was a partner. Alcatel having the full backing of the French government, the CD 900 scheme had the favour of the DGT director in charge of industrial policy, to such an extent that he was opposed to the award of a second contract to LCT for SFH 900. LCT, a recently nationalised ex-ITT company, later to be bought by Matra, was still considered to be "foreign". I spent weeks arguing that the purpose of the Franco-German programme was to test different radio techniques and that both projects were based on widely

---

[6] Meeting report of GSM#6 section 8 in the attached CD ROM folder A1, document GSM 76/84 submitted by France and Germany in Folder A2. A progress report with more details was given to GSM#7 (document GSM 6/85).

different technical approaches which had to be investigated. In May 1985 Jacques Dondoux eventually followed my advice and the four contracts could proceed.[7]

When they were ready for signature an agreement between the DGT and DBP was elaborated to deal with the various commercial implications: sharing of the cost, ownership of the results, etc. This agreement started by defining the objectives of the co-operation:

> The French PTT and the Deutsche Bundespost have decided to introduce before the end of the present decade a common digital cellular radio communication system based on a European standard the definition of which is presently under way in CEPT. This agreement relates to the first phase of the work to be carried out to reach this objective.

> In this first phase both parties will support the work of CEPT with a joint experimental programme aiming at a comparative evaluation of some digital radio transmission techniques among which CEPT will have to make a choice before the end of 1986.

The agreement was signed by correspondence in July-August 1985 by Jacques Dondoux and Waldemar Haist. The total cost was 18 million ECU.[8] The text stipulated in particular that:

1. the expenses would be shared evenly between DGT and DBP;[9]
2. the contracts would contain harmonised intellectual property clauses giving DBP and DGT the right to exchange all technical information between them, allowing all technical information relevant to the elaboration of the specifications of the European system to be passed to CEPT and committing manufacturers to grant free and non-exclusive licences on possible patents essential to the implementation of the European standard to any party belonging to CEPT.

An annex also described in broad terms the procedure to be applied for a comparative evaluation of the techniques demonstrated by the different projects. It is interesting to note that the principles set forth there were later endorsed by the GSM group when they had to conduct this comparative evaluation.

2.3.4.1 Field Testing in Paris

The demonstrators were delivered during 1986 and tested according to the individual contracts. The comparative evaluation mentioned above was in fact made in the GSM framework. GSM WP 2, in charge of the radio aspects of the future system, defined the methodology. It included field tests. At the GSM#11 meeting (Copenhagen, June 1986) Klaus Spindler proposed that all demonstrators, including the four Franco-German and others built by Elab, Ericsson, Mobira and Televerket, thus a total of eight, be tested at the same location under the supervision of the same team of engineers. He suggested CNET in Issy-les-Moulineaux, near Paris, as the place where the required infrastructure and expertise was more easily available.

---

[7] It is interesting to note that it is in Singapore that the decision was made. Jacques Dondoux, the DGT director in charge of industrial policy in DGT and I were there on the occasion of the ITU Asia Telecom forum. At a cocktail party Jacques Dondoux took us aside to resolve the dispute and after hearing our views decided to let the SFH 900 contract go.

[8] Equivalent to 18 million Euro.

[9] As the cost of the contracts awarded by DGT was more than the cost of the contracts awarded by DBP this resulted at the end of the programme in a large payment by DBP to DGT.

An additional advantage was that the Permanent Nucleus of the GSM group, led by Bernard Mallinder, was being installed in Paris, only about 2 km away from CNET, and could therefore be associated with the testing activities.

The results of the tests have been reported by Thomas Haug in the preceding section. If one considers more precisely the impact of the Franco-German co-operation two remarks can be made. The first one is that the four Franco-German demonstrators were certainly among the better engineered, as the project had been properly funded and managed in the framework of formal procurement contracts. Because of this they have certainly contributed to establishing the necessary confidence in the feasibility of a digital solution. The second is that the SFH 900 project has played a vital role in demonstrating the value and feasibility of slow frequency hopping combined with narrowband TDMA. Narrowband TDMA was eventually selected, rather than wideband TDMA as used in CD 900, because it was a technology easily accessible to most manufacturers. It also offered a greater flexibility with respect to spectrum usage and had lower requirements on signal processing in the terminals. But it is less powerful to counter the effect of either multipath propagation or co-channel interference. This why slow frequency hopping was needed. With slow frequency hopping narrowband TDMA can effectively compete with a wideband solution.

2.3.5 Extension to Italy: the Tripartite Co-operation

Extending the Franco-German experimental programme itself to other countries would have raised problems, and would have delayed the process. But most of the other components of the Franco-German co-operation agreement could easily be extended to other European countries. The first one to join was Italy. A tripartite agreement was signed on 20th June 1985 by Jacques Dondoux for DGT, W. Florian for DBP, D. Gagliardi for the Italian Ministry and P. Giacometti for SIP.

The stated objectives were to jointly support the work of the GSM group, to place manufacturers of the three countries in a position to resist competition from extra-European manufacturers and to set up co-ordinated plans for the earliest possible implementation of a compatible service covering the three countries. Two areas for co-operation were identified: the co-ordination of R&D activities and the co-ordination of operation plans. With regard to the first one the agreement included the same clauses on the circulation of information and intellectual property conditions of the R&D contracts as in the Franco-German agreement.

One annex described the work undertaken by DBP and DGT in the framework of their joint experimental programme. Another one described the various R&D activities of SIP.

Compared to the Franco-German agreement, this agreement was of course lacking the substance of a joint R&D programme but it went further in two directions: co-ordinating the views of the three countries on the issues debated at CEPT level, and co-ordinating the future implementation of the standard, something which later became the mission of the GSM MoU group of signatories.

2.3.6 Extension to the UK: the Quadripartite Co-operation

Very soon after Stephen Temple succeeded in making the UK join. This took the form of an additional protocol to the tripartite agreement signed in April 1986 by John Butler for the

DTI, Gerry Whent for Racal-Vodafone Ltd and John Carrington for Telecom Securicor Cellular Radio Ltd. Racal-Vodafone and Telecom Securicor were by then two successful analogue cellular operators. That they expressed a commitment to the future European digital standard was certainly an encouragement to those working on GSM. At the same time a new annex was attached describing the relevant R&D activities in the UK. It was certainly too late to expect new technical concepts to emerge from these activities. The contribution of the quadripartite co-operation has been indeed at a different level. Four examples given below illustrate some of the different areas in which it had the most positive impact.

2.3.6.1 The Top Down Study

The R&D activities listed in the annex to the quadripartite agreement included the construction of a TDMA test bed, used to assess different speech coding and modulation techniques. But the most remarkable contribution of the UK was probably the "Top Down" study conducted by Plessey and funded by the DTI. In 1986 the two UK TACS networks were in operation and met with a tremendous success. The various market actors had thus very little ground to immediately commit themselves to a new Pan-European approach. The study helped them acquire support for the GSM work. It also provided a firm basis for the formation of the consensus on the selection of the radio and speech coding.

On this subject the study made the point that the new second generation system should be "as good as TACS in six respects, and better in at least one respect of:

1. subjective voice quality;
2. base station cost;
3. mobile cost;
4. spectrum efficiency;
5. viability to support handportables;
6. new services."

These criteria were not new to GSM. Most of them were already included in a document that Stephen Temple had tabled at a GSM meeting in 1985[10].

In addition a legitimate UK requirement was stated which was "the ability of the second generation system to co-exist or overlay with the existing and planned TACS network for a period of approximately 10 years".

These criteria thus served two purposes. They helped build the confidence of the British industry in GSM and, the following year, when GSM had to make the difficult choice among the proposed radio technologies, and finally decide to go digital, these criteria were indeed used as benchmarks and were extremely useful in helping to arrive at a consensus.

2.3.6.2 The Creation of the GSM Permanent Nucleus in 1986

Another value of the quadripartite co-operation is that it created a forum where the four countries could co-ordinate their views on some issues discussed in the GSM group. The four Nordic countries had done so since the creation of the GSM and this certainly had had a positive impact on the GSM decision making process.

An instance in which the quadripartite co-ordination helped to achieve a decision is the

---

[10] GSM 10/85

location of the Permanent Nucleus of the GSM, i.e. the team of full-time experts then about to be set up to manage the production of the GSM specifications. According to the CEPT rules one of the CEPT members had to offer the required office space and equipment, secretariat staff, etc. at no cost to the other members. A difficulty arose as there were two competing offers, one by the Danish telecoms to house the GSM Permanent Nucleus in Copenhagen and one by the French DGT in Paris. The first offer was backed by the Nordic group and the second one by the quadripartite group. At the same time the UK offered to second Bernard Mallinder as the co-ordinator of the permanent nucleus and Germany offered Eike Haase as deputy co-ordinator. The choice of Paris versus Copenhagen was the subject of numerous telephone meetings over a period of nearly 6 months. Thomas Haug did not press to have a quick decision as the question had been raised early enough. Eventually the determination of the quadripartite group helped to achieve a consensus on Paris, which was formally endorsed by the GSM group at GSM#10 (Athens, February 1986). When it was done I shook hands with Marius Jacobsen the head of the Danish delegation who said: "Thank you for a good fight". I think that more than anything else it had been a good exercise in consensus formation, an area where the GSM had still a lot to learn, and actually learned a lot in the following years.

2.3.6.3 The Resolution of the Madeira Dispute in the First Half of 1987

Section 2 has reported about the dispute which arose at the Madeira meeting of the GSM Group (16th–20th February 1987) where France and Germany supported the broadband TDMA approach against the majority of the GSM delegations from 13 other European countries who favoured the narrowband one. The earthworks for resolving the dispute had to be done in France and Germany. Armin Silberhorn, who had then taken over from Klaus Spindler as head of the German GSM delegation, and myself in France had to convince our high level management that there was no alternative to rallying the European majority.

On the French side, this was quite difficult. Marcel Roulet had succeeded Jacques Dondoux. It was difficult for him to go against Alcatel who were pushing their CD 900 concept even more energetically since Alcatel had bought the ITT telecom activities with the consequence that SEL, the inventor of CD 900, had become an Alcatel company.

The narrowband solution had been favoured on both the French and German sides at the working level represented in Madeira, but there were clear directions to support the wideband solution due to preferences at the political level.

After the Madeira meeting the German minister ordered a review of the position taken by the technical level. This review was carried out under the Chairmanship of the President of the Technical Engineering Centre of DBP (FTZ) in a 2-day meeting by the Research Institute of DBP on 24th and 25th February in Lahnstein (near Koblenz). The narrowband TDMA solution was confirmed as the technical and economic optimal solution. This result was presented to the minister Schwarz-Schilling on 17th March. The minister endorsed the result. This meant giving up the broadband solution in order to reach the common single European solution.

On 18th March I hosted a "tripartite" meeting of the French team, the German team led by Armin Silberhorn and an Alcatel delegation (France and Germany) led by Philippe Glotin. After an intensive discussion the French and German teams confirmed their view, that narrowband TDMA is superior to the broadband solution considering technical, operational and economic criteria.

The two Director Generals of telecommunication, Marcel Roulet and Waldemar Haist met in Paris on 25th March. Marcel Roulet requested, that the trials should be continued or that as a fallback position the development of both systems should be continued and the final decision be postponed to December 1987.

Finally Christian Schwarz-Schilling informed the French minister Gérard Longuet on 3rd April that Germany would not continue to support the wideband scenario and would also not support the proposed additional trials, since they would lead to a delay of the whole program. He invited his French colleague to propose a solution based on the narrowband scenario. Minister Schwarz Schilling also talked to the British minister Geoffrey Pattie on the same day. They agreed that a narrowband solution should be supported and that the further discussions should be carried out in the quadripartite framework.

After his conversation with Christian Schwarz-Schilling, Gérard Longuet did not pass any message to the DGT. At this time he took more interest in the award of a second analogue cellular licence than in the work on GSM. The final inflexion in the French position came a few days later when Alcatel agreed to accept the solution supported by the majority in the GSM group. The decision was taken at a meeting in the Alcatel (ex-ITT) headquarters in Brussels on 9th April 1987. It is interesting to note that it was influenced by the European Seminar on Mobile Communications organised in Brussels by the French Société des Electriciens et Electroniciens on 7th–8th April.[11] Various presentations on markets, technology and the work of the GSM Group were made by speakers from different European countries. It ended with a panel discussion on the way forward for Europe, which quite naturally centred on the resolution of the Madeira dispute. Philippe Glotin, head of mobile communications in Alcatel, took part in this discussion and was impressed by the determination of the majority of European players to go ahead on the basis of the GSM preferred solution, with or without Alcatel. When he reported this the next day, Alcatel decided to stop a hopeless fight.

For the final negotiations a quadripartite meeting at Director General level took place on 22nd–23rd April in Bonn. This freed the way to the final agreement.

On 5th–6th May a quadripartite meeting took place at the working level in Paris. A technical contribution to CEPT GSM WP 2 and a draft for the common declaration of the four ministers were agreed. Furthermore, the creation of a memorandum of understanding for the introduction of the service in 1991, which would be open to other countries was discussed. The Chairman of CEPT GSM and the head of the Swedish delegation in GSM participated as observers.

The signature of the common declaration took place in Bonn on the 19th May 1987. Several points in the declaration are interesting.[12]

Probably the more important strategic item was the confirmation that the Pan-European digital system should be built on the basis of narrowband TDMA with the agreed technical improvements and that the four countries would take every necessary measure to secure the opening of a service in their countries in 1991. It was further recognised that it was vital for

[11] The meeting was organised by Jean-Paul Aymar who was in charge of mobile communications in SEE, assisted by Didier Verhulst who had been the first chairman of GSM WP 2. Among the authors of presentations were W. Fuhrmann, Fred Hillebrand, Bernard Mallinder, Stephen Temple and Didier Verhulst.

[12] There were two originals, one in English and one in French. I have kept the French one. The English version and a German translation can be found in GSM Doc 68/87.

the success of the European system that operators jointly develop co-ordinated plans for its implementation. Furthermore, they were invited to record these plans in a Memorandum of Understanding to be signed at the latest in September 1987 and to be opened for signature by other CEPT countries. This article which had been suggested by Stephen Temple has indeed been the foundation stone of the GSM MoU, now known as the GSM Association.

It is also worth mentioning that the text also dealt with two technical points, the modulation method and the use of slow frequency hopping.

The narrowband TDMA concept defined by the working assumptions of the Madeira meeting included a modulation method proposed by Norway. But further studies in WP 2 had shown that another approach might be preferable. To make sure that this possibility would be fully explored the French engineers had suggested to add to the reference of the Madeira narrowband TDMA concept in the common declaration the words:"enhanced in the area of modulation and coding to provide the greatest flexibility in receiving equipment implementation". WP 2 actually decided in their May meeting to adopt the GMSK modulation scheme. GMSK is based on a Japanese patent by an engineer of NTT and our Japanese colleagues later were delighted to hear that they had contributed a component of the GSM technology.

The working assumptions of the Madeira meeting stated that slow frequency hopping (SFH) would be used as an operator's option. This meant in particular that SFH had to be specified, that a SFH capability had to be included in all mobile stations, but it had not been explicitly written. The French engineers were worrying that this may not be enforced if a majority of operators decided that they would not use SFH in their network. Therefore, they insisted that a mention of SFH would be made. It was actually inserted in the article asking operators to agree on an harmonised approach to the implementation of network features.

The declaration was signed by the ministers, or their representatives. In France, Germany and Italy the ministers could also commit the state owned monopoly operators. This was not the case in the UK and therefore the British operators also signed the declaration. That way they eventually became firmly committed to the introduction of GSM.

2.3.6.4 Advertising GSM: the Hagen Seminar in 1988

After the signature of the quadripartite declaration there was a huge amount of work to complete, essentially produce the detailed specifications and ensure their adoption as standards by the market actors. With respect to this second objective, there was a well-defined strategy to enrol European operators in the proposed memorandum of understanding. But it was also necessary to encourage manufacturers to develop products based on the GSM specifications. A first step was to make them known. During the GSM meeting in Brussels in June 1987 we started discussing the organisation of a seminar to be held in 1988 where a comprehensive presentation of the GSM specifications would be made.[13] It was agreed that it would take place in Germany and Armin Silberhorn was put in charge.

There was of course a more immediate opportunity as Renzo Failli had organised, with the support of the quadripartite group, a conference on Mobile Radio in Venice from the 1st to the 3rd July 1987. The venue was prestigious. It was the Foundation Cini in San Giorgio Maggiore, formerly a Benedictine convent. Although the scope of the conference was much broader, there were numerous presentations related to the work on GSM. I was

---

[13] Doc. GSM 187/87

surprised to see there a group of top scientists of Motorola led by Dr Mikulski who had made the trip from Motorola headquarters in Schaumburg, Illinois. I had seen them in 1982 on a fact finding mission on the US analogue cellular standard, AMPS. I commented that five years earlier we had come to learn from them and now it was the other way round. They were not too happy about this comment. Motorola accepted GSM as a standard for Europe, but strongly opposed the idea of having it used elsewhere. The presence of this delegation at least proved that they were taking it seriously. It was a good indication of what was going to happen in the following months.

At the same time, in Germany, Armin Silberhorn had difficulty organising the proposed seminar. The best solution would have been to have it managed by the German engineering society, VDE/ITG, but they said that it was too late to plan any event in 1988. Fortunately I was then contacted by Prof. Ludwig Kittel of the Fernuniversität of Hagen, a city of 200 000 inhabitants in Westphalia. The university of Hagen was planning their 1988 annual event, the "Dies Academicus 1988". The theme was Franco-German politics and Ludwig Kittel invited me to speak about the Franco-German cooperation in the field of digital mobile radio. I agreed to give a speech on " The human factors in Franco-German cooperation" and reported this to Armin Silberhorn who immediately saw the opportunity to hold our seminar in Hagen in October 1988, in conjunction with the Dies Academicus. There was in Hagen a beautiful multipurpose building, the Stadthalle, which could be converted very quickly from a congress centre into a concert hall or a banquet room and could accommodate more than 1000 participants. When I saw it I had no hesitation. Ludwig Kittel volunteered to take care of all details, with the support of the regional Bundespost directorate and of an organising committee composed of himself, Armin Silberhorn and myself. We had frequent meetings in Hagen during the spring and summer and I came to know the city quite well. The programme of the Digital Cellular Radio Conference, as we decided to call it, was finalised during the GSM meeting in Vienna in April 1988[14] and we started advertising. We soon got more than 600 registered participants.

Even with this big attendance, the conference went very smoothly, thanks to the efficiency of the Bundespost team and to the careful planning of Ludwig Kittel. There was a comprehensive set of lectures on the various technical aspects of the GSM by people who had contributed themselves to the selection of the technical solutions. This was the time at which the discussion of Intellectual Property Rights was the most active, in particular with Motorola who, in the meantime, had filed a set of over 100 patent applications on some of the features included in the GSM. For this reason one additional purpose of the seminar was to reveal all technical information so that it would not be possible to file more patent applications on GSM technical solutions. To this end we had even decided that participants could buy copies of the draft specifications. The French secretary of the Permanent Nucleus, Sylviane Poli, had come from Paris to run the sale office. The first afternoon we were asked to close the conference strictly on time as the main auditorium had to be arranged for a concert in the evening. It was a Richard Strauss concert by the Hagen Symphony Orchestra. They played the Metamorphoses for 23 string instruments and the Alpine Symphony. Finally the famous Gundula Janowitz sang the Last Four Songs for high voice and orchestra. She had rehearsed in the building in the afternoon and from time to time her beautiful voice in the distance had distracted us from the technical presentations. The second afternoon again we were asked to close the presentations on schedule, as the auditorium had

[14] Doc. GSM 120/88

to be arranged for the conference dinner in the evening. I have a vivid recollection of this evening as I had the mission to give the dinner speech. There was a band playing continuously and I had been told to pick the right moment, at the time dessert would be served, go the band, ask them to stop playing, pick a microphone and deliver my speech. Getting the attention of about 600 diners had seemed to me a formidable task, but it was a good audience. I explained, and I think everybody agreed, that we had come to the crucial point where the GSM fathers were handing over the result of their work to operators and manufacturers to develop products and services and offer them to their clients. On the last day there was a panel discussion led by a journalist from the Financial Times. Stephen Temple was one of the panellists and he made a short presentation the title of which was: "Selling GSM to the world". When he had proposed this subject I had thought it was a bit premature, or too arrogant. Later we understood that it was right to raise this question, even if this was upsetting our friends in Motorola.

Chapter 2: The Agreement on the Concepts and the Basic Parameters of the GSM Standard (mid-1982 to mid-1987)

Section 4: The GSM Memorandum of Understanding – the Engine that Pushed GSM to the Market

Stephen Temple[1]

2.4.1 Introduction

In the 1980s there was a significant transformation of the European telecommunications market. The decade began with strongly national focussed structures in place. The PTT (Posts, Telephone and Telegraph Administration) was at the centre of power of this structure. In some EU countries the PTT was a government department. In other countries it was a state owned industry sponsored by a ministry that worked closely with it. In the larger countries of Europe the PTT had an important role in fostering the local telecommunications manufacturing industry. A technology initiative would usually comprise a close co-operation between the national PTT and one or more of the major national manufacturers. The partnerships often involved the PTT providing R&D funding and entailed either a guaranteed supply contract or the strong expectation of one.

There were examples of good European and international co-operation in telecommunications standards. The examples were usually driven by the expanding and profitable international telecommunications traffic. However there were also examples of technical standards being used as technical barriers to trade. For example, an effort would be made by the government in one European country to use a technical standard for a new technology to help secure a lead for its supply industry. The response from other countries would be to select a different technical standard. The reason for this was to thwart the industries in the first country from getting any sort of first mover advantage. The worst example of this was the videotext standards in Europe. Cellular radio was also a victim of these national industrial

[1] The views expressed in this section are those of the author and do not necessarily reflect the views of his affiliation entity.

tensions. Prior to GSM, the efforts at the European level to arrive at a common technical standard or even frequency band of operation had been a failure. There was not even any pretence at the global level, in the International Telecommunications Union, to seek a common technical standard for cellular mobile radio.

The UK can claim some credit for breaking the mould in respect of the national PTT and national supplier relationship. The government of Mrs Thatcher had put privatisation of British Telecom (BT) at the top of its political agenda for telecommunications. This inevitably shattered the bond between BT and its local (UK) supply industry. BT from that point on would only buy from UK owned companies if they had world class products at the lowest prices. The UK telecommunications equipment supply market rapidly became open and very competitive. The second political priority was to provide BT itself with competition. Cable & Wireless was given the chance and Mercury was born to take on BT in the wire-line market. In cellular radio the chance was given to Racal and Vodafone emerged.

The UK Government's approach to the promotion of competition in the telecommunications market was not helpful in trying to achieve a common European cellular radio standard. In fact it started off by making it much more difficult. The overwhelming pressure to enter the market quickly did not provide the time for Europe to achieve a common European standard for its first generation analogue cellular radio networks. The UK's competitive telecommunications market did make one contribution. This was to inject into the European strategic thinking in standards making a strong dose of market realism. However, this alone would not have got Europe the world lead in mobile radio that it enjoys today. What secured for Europe this leadership was the combination of:

- The technology development efforts of France, Germany, Sweden and Finland;
- Efforts of the French and German operators to plan a next generation system for a mass market;
- The very positive market take-up of cellular radio services in the Nordic countries;
- The effort that had to be made by the DTI to bridge between its European partners and its domestic competitive players Cellnet and Vodafone;
- A shrewd move by the Commission to table a directive on safeguarding the frequency bands for a Pan-European cellular radio system;
- The close working relationship that the GSM group achieved between key national officials
- A slice of good luck and well judged timing.

All of these positive forces coalesced around the "GSM Memorandum of Understanding" for introducing the GSM system by a common date. The GSM Memorandum of Understanding (MoU) was a product of its time. For this reason some background of the happenings prior to the Memorandum being drawn up is essential. The story begins where so many imaginative European initiatives of the time used to begin. A summit between the Heads of State of France and Germany was held in November 1984. It brought to fruition discussions between PTT officials for a Franco-German initiative on a new digital cellular radio technology.

2.4.2 Getting GSM to Market – The 1994 French and German Perspective

The strategy of the French and German Governments was announced to the GSM Group in late 1984. Their representatives at the GSM Plenary in October/November 1984 in London explained that they had given up the project of a Franco-German S900 analogue system and intended instead to focus on a Pan-European digital system. The two governments would fund the development of digital technology with their respective industries in some form of collaborative programme and trials would be carried out. The CEPT organisation would be expected to produce a technical standard around the result and PTTs needing to expand their capacity in the 900 MHz range of frequency channels would use the CEPT standard.

There were some subtle sub-texts to the agreed approach by the two political Heads of State. The German Research Ministry had in mind that the technology would be based upon work they were funding with SEL. This was a broadband Time Division Multiple Access (TDMA) technology. The French PTT were doing their own R&D and at this stage were very open minded on the choice of technology. (After Alcatel bought SEL the French Government favoured the SEL system). The political levels that set up the 1985 Franco-German technology co-operation were not aware of just how many dialects of digital technology were in the process of development in research laboratories around Europe.

There were also significant differences at the commercial level between the telecommunications operators of the two countries. At the root of these differences was the spare network capacity that the German and French operators each had in hand to meet future growth of their mobile radio customer base. The Germans had just introduced their C-System. They estimated that they had enough capacity to see them through to the turn of the decade. The French PTT on the other hands had put into service a system of much more limited capacity. Their internal forecasts were that they would need a new system to be brought on-stream perhaps as early as 1988. Thus, the Franco-German strategy of 1984 was essentially running on the premise that the French PTT would be the first to bring a new technology into serious commercial use, this might be paralleled by Germany as a token German network and other European PTTs might follow in due course. This assumption would have left it to the French PTT to bear the main brunt of bringing new immature technology into service.

2.4.3 The Perspective in 1984 of other European Countries

It was not evident in early 1984 that the Franco-German strategy would intercept the path the rest of Europe was on. The perspectives from the UK and Nordic countries were quite different from those of France and Germany.

The Nordic countries were just in the process of introducing their new NMT900 analogue cellular radio system. By any measure the Nordic countries led the rest of Europe in the development of a cellular radio service. Their customer penetration figures inspired the rest of Europe and gave confidence to invest in the service. Whilst the Nordic countries had some excellent research in digital cellular radio there were no serious attempts at plotting a strategy to introduce the technology into service. The focus was in persuading customers to move from the highly popular but rapidly congesting NMT450 systems to the new NMT900 system that provided a lot of spare network capacity to accommodate growth in the number of customers. This would meet foreseen customer demand for many years.

In many other European countries those responsible for running mobile services in their PTTs were having a hard job persuading their fixed wire-line dominated organisations to make investment in mobile radio a high enough priority. There was no evidence of any serious studies on the introduction of digital cellular radio services from any other quarter in Europe. It was left to a few individual engineers who attended the GSM meetings to form their own views on the importance of digital cellular radio technology.

Within the UK competition had started. The customer was very much in the driving seat. Cellnet and Vodafone brought their TACS analogue cellular radio systems into service in 1985. The UK operators viewed 1993/94 as the next likely opportunity point for introducing any new technology based on the demand for additional capacity. One forecast made by Cellnet in late 1984 stated that it would not prove feasible to introduce a new European standard until after the turn of the century. This dispersion of dates when the "natural" window of opportunity to introduce a new network technology would arise in different European countries was a crucial factor in any strategy for introducing a Pan-European system. In 1984 little thought had been given to this factor. One reason for this was the priority the PTT tended to give to the needs of its supply industry over that of its customers. The monopoly enjoyed by the PTT enabled them to control the time when new technology was introduced. The customer was often simply made to wait.

2.4.4 Why the MoU Emerged from the UK

In the 1982–1984 period a cultural divide opened up between the UK and the rest of Europe in the telecommunications field. The UK government was in the business of promoting open markets and competition. It had turned its back on "picking winners" amongst emerging technologies. The concern of other governments in Europe remained to promote the strength of their national manufacturing champion. This was a part of the national consensus in these countries. At the time, some viewed competition in telecommunications as an idiosyncrasy of the British. It fell to the Department of Trade and Industry to build bridges between the two ideologies. The single European market was just emerging on the EC agenda. The UK had the opportunity to engage and shape opinions. An open and competitive market across the whole of Europe would bring huge benefits not just for the UK but for the whole of Europe. This in turn would make a powerful contribution to world trade.

It was this space created by the events at the time that put the DTI in a unique position. It understood the workings of a competitive cellular radio market. Its positive engagement with Europe provided insight on the potential of digital cellular radio technology to provide consumer benefits and achieve a mass consumer market. This was the stimulus to think ahead on how the GSM standard could be brought to market. The French and Germans had already thought the game plan through of linking research activities to the process of standardisation. It fell to the UK to work out the strategy to move from the process of standardisation to the process of launching, on a commercial basis, a new technology into an emerging competitive European cellular radio market.

It was evident that the emergence of a technical standard would not in itself create a market or force cellular radio operators to invest in new networks. Most were still investing heavily in the analogue networks that they already had recently introduced. A mechanism had to be found to capture the drive towards digital cellular technology being created by the French and German Governments and harness, at the same time, the dynamism of a competitive market of the sort emerging in the UK. The key was to identify exactly what drove a cellular radio market forward and build the strategy around that "driver".

2.4.5 What Drove a Cellular Radio Market Forward?

A key element of the DTI analysis of the mobile radio market was that it was a two-tier market. The cellular radio operator provided the "service" that customers wanted to buy. The manufacturing industry supplied the equipment to enable the operators to mount the service. The availability of network equipment and cell phones alone would not in itself create a new market. An entity had to raise the capital, organise the human resources to find base stations sites and roll out the network. Efficient retail distribution channels had to be organised for the cell phones. IT systems had to be in place to register customers, collate their calls and regularly bill them. The service had to be sold to the public. These were the essential functions provided by the cellular radio operators. The cellular radio operators were the key players needed to drive a new network technology into the market.

Another emerging trend observable in 1984 was the popularity of the hand portable cellular phone. Prior to this, the cellular radio was known for good reason as a "car phone". There were practical factors behind this. The car battery was large enough to supply the power-hungry early cellular phones. The aerial mounted on a car was more efficient than a hand portable aerial. This provided the power to cover the typical distance to the local base station. Physical size was not an issue for the car phone. The bulk of the equipment could be accommodated in the boot of the car. The very first cellular mobile hand portables were certainly bulky. One became affectionately known as "the brick". The first sets small enough to be put into a (large) pocket achieved the size reduction by simply reducing the size of the battery. This had serious disadvantage for the customer. The battery only lasted 5–6 hours before it needed to be recharged. Even so, 6 hours was only achieved if the user didn't make or receive any calls. If the set was used, the transmitter drew off additional energy from the battery and the time before the battery needed to be recharged was significantly shortened.

In spite of these severe constraints the popularity of the hand portable cell phones was rising. GSM had recognised hand portables in its terms of reference as early as 1982. In 1986 the number of hand portable cell phones only represented 16% of the total population of cellular radios but was 40% of all new connections. Improved designs were starting to emerge. The market trend was recognised in the both the Nordic and UK markets. Pressure from these countries elevated "ability to support hand portable" cell phones to one of the key six CEPT criteria for evaluating the proposed GSM technology. This trend had signifi-cant implications for bringing the GSM technology to market. The hand portable was pushing cellular radio in the direction of a mass consumer item. As such, the rules of the consumer electronics industries had to be applied. First and foremost, huge volumes were essential to interest and excite the semi-conductor industry. The role of the chip manufac-turers was to invest in advanced chips to allow the size of cellular mobile phones to be shrunk and their thirst for electrical power to be dramatically reduced. Their large R&D costs for doing this, in turn, had implications for the size of market that would be necessary to drive such volumes. A view emerged that a single European country, no matter how large, was not of a sufficient size alone to create the size of market needed to lift off a new immature technology. This was an important realisation and a turning point in the European telecommunications market.

These interdependencies between the different parts of the industry had to be rationalised into a structured road map that all parts of the industry understood and responded to.

2.4.6 The European Strategy to get GSM to Market

The road map to get the GSM technology to market required four parallel planes of activity:

The top plane was the political level. The European telecommunications market in 1985 was a heavily regulated market divided strongly along the lines of the member states. The political will had to be generated to make an agreement on GSM to happen. This was particularly important since markets were all out of phase. The French and German Heads of State agreement of November 1984 had started the process. The UK made its contribution through facilitating a statement in the report of the 1986 meeting of the European Heads of State. This created the political support the European Commission needed to table the directive reserving frequency bands for the GSM technology. The opening up of a new range of frequencies for cellular radio itself is a powerful stimulus for cellular radio operators. One of the neatest tricks of industrial strategy in a market economy is to link the release of new spectrum to the market with a new technology. Since nobody in the market is obliged to ask for the frequencies there is no coercion to use the technology. It is a voluntary act by the market actors. Where the market actors themselves have been directly involved in the selection of the technology the voluntary approach is re-enforced. The net effect is to channel the market drive towards the common technology and widen the public good.

This approach contrasts with forcing market actors to reverse direction on the technology that they had already chosen. The European Commission together with some member states tried this with the MAC technology for broadcasting satellites in the late 1980s. It also contrasts with the approach of Governments abrogating any responsibility and leaving it to "market forces". Where no player is dominant in the market that market will inevitably fragment amongst a number of technologies. This happened in the US in the 1980s when the FCC elected not to support a unique cellular radio standard, including the one from its own industry standards body. Its policy was to leave the choice of the digital cellular radio technical standard to the market. This FCC decision not to come behind a single technical standard in the mid-1980s lost US industry the premier position for the supply of cellular radio technology. A more serious consequence arose from the FCC's decision not to embrace the GSM technology and to allocate the same 900 MHz frequency bands for new digital cellular radio that were to be used in Europe. This would have involved moving some unrelated radio services to other radio spectrum. This is always difficult to do and so it is easy to see why the FCC took the view that they did. But the consequence was to leave its business community (and indeed all US citizens) with inferior communications tools when travelling abroad. Whilst the European business man abroad in the 1990s could enjoy sitting in a warm restaurant making a GSM call home, a US traveller would be seen queuing in the street in the pouring rain outside a public payphone with a telephone card in hand. This slightly exaggerated imagery sums up a true disparity of mobile radio support to the global travellers from the respective regions over the past decade. Only recently have the handset manufacturers closed the gap with multi-standard multi-band cellular radio hand portables. The European political leadership of the 1980s moved Europe ahead of the game by getting the right balance.

The second plane of activity was to obtain the commitment of the cellular radio operators to purchase the new networks and open a service on a common date. This was a crucial commercial agreement. Vague promises to open a service at some future time were simply not good enough. It was estimated that at least three large markets had to come on stream at the same time to generate enough economies of scale. The risks were too high to make it tenable for one operator to try and lift an immature technology on their own. The synchronised introduction into service by a number of European operators was needed to kick-start the whole industrialisation process. To launch the GSM network was by no means an assured success. Far from it, the analogue cellular radio technologies were advancing fast. Economies of scale were building up behind the NMT 900 (used in the Nordic countries and Netherlands) and TACS standards (used in the UK, Italy and Austria). Hand portable cell phone prices were falling rapidly. In the silicon driven world emerging in the 1980s nothing bred success like success. Nothing was more likely to lead to rapid death than for a new technology to find itself on a downward spiral of low volumes leaving prices high that in turn left volumes low. There had to be a commitment by the cellular radio operators to invest quite large sums of money in GSM networks of a geographic size to attract enough customers. Even better was to have this under competitive conditions, where operators would speed up the process of introducing GSM networks to get ahead of the competition.

It was also realised that a number of European operators using networks to a common standard would create the conditions to generate entirely new revenues from European roaming. Whilst it was not economical to build a cellular radio network solely for those travelling between European countries – the mere fact of having networks on an identical technology would allow these new revenues from roaming to be generated at almost zero incremental capital cost. Again tribute has to be paid to the example set by the Nordic countries that pioneered roaming between their respective countries. The concept of European roaming came at the same time as the political initiative for the single European market. Indeed cellular radio roaming across the European Union was probably one of the best examples of the benefits of a single market without frontiers.

The third plane of activity was the technical standardisation effort. Failure to agree a common technical standard would leave Europe with the continuation of its fragmented cellular radio markets. The standardisation process created the focus for the R&D efforts of the supply industry. The standardisation process started in CEPT and was later absorbed into the European Telecommunications Standards Institute. It essentially served three functions. The first was to produce technical specifications that would allow any cellular mobile radio to function with any cellular radio network. It also provided a place of mediation between the buyers of networks and suppliers of networks. A third task was to design substantial parts of the technology and its architecture. It is probably quite unique in standards making history for a system of this size and complexity to have been designed "in a committee". The insight into this remarkable achievement of technical leadership is to be found elsewhere in this book.

Finally the fourth plane of activity was the industrialisation by the supply industry. They needed to see a market and have the confidence to invest large sums of money to make it happen. In fact three industries had to be galvanised. The major systems companies had to develop the core network software. The equipment suppliers had to play their role in producing base stations and mobile equipment in good time. Finally, the semi-conductor industry was pulled behind the equipment manufacturers to shrink the complex new technology into a few a chips drawing as little electrical current as possible.

One significant difference between the GSM technology and the alternative US technical standards of the time was that the US digital cellular standards were only radio interfaces. GSM was both a radio interface and a well-defined intelligent network sitting behind it. This was a reflection of the different standards making structures in the US and Europe at the time. In the US the telecommunications network standards were made in the T1 Committee and the radio standards in the Telecommunications Industry Associations. In Europe first CEPT and then ETSI had both functions under the same roof. Europe therefore had the standards machinery to embrace a coherent set of technical standards for the complete cellular radio platform. The process of standards making in ETSI included both the cellular radio operators and the supply industry. This allowed the process of industrialisation to proceed in parallel with the standardisation effort.

2.4.7 The Construction of the MoU[2]

At one level the MoU was a simple concept. "We the undersigned cellular radio operators agree to purchase our next generation of cellular radio network to a common agreed technical standard. Furthermore we agree to open service in 1991." The first commitment created the market that allowed the industry to plan its investments. The second commitment of the common date generated the vital momentum. It also reduced the risks for all the parties involved.

The choice of 1991 was a fine judgement by all concerned. It put pressure on those with a pressing market requirement, namely France, to delay their entry into the market with new capacity until the supply industry was ready to deliver the new technology. Here is one area where history could not easily repeat itself. France should have launched a new network in 1988/89 to meet its growing customer base. It sacrificed the short-term needs of its customers for the longer-term benefits of a digital Pan-European technology and the requirement to position its industries to benefit from a new technology. In the very competitive cellular radio market in France today it would be difficult to imagine this happening again. On the other hand countries like Italy and the UK did not need the additional capacity that GSM offered until 1993. The 1991 date represented a different sort of sacrifice. The operators had to advance their investment by a few years. The emergence of new market entrants such as Mannesman in Germany, SFR in France and Omnitel Pronto Italia in Italy created additional pressure for an early date.

The date also had to be credible for the supply industry. They had to believe it and play their part in making it happen. Had an earlier date been chosen then the supply industry would not have had time to develop the system and components. GSM might have been damaged by broken delivery promises. Had a later date been chosen then the market pressure may have become intolerable and operators sought political dispensation to introduce an "interim" technology to market. This would have blown away the initiative. The date of 1991 proved to have been a good compromise.

The next feature of the MoU was the balance between what went into the agreement and what was left out. In a competitive market operators had to be free to exercise their own commercial judgements on a whole range of issues. However, there were some areas where the operators had to be prepared to bind themselves to common decisions. The technical

---

[2] The text of the GSM MOU can be found on the attached CD-ROM in folder D.

infrastructure and protocols to support roaming was one such area. Great care was taken to get this balance right. The agreement had to look to the future of a fully competitive cellular radio market and not model itself on the cartel-like features of the old CEPT. Yet the telecommunications operators who were still part of the old CEPT way of thinking had to be comfortable with the MoU. The DTI was uniquely qualified in the 1986 time period to understand both worlds and find a framework that would work for both.

One element of the MoU was a break with the prevailing CEPT culture of the time. It stipulated that "operational networks *shall* be procured by the signatories with the objective of providing public commercial services during 1991". This made sense. If all the operators were under the same time constraint and taking on the same level of risk they were far more likely to converge on the same decisions. Pressure was to come later from some CEPT countries to soften this clause. They wanted to sign the MoU and enjoy the status of being in the vanguard of European technological development but then procure their systems at a later more convenient date. They pressed for the word "shall" to be changed into a "best endeavour" language. This was firmly resisted. As a result the GSM opportunity was brought to the top of the agenda of the boardrooms of the European PTTs and minds were crystallised.

The chosen instrument of a "MoU" also was a matter of getting the balance right. CEPT traditionally expressed its agreements in the form of a CEPT Recommendation. A "Recommendation" was something that could be ignored when it became inconvenient to a national authority. Using this instrument could have been interpreted by the supply industry as being too weak. There were too many CEPT Recommendations not being followed. Industry was being asked to make enormous risk investments to bring the technology to market. Something stronger was needed. Another option might have been a commercial contract. The risk with a commercial contract was to have got drawn into far too much detail. The MoU allowed the essentials to be described in plain language. The objectives and obligations could be set out in a clear and understandable way to all the parties who depended on each other to make the GSM project a success. The detail could be defined later by the signatories to the MoU bound together by a common objective.

The focus of the Memorandum at the time of drafting was very European centric. A provision was put in for signatories to work co-operatively to provide advice and other appropriate support to operators in other parts of the world who were considering using the GSM standard. But the pressure for this to be included was largely coming from the supply industry partners. Vodafone was possibly the main exception amongst the European cellular radio operators. They were well in the vanguard of having a well-defined global vision. But for the other operators and governments the prize was a Pan-European cellular radio area stretching from Athens to Stockholm, from Bonn to Madrid and from Rome to London. This resonated with that huge project of the 1980s – the single European market. The globalisation agenda was still to come to the fore. In 1986 it was a "nice to have feature" that nobody could take exception to. There was also a feeling at the time that a Pan-European system would itself be a huge accomplishment. To set global ambitions, at that point in time, would have seemed over ambitious.

2.4.8 The MoU and its Impact on Intellectual Property Rights

The only clause that went into the text that was to have explosive consequences later was a clause calling for the signatories to co-ordinate their intellectual property right policies.

Specifically, those that affected the "open" interfaces defined in the GSM technical standard. Whilst the seeds for this conflict were sown in the Memorandum, the explosion actually took place in the European Telecommunications Standards Institute a few years later.

The intellectual property rights clause was included at the request of the French and German governments. The two governments had funded a joint R&D programme with a view to producing an open standard. They had put in their R&D contracts a condition that any intellectual property rights generated from their funded R&D programme needed for the technical standard must be licensed to third parties "royalty free". Since both governments were paying for the R&D trials it was entirely reasonable that they could set down the terms for dealing with any intellectual property rights generated as a result of the trials. However, they did not want imbalances occurring in the GSM standard where French and German companies were required to licence their patents royalty free but industries from other countries could charge for any of their patents included in the technical standard. The plan was to use the procurement contracts of the cellular radio operators who were signatories to the MoU to enforce royalty free conditions on any company who was successful in winning contracts.

This infuriated a number of industrial companies. Motorola led the charge and were somewhat unfairly characterised as the villains by a number of European Operators. In fact a number of European companies were equally upset but kept their heads down since they had been parties to the funded R&D programme. What annoyed all of them was that cross licensing of intellectual property rights is an important part of modern high technology industry. Enforced royalty free licensing clauses tipped the balance in favour of companies not investing any money on R&D. In the end one of the cellular radio operators broke rank and placed a procurement contract without the offending intellectual property rights conditions.

Life moved on. But the issue did not go away. It was carried into the European Telecommunications Standards Institute together with the GSM Group itself. Since the telecommunications world had lived with intellectual property rights without undue problems the decision taken by the ETSI management (which by then I was a part) was to park the GSM issue and try to find an intellectual property rights policy for all areas of ETSI's activity. The expectation was that the friction generated in the GSM Group would be damped down in a much wider industry group. The opposite turned out to be the case. The GSM generated tensions in this area ignited the entire European telecommunications industry. A major row ensued.

The ETSI membership split into three factions. The bulk of the European telecommunications industry (operators and manufacturers) supported an ambitious "patent pool" concept based upon "licensing by default" of intellectual property rights caught up in a public standard. A company kept its rights to exclusively exploit its own intellectual property rights, providing it made this clear to ETSI before work began on a new standard. Where no such declarations were made, members were obliged to licence any of their intellectual property rights caught up in that new standard on "fair and reasonable" terms.

A number of US multinational companies led by Motorola and Digital strongly opposed any such restriction on their intellectual property rights licensing freedom.

The third faction was the Italian industry led by the main telecommunications operator STET but widely supported by the rest of Italian industry. STET took what many in ETSI viewed as an extreme position, favouring those not having any intellectual property rights.

Agreement was nearly reached by ETSI on the ambitious scheme but the Italian members of ETSI held out too long for too much. Pressure from the US trade authorities built up behind the position of Motorola, Digital and others. The European Commission's internal view split between the directorates responsible for intellectual property rights, international trade and telecommunications. In the end, the "middle ground" consensus broke in ETSI and the US view prevailed.

One of the consequences of this outcome is to leave buried an unresolved conflict when complex technical standards are enforced upon the market as regulations. Such standards inevitably have patents buried in them. The owners of these patents are not only granted a monopoly through the granting of the patent itself but competitors are precluded by the regulation from finding alternative non-infringing solutions to enter a market. One result is that the old "national" technical barriers to trade abolished under the World Trade Agreements are effectively replaced by regulatory enforced "private" technical barriers to trade. There is some anecdotal evidence that this happened in the early stages of the GSM handset market.

2.4.9 Other Issues of Interest in the Memorandum

During the drafting of the Memorandum there was a debate on whether the agreement was for signature by the cellular operators only or by governments as well. The DTI had in mind a purely commercial operator's agreement. The Italian Ministry insisted that governments should also sign. The French and Germans specifically wanted the DTI to sign. The public argument in 1987 between Vodafone and BT/Cellnet over some frequency channels Vodafone had been given temporary use of was fresh in their mind. They found it difficult to comprehend that competitors could still co-operate on a self-enlightened basis. For them the DTI represented some necessary glue for the co-operation. This was agreed subject to carefully separating out the respective obligations. The commercial operators were to carry the commercial risks and responsibilities. The government's role was limited to appropriate supportive actions such as technical standards. The intention was that the government representatives should not be able to influence the commercial decisions.

The voting provision on technical issues in the Memorandum reflected another watershed of European telecommunications policy. Everyone was appalled that a dispute on the choice of digital cellular radio technology had to go all the way up to the level of ministers to resolve it. The Germans in particular insisted that a better way had to be found to resolve the sort of dispute over the selection of a technology that had just taken place. Others were also fed up with the principle of unanimity that allowed one delegate to block the work of an entire group. As a result a provision was put in for "weighted" national voting on technical issues. The "weighting" was modelled on the European Union model, where the number of votes tended to reflect the size of the country. However, the weighting values left the smaller countries with a lot of influence. It signalled for the first time the willingness of the powerful European PTTs to cede their sovereignty over technical standards. It laid down the foundation for the voting arrangements in the European Telecommunications Standards Institute. National voting was to come under pressure later in ETSI from the European supply industry. There is no doubt that, if such a memorandum was to be drafted today, voting rights would be given to individual cellular radio operators. However, in 1987 it was a pragmatic solution for the GSM MoU and the European Telecommunications Standards Institute to have identical voting rules on technical issues. It eliminated any risks of the lock-up between the Memorandum signatories and ETSI over technical issues as the complex GSM standard was being drafted.

Both these issues were resolved at one meeting of officials from the four countries (Germany, France, Italy and the UK) plus Sweden. Indeed the entire MoU was agreed at this one meeting based upon a draft submitted by the DTI. This reflected a very close working relationship built-up between the key officials that attended the GSM meetings over the 1984–1987 period.

2.4.10 The People who Created the MoU

The GSM was an unusual forum in standards making terms. Even within CEPT it had a special status (the original "S" in GSM). As a result it was self-contained. It handled the politics and commercial considerations alongside the purely technical function of making standards. Thomas Haug from Sweden skilfully chaired the group. His method of dealing with the political "hot potatoes" was to have informal "coffee break" meetings with senior officials from the leading countries in CEPT. For the most part this comprised Phillip Dupuis from France, Armin Silberhorn from Germany (who took over from Klaus Spindler), Renzo Failli from Italy and myself representing the UK. The reason for this "group of four" was that in parallel with the CEPT, a "quadripartite agreement" had been set up at a political level for support for the R&D and standardisation activities. This "quadripartite agreement" began its life as the Franco-German partnership announced in 1984. It was extended in 1985 to include Italy (principally the Telephone Company SIP). After some diplomatic gymnastics this had been extended to include the UK by 1986. The "group of four" senior officials essentially drove the strategy. It was entirely appropriate that I first floated the idea for a MoU in this forum.

In early 1987 the GSM project was getting a very high level of public and political visibility due to the row over the choice of technology. The CEPT found itself impotent to resolve the disagreement that emerged at the Madeira (Funchal) meeting of the GSM. CEPT worked by unanimity. The only way the impasse could be broken was through political means.

The ministers from the quadripartite countries provided the only forum that could effectively deal with the impasse. The meeting of the four ministers or their very senior officials was chaired by the German minister, Dr Schwarz-Schilling and took place on the 19th May 1987. Senior officials had prepared the way in the previous week. They hammered out a deal on the technology. The need for an Operators Agreement in the form of an MoU was also agreed. The meeting under the German minister confirmed the deal on the choice of the technology. There was then a short speech from each party. It fell to the UK minister, Mr Geoffrey Pattie, to raise the need for a commercial operator's agreement. The ministers agreed to instruct their officials to have such an agreement drawn up in the form of a MoU ready for director-generals to sign by September 1987. The signing of a four-country agreement on the technology deal then followed. The foundation of the GSM technical standard had been laid.

At the end of July 1987 the Italians had arranged a large technical conference on digital cellular radio in Venice. I agreed with Armin Silberhorn, Phillip Dupuis and Renzo Failli to take advantage of the event to draw up the MoU. We all agreed we should extend the co-operation at this critical stage to include Sweden. Through them there was a good chance of bringing in the Scandinavian countries. Senior representatives from our two cellular radio companies accompanied me. Peter Carpenter represented Cellnet although strictly speaking he was from BT. John Peett represented Vodafone. Renzo Failli had arranged for us to meet in the head office of the local Telephone Company SIP. They were housed in a magnificent old Venetian building, which they were in the middle of renovating. This blend of modern telecommunications company headquarters and a very historic building made an appropriate setting for an MoU that was bridging the old and the new structures in European telecommunications. I tabled a draft and we rapidly worked through it in the director's office, resplendent in dark oak panelling.

In the run up to the CEPT director general's meeting in Copenhagen there was speculation on how many operators would sign the MoU. The Memorandum only came into force when the "weighted" votes of the members exceeded 30. At that time a large European Community member had ten weighted votes. In other words the operators from three large countries or some combination of smaller countries and one or two large countries had to sign to make the Memorandum valid. The Scandinavian countries were known to be vacillating. They didn't need the capacity of a new system until well into the 1990s. On the other hand they had hitherto led Europe in the development of cellular radio. The idea of the large central European countries stealing the lead with the new technology would be a blow. The prevailing speculation was that the operators from between four to eight countries might sign.

The director general's meeting had been ostensibly called to discuss the establishment of a European Telecommunications Standards Institute. The current chairman of the CEPT Telecommunications Commission, Mr Loenberg, had agreed to also include the Pan-European digital cellular radio item. It took most of the morning to deal with a last minute argument on whether French (the official language of CEPT) or English (in which the Memorandum had been drafted) should be the official language of the Memorandum. This was solved by making both the English and French languages equally authoritative. The CEPT chairman then asked who intended to sign. Everyone was truly amazed at the number of hands that went up – 13 operators from 12 countries. The Spanish operator had still to receive authorisation from his board. The Memorandum was tidied up and the sheets prepared for the director generals to sign. By lunchtime I thought the job was done but not quite. The Portuguese delegate announced that he had just got authorisation from his company to sign. The Memorandum was launched with 13 operators from 14 countries. A Spanish signature was only a few weeks away. The GSM bandwagon had been launched in grand style.

2.4.11 The First Few Meetings of the MoU

The first meeting of MoU signatories was to be chaired by a representative nominated by Germany. We all pressed the German Administration hard for this to be Armin Silberhorn. He had been a representative on the Intelsat board of governors at an earlier stage of his career. This gave him a good background to get the MoU meetings on a businesslike footing. An early sign that the MoU was functioning as a market led organisation was the way Vodafone representatives were made welcome as equal partners by the other operating companies. Mercury experienced more difficulties in CEPT groups.

The only government representatives at these early meetings were from the Italian ministry and the DTI. Whilst the Memorandum had carefully delineated the role of government representatives from that of the cellular radio operators, sometimes the boundary was not so clear cut.

An exciting dimension of the GSM system was the opportunity it gave to businessmen and ordinary citizens to travel around Europe and use their cellular radio wherever they were. At one of the early MoU meetings the tariff experts from the signatory operators started to discuss the commercial principles on which this would be based.

A proposal came from several operators to apply the model used in many European countries for paging networks. Here the customer would pay a subscription according to the number of paging "cells" over which he or she wanted to be paged. To be paged over one cell covering one city would be the lowest subscription. To be paged over two cities would cost double the subscription and so on. To be paged over the whole of a country would cost the most money. The idea was that a GSM customer would contract with his or her cellular radio operator to have a service only in the home country for the lowest subscription. Then a subscription supplement would be paid for each additional country they wanted a service in. Thus, to have a roaming service over the entirety of Europe would be likely to cost the consumer a substantial amount of money.

The idea appalled me. Throughout my numerous visits to countries around Europe to overcome the difficulties of making GSM a reality I always found myself back in the immigration queue for UK and EC citizens at Heathrow airport. In those days the queues were very long. I could see a potential market for GSM phones standing in front of me. They would all want to use a GSM phone if it could be used anywhere in Europe. It contributed to my motivation to make GSM succeed. When I heard the paging model being elaborated in the MoU group I could see that vision slipping away. In any business organisation the pressure would always be on staff to minimise the number of countries they took a roaming subscription for. This was no less true for civil servants. Brussels maybe but forget anywhere else. It would certainly be true for tourists. How many people would go through the hassle of buying a temporary roaming subscription just prior to travelling? Who would bother to take their cellular phone with them on their holidays? It also had the perverse characteristic that an Italian who was put off buying a roaming subscription for UK by a high subscription rate set by an Italian operator was actually losing call revenue income for the UK operators. I believed that the net effect of applying a "paging type" tariff regime would be to depress the overall level of roaming across Europe. I expressed my dismay to Armin Silberhorn during the next coffee break. To my great relief I found that he shared my concerns. So we decided on a joint intervention.

When the MoU signatory meeting re-convened I asked for the floor. I said that all the efforts that Europe's engineers had made in GSM to remove the technology barriers to roaming was now being replaced by the tariff experts with commercial barriers. My arguments were couched in forceful terms. The speech cumulated in a simple proposition – one subscription, whole of Europe! As I made this point I banged the table slightly too hard and all the coffee cups around the table jumped up. As they landed the noise startled all the delegates. There was a shocked silence. Right on cue Armin Silberhorn came in. He admonished the experts for their backsliding. The proponents for paging tariff principles were vanquished. There was no further discussion. From that point on it was "one home subscription – whole of Europe roaming".

This decision has been questioned by some of those joining the MoU later. Others might reasonably have argued that tariff principles were a purely commercial matter between operators. It was arguably well off-limits for a government representative in the MoU to have intervened on the matter. A speaker at one of the early GSM congresses in Cannes said that the decision was over simplistic and lost some flexibility for differentiating tariff packages. I happened to be chairing that conference session. It gave me some food for thought.

Many years later I read in a newspaper of some German climbers in the French Alps who had been buried by an avalanche. France Telecom engineers had located them by triangulating on their GSM cellular radio calls. I like to think that but for GSM and the "one subscription, whole of Europe" those climbers would not have bothered to carry their cellular phones. Over the years the benefit of roaming has extended to most of Asia, the Middle East and Africa. This has thrown up other such dramatic news items. One was a reporter in a war wracked Asian country. He leapt into the undergrowth under fire from rebels. He was able to phone the authorities for help on his GSM mobile. A more recent example was a girl on board a boat that had lost its power and was drifting in stormy waters off the coast of Indonesia. The ship to shore radio had broken down. She sent a Short Messaging Service (SMS) message to her boyfriend half way around the world in England to raise the alert of her plight. Information of the plight was relayed around the world to the Indonesian authorities who launched a rescue.

The world's media picked up on the SMS aspect of this event. Everyone loved the idea of how SMS helped the boyfriend provide assistance from thousands of miles away to his girlfriend who was in difficulty. The media did not stop to marvel how it happened that the complex technology used to link both ends of the cellular radio call was absolutely compatible. There was not a second thought on the fact that when she switched on, the local GSM operator instantly accepted her as a valid customer. They may not have been aware that, just as soon as her mobile handset locked to the nearest base station, the local Indonesian GSM network and her UK home network had automatically sent a series of messages back and forth. Within a few seconds her UK operator had confirmed she was their customer and sent a complex set of data to ensure the security and privacy of all the local calls she may have wanted to make. The trust existing between the GSM operators on different sides of the globe that the call would be paid for is now assumed to be just all part of the service.

Even without such dramatic examples, there have been numerous times when the GSM hand portable has been a boon to phone home, reschedule a travel itinerary that is going wrong or maintain contact with the office. The very first article in the memorandum sets the objective of establishing an international roaming service whereby "a user provided with a service in one country by one of the network operators can also gain access to the service of any of the other network operators in their respective countries". Whatever the commercial arguments, the evidence is overwhelming that the consumer has been well served by the GSM international roaming. It is heartening that Europe has been able to share the benefits of this service with the citizens of so many other countries in Asia, Africa, Australia and other locations around the globe. It is a clear example where one technology being better that another is far less important than the major customer benefit of simply using the same technology. It is a lesson that some companies and even countries have still to learn in the cellular mobile radio market.

2.4.12 Was the GSM MoU a Vital Ingredient in GSM's Success?

A fair test of what the essential ingredients of the GSM success were is to remove each of the candidate ingredients and ask the question – without that ingredient would GSM still have succeeded? The technology was the raw ingredient but there were a number of candidate technologies. The directive reserving the frequency channels for the GSM technology certainly passes this test. It channelled Europe's market along a common technological path. Competition has to be mentioned as an essential success factor. Companies like Mannesman and Orange kept the incumbents on their toes. But the sheer weight of the whole of Europe pulling in the same direction and focussing its technological skills on a common market led project was the most vital success ingredient of them all. If today the CEO of any of the largest US technology companies were to be asked – in which area of technology is Europe ahead of the US they would all immediately say "mobile radio" and then hesitate to think of another example. This was another achievement of the MoU.

Chapter 2: The Agreement on the Concepts and the Basic Parameters of the GSM Standard (mid-1982 to mid-1987)

Section 5: The Role of the Commission of the European Communities

Philippe Dupuis[1]

2.5.1 The Early CEC Initiatives and Doubts

In the early 1980s the Commission of the European Communities took a close interest in telecommunications and information technology. This led to the creation of a taskforce which soon evolved into a general directorate, DG XIII, led by Michel Carpentier. When DG XIII officers heard about the work of the GSM Group, two of them Christian Garric and Jorgen Richter came to visit me in Paris to learn a little more. They were obviously interested. Christian Garric had a lot of sympathy for our efforts and later tried to help resolve the 1987 dispute. Jorgen Richter was put in charge of the mobile telecommunication sector in DG XIII and followed the GSM work for many years, even sitting at several GSM meetings as an observer.

In the same time frame a group was created called SOGT (Senior Officials Group for Telecommunications) consisting of representatives of the telecommunication administrations of the member states to advise the CEC on policy issues. SOGT created a working group called GAP (Groupe d'Analyse et de Prévision). It was chaired by a French colleague, Jean-Joseph Viard. GAP produced a first report on ISDN and then started to work on mobile telecommunications. I took part in the meetings on mobile communications. The CEC was then attempting to define guidelines for the European harmonisation. Ideally they would have liked to influence the work programme of the GSM. This was, however, difficult as the CEPT geographical basis was broader than the European Communities. Nonetheless they attempted to impose their vision that the mobile network should be an extension of the ISDN. Of course

---

[1] The views expressed in this section are those of the author and do not necessarily reflect the views of his affiliation entity.

the GSM group itself was trying to re-use the ISDN concepts to the greatest extent possible but had a slightly different vision in which the mobile network was interworking with, rather than integrated into, the ISDN. There was indeed no other way to proceed as some ISDN specifications were either not available or not fully harmonised on a European basis, or did not provide essential functions (e.g. mobility management and roaming). The GSM Group thus followed the spirit rather than the words of the GAP recommendations and eventually managed to stay independent from the "political" forces represented by the CEC. It is thus unfair to accuse it of having been "politically driven".

During the GAP meetings it also appeared that some of the DG XIII officers had doubts as to whether the work of the GSM group would eventually produce a viable and successful cellular communication standards, although this opinion was not explicitly expressed. The CEPT, a gathering of the historical telecommunications administrations operators did not have a good reputation in Brussels circles. CEPT was working on the basis of consensus which sometimes made it difficult to reach a final agreement. It was also criticised for "overspecifying", thus blocking innovation. Finally the ambition of the GSM Group was to introduce a digital standard and some people were not convinced that it was actually feasible.

Those who did not believe in the success of the GSM Group had ready-made contingency plans. In 1986 two 900 MHz analogue cellular standards were operating successfully in Europe, TACS in the UK and NMT 900 in the Nordic countries. If the GSM Group had failed to deliver a workable solution then either one could have been adopted as the "Pan-European" cellular system. In the competition which would have developed NMT 900 would have led. It had been developed in Europe, although outside of the EC countries. But Denmark had since joined the European Communities. Jorgen Richter was Danish.[2] The Danish representative in the GAP indeed stated that Denmark was perfectly happy with NMT 900 and did not need the GSM system. Denmark would, however, he said, build a GSM infrastructure for roaming subscribers. He also said that a danger of the GSM system was that it would be like a "ripe plum for the Japanese manufacturers to pick".[3]

The doubts expressed in the GAP were just an echo of those appearing in magazines, at conferences or in the report of studies made by international consultants. Typical statements were that the national delegations in CEPT would never succeed in reaching an agreement on a common technology, or that the system would be over-specified, or simply that a digital cellular radio technology would never work, or would require bulky terminals. The GSM Group was accused of being politically driven, of producing a "white elephant" as well as of being driven by "technology push" rather than "market pull". Whether such criticism was really spontaneous or inspired by the supporters of the analogue systems is difficult to determine. Opinions expressed in the communication satellite arena were not more positive.

---

[2] In a presentation given at the "8e Journées internationales de l'IDATE" (J.A. Richter: ISDN and Mobile Services, Proceedings of the 8e Journées internationales de l'IDATE, p. 67, Montpellier, November 1986), Jorgen Richter gave a summary of the GAP report. During the discussion which followed there was a question about the signalling requirements for the support of the mobility management in the mobile network. As an answer, to the great surprise of the audience, he just stated that the solution which the GSM Group had adopted would never work.

[3] This of course never happened. When the GSM specifications were released European manufacturers were actually at advantage because they had been associated with the elaboration of the specifications, either when taking part in the various R&D programmes or later through their participation in the GSM group or its working parties which were progressively opened to them starting in 1986, thus anticipating the future ETSI regime.

I remember making a presentation of the work of the GSM Group at a colloquium on mobile communications organised by the European Space Agency in 1986. The comment of the session chairman was that the GSM approach would never work and that only a satellite based system could provide a European wide service. Then he gave the floor to representatives of the American Mobile Satellite Consortium who were proposing to extend their service to Europe.[4] The same negative conclusion was reached in a thesis on mobile communications written the same year by two students of Ecole des Mines,[5] one of the top scientific and management schools in France.

2.5.2 CEC Contribution to a Single Pan-European Cellular Telecommunication System

Despite this lack of faith in the GSM Group the CEC gave a very positive support to the introduction of the GSM on the European market. A few administrative or regulatory prerequisites had to be satisfied for the introduction of a Pan-European cellular telecommunication system, be it digital or analogue, such as the availability of common frequency bands, the mutual recognition of the type approval of mobile terminals as well their free circulation.

In countries where 900 MHz analogue systems were operated they met with a tremendous success and the operators were requesting more spectrum. There was thus a risk that not enough spectrum would be kept aside for the initial introduction of the GSM system. A directive adopted in June 1987[6] requested member states to reserve the 905–914 and 945–959 MHz bands or equivalent bands "for a public Pan-European cellular digital mobile service by 1st January 1991" as well, in a second stage, to free the whole of the 890–915 and 935–960 MHz bands. The word "digital" had been added in the final text as the GSM Group had then selected a digital radio technique a few weeks before.

On the mutual recognition of the type approval of mobile terminals the CEC co-operated with the CEPT to implement a comprehensive scheme including the TRAC (Technical Recommendations Approval Committee), the NETs (Normes Européennes de Télécommunications), as well as a network of testing laboratories. The domain covered by this scheme was of course broader than mobile telecommunications. It was in particular intended to also regulate the ISDN. The difficulty was there again that the CEPT had a broader geographical basis than the European Community.

The free circulation of mobile terminals was easier to establish as it was just a radio regulatory matter for which the CEPT was fully competent.

The role of the CEC in supporting the adoption of the GSM in its member states through these various actions must be acknowledged. The GSM was expected to reach the market in 1991. Whether this date would be met and whether GSM would be able then to compete effectively with mature analogue systems could be questioned. Looking at the success of the cellular service in the UK and in the Nordic countries, there was thus at the end of the 1980s still a strong business incentive for telecommunication mobile operators in the other countries

---

[4] The American Mobile Satellite Consortium project was eventually cancelled a few years later.

[5] This thesis, by Marie-Pierre Boisseau and Stéphane Baugé, is probably still available at the Bibliothèque de l'Ecole des Mines, Paris.

[6] Council Directive of 25th June 1987 on the frequency bands to be reserved for the co-ordinated introduction of public Pan-European cellular digital land-based mobile communications in the European Community (87/372/EEC).

to also introduce a TACS or NMT 900 service at the risk of jeopardising the future market and spectrum for GSM. TACS networks were introduced for instance in Italy and Spain during this time frame. The frequency directive and the political pressure exercised by the CEC were then extremely useful to ensure that a sufficiently large initial market would remain open to GSM.

Chapter 3: The Detailed Specification Work Leading to the GSM Phase 1 Standard used for the Opening of Service (1987–1991)

Thomas Haug[1]

Mid-1987 was a milestone in the work of GSM. I think everybody in the group felt, despite what they might have felt earlier, that the agreement on the basic parameters in the air interface marked the turning point in the work. Another very important event took place in 1987 after years of preparations, i.e. the Memorandum of Understanding (MoU), signed by – initially – 14 operators who made a firm commitment to implement the system by 1991. This is described by Stephen Temple (Chapter 2, section 4), and I shall not go into it here. Suffice it to say that without a commitment on the part of the operators along those lines, there would hardly ever have been a GSM system in the real world, only on paper, simply because the manufacturers would never have dared to invest the huge sums in hard- and software work required to implement the system. The first set of specifications (or "Recommendations" as we used to call them at that time, keeping in line with the ITU and CEPT terminology) used for tendering is seen in the document GSM 31/88 in the acompanying CD-ROM. Another example of our intention to make the work better known was the Hagen seminar in October 1988 as seen in the documents GSM 187/87 and GSM 120/88. This activity is covered in Philippe Dupuis's chapter and I need not go into details here.

One change in the work rules of CEPT was very beneficial to GSM. Originally, CEPT had always worked as a very closed organisation, not open to outside bodies, and its results were not made public (although they often leaked out). Therefore, many CEPT standards were known to the manufacturers only if a PTT wanted to use a particular standard in connection with a contract for delivery of equipment, and usually they had very little influence in the preparation of standards. In December 1986, however, the UK representative (Stephen Temple, DTI) proposed to CCH that in the future, each delegation to CEPT working groups could bring along up to two representatives of their national manufacturing industry as technical advisors. The argument presented was that the situation had changed

[1] The views expressed in this section are those of the autor and do not necessarily reflect the views of his affiliation entity.

since the early days of CEPT, which was conceived in the days of the monopolies, and CEPT could no longer work in a secluded environment. I think it was natural for DTI to propose this since the UK was the country where the break-up of the telecom monopoly first became a reality. The change was implemented by GSM for the first time in June 1987, as it was felt that the crucial meeting in Funchal in February 1987 should be handled by the CEPT members alone.

A number of activities were gradually moved from GSM to the MoU forum, above all issues of an operational and commercial nature, and also the IPR questions, which eventually became a very difficult area to deal with. The problems were not yet solved when the Phase 1 version of the system was opened.

The technical work of GSM during this period was characterised by an ever increasing workload as more and more problems concerning details became apparent. It also became clear that the group had to rely on its own competence and if necessary its ability to enlist qualified assistance which could work under the leadership of the group, the subgroups, or the Permanent Nucleus (PN). The original idea of CEPT was that GSM should as far as possible "subcontract" work to other, existing groups in CEPT, but the way in which those groups worked was not usually suitable for the time schedule of GSM. Therefore, GSM and MoU had to start a number of activities in order to deal with those questions.

The setting up of ETSI in 1988 did not have an immediate effect on GSM, which remained a CEPT working group until March 1989. The work then proceeded by and large as before, but with the difference that many more delegates took part in the meetings since ETSI was open to all manufacturers. This was in some respects a great advantage, since many of the new members contributed a great deal to the work. The PN changed its name to Project Team 12, but stayed in Paris and went on with its work as before. The PN had a very important task to fulfil, and did it brilliantly. One of its most important tasks was to keep an eye on the various specifications which without exception went through a series of modifications, and it became a substantial burden for the staff at the PN. An efficient tool for keeping the files updated was the system of Rules for Change Requests, most of which came through the subgroups. In some meetings of GSM (SMG) there were more than 150 changes to the specifications, and all those which were approved, which means most of them, had to be worked into the specifications as soon as possible. The tools used today, when everybody has e-mail and a PC, were mostly unknown at that time, so everything had to be copied on paper and distributed in the meetings or shortly after.

One change was that the responsibility for the circulation of the security algorithms was moved to the MoU forum. Although this concerned a subject which was more technical than commercial in nature, it was considered preferable to secure the confidentiality of the algorithms, since ETSI was to be a totally open body, and it was then felt that the subject belonged to the operators' group. The specification of the SIM, however, clearly belonged in the ETSI sphere, since the manufacturers were members of that body.

Of course, there was an immense amount of work on details to be done in the years that followed, as always in a large project. In addition, the technology of miniaturisation developed quickly as we went along. Therefore, the emphasis on hand-held terminals, which is so natural today, made several changes to our specifications necessary, since we had originally based much of our work on car installed terminals, with quite different parameters in output power, current consumption, volume and weight. However, the foundation that was laid in the spring of 1987 was very firm, and despite many struggles, the goal of our work was never in doubt. It would be tedious to go through all the details of the meetings – those who are interested can study the reports of the meetings, contained on the CD – but as we all hoped, more and more bits fell into place through the following years, and the aim of the MoU to open the system by 1 July 1991, was felt to be within reach, provided a liberal interpretation of what was meant by "the system" was acceptable to the operators. In other words, it was seen that a system with a full set of features could not possibly be opened at that date, but a limited one with only the basic features could. That was what is generally termed "Phase 1". In order to do that, a freeze of the specifications had to be implemented, so that after 1990, changes should not be made to the specifications of Phase 1 unless it was absolutely necessary. However, since there was improvement and elaboration of the features going on all the time, it was necessary to start work on specifications for Phase 2. One very important objective was then to secure the compatibility between the two phases. To a large extent, this was achieved, but in some cases it proved impossible.

One problem which was not wholly within the responsibility of the committee, was the issue of type approval. This question was obviously of very great importance for the implementation of the system, in particular because the Pan-European character of the system necessitated mutual recognition of type approval between all the participating. Consequently, type approval in one country would therefore be a final stamp of approval and it would be very difficult to withdraw the approval once it had been granted. To work out the requirements for type approval was the task of the committee, while the MoU followed the development of the test equipment itself. This split in the responsibility may have been one of the reasons for the delay which the manufacturers and the operators complained about, and for a long time, the terminals had to be sold with only an interim type approval. The work on the specifications is further described by Remi Thomas and David Barnes in Chapter 17.

At the same time, there was a new task allocated to GSM. The UK expressed a need to have a Personal Communication Network (PCN) for the 1800 MHz range in place quickly. ETSI found after studying the question that since there already was a system specification available which was almost completed and could be considered stable, namely that of GSM, an update of that specification to the PCS frequency range would be the fastest way to meet the UK demand, and so, the 1800 MHz GSM was born. The change mainly involved changes in the radio parameters, channel numbering and so on, but basically, the new 1800 MHz system was a GSM system.

The first Phase 1 network to be opened was an experimental one in Finland, an event which occurred on 1 July 1991, as stated in the MoU.

In the last meeting of GSM (GSM#32, 30 September – 10 October 1991) a decision was taken to go into a new study area, i.e. the UMTS system, aiming at the next generation of mobile systems to be implemented after the year 2000, and a new Sub-Technical Committee was created for that purpose. Stein Hansen of Norway was appointed chairman of that group[2].

With the decision of ETSI to allocate the work on the UMTS to GSM, it was also decided that the name of the group should be changed to SMG as of January 1992. After meeting No. 2 of SMG in Oostend in April 1992, the author of these lines went into retirement and Philippe Dupuis of France Telecom took over the chairmanship of SMG.

---

[2] GSM#32 meeting report in the attached CD ROM folder A1, input document GSM 359/91 in folder A2.

Chapter 4: Consolidating GSM Phase 1 and Evolving the Services and Features to GSM Phase 2 in ETSI SMG (1992–1995)

Philippe Dupuis[1]

4.1 General

When the agreement on the selection of the GSM technology had been achieved some people thought that the rest would be easy. We knew however that producing a set of specifications that would ensure the interpretability of mobile stations produced by any mobile terminal manufacturer, and network infrastructure produced by any manufacturer of network infrastructure, or the interoperability of different elements of the network infrastructure produced by different manufacturers, would be a formidable task, particularly in the relatively short time frame agreed. It is thus not surprising that, when the first set of specifications was released in 1990 with a total volume of about 5000 pages, it was incomplete and contained many imperfections or errors. It was thus decided to call it phase 1 and to immediately start working on a second release called phase 2. After the testing of pre-operational phase 1 networks in 1991, followed by their commercial opening in 1992, the work of the GSM group focused on the development of the phase 2 specifications which actually lasted until 1995.

Several new specifications had to be written, but more importantly, all existing specifications had to be reviewed and checked for completeness, correctness and consistency as well as their ability to evolve in the future. The accent was on the management of the documentation. Precise procedures had to be put in force. This work was possibly less glamorous than the work done during the previous period but it was certainly equally important to the success of GSM, particularly outside of Europe. Australia had been one of the first non-European countries to adopt GSM and some Australian colleagues were sitting at our meetings to make sure that we would deliver a workable product. The volume of the work on phase 2 was formidable. In addition we invented a concept for further evolution, known as phase 2 + , and

---

[1] The views expressed in this section are those of the author and do not necessarily reflect the views of his affiliation entity.

had been put in charge of early work on UMTS. The work which took place in parallel with these two other areas is reported in the following chapters. Needless to say there was never a dull moment at our meetings.

At the same time the GSM group had become an ETSI technical committee, it was renamed SMG and I took over from Thomas Haug as chairman in spring 1992. SMG had to cope with the new ETSI environment. This was an additional challenge. While it entailed some bureaucratic burden the integration within ETSI has had some very positive effects as it enabled us to have fruitful exchanges with colleagues in other fields on the specification and testing methodology and opened new channels for mobilising the necessary resources. The old Permanent Nucleus (PN) for instance had been transferred to ETSI headquarters as a "project team". It became PT12 and was funded by the ETSI budget, through which some financial support from the CEC was made available. The various ETSI forums also provided opportunities to advertise our work and to attract contributions from companies represented in ETSI.

This section shows how a second GSM battle was won thanks to the excellence of the work done in the various sub-technical committees (STCs) of SMG, where hundreds of engineers from the European industry contributed their expertise, time and efforts. This whole process was supported by PT12. They integrated all the contributions and checked them for consistency.

4.2 SMG Within ETSI

ETSI had been created in 1988. The initiative came from the CEPT as an answer to the wishes of both the CEC and the European Free Trade Association (EFTA) who considered European standardisation in telecommunications as an important step towards the establishment of a single market.

ETSI pioneered the concept of involving in the creation of standards all concerned parties. In 1993 there were 316 full members from 25 countries comprising administrations, national standards organisations, public network operators, manufacturers, research bodies, service providers, users, etc. It was thus a large organisation and had to be ruled with well defined procedures. It was governed by a General Assembly (GA), a Technical Assembly (TA) and a Director General assisted by a permanent secretariat based at ETSI's headquarters in Sophia Antipolis, in the South of France.

The SMG chairman had to report to the TA. The meetings of the TA, usually in Nice, were actually prepared by meetings of the Technical Committee Chairmen (TCC) which took place in Sophia Antipolis. Thomas Haug, because he had been in charge long before ETSI existed, could take the liberty not to attend all of these meetings.. One factor which made things easier for him and later for me was that in the early years of ETSI the chairman of the TA was Stephen Temple who was of course convinced that ETSI had to support effectively the work of GSM/SMG. The Director General, D. Gagliardi, as former chairman of the CCH in CEPT also had some sympathy.

It was obvious however that this situation would not last for ever and that I had to attend the TCC and TA meetings on a regular basis, and "play the game".[2] One issue in the TA was the approval of our work programme. It was very difficult for SMG to comply with the proce-

---

[2] TDoc SMG 31/94, TDoc SMG 179/94 and TDoc SMG 508/94, chairman's reports to SMG describe typical issues discussed at Technical Assembly meetings.

dures as our work programme included a large number of items and was constantly updated. I was often criticised for this. Some of the ETSI members, mostly manufacturers, were indeed concerned that SMG was going too fast and that it was difficult for them to follow in developing the products. Because of this they would have liked to be able to control our work programme at the TA level. They were typically those who had not initially believed in the success of GSM and were then behind schedule in the development of GSM products.

Another important issue at TCC and TA meetings was the funding of the "Project Teams", the teams of experts who worked fulltime at Sophia Antipolis to assist the Technical Committees in the development of specifications. We had been able to keep the same level of support that we had before with the CEPT Permanent Nucleus. With a total at some times of up to ten full-time experts SMG was among the largest spenders in ETSI and I was also criticised for this. Another criticism was that we were using our PTs for the programme management, while the rule was that they should deal only with the drafting of specifications. The running cost of the PTs was paid by the ETSI budget itself funded by the members fees, or by special financial contributions from ETSI members interested in a particular project. The major part of such voluntary funding came from the Commission of the European Communities (CEC) interested in specifications needed for mobile station type approval testing. At the most difficult time the GSM MoU and some manufacturers also contributed financially to make sure that PT12 would stay alive.

The funding of PTs working in the SMG area was thus an annual headache for the chairman of SMG and the co-ordinator of PT12, without mentioning the experts themselves who often lived in Sophia Antipolis with their family, and had to know in advance whether their contract would be renewed.

It led to endless and passionate discussions with Peter Hamelberg who succeeded Stephen Temple as TA chairman and with the new director general, Karl-Heinz Rosenbrock. With the success of GSM both in Europe and outside of Europe they realised the importance of our work and things were easier. However, at the beginning I had to fight hard. I still remember an argument with Peter Hamelberg in which he had said that we were possibly overdoing it when seeking a full European harmonisation, as he thought that mobile users were mostly using the service in their home country. I told him, using words that Thomas Haug had often used in the past, that he had yet to "see the light". A few months later he started using a GSM handset and called me to say that he had "seen the light", which made our collaboration much easier.

Another problem we had with the ETSI rules was that, when GSM was adopted in many countries outside of CEPT, these countries wanted to send delegates to the SMG meetings. This had started with Australia. Later representatives of the American PCS 1900 operators became regular participants, and contributors, in the SMG meetings. The objective of this participation was of course to fully align the GSM and PCS 1900 specifications, with a view to the later introduction of tri-band operation. We also had occasional delegates from Russia and Hong Kong. ETSI meetings were in principle restricted to ETSI full members, belonging to CEPT countries, and therefore, for each of them, I had to negotiate exceptions with the ETSI secretariat. ETSI at large was however proud of the GSM world-wide success and tried to become more open to non-European participation.

4.3 Co-operation with the GSM MoU

SMG also had a close relationship to the GSM MoU Association. The chairman of SMG

presented a report at each MoU plenary meeting.[3] Having chaired the GSM MoU in 1988 I
was even invited at one time to sit in a small GSM MoU steering group including the past
chairmen. A representative of the GSM MoU also took part in SMG meetings. Initially the
MoU chairman attended our meetings, or was represented by John Moran, the MoU Secretary
based in Dublin. Later Fred Hillebrand became the full-time GSM MoU technical executive
and was put in charge of the co-operation with SMG.

This co-operation included several areas. Firstly, the operators in the GSM MoU co-
ordinated their views on the introduction in GSM of new services or features and the GSM
MoU thus provided a most important input to SMG on what to include in the work
programme. The GSM MoU was also in charge of some "administrative" matters primarily
concerning the operators, such as allocating various network or base stations codes, as well as
managing the circulation of the security algorithm. The GSM MoU had also accepted to fund
the development of the type approval test equipment, which required a strong co-ordination
with SMG itself in charge of the elaboration of the test specifications. This co-ordination is
described in more details below. The GSM MoU was also in charge of developing specifications of interest to them only, such as the one concerning the exchange of charging information between networks for roaming users, but this was clearly a separate area.

During this period the GSM MoU grew from a European to a world-wide organisation but
the excellent co-operation spirit remained what it had been in the early days.

4.4 SMG Working Structure in 1993

4.4.1 Working Structure

Like the 11 other TCs of ETSI, SMG had set up some subcommittees (STCs), actually six of
them. They are represented in Figure 4.1 below. In addition to them, two expert groups dealt,
respectively, with test specifications and speech coding. All groups, except SMG5 were
contributing to the GSM standard, SMG5 being dedicated to the co-ordination of the work
on UMTS.

4.4.2 PT12

Possibly the most important element of the working structure was PT12, the successor of the
PN who had moved from Paris to ETSI headquarters in Sophia Antipolis in the South of
France. It used to be known in SMG as PT12 although in ETSI's terminology it was a set of
Project Teams (PTs) contributing to the work programme of SMG, each one being assigned a
specific task, as represented in Figure 4.1.

This aggregation of PTs consisted of seven experts seconded to ETSI by ETSI members.
Each of them was in charge of assisting one STC, organising the meetings, keeping track of
the work programme and documentation. One of the PT12 experts acted as co-ordinator. The
first co-ordinator had been Eike Haase, who had joined the initial PN in Paris as deputy to
Bernard Mallinder. When he left in 1992, Jonas Twingler of Sweden took over until 1995, and
then Ansgar Bergmann from Germany. The role of the co-ordinator was fundamental. He was
in the best position to propose improvements to the working methodology. He had at the same

---

[3] TDoc SMG 278/93, TDoc SMG 115/94 and Tdoc SMG 125/96 are examples of such reports.

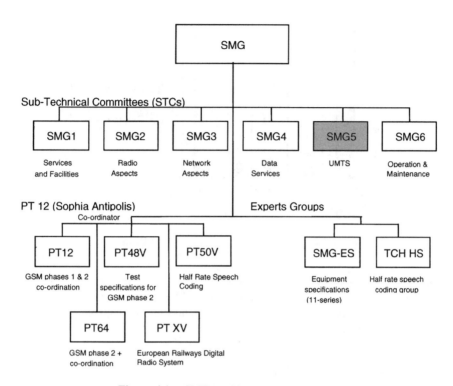

Figure 4.1   SMG working structure in 1993

time the responsibility of selecting the individuals proposed for the SMG PTs and maintain a good team spirit. For the period from 1992 to 1995 which is considered in this section one must particularly acknowledge the work done by Jonas Twingler on GSM phase 2.

4.4.3 Working Procedures

In SMG, like in other TCs of ETSI the detailed work was done in the STCs. SMG itself was thus mostly co-ordinating the work of STCs, defining their work programme, approving their deliverables and deciding about policy issues.

The GSM standard consisted then of 140 specifications and ten reports. The primary responsibility of each of these deliverables was allocated to a particular STC and to a rapporteur. Once the draft had reached a sufficient state of completeness and stability it was submitted to TC SMG for approval. From then on, other STCs based their work on this draft. This implied that any change must be brought to the attention of, and approved by, TC SMG. Proposed changes were presented in documents called Change Requests (CRs) complying with a precise format defined by PT12.

To give an impression of the amount of work, Figure 4.2 indicates the number of CRs approved by TC SMG at its Meeting Nos. 4–8. One notes the steady decrease of the number of CRs relative to GSM phase 1, corresponding to a maintenance phase in which they were mostly of an editorial nature and related to testing. For GSM phase 2 the number of CRs reached almost 300 at the SMG Meeting No. 8 at the end of 1993. Later it fluctuated around

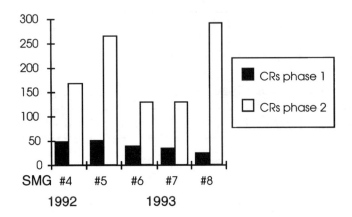

Figure 4.2   Number of CRs approved at SMG meetings

200 CRs per meeting. Of course not each of them was actually presented and discussed at the meeting. The fact that, despite this, any participant could object to it was considered as part of the "democratic" process as it allowed a small organisation not capable of sending representatives to all the STC meetings to have a say in the final decision.

4.5 The Phased Approach Concept

The concept of the phased approach had been introduced during the course of the development of the GSM specifications. In 1990 it became apparent that all the intended features could not be specified in time for the commercial opening of the first GSM networks in 1991. It was also recognised that there was still room for improvement in many areas of the signalling and procedures, as well as weaknesses to be corrected and enhancements to be included for the future proofing of the system.

It was then decided to finalise a self contained version of the specifications supporting only a subset of the planned services and called GSM phase 1, with phase 2, the complete, and at this time one thought final, version of the GSM specifications, intended for release some time later.

4.5.1 Cross Phase Compatibility

A major subject for debate was of course upward compatibility. The main requirement was that phase 1 services should remain available to phase 1 mobile stations accessing a phase 2 infrastructure. The obvious way to achieve this was for the phase 2 infrastructure to be able to identify a phase 1 MS and, once it had done so, to revert to phase 1 signalling.

Such a solution was of course feasible because added complexity on the infrastructure side was not considered to be an obstacle. On the MS side on the contrary all efforts were then made to limit complexity and such an approach could not be used. It was therefore agreed that the reverse upward compatibility objective, i.e. that the phase 1 services be available to a phase 2 MS in a phase 1 infrastructure, could necessitate some upgrading of such infrastructure.

4.5.2 Transition

In the networks the transition from phase 1 to phase 2 was expected to proceed in a gradual way and indeed started very early as most of the products designed after 1992 were already compliant with some parts of the phase 2 specifications.

On the MS side however, a clear distinction between a phase 1 mobile station and a phase 2 MS was needed as the infrastructure had to give a different treatment to both categories. There has thus been a stepwise transition at the date at which the first phase 2 mobile station entered a GSM PLMN. This of course was closely linked to the initiation of the phase 2 type approval and took place only in early 1996.

For the reverse case of a phase 2 MS operating in a phase 1 network, report GSM 09.90 released in March 1993 had given guidance to the early PLMN operators as to the modifications which should be introduced in their networks to remove possible incompatibilities. This report was remarkably short showing that it was not a difficult issue.

4.6 The Work on the Core Specifications

4.6.1 Phase 1 Maintenance

Before going into the work on phase 2 let us recall that SMG had also the responsibility of the maintenance of the phase 1 specifications, e.g. to correct errors or ambiguities that would be discovered, normally this concerned the test specifications. One interesting issue came at the time Slow Frequency Hopping (SFH) was activated in the networks. Although all mobile equipment has been tested for SFH it was discovered that certain units did not behave well. When it was discovered that a first mobile equipment manufacturer had put faulty units on the market, the other manufacturers were rather inflexible and did not want to seek a solution that would avoid recalling such mobile units. Very soon after it was discovered that most units on the market had problems. Then with a remarkable co-operation spirit, mobile equipment manufacturers, base station manufacturers and operators agreed to jointly identify solutions which could be implemented in the base stations to minimise the inconvenience. Report GSM 09.94 was presented on this subject at SMG#15 in July 1995. That competitors could work together to jointly correct their respective mistakes is another demonstration of the value of open standardisation forums.

4.6.2 Phase 2 Additional Services

It is often thought that the GSM phase 1 specifications contained a drastically reduced set of services. This is not true. The phase 1 specifications indeed included the basic voice services, almost all data services initially planned, fax services, a set of supplementary services, such as call forwarding and call barring, and even Slow Frequency Hopping (SFH). The support of multiple encryption algorithms, to suit the requirements of the law enforcing agencies, was even added later. The most likely reason why fax and data services were offered, and features such as SFH activated much later, is that operators and manufacturers attempted to spread the huge development efforts and costs required by the introduction of GSM over time. They also had to concentrate all their available R&D resources on the development and debugging of the basic telephony service.

The inclusion of additional services was thus not actually the main objective. New phase 2 services, however, included now popular supplementary services such as *calling line identi-fication*, *call waiting,*as well as *multiparty communications*and *closed user group*. It is also worth noting that many new or improved functions were defined for the Short Message Service (SMS) as well as for the Subscriber Identity Module (SIM).

4.6.3 Optimisation

Optimisation aimed at the correction of errors or limitations discovered in the phase 1 specifications, as well as the improvement of some key functions such as cell selection and re-selection for a Mobile Station (MS) in idle mode, and handover. It also included the merging of the GSM and DCS 1800. When DCS 1800 had been introduced in the UK the specifications were urgently needed and for this reason were managed separately from GSM phase 1, leading to divergence in some areas. In phase 2 it was decided to fully merge both, thus paving the way for the future dual band operation.

4.6.4 Future Proofing

This concerned in particular the addition of *hooks*intended to allow the later introduction of new elements, in the same way the introduction of multiple encryption algorithms has been made possible.

4.6.5 Specification Methodology

Such a huge amount of work could not have been achieved without a clear methodology. SMG 3, the STC in charge of network aspects, particularly under Michel Mouly's chairman-ship was more particularly involved in this.

Initially there had been a debate as to whether GSM would follow the OSI layered model. Some argued that it would make the signalling more complex but eventually GSM followed the OSI model. More apparent at our meetings was the use of the Stage 1, 2 and 3 methodol-ogy. To illustrate it in simple terms, if SMG was asked to specify a new service, the work would start with the drafting of a precise functional specification, which was called Stage 1, normally under the responsibility of SMG 1. On the basis of this specification one would then identify the required information exchanges between network elements, which were the subject of the Stage 2 specification. Finally SMG 3 would produce the detailed specification of the corresponding signalling messages known as the Stage 3 specification. This was the basis on which SMG and its STCs worked as a well-oiled machine.

4.6.6 The Half-Rate Speech Codec

When the speech codec, later known as "full-rate speech codec", was adopted in 1987 one believed that a more efficient speech codec with a much lower bit rate would soon be feasible. Radio channels were thus designed in a way which would allow the later introduction of a half-rate speech codec, thus doubling the capacity of the networks, as well as the spectrum efficiency of the system. It was thought that half-rate speech was needed to make GSM competitive on the world market.

The rationale behind this plan was that the progress in the state-of-the-art of speech coding would produce more sophisticated coding algorithms achieving a greater data compression. Such algorithms would of course require more computing power and memory space but the progress of microelectronics allowed the packing of more of this into a GSM handset. Because this evolution did not proceed as fast as expected, or the GSM requirements were to difficult to meet, workable solutions emerged only in 1992. A pre-evaluation was conducted which in early 1993 concluded that the two best proposals originated from Motorola and ANT. Both companies then proceeded to fully develop their proposal and delivered a hardware implementation as well as a C-code description of their algorithm for the final round of tests. Both candidates had a complexity of 3–4 times the complexity of the full-rate speech codec. It should be noted in this respect that SMG had decided that half-rate mobile stations should have the full-rate capability, and the codec had thus in fact to be "dual-rate".

An argument arose at SMG#8 in October 1993 as Motorola had been late in delivering these elements but they were forgiven. The big argument came at the next meeting, SMG#9 in Nice in January 1994.[4] The results of the subjective tests indicated that the Motorola codec was better that the ANT one but it did not meet all the agreed requirements. Both candidates had also poor performances in high acoustical background noise and in tandeming conditions, i.e. on mobile-to-mobile calls. It was thus not possible to make a decision based on the previously agreed rule of the competition. There was a strong Motorola delegation led by Dr Mikulski from Motorola headquarters in Schaumburg, where I remember having met him as early as 1982. There was certainly among us a willingness to reach a conclusion on a matter which had been on the agenda for too long. On this basis we could possibly have forced a decision to adopt the Motorola codec, but Dr Mikulski demonstrated his co-operation spirit in agreeing with Dr Guels of ANT that both companies would work together to develop an improved version of the Motorola codec. Eventually its specifications were approved by correspondence in February 1995.

What happened later is another story. To my knowledge there has thus been no large scale implementation of the half-rate speech codec. The market decided that there was more straightforward solutions to increase the capacity of the networks, in particular the introduction of micro-cells. One reason may also have been that a high quality full-rate codec was adopted by SMG soon after as we will see in Chapter 5.

4.7 The Work on the Test Specifications

4.7.1 Back on the Type Approval of Phase 1 Mobile Equipment

The GSM phase 1 specifications were frozen in 1990, in order for the first networks to start service in mid-1991. Commercial operation was, however, postponed until spring 1992, because of the delay in the development of the mobile stations and test tools for their type approval. Eventually type approval started in early 1992, but it was considered only as an interim type approval. A new European type approval regulatory regime had indeed been defined, to be enforced starting on the 1 January 1994. GSM phase 1 full type approval had thus to be based on the European Directive on telecommunication terminal equipment (Directive 91/263/EEC) and at the same time to include a more comprehensive set of tests.

---

[4] TDoc SMG 524/93 deals with the decision making procedure to be followed at this meeting.

In 1992 and 1993 SMG produced the corresponding test specifications. The final version of the corresponding Technical Basis for Regulation (TBRs) was approved at an extraordinary meeting held in Paris on the 28 May 1993. These TBRs were later incorporated into the Common Technical Regulations (CTRs) 5 and 9, the first ever CTRs of the new regime, adopted on the 28 September 1993.

4.7.2 The Development of the Test Tools

To ensure the full interoperability of mobile equipment produced throughout the world it was important that all type approval laboratories use the same test equipment, at least that there exists a set of reference test tools.

SMG was not in charge of the development of such testing tools, be they hardware or software. The responsibility of SMG ended with the delivery of the test specifications. Of course, the development of the specifications had to be done in close liaison with the entities in charge of the development of the test tools, e.g. giving the higher priority to the development of the test most likely to be performed for type approval.

For GSM phase 1 the initiative had come within the GSM MoU where a group of operators and administrations agreed to fund the development of the System Simulator (SS) used for type approval testing and entered a contract to this purpose with Rhode & Schwartz of Germany. The corresponding phase 2 activities were again initiated in the MoU even if there was an increased involvement of other actors, such as the test houses, and the CEC who provided part of the necessary funding.

4.7.3 Issues Related to Type Approval Testing

Several difficulties as well as other important issues were debated regarding type approval testing.

One difficulty was that tests of a quite different nature were required. The verification of the radio characteristics on one side was mostly a matter of hardware, while on the other side the tests of the signalling protocols were essentially a matter of software. In GSM phase 1 it was attempted to use a unique system simulator for all tests. This was a challenge in a way that the expertise on radio hardware and signalling protocol was in general located in different companies. Therefore for GSM phase 2 it was decided to use several machines.

One issue was the depth of testing. The appropriate depth of testing for type approval is a matter of judgement as it is known that no finite set of tests can guarantee the full conformance of the implementation of a complex protocol. Operators were in favour of a large number of tests as they expected that it would save them future trouble in the operation of their networks. Manufacturers were not far from the same position as they were afraid to have to recall later large quantities of mobile stations needing a software correction. But regulators were just interested in the verification of "essential requirements", such as user safety, electromagnetic compatibility, effective use of the spectrum, etc. This was the subject of endless discussions leading to thinking that in the future it would be necessary to create an "industry certification" complementing the regulatory type approval.

Another issue concerned the extent to which the development of the test specification should be based on the use of the Tree and Tabular Combined Notation (TTCN) test description language. TTCN had been developed in ETSI with the support of the CEC and its use was actively encouraged by both.[5] A problem was that we had deadlines to meet and there were few experts familiar with TTCN. It was also felt that the use of TTCN would pay off only when TTCN compilers would be available to produce executable test software directly from a TTCN test description. This again led to passionate arguments between the involved experts in the GSM MoU and SMG.

Eventually the CEC had the last word as they were funding part of this work and SMG went to TTCN. A PT was created to convert some of the layer 3 tests from prose to TTCN. SMG#16 in October 1995 released the phase 2 mobile station test specification, GSM 11.10, consisting of three parts: prose specifications in Part 1, Implementation Conformance Statements in Part 2 and layer 3 TTCN test cases in Part 3. Part 3 was of course available as a computer file. This allowed the phase 2 type approval to start in early 1996. This result had been achieved thanks to the patience and dedication of many experts, more particularly David Freeman, John Alsoe and Rémi Thomas who successively chaired the group in charge of developing the test specifications. This group eventually became a full STC, SMG7.

---

[5] SMG TDoc 145/93 from Tele Danmark supports the use of TTCN for protocol testing.

Chapter 5: Evolving the Services and System Features to Generation 2.5 by the GSM Phase 2+ Program

Section 1: The GSM Phase 2+ Work in ETSI SMG (1993–1996)

Philippe Dupuis[1]

5.1.1 The Invention of Phase 2+

5.1.1.1 The Invention of the Phase 2+ Concept

In 1992 SMG had to stop adding new items to the phase 2 work programme. It was nevertheless clear that there would be something after phase 2. Some proposed to call it "phase 3". This would of course have later caused some confusion with third generation. But the actual reason why SMG rejected this expression is that it would have suggested a phase 2/phase 3 transition similar to the phase 1/phase 2 transition, while it was thought that, for this further evolution, one should aim at a full cross phase compatibility. It would not have been indeed acceptable to impose a retrofit or upgrading to all networks each time a category of mobile equipment with novel features is introduced to suit a particular market requirement. As explained in Chapter 4 such a full cross phase compatibility required more complexity in the mobile equipment but this was now feasible in most cases. Even in phase 2, a trend in this direction existed with the decision to have mobile equipment with multiple encryption algorithms or multiple speech codecs. The expression "phase 3" having been rejected different periphrases were used such as "beyond phase 2", or "the further work programme". Then at the beginning of 1993 PT12 invented the expression "phase 2+" which was immediately adopted without debate. It appeared for the first time in the meeting report of SMG#6 (Reading, March 1993).

Phase 2+ was considered to be open ended. Its objective was to allow GSM to adapt to new

---

[1] The views expressed in this section are those of the author and do not necessarily reflect the views of his affiliation entity.

market requirements coming from users, operators or manufacturers and resulting from growing users expectation or from the progress in microelectronics which, for instance, makes it possible to introduce terminals with enhanced features and functionality.

In 1996 it was decided that phase 2+ would be organised in annual releases. The objective here was to maintain a full internal consistency of the GSM core specifications and test specifications throughout the various stages of their evolution. At SMG#17 (Edinburgh, January–February 1996) an ad hoc group composed of Ansgar Bergmann, Simon Pike, Rémi Thomas and Jonas Twingler made proposals in this direction[2] which were agreed at SMG#18 (Bonn, April 1996). The first release was the 1996 release.

5.1.1.2 The Content of Phase 2+

Phase 2+ in March 1993 was just an extension of phase 2.[3] The phase 2+ workplan contained new supplementary services such as "Completion of calls to a busy subscriber", "Malicious Call Identification", "Compression of user data". One work item, "Three-Volt technology SIM", was a consequence of the progress of microelectronics, as is explained below. There was nothing really revolutionary even if other items were already more future oriented such as "Extension to the SMS alphabet" because of the adoption of GSM in the Arab countries, and an expected similar move in China. "Service to GSM handportable Mobile Stations in fast trains" was also triggered by the growth in the use of GSM in Europe. "Operation of Dual Band GSM/DCS by a single operator " was already a precursor of the future work on dual-band operation. Finally there were other work items required by actors in other areas, such as GSM-DECT interworking or the support of Universal Personal Telecommunications (UPT). This last category of work items never resulted in market products as eventually GSM networks themselves succeeded to serve the various underlying user requirements.

It is only at SMG#6 that a process was started through which phase 2+ would acquire its full dimension. It started with a document[4] tabled by Nokia suggesting that we could evolve GSM beyond what we had previously envisaged. This document was only 2 pages long and entitled "GSM in a future competitive environment". When it was first circulated everybody was puzzled by the title and read it. Several participants immediately expressed their disagreement very loudly. The document was indeed putting in question all the mainstream ideas about the following generation, or UMTS. It was for instance proposing to adopt improved speech coding algorithms, to introduce higher bit rate data services, etc. Many of the meeting participants therefore objected that what was proposed in the document actually belonged to the third generation and said they were very much against transforming GSM into a 2.5 generation system. Passionate discussions took place during the coffee breaks. Eventually we had to agree that all this was making sense. When the document came for discussion and was presented by Heikki Ahava it received significant support. Following a course of action which had been taken in similar circumstances in the past it was proposed to arrange an extraordinary meeting to discuss the matter further. I had a different idea. I thought that what was needed was brain storming rather that quick decisions and suggested to hold an open workshop to which we could invite experts from non-ETSI

---

[2] Report of the ad hoc group on working procedures, TDoc SMG 173/96, January 1996.

[3] The first phase 2 + work plan was produced by PT12 for SMG#7 (June 1993), see Tdoc SMG 475/93 and 517/93.

[4] GSM in a future competitive environment, TDoc SMG 234/93, March 1993.

companies. Nokia was put in charge of arranging that workshop in co-operation with the SMG chairman. I remember reviewing with Heikki Ahava during the following coffee break a list of possible topics and speakers on each of them and adding to the list the integration of Intelligent Networks (IN) concepts in GSM.

5.1.1.3 The Helsinki Workshop in October 1993

The workshop on "GSM in a future competitive environment" took place on the 12–13 October 1993 in Helsinki. It attracted 64 participants including some from organisations not belonging to ETSI. The report[5] was submitted to SMG#9 (Nice, January 1994).

It is difficult to summarise the results as proposals addressing a large number of different areas were discussed. Most of them were not entirely new. However, putting them together in perspective gave a striking effect. It made it clear that it was possible to design an evolutionary path from GSM to the next generation of mobile communications. This was indeed what Nokia had in mind.[6] They thought that the mobile communications industry, having invested so much in GSM, could not one day abandon it to adopt an entirely new system, as UMTS was then expected to be. The point was made that in other regions of the world more consideration was given to an evolutionary approach. In this respect an interesting paper was presented by Dr Tiedeman of Qualcomm. I had invited him because he had inspired in the ITU a paper from Korea suggesting the association of Qualcomm's CDMA radio technology with the GSM platform. This could have been a part of another evolutionary path towards a single world standard.

Looking at the different evolution areas I was impressed by the fact that different features that we had thought of to materialise only in UMTS[7] could be implemented in GSM, such as high quality speech, the integration of IN concepts, the integration of satellite and terrestrial mobile communications, etc.

On the first day a dinner was hosted by Sari Baldauf of Nokia whom we had met in similar circumstances at a GSM dinner in Helsinki in 1988. Her belief in the success of GSM and the growth of mobile communications was certainly an encouragement to all of us.

5.1.2 Major Phase 2+ Work Areas

It is not easy to make a comprehensive presentation of the work done in 1993–1996 on phase 2+. Work was undertaken on so many different items that it would be too long to list them all. In many cases also the results materialised some time later. In this section we just give some indication of some of the work areas which are representative of the diversity of the content of phase 2+ or of the general trend in GSM evolution.

5.1.2.1 3-Volt SIM

This is a typical example of a technology driven evolution. As the reader knows, the SIM is one of the most interesting elements of GSM. It carries all user specific data and can be inserted in any type of Mobile Equipment (ME).

[5] TDoc SMG 2/94
[6] TDoc SMG 234/93
[7] See Chapter 8, Section 1.

For this purpose it was necessary to specify the SIM/ME interface and this was done in 1990 on the basis of the current 5-Volt technology. Having left this unchanged would have prevented the manufacturers of mobile terminals from exploiting the benefits of the new 3-Volt technology, in particular a lower power consumption. In 1995 SMG thus specified a new interface and, more importantly, reached a consensus on transitional arrangements.

5.1.2.2 New Speech Codecs

The introduction of new speech codecs was needed to enable GSM to offer a speech quality fully comparable to the quality of the fixed networks and to effectively compete against them. It was decided to undertake preliminary studies for the introduction of an Enhanced Full-Rate (EFR) speech codec already at the SMG#9 meeting. Very soon after this, introduction was urgently required by North American PCS operators who were planning 1900 MHz networks using the GSM based ANSI J-STD-007 standard. Several manufacturers, including Alcatel, Ericsson, Motorola and Ericsson tabled a proposal which was adopted at SMG#16 (Vienna, October 1995). In Europe the DCS 1800 operators were the first to implement it.

5.1.2.3 Interoperability within the GSM Family and Multi-band Operation

In 1995 the GSM family consisted of GSM at 900 MHz, DCS1800 at 1800 MHz, which was then renamed GSM 1800,[8] and of the recently adopted ANSI standard J-STD-007 for American PCS systems at 1900 MHz.

The early form of interoperability between GSM and DCS 18000 networks was "SIM roaming". By inserting a GSM SIM in a DCS 1800 mobile equipment, or vice versa, an user could indeed access networks of the two categories. The same procedure was also extended to the US 1900 MHz networks. To allow this form of interoperability the network specifications of the three variants had only to be aligned to the greatest extent possible. For GSM and 900 and GSM 1800 this was part of phase 2. For J-STD-007 it was decided that European and American experts would co-operate to remove all possible incompatibilities.

Later manufacturers developed 900–1800 MHz dual-band mobile stations, followed by 900–1800–1900 MHz tri-band mobile stations. This required in addition the use of a unique radio channel numbering plan, a point that T. Ljunggren had already addressed in a presentation at the Helsinki seminar in 1993.

Beyond this it was recognised that, in Europe, it would be possible to operate mixed 900–1800 MHz networks in which dual band mobile stations would switch bands as often as required, even on the occasion of a handover. This perspective was quite attractive to 900 MHz operators who could gain capacity in high density urban areas as well as to 1800 MHz operators who could benefit from international roaming onto 900 MHz networks. For regulators it provided a way to harmonise the spectrum allocations of the different competitors. It required a further set of specifications. Thanks to the dedication of T. Ljunggren to this subject they were adopted at SMG#15 (Heraklion, July 1995). A few weeks later a live

---

[8] The expression DCS 1800 had initially been used. It was proposed by the Mobile Expert Group in 1990 following a request of some manufacturers who thought that GSM 1800 would deprive them from any flexibility in IPR negotiations. The 1800 MHz operators also preferred it as they hoped to be able to offer a more advanced set of services.

demonstration of a dual-band handover was made in Stockholm by Telia using a prototype dual-band mobile station built by Motorola.

5.1.2.4 CAMEL

In 1993 some GSM operators already offered customised services. In most cases their users could not access these services when roaming in a foreign network. The reason was that service customisation required the implementation of some IN concepts and this implementation had then been carried out by the different GSM infrastructure manufacturers on the basis of proprietary solutions. The answer was obviously to integrate IN concepts into GSM in an harmonised manner. The far reaching implications of this issue were obvious. This is why I had insisted on having a first discussion on the subject at the Helsinki workshop.

In 1994 SMG1 proposed a first step in this direction in a work item called Customised Application for Mobile Enhanced Logic (CAMEL). But operators in the GSM MoU had eventually understood the value of the IN approach and were proposing to go even further. A joint SMG/MoU workshop was then called to discuss the various proposals and ideas. The organiser on the GSM MoU side was Michael Davies of BellSouth in New Zealand. It was entitled "The evolution of GSM towards IN" and took place in Brussels in February 1995. I had invited Nicola Gatti of Telecom Italia, the chairman of the NA6 group responsible for IN within ETSI, to take part. Ambitious proposals were made including a service creation environment for mobile operators.

This integration of IN concepts was a formidable task. It was not possible to rely on the work done for fixed networks because it did not include the mobility component. It was therefore undertaken to enhance the GSM Mobile Application Part (MAP) with IN components, rather than using the Intelligent Network Application Part (INAP) under development for the fixed networks. All this took time and only a first phase of CAMEL was included in the 1996 phase 2+ release. In the meantime operators and manufacturers could not wait and continued to introduce services based on solutions which were not standardised, or "quick and dirty" according to an Ericsson colleague. But a trend had been set which was eventually going to bring GSM further along the evolutionary path towards third generation.

5.1.2.5 GPRS

All GSM data services were initially based on circuit switched solutions and consequently charged on the basis of the connection time. The attraction of packet based data services is the ability to avoid the connection set-up time and to be charged on the basis of the amount of data transferred, irrespective of the connection time, which makes it possible to keep a permanent connection.

In the early days of GSM, probably around 1988, two companies IBM and Motorola had suggested that the GSM should include packet mode data services. This was rejected. As GSM was based on a circuit switched architecture it was not so easy to accommodate packet mode services.

In 1992–1993 SMG was again under pressure to introduce packet mode services in GSM, both by the CEC who had a special interest in road transport telematics applications and by the UIC (Union Internationale des Chemins de fer) who were about to select GSM as the technology on which applications for the European railways would be developed. Another reason for SMG to start working in this area was that CDPD, a packet mode service, had been introduced in some US cellular networks. A packet mode service was therefore needed for GSM to be competitive on the world scene. The General Packet Radio Service (GPRS) was then adopted as a phase 2+ work item. Work was initially expected to be completed in 1994 but later it was recognised that it could not be finalised before 1996. In the meantime another objective of GPRS had emerged which was to provide an efficient access to the Internet or other IP networks. This has now become the major stake of GPRS.

More details can be found in the dedicated description in Chapter 7, Section 8.

5.1.2.6 SIM Application Toolkit

Initially called "proactive SIM" the aim of the SIM application toolkit is to take advantage of the unused computing power available in the SIM. To do this a major obstacle was that the protocol used at the SIM–ME interface includes commands from the ME to the SIM, not vice versa. This protocol had thus to be expanded to allow for instance the SIM to control the display of information on the ME screen (e.g. a menu) or the transmission of short messages (SMS). With this it becomes possible to run in the SIM a simple application allowing for instance a mobile user to access a banking server via automatically generated SMS and perform simple transactions. One of the first applications of this technology has indeed been developed by Cellnet and Barclays bank.

5.1.2.7 Extension to the SMS Alphabet

The first countries who adopted GSM and did not use the Latin alphabet were the Arab countries. This was in 1992 and therefore work in this area started early in SMG4. Initially we were following the work in CCITT which was expected to produce alternative alphabets. Progress was very slow. But in 1995 we discovered that the ISO (International Standards Organisation) had almost completed the development of a Universal multi-octet Character Set. The basic plane consisted of a set of more than 65 000 two-octet characters known as Unicode. It was then just a matter of a few meetings for SMG4 to finalise a specification allowing short messages to use either our initial alphabet, then called the default alphabet, or Unicode. Of course with two-octet characters it meant that the maximum length of a message was 80 instead of 160 characters. But it was understood that it would be easy to implement a compression algorithm. My major concern at this time was not to disappoint our Chinese colleagues who were then building GSM networks at full speed. It was solved immediately and I still remember an MoU plenary meeting at which representatives from Hong Kong, China and Taiwan came to me together to learn more about this solution. I also remember that a few months later one of the leading GSM mobile handset manufacturers gave live demonstrations in Beijing of the transmission of SMS in Chinese ideograms. This story also illustrates the convergence of telecommunications and information technology. We were expecting a solution to come from the telecommunications world and it came from the computer industry.

5.1.2.8 DECT-GSM Interoperability

Following the model of the British CT2, DECT had been intended as a radio technology which would be simpler, and hopefully cheaper, than GSM and be suitable for cordless phones and for short distance radio communication services either public, as the UK Telepoint, or Wireless PABXs at industry or business sites. Many ideas emerged about possible forms of interworking between DECT and GSM. Within ETSI, SMG was asked to develop the necessary specifications. This never went too far as there was never a clear statement of the functional requirements. Other factors contributed to lessen the interest of DECT for short range radio services. One was that DECT did not include any feature to combat the effect of multipath, and even in short range applications this was a severe limitation. Eventually also one of the reason for using DECT, which was its superior speech quality, disappeared when GSM adopted the EFR. DECT is now in use in cordless telephones and there exists even a combined GSM-DECT handset which enables users at home to receive or originate calls either on the GSM network or on their wire telephone line. But this particular application does not involve any form of interoperability. DECT-GSM interoperability is thus another example of work undertaken without resulting in the successful introduction of market products. The main reason is that GSM alone could meet most of the requirements.

5.1.2.9 Support of UPT

A similar example is the work undertaken for the support of UPT. UPT was a concept invented in the 1980s in which a user would receive a "personal" user number. A call attempt using this number would be re-directed towards the current location of the user, either in the PSTN, ISDN, a GSM network, etc. SMG was asked to develop specifications for the support of UPT in the GSM networks. Again the large adoption of GSM by telecommunications users and the broad coverage of GSM networks made UPT lose its interest, as the basic requirement could be met more simply by just using a GSM handset.

Chapter 5: Evolving the Services and System Features to Generation 2.5 by the GSM Phase 2+ Programme (1993–2000)

Section 2: The GSM Work in ETSI SMG from April 1996 to July 2000

Friedhelm Hillebrand[1]

5.2.1 GSM Specification Work to Meet the Market Needs

The market provided tough challenges to the specification work. The explosive growth in users, networks and countries covered[2] called for new services, improved quality of service, higher security and capacity. Major efforts were needed to secure the integrity of the GSM specifications at the global level. The take-off of the Short Message Service (SMS) and data services and the potential of mobile Intranet and Internet access called for an accelerated GSM evolution.

The standardisation work needed to support the high growth by providing

- new services
- higher quality of service
- higher capacity
- higher security

The wide global acceptance of GSM required special attention to maintain the compatibility and integrity of GSM world-wide.

---

[1] The Technical Committee SMG in ETSI responsible for GSM and UMTS ceased to exist at the end of 31 July 2000, since all GSM and UMTS specification work has been transferred to 3GPP. I was elected SMG Chairman in April 1996 and was twice re-elected. The views expressed in this section are those of the author and do not necessarily reflect the views of his affiliation entity.

[2] For exact figures see Chapter 1, Section 3.

5.2.2 The Four GSM Releases: 96, 97, 98 and 99

ETSI Technical Committee SMG (Special Mobile Group) produced four major releases of the GSM Technical Specifications during the years 1996–2000. They cover nearly the complete GSM Phase 2+ program, the continued evolution of the basic phase 2 system. The four specification releases were: Release 96, 97, 98 and 99. The core specifications were completed at the end of the year which gave the name to the release. Often smaller parts could only be approved in the first quarter of the following year.

Typically 6 months after the completion of the core specifications the necessary operation and maintenance specifications were completed. Typically 1 year after the completion of the core specifications the Mobile Station test specifications for type approval were completed. Stabilisation lasted typically around 4 years, depending on complexity of the tested features, the number of tests, the date of implementation of test tools and arrival of mobiles in the market.

A surprise was the ever growing flow of innovative new work items, which demonstrated the vitality and evolution potential of the GSM platform. Nearly all work items were usable in UMTS. Many were critical for the UMTS success (Table 5.2.1).

5.2.3 Selected GSM Phase 2+ Work Areas

The details of the production of the four releases can be found in the SMG plenary meeting reports from Plenaries SMG#19–SMG#32. These reports are contained as reference documents in the attached CD-ROM. Each meeting report provides a snapshot of all GSM and UMTS activities at the time of the meeting. In order to illustrate the development over time the following sections provide a chronological report about selected work items. The issues are presented from the perspective of the plenary as the highest decision making body.

5.2.3.1 The Inquiry of the European Commission into "SIM Lock" and the Legal Review of the Standardisation Results

The European Commission started in early 1996 an ex-officio investigation into "alleged anti-competitive conduct" by several manufacturers, operators, ETSI and the GSM MoU Association. The subject investigated was the so called "SIM lock" feature and its use.

ETSI SMG had elaborated a specification GSM 02.22 in response to operators' and manufacturers' requests in order to avoid a fragmentation of the market by proprietary solutions. The feature allowed the firm coupling of to one mobile equipment to one SIM, so that this handset would work only with this SIM. Operators wanted to use this in order to protect their commercial interest for subsidised handsets or in case of leased handsets.

Very comprehensive material was submitted to the Commission explaining the functions and possible applications as well as details of the standardisation process. Three SMG plenary meetings and an ETSI General Assembly dealt with the matter.

During the investigation SMG amended its draft standard, deleting certain wording relating to the use of the SIM lock without changing the functionality of the feature. As the discussions became more heated, SMG suspended the work on 27 June 1996 until the completion of the investigation at the end of July 1996.

Finally a meeting was granted by the Commission on 25 July. The Chairmen of SMG and

Table 5.2.1   Overview of the GSM Releases 96–99

GSM Release 96 Core specifications completed in December 1996 contains 26 work items, e.g.:
14.4 kbit/s data transmission (including $n \times 14.4$ kbit/s)
SIM ME personalisation (including a review under competition law by the European
Commission)
CAMEL phase 1 (service creation and portability based in IN)
EFR (Enhanced Full-Rate Codec) (taken over from ANSI T1P1)
HSCSD (High Speed Circuit Switched Date)
SIM toolkit
Support of Optimal Routing phase 1
ASCI (Advanced Speech Call Items) phase 1: functions for workgroups to be used by the European
railways
GSM Release 97 core specifications completed in March 1998 contains 20 work items, e.g.:
CAMEL phase 2: additional service creation tools based on IN
GPRS (General Racket Radio Service) phase 1
CCBS (Call Completion to Busy Subscriber)
ASCI phase 2 (Advanced Speech Call Items)
SPNP (Support of Private Numbering Plan)
SMS enhancements
SIM security mechanisms for the SIM toolkit
GSM Release 98 core specifications completed in February 1999 contains 30 work items, e.g.:
AMR (Adaptive Multi-rate Codec)
EDGE (Enhanced Data Rates for GSM Evolution): basic functions
FIGS (Fraud Information Gathering System)
MNP (Mobile Number Portability)
MExE (Mobile Application Execution Environment)/WAP phase 1
TFO (Tandem Free Operation) phase 1 (in-band signalling)
CTS (Cordless Telephony System)
GSM Release 99 core specifications completed in February 2000 contains new services, e.g.:
SMS Advanced Cell Broadcast Service
MEXE/WAP phase 2: Mobile Station Execution Environment
CAMEL phase 3
GPRS phase 2: General Packet Radio Service
EDGE phase 1: Enhanced Data Rates for GSM Evolution
GSM400: GSM in 450 and 480 MHz bands
LCS: Location Services (R98, completion in 1999)
Quality enhancements, e.g.:
TFO (Trancoder free Operation) phase 2: enhancements and out of band signalling
AMR (Adaptive Multi Rate Codec): enhancements
Security enhancements, e.g.:
Signalling System No. 7 Security Review
IMEI Security: stricter principles
A5/1: use of full key length

SMG1, the ETSI legal advisors and representatives of the operators and manufacturers Associations met representatives of the European Commission. The main results were:[3]

- The Commission had sent a letter to the manufacturers on 22 July allowing the manufacture of mobiles with a SIM lock provided that a simple unlocking by the user is possible
- The Commission agreed that ETSI could continue standardisation provided an unlocking feature is included (it was confirmed that this was already the case)

ETSI issued a press announcement about the continuation of the work on 14 August 1996.[4]

Based on this experience I asked the legal advisor of ETSI to perform a legal review of all SMG output documents under competition law aspects. The result showed that the work was in line with the requirements of competition law.[5] It identified a part of a single sentence in one Specification which might have lead to problems. This text was deleted.

5.2.3.2 New Services

5.2.3.2.1 Customised Applications for Mobile Enhanced Logic (CAMEL)

The basic operator requirements to find a means of services customisation in order to differentiate themselves in competition and the creation of the work item for CAMEL is described in Chapter 5, Section 1, paragraph 2.4.

It turned out that the CAMEL concept of offering services creation and portability based on Intelligent Network (IN) concepts was such a formidable task that it had to be developed in phases. The service requirements of CAMEL phase 1 was approved by SMG#19 in June 1996. These service requirements contained basic mechanisms: trigger detection points, event detection points and operations of the Intelligent Network Application Part (INAP) protocol.

Regarding the protocol specification a discussion emerged between the ETSI Technical Committees– SMG and SPS (signalling protocols and switching). SPS was responsible for all protocols in ISDN. This included Signalling System No. 7 used in ISDN. The key mobility handling protocol in GSM is MAP (Mobile Application Part) a high level protocol using the transport capabilities of Signalling System No. 7. In the past SMG had produced stage 2 MAP specifications (architecture aspects) and SPS stage 3 MAP specifications. SPS had handed back the MAP stage 3 to SMG in 1995, since they saw INAP, the Intelligent Network Application Part, as the main avenue into the future also providing mobility management in broadband-ISDN and UMTS. CAMEL, however, needed to use some existing INAP functions and to create new INAP functions. SPS wanted to take over this work. SMG however felt that it was so deeply connected to MAP and the rest of the GSM work that it could not be separated. Another mismatch was the timing. The SMG demand was much more urgent than the demand for the INAP development, which was driven by the fixed network demand and was aligned with the ITU INAP development.

After a longer dialogue between SPS and SMG the following solution was found: SMG3 produces CAMEL stage 3 specifications under the heading CAMEL Application Part (CAP). SPS takes this work and mirrors it into the ETSI INAP specification, which is aligned with the ITU INAP. SPS would introduce as much of the CAP material into INAP as possible. The

---

[3] SMG 567/96.
[4] SMG 570/96.
[5] SMG 765/96.

serious background of this difficult balancing act was, that major manufacturers wanted to maintain fixed-mobile convergence and a single generic CAP/INAP platform in the core networks. To achieve this co-operation a lot of discussion between the Sub-Committee Chairmen (Michel Mouly, SMG3 and Hans van der Veer, SPS3) was necessary. But it would not have been successful without the support and help of Dieter Kaiser, the SPS Chairman and the constructive dialogue we had.

The ETSI reform had offered as a new means of co-operation within ETSI subcontracts between Technical Committees. SMG and SPS agreed on a subcontract for CAMEL phase 1. This was approved by SMG#21 in February 1997.[6] It was extended to later CAMEL phases. This was the first use of this innovative concept of the ETSI reform.

All specifications for CAMEL phase 1 were approved by SMG#21 in February 1997.

A first discussion on a feature list for CAMEL phase 2, which should follow 1 year after phase 2, was held by SMG#20 in October 1996. A very comprehensive list of features was elaborated. This initiated a very controversial discussion on the timing of phase 2 in February 1997 (SMG#21). Several operators wanted the very comprehensive phase 2 in Release 97. Several manufacturers pleaded for phase 2 in Release 98 in order to have sufficient time for a proper standardisation. The debate became heated. Both parties supported their case by documents. It became impossible to reach a consensus decision between the two alternatives. A show of hands resulted in a clear majority of operators who wanted CAMEL Phase 2 in Release 97 (10:3) and an equal split amongst manufacturers (4:4). In the discussion T-Mobil had made a compromise proposal for a way forward.[7] Based on the T-Mobil proposal and the discussion SMG agreed on a way forward to define a subset of the existing comprehensive phase 2 list as CAMEL phase 2 in Release 97 and to work withSMG1 and 3 towards that goal and to review the situation in the next plenary.[8] This proposal was approved. It opened effectively a race of members contributions against a fixed schedule.

SMG#22 in June 1997 was able to approve the feature list for CAMEL phase 2 and a subcontract with SPS for CAMEL phase 2.

At this meeting Alan Cox, SMG1 Chairman, proposed to introduce SPNP (Support of Private Numbering Plan) by CAMEL only.[9] This showed the potential of the CAMEL concept to end the standardisation of a never ending sequence of supplementary services by using the service creation potential of CAMEL. It was also reported that FIGS (Fraud Information Gathering System) specified by SMG10 would be based on the CAMEL platform.

The work was complex and needed an elaborated work item management. SMG#24 in December 1997 approved a revised version of GSM 10.78 "CAMEL Project Scheduling and Open Issues" as an excellent example of a project monitoring specification.[10] CAMEL phase 2 was completed as part of Release 97, and CAMEL phase 3 as part of Release 99.

[6] SMG 183/97.
[7] SMG 205/97.
[8] SMG 237/97.
[9] SMG 470/97.
[10] SMG 916/97.

5.2.3.2.2 Mobile Station Execution Environment (MExE) and Wireless Application Protocol (WAP)

SMG#22 agreed in June 1997 to establish a small project team to elaborate a work item description backed by a feasibility study on the mobile station application execution environment (MExE). This work item intended to use the intelligence of the mobile station to enable a comfortable Internet access.

In parallel the WAP Forum was established by a small number of companies. It became obvious that the work of this forum was of great relevance to the MExE work item. But the forum was not fully open to all companies. SMG#23 approved in October 19997 the work item and a feasibility study under the condition that the relationship between the SMG work and the WAP consortium is clarified before WAP is taken into account. At SMG#24 in December 1997 the work item description[11] was revised. But no full clarification regarding the co-operation with the WAP Forum could be achieved. The complaints of two companies continued at SMG#25 in March 1998.

SMG4 was asked to organise a technical workshop to review the WAP Forum results. This workshop came to the conclusion that WAP has the potential to fulfil the MExE requirements. SMG4 proposed a co-operation method between SMG and the WAP Forum. SMG4 and 9 should liase directly with the WAP Forum. The Chairmen of these groups should explore with representatives of the WAP Forum the most appropriate way to standardise WAP. It was clarified that (parts of) the WAP documents should be approved directly as SMG documents. SMG#26 agreed in June 1998 that SMG4, 9 and 12 should develop a paper on working methods between SMG and the WAP Forum, which could be approved by SMG and the WAP Forum and form the basis for a co-operation agreement between the WAP Forum and ETSI. The ETSI Board was informed.

This cleared the way for a constructive working relationship between SMG and the WAP Forum, which lead to the completion of MExE at SMG#29 in June 1999 as a Release 98 work item. The openness issue of the WAP Forum had been sorted out in the background. The successful process between SMG and the WAP Forum was certainly a catalyst for this.

5.2.3.2.3 GSM Cordless Telephony System (CTS)

The idea to use standard GSM for wide-area mobility and a "cordless" solution at home or in the office attracted support. The first attempt lead to a dual-mode DECT/GSM handset, which could operate on GSM and DECT.[12] There was very little interoperation between GSM and DECT. The work was mainly done in ETSI Project DECT and accompanied by SMG.

The next push to deal with such a concept came from the world market. The American National Standards Institute (ANSI) had specified a concept for wide-area mobility in standard cellular mode (ANSI136 TDMA) and for a cordless mode at home or in the office using cellular channels not occupied by the public mode with very low transmission power.

A GSM solution was proposed under the name "GSM Compatible Home Base Station System" by Ericsson to SMG#21 in February 1997.[13] The idea was to provide a cordless

---

[11] SMG 1032/97.
[12] See description in Chapter 5, Section 1.
[13] SMG 87/97 and 88/97.

functionality to a standard GSM mobile station with minimum impact on it (update of software only). The home base station would be connected to the PSTN or ISDN.

This proposal created a lively discussion by operators and regulators mainly focussed on frequency usage matters. The immediate killing of the work item could only be prevented by the proposal to enter into a feasibility study phase to analyse an agreed catalogue of relevant questions.[14]

The study was undertaken by an ad-hoc group formed of SMG1, 2 and 3 delegates. They completed the feasibility study and a work item description which proposed a phased approach starting with Phase 1 (speech and DTMF). These results and a name change to "GSM Cordless Telephony System" (CTS) could be approved already atSMG#22 in June 1997.[15]

At SMG#23 in October 1997 progress was reported and a majority view expressed as guidance for the work: The CTS operation should be tied to a GSM subscription and operate in the spectrum of the home network of that subscription. A completion of CTS phase 1 was foreseen in Release 98.

During late summer 1998 the interest of Ericsson as the leading company disappeared. Alcatel jumped in and provided the work item manager and most rapporteurs. All specifications of CTS phase 1 were approved by SMG#28 in February 1999 and by SMG#29 in June 1999. The work item is a part of Release 98.

The success in the market remains from my point of view a bit doubtful, since this solution requires two subscriptions (GSM and ISDN) with monthly fees. The competition comes mainly from the GSM charges. The operators are lowering the call charges and introducing local tariffs.

5.2.3.2.4 Interworking between GSM and Mobile Satellite Services

Several consortia had planned mobile satellite services for small hand-held terminals in the late 1980s and early 1990s (e.g. Iridium, ICO, Globalstar, ACeS, Odyssey, ASC). Originally the idea had been to compete with terrestrial cellular networks. But during the early development of these systems it was realised that cost and time to market was very critical. Therefore they all decided to use a standard GSM core network with the necessary minimal modifications required for satellite operation. GSM was the only complete available system standard and GSM network components were in volume production in the early 1990s. Even by using GSM core networks the cost remained high. In addition it became clear that mobile satellite services would not be able to provide in-building coverage. Therefore these parties repositioned their service as a complement for terrestrial cellular services in areas which could not be covered economically by terrestrial means (e.g. deserts[16] or oceans). This lead to the desire to offer roaming between MSS and GSM based on dual-mode/dual-band terminals and dual-mode/dual-band of operation in the networks. For this purpose most mechanisms existed already in GSM.[17] The mobile satellite services operators approached the GSM Association (GSMA) in order to get contacts to many GSM operators and the GSMA Perma-

---

[14] SMG 287/97.
[15] SMG 382/97.
[16] In Australia GSM covers 95% of the population, but only 5% of the territory.
[17] See Chapter 5, Section 1, paragraph 2.3.

nent Reference Documents needed for this purpose. They concluded co-operation agreements which lead to full membership.

GSMA proposed to SMG#19 in June 1996 a work item dealing with GSM/mobile satellite services interworking and offered to find a work item rapporteur in PT SMG.[18] The work item was approved by SMG#19 in June 1996.

During our bilateral discussions between Technical Committees SMG and SES (satellite and earth stations) a work split was found for the work item.

A point of special interest for the mobile satellite services operators was a defined space in the SIM directories. This was elaborated by SMG9 and approved by SMG#21 in February 1997. The other necessary small changes were elaborated by the Technical Committees SMG and SES in due course. Therefore all mobile satellite services operators mentioned above were able to offer roaming with GSM operators.

But the very high cost burden made the future life of mobile satellite services operators very difficult as we know today.

5.2.3.2.5 GSM Number Portability (MNP)

Identification of Requirements and Agreement on a Process
The European Commission prepared a mandate for ETSI to standardise a solution for number portability within GSM. I learned this during a visit of the Commission in autumn 1996. The requirements of the Commission were to be implemented in all European GSM networks by January 2000. I informed the Technical Committee SMG and provided the draft mandate of the Commission[19] for ETSI work to SMG#20 in October 1996.[20] SMG agreed that in order to gain time for the work the official arrival of the mandate should not be awaited and that SMG1 should start the work immediately.

In SMG1 a controversial debate arose and the work was not started since the status of the mandate and the commercial basis of number portability was seen as unclear. The GSM MoU Association tried to establish their views. During SMG#21 in February 1997 an ad-hoc group studied the matter. The main conclusion was, that more regulatory and commercial guidance was needed for firm conclusions. However, a work item description for local number portability required by the Federal Communication Commission in the US was approved. In the discussion a common solution fulfilling both the American and European requirements was favoured. It was noted that technical work needed to start urgently, but that a commercial framework is needed prior to a technical solution.

After the meeting it became known, that the Dutch regulator requested an implementation by 1 January 1999. The Dutch actors did not like an isolated activity and offered their results to SMG. Some work had started in the UK. The GSM MoU Association's European Interest Group started to work. The Commission mandate was not yet officially sent. Based on this

---

[18] SMG 461/96.

[19] ETSI is a voluntary standardisation organisation. It is organised as a non-profit association under French law. There is an agreement between ETSI and the European Commission. The Commission is entitled to award mandates to ETSI for work the Commission needs, mainly for regulatory purposes. ETSI is committed to fulfil such mandates. The Commission finances this work. The Commission grants mandates in the field of telecommunication standards to ETSI on an exclusive basis.

[20] SMG 684/96.

information I asked the SMG ad-hoc group to prepare an input to SMG#22 in early May 1997.[21]

At SMG#22 in June 1997, the ad-hoc group reported[22] and proposed SMG to task SMG1, 3 and 10 to study the matter under SMG 1 co-ordination and to establish liaison with other ETSI groups and the GSM MoU Association. A liaison statement from the GSM MoU Association[23] requested to pick up the work with urgency. The Dutch situation and requirements were described to SMG.[24] Enabled by this level of built-up requirements SMG#22 agreed in June 1997 to start the work on GSM number portability as a matter of urgency and to agree to consider the number portability with other networks in the future. The ad-hoc group presented an action plan[25] proposing a feasibility study, which should identify the regulatory requirements, consider charging aspects and a phased introduction and as one of the first actions of an open workshop. This action plan was approved by SMG#22. This decision provided the way forward after a very difficult phase of clarification of regulatory and other requirements.

The Standardisation Work for GSM Number Portability

SMG#23 received the report[26] of the workshop on GSM number portability in October 1997. It proposes a feasibility study by the end of 1997 and the actual standardisation work during 1998. A work item description had been elaborated by SMG1. It was approved.[27] A preliminary stage 1 specification (i.e. the service requirements) was presented for information. SMG#24 in December 1997 approved the stage 1 specification, which serves as "polar star" for the elaboration of the technical solution.

SMG#25 received a progress report in March 1998 from SMG3-SA (system architecture) indicating that two main alternatives had been identified which needed further study.

During the second quarter of 1998 the work of SMG12 system architecture got into difficulties. The recommendation of the European Commission, that number portability in general should be implemented soon, lead to requirements by some national regulators to implement the GSM number portability very soon. Other regulators had no such requirement. The differences in national requirements and regulations (including timing) did not allow just one solution. SMG12 proposed therefore two solutions for circuit switched applications:

- A signalling relay function solution
- An IN based solution

On the other hand it was stressed that a unified solution for the GSM number portability requirements is strongly desirable to avoid interworking problems between different portability clusters of networks.[28]

The debate at SMG#26 in June 1998 was between those who wanted a solution very soon and saw less value in international and fixed to mobile NP interworking and those who placed great value on interworking within GSM and with other networks and less value on a very

[21] SMG 322/97.
[22] SMG 463/97.
[23] SMG 420/97.
[24] SMG 510/97.
[25] SMG 547/97.
[26] SMG 724/97.
[27] SMG 867/97.
[28] Report of SMG12 in SMG 298/98.

early availability. The first group supported the solution based on signalling relay function and the second group supported the IN based solution.

The difficulty in the decision situation had been created by an insufficient harmonisation of the requirements on the standardisation. Therefore, the only way forward for SMG was to approve both solutions mentioned above and to produce specifications for both solutions for all GSM services.

SMG#28 approved the stage 2 specification dealing with fundamental architecture aspects in February 1998. The detailed stage 3 specifications were approved by SMG#29 in June 1999.

5.2.3.2.6 New High-Speed Data Services

The explosion in processing power of laptops and in the use of the Internet requested higher data rates than the 9.6 kbit/s provided by the GSM system in the beginning (see Chapter 13, Section 1) or 14.4 kbit/s specified later by ANSI T1P1 and endorsed by ETSI SMG. An evolutionary concept is provided by High Speed Circuit Switched Data Services (HSCSD), and by the General Packet Radio Service (GPRS). Enhanced data rates for the GSM Evolution (EDGE) enables HSCSD and EDGE to offer even higher data rates.

High Speed Circuit Switched Data (HSCSD)

HSCSD is not an appealing abbreviation and the service is not regarded as "sexy". But it is a real high speed service which can be easily implemented. The basic idea was that the growing processing power of the digital signal processors in the terminals would allow the processing of more than one time slot with the same hardware. This would allow the combination of the bit-rates of the slots and offer users $n \times 9.6$ or $n \times 14.4$ kbit/s, the rate of one GSM full-rate traffic channel. This can be easily implemented in terminals and networks.

HSCSD was driven by Nokia contributions. The work was completed as part of GSM Release 96. It was approved atSMG#21 in February 1997. Some small alignments with ASCI and GPRS were endorsed atSMG#23 in October 1997. Some error corrections were approved in the two following plenary meetings. These error corrections were detected by the initial product development. The number of corrections was very low.

The Damocles sword hanging over HSCSD during standardisation was the question, whether operators would charge n times the charge of a phone call for n times the rate of the basic traffic channel. In the meantime some operators realised that more user friendly charging solutions are needed for HSCSD. This was relatively easy for some "spectrum rich" operators. But another aspect was also helpful to enable such solutions. In the fixed network a circuit switched connection occupies the full resources for the time of the call. In a GSM network, however, with its clever implementation the call needs the full resources at the radio interface only when it transmits data. When no data are transmitted the feature DTX (discontinuous transmission) stops the use of radio resources. Slow frequency hopping leads to reduction of the interference. Therefore, the capacity occupied sinks, when no data are transmitted.

Much time has been lost in standardisation with the complexity needed for high rates requiring more than one 64 kbit/s channel in the fixed part of a GSM network and the interworking to the ISDN. It would have been more efficient to limit the standard to bit-rates of up to 64 kbit/s in the fixed part. At the time of writing several operators have introduced HSCSD with innovative charging solutions and bit rates of up to about 42 kbit/s. The service is popular, e.g. for a fast e-mail download.

General Packet Radio Service (GPRS)

Complexity of the Issue   The motivation and demand for GPRS is already described in Chapter 5, Section 1, paragraph 2.5. A detailed report can be found in Chapter 16, Section 3. The standardisation work started in 1994. The initial ideas were to complete the GPRS standards in 1995. But it soon turned out that GPRS had major impacts on GSM. Substantial changes were needed in the radio subsystem. In the core network an overlay network needed to be developed.

A new higher layer radio transmission mechanism needed to be embedded into the lower layer capabilities of the GSM radio interface. It allows the use of either one or several time slots or a complete radio carrier for packet transmission. This provides a data rate of about 100 kbit/s which is shared by several users. Therefore to my mind the main advantages of GPRS is "always on" and "charged according to use". "High speed" is not in the foreground of my thinking.

In the core network a packet overlay was needed. Two new types of nodes were defined:

- Serving GPRS Support Node
- Gateway GPRS Support Node

GPRS Phase 1 Specification Work for the Basic Functions   The work was so complex that it was split into several phases. The standardisation for GPRS phase 1 provided the basic packet transmission and switching functions within the GSM network. This phase lasted from 1994 to 1997. GPRS phase 1, which had started in 1994, became a feature of Release 97. The timing was over-ambitions. It became obvious from the volume of change requests which flew back from the initial implementations by all major manufacturers (see Chapter 16, Section 3, paragraph 6).

Therefore the stabilisation of GPRS phase 1 became a priority over the elaboration of enhanced features. This situation was not allowed to continue with phase 2 for Release 98. Instead GPRS phase 2 had to be postponed to Release 99. This is a good example for the case that an over-ambitious target does not lead to an acceleration of the work, since changes follow suit.

GPRS Phase 2 Provides Enhancements of Phase 1   Examples are:

- point to multipoint services
- real time services in the packet domain
- enhanced quality of service support
- advanced charging and billing: advice of charge, hot billing, pre-paid
- GPRS to mobile IP interworking
- enhanced access to Internet service providers and intranets
- FIGS (fraud Information Gathering System) on GPRS
- adaptations for the use in UMTS

Enhanced Data Rates for the GSM Evolution (EDGE)

EDGE is a new GSM radio technique for higher bit rates: It re-uses the GSM radio channel structure and TDMA framing and introduces new modulation and coding and the combination of timeslots.

EDGE phase 1 in Release 1999 is applicable to GPRS and HSCSD:

- EGPRS: single and multislot packet switched services: 384 kbit/s up to 100 km/h, 144 kbit/s up to 250 k m/h
- ECSD: single and multislot circuit switched services up to 64 kbit/s (limited by the fixed part of the network)

The ANSI 136 TDMA community plans to have EDGE as their 3G solution and had two major additional requirements for the deployment possibilities: the use of the 800 MHz band and a low amount of initially available spectrum (1 MHz). The latter requirement lead to the development of EDGE Compact.

EDGE phase 2 will be part of Release 2000; it provides enhancements, e.g. real time services.

5.2.3.3 Enhanced Speech Quality

5.2.3.3.1 The Strategy

There had been some dissatisfaction with the speech quality of the half-rate codec and the enhanced full-rate codec under difficult radio conditions, since they do not achieve wireline quality under these circumstances.

The Speech Quality Strategy Group (SQSG) had been set up by SMG#16 in October 1995 with the task of elaborating a strategy and an action plan for new GSM speech codec(s) and for effecting enhancements to other aspects of end-to-end speech quality. The group delivered their final report to SMG#19 in June 1996.[29] It contained the following proposal to avoid a proliferation of codecs:

- To develop a single integrated codec system providing "wireline" quality at half- and full-rate modes under a wide range of operating conditions.
- To provide a real-time adaptation which selects the bit-rate to provide the best combination of capacity and quality possible.

This solution was called AMR (Adaptive Multi-Rate codec system). To achieve the targets a 15-month feasibility study was initiated. In addition a new subgroup on end-to-end performance was proposed. For an integrated management the establishment of a new Sub-Technical Committee SMG11 with three subgroups was proposed.

The program found a strong interest in the GSM MoU Association and the North American GSM community. The strategy, the work program and the revised organisation were endorsed by SMG#20 in October 1996.

5.2.3.3.2 The Path to AMR

The AMR study phase report was presented to SMG#23 in October 1997.[30] The report predicts a higher robustness in full-rate mode, a quality improvement in half-rate mode

---

[29] SMG 447/96.
[30] SMG 740/97.

and a capacity improvement compared to the full-rate codecs. In addition the concept of wideband AMR (up to 7 kHz audio instead of 3.4 kHz) was introduced.

It was very difficult to find a compromise way forward between those who wanted the advantages of the narrowband AMR as soon as possible and those, who wanted to add the wideband capabilities. SMG#23 agreed the following strategy:[31]

- To start the narrowband AMR work immediately
- To assess the feasibility of a wideband AMR and decide later about the introduction/ integration based on more information

After intensive work in SMG11 the results of the narrowband AMR qualification phase was completed and reported to SMG#26 in June 1998. It included performance results and a proposal for a short list of the codec systems to be used. After a difficult discussion and a show of hands a shortlist was agreed.

In a subsequent narrowband AMR selection phase a proposal for a decision on a candidate selection was made to SMG#27 in October 1998. After a very intensive discussion SMG#27 unanimously selected the codec system for the AMR. After that an optimisation and detailed specification phase took place. These were completed as part of Release 97.

As asked by SMG in October 1997, SMG11 made a feasibility study on a wideband AMR providing an audio band width of 7 kHz on a GSM full-rate channel (22.8 kbit/s gross rate).[32] While current GSM codecs achieve a good performance for narrowband speech (up to 3.4 kHz), the introduction of a wideband speech service audio (up to 7 kHz) would improve the naturalness in voice. The study performed showed that the target is feasible. The possible higher bit-rates of EDGE and the possibilities of UMTS were also considered. Based on the feasibility study SMG#29 in June 1999 approved the work item. The work was jointly performed by SMG11 and 3GPP TSG S4. Nine proposals were received. The definition was performed in two stages: qualification and selection. The final decisions were taken in 3GPP.

5.2.3.3.3 Transcoder-free Operation (TFO) to Improve the Quality of Mobile-to-Mobile Calls

With the growth of the GSM user numbers, the number and share of mobile-to mobile calls grew. A major quality limiting factor was the transcoding from GSM full- or half-rate to ISDN coding in the GSM core network and from ISDN to GSM full- or half-rate. In order to eliminate this double transcoding, work called initially "tandem-free operation" and later "transcoder-free operation" (TFO) was initiated in 1996. Initially tests were performed in order to verify the quality gain (reported to SMG#20 in October 1996). There was a strong interest in having TFO soon (SMG#22 in June 1997). In order to reach an earlier implementation a TFO phase 1 with in-band signalling was defined. Results were presented at SMG#23, 24, 25, 27 and 28 in February 1999. TFO phase 1 became part of Release 98. TFO phase 2 (out-of-band signalling) was completed in Release 99.

---

[31] SMG 860/97.
[32] SMG P-99-42.9

5.2.4 Work Management

5.2.4.1 New Plenary Format

Before 1996, the plenary did not allocate sufficient time to UMTS. The new change request procedure (see paragraph 4.5) and more focussed debates on all subjects gained time for the adequate treatment of UMTS.

Then there was not sufficient time for consultations between the delegates and the delegates and their companies in the case of critical controversial items. Therefore the treatment of controversial items was reorganised. In these cases all arguments were presented at the plenary. Then I tried to reach a decision. I frequently used an indicative voting by a show of hands. If this did not lead to an agreement, this issue was declared a postponed item, which would be treated again on the last day of the meeting in a session dedicated to postponed items. One delegate was made responsible for trying to reach an agreement during the plenary week outside the main meeting and to report in the special session on Friday. In this special session only reports supported by a document were accepted. This allowed firm decisions which were well documented. Normally we had between 20 and 50 postponed items. It was nearly always possible to reach a consensus decision, since this process created sufficient time for consultation during the week.

This lead to a new plenary format:[33] GSM was treated from Monday morning to Wednesday noon, UMTS from Wednesday noon to Thursday evening. Friday was reserved for postponed controversial issues, which could not be resolved earlier in the week. This new format was already introduced at SMG#19 in June 1996.

5.2.4.2 Work Item Pruning

Often work items are introduced by members with great enthusiasm. Later sometimes the initial supporters lose interest. Often others step in. But there are also cases that work items fall into a "sleep mode". In order to avoid unnecessary administrative overhead but perhaps more to generate new interest, I proposed a regular review of all work items and a deletion of sleeping work items. SMG#20 in October 1996 performed the first deep discussion on this and deleted 20 work items. This was about 10% of the existing data base of all not (yet) completed work items. Such exercises were repeated regularly.

5.2.4.3 Number of Documents and Change Requests

There was an ever growing flood of documents. In GSM#16 in December 1987 the first set of specifications for tendering was approved, the plenary needed 150 000 copies. In my Chairman's period we needed between 400 000 and 600 000 copies per plenary. Copying cost was the most costly item of the hosts.

The number of temporary documents grew over time mainly caused by the number of specifications and the Change Requests (see Chapter 20) (Figure 5.2.1).

The first specification release for tendering was produced in 1988. From 1988 onwards a wave of change to the existing specifications created large numbers of documents. The peak in 1997 was driven by GSM phase 2+ and the initial UMTS specification work. The 1999 and 2000 figures exclude UMTS, since it was transferred to 3GPP. The 1999 and 2000 figures are

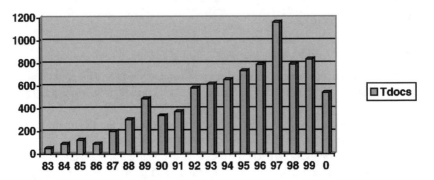

Figure 5.2.1   Number of temporary documents per year in the GSM and SMG plenary

blown up by the wave of changes in GPRS. 2000 covers only half a year, since all remaining GSM work was transferred to 3GPP.

Included in these documents was a large number of Change Requests (CRs) (normally there are several Change Requests in one Temporary Document) (Table 5.2.2).

Table 5.2.2   Change requests

Year	Number of CRs	Comments
1996	887	
1997	1187	
1998	1014	
1999	2194	Without UMTS (transferred to 3GPP)
2000	793	Until the end of July (closure of SMG)
Total	6075	

5.2.4.4 Electronic Working

These volumes requested the intensive use of electronic means in the production. But there was strong resistance by some delegates against, using in the approval process, electronic versions only since there were incompatibilities between different computers which could corrupt the meaning of a document. This problem became better over time. A breakthrough happened at SMG#21 (February 1997), when three thick detailed test specifications were not copied for all delegates. It was agreed to avoid the copying of thick test specifications under the condition that every delegate was entitled to request a paper copy at the meeting secretariat. This process slowly expanded and from the beginning of 2000 onwards no paper copies were provided to the delegates.

Another landmark in electronic working was the distribution of the meeting report, updated specifications and temporary documents on a CD-ROM after each plenary. The premiere was SMG#21 in February 1997.

Then during 1999 the first local area networks appeared in the SMG Plenaries. A main driver of this development was Kevin Holley the Chairman of SMG4 (data services).

5.2.4.5 New Change Request (CR) Procedure

The large wave of CRs had eaten up more and more meeting time by the presentation and sometimes detailed discussions in the first half of the 1990s in the SMG plenary. When I came into office I wanted more meeting time for strategic issues (mainly UMTS) but I wanted also to avoid the repetition of discussions held in the subgroups in the plenary. I proposed therefore immediately before I came into office as Chairman to distinguish between two classes of CRs:[34]

- "Strategic" CRs should be presented individually, discussed and approved.
- "Non-strategic" CRs should be submitted as documents and be approved without presentation and discussion.

The class should be selected by the subgroup, but each plenary member should have the right to request a presentation and discussion of "non-strategic" CRs. This concept was endorsed on a trial basis by SMG#18 in April 1996 and confirmed later.

It turned out that about 90% of all CRs were "non-strategic". This saved considerable meeting time. In addition this process provided an implicit delegation of decision power to the subgroups, since 90% of all CRs were effectively approved by them.

5.2.5 The Global Co-operation was Intensified and Re-structured to Secure the Integrity and Consistency of the Specifications

The global co-operation was re-structured to secure the integrity and consistency of GSM and UMTS world-wide by four major measures:

- Co-operation agreement between ETSI and the GSM MoU Association
- New working-together for GSM between ETSI SMG and ANSI T1P1
- Integration of all Chinese GSM requirements
- Close co-operation with the ANSI 136 community to evolve EDGE and GPRS as a common technology

5.2.5.1 Co-operation Agreement between ETSI and the GSM MoU Association

During the years from 1987 to March 1989 a very close liaison between the standardisation group CEPT GSM and the GSM MoU group existed, since regulators and network operators were members of both groups. Often the heads of delegations were the same persons.

After March 1989 the standardisation work was transferred to ETSI. There also the manufacturers were members. Since the strategic decisions had been made, ETSI GSM (later SMG) and the GSM MoU group worked relatively independently during the more detailed work from 1989 to the mid 1990s.

ETSI discussed and agreed new models for standardisation work in the strategic review by the High Level Task Force. A far reaching autonomy was agreed for Technical Committees on all technical matters. The concept of ETSI Partnership Projects was created. This was an activity when there is a need to co-operate with an external body and where such co-operation

---

[34] SMG 283/96.

cannot be accommodated within an ETSI Project or Technical Committee. This model had been designed within ETSI with the GSM and UMTS work in mind.

Within the GSM MoU group there was a great dissatisfaction that no non-European GSM operators could participate in the GSM work in Technical Committee SMG in ETSI. In addition there was displeasure with the influence of the ETSI General Assembly and Technical Assembly on the GSM standardisation work in Technical Committee SMG. The membership of Technical Committee SMG and GSM MoU Group comprised companies committed to the GSM work. Whereas the ETSI GA and TA was open to the general ETSI membership and every member could have influence or interfere in the GSM work.

The professional technical support (PN, PT, MCC, see Chapter 16) used for project management, consistency checking, document management, etc. was a key element of the working methods in Technical Committee SMG. No other technical body in ETSI used such a support function. Regularly delegates in the ETSI Technical Assembly questioned the need of the technical support for the GSM work. The annual budget approval in the ETSI Technical Assembly and General Assembly was a regular big fight. In addition the influence of the Technical Assembly in technical matters of Technical Committee SMG led to difficult debates (e.g. work item approval).

In the GSM MoU group remedy was sought. There was one group who wanted to transfer the GSM and UMTS work to an ETSI Partnership Project, a concept newly invented in the ETSI reform for cases with much non-ETSI participation. Another group, where I was a leading member, was appealed by several elements of the ETSI reform (opening of ETSI for non European organisations as Associate Members, autonomy of Technical Committees in technical matters, recognition of the PN concept as a valid working method in the ETSI Directives). This group thought that the existing problems could be solved by using these means. After a very intensive discussion in the GSM MoU group the second way forward did win the support of the Chairman and the majority.

In order to stabilise the situation I drafted, in consultation with Karl Heinz Rosenbrock, the ETSI Director General, a co-operation agreement between ETSI and the GSM MoU group, which was endorsed by the GSM MoU ex-chairmen group (a Steering Group) and the ETSI Interim Board and then approved by the plenary of the GSM MoU group in Atlanta in 1996 and the ETSI General Assembly.[35]

In the "considering" section it recognises the role of both organisations and their contributions to GSM: ETSI standards and GSM MoU Permanent Reference Documents on services, charging /accounting, international roaming, security and fraud. It confirms relevant elements of the ETSI reform. In the "agreement" section on information document exchange was agreed. GSM MoU is entitled to send observers who can submit documents and have the right to speak to relevant ETSI Technical Committees (i.e. mainly SMG). GSM MoU members outside of Europe got access to all GSM documents without additional payments. GSM MoU contributes a substantial fixed sum to the SMG Project Team budget. It was further agreed to "make any effort necessary in order to maintain the integrity of the GSM standards by close liaison with ANSI...".

This co-operation agreement formed a stable environment for the work of Technical Committee SMG for some time. It confirmed the leading role of the GSM MoU group

---

[35] SMG 479/96.

regarding services requirements. It allowed a limited access of non-European GSM operators. It was agreed in the framework of the ETSI reform principles:

- Openness of ETSI for non-European organisations as Associate Members.
- Autonomy of the TC in all technical matters.
- PT as a recognised working method funded by ETSI and contributions from GSM MoU, CEC, etc.

Difficulties within ETSI arose when several key elements of the ETSI reform were forgotten by certain ETSI delegates and the spirit of the old Technical Assembly and the related debates on competence, technical support and funding resurrected.

But the global acceptance of GSM continued and when a global UMTS was emerging the consistency and integrity of the GSM and standardisation work needed a new globally open organisation (see Chapter 6, Sections 3 and 4).

5.2.5.2 New Co-operation for GSM between ANSI T1P1 and ETSI SMG

A new working-together method for GSM specifications between ANSI T1P1 and ETSI SMG was introduced in 1996/1997 (see also Chapter 6). Before this time there were two independent sets of Technical Specifications with different scope and structure for GSM900/1800 and GSM1900 (used in the US). This bore the risk of incompatibilities. In addition there were differences in services and features caused by the fact that the development speed was different. This impacts on the roaming between the two GSM parts of the GSM world.

Based on a initiative by Ansgar Bergman (the leader of PT SMG) and me, ANSI T1P1 and ETSI SMG agreed to merge the two independent sets of specifications into one common set and to evolve it using an innovative co-ordinated working method by both committees. Each work item was approved by both committees. A lead committee for the work was agreed. The other committee accompanied the work by review and comments. The results were approved in both committees and incorporated into the common technical specifications.

The agreement was made in 1997 and then implemented in steps. This measure improved the integrity of GSM between the US and the rest of the world. But the working process was not very efficient, since great efforts from both sides were needed for co-ordination.

5.2.5.3 Integration of all Chinese Requirements to the GSM Specifications

The integration of all Chinese requirements to the GSM specifications was essential, since it was visible that China would become a very large market. The Chinese network operators had implemented large GSM networks based on the existing ETSI specifications. In order to get some specific additional requirements fulfilled, they had started independent specification work on the MAP (Mobile Application Part), the key protocol in the GSM core network subsystem. This had been done without contact with the GSM development in SMG. SMG took the initiative and agreed with the Chinese authorities, that they could introduce their requirements fully into the SMG work process and that they could participate fully in the work. This was implemented in 1997.[36] This measure secured the integrity of GSM between China and the rest of the world.

---

[36] See meeting report of SMG#22, p. 4.

5.2.5.4 The Co-operation with the ANSI 136 TDMA Community

A close co-operation with the ANSI 136 TDMA community was essential to narrow the differences between the GSM community and the second largest community in terms of user numbers. In order to realise synergies they chose EDGE and the GPRS evolution as their solution for the third generation. This was enabled/supported by an intensive co-operation between SMG and UWCC. A framework agreement was concluded between ETSI and UWCC. SMG took the UWCC requirements into consideration in the development of EDGE and GPRS. This included a version EDGE Compact, which could be implemented with very little spectrum and a 800 MHz EDGE version. UWCC delegates took part regularly in SMG meetings.

5.2.6 Conclusions

The four GSM Releases 96, 97, 98 and 99 produced by Technical Committee SMG during the period from 1996 to 2000 contain nearly the complete GSM phase 2+ program. There are new services and network features as well as the evolution of existing services and features. They improve the services portfolio, the quality of service, the capacity and security of GSM networks. Enhanced customisation and portability is provided by CAMEL. Internet in mobile phones is enabled by WAP/MEXE. Packet network Internet access is provided by GPRS: more than 64 kbit/s, always on-line, charging per usage. High speed Internet access up to 384 kbit/s is realised by EDGE. HSCSD provides high speed high throughput data transmission.

The quantity of innovations leads to a change in quality. GSM mobile telephony networks will be transformed into mobile Internets by implementing these four releases. The GSM phase 2+ program has achieved the major goals of the third generation already and has evolved, therefore, GSM from a second generation system to a generation 2.5 system.

With these achievements Technical Committee SMG had fulfilled its mission within ETSI. The global market success of GSM and the need to secure the integrity of GSM world-wide and the cohesion with UMTS called for a global specification organisation to perform the further evolution of the GSM specifications in one structure with the UMTS work. The work was therefore transferred to 3GPP and SMG was closed at the end of July 2000.

Chapter 5: Evolving the Services and System Features to Generation 2.5 by the GSM Phase 2+ Programme

Section 3: GSM Railway (GSM-R)

Ansgar Bergmann[1]

GSM is not only by far most successful system for public mobile cellular networks: European railways are currently introducing GSM as the harmonised platform for their (non-public) mobile telecommunication networks. The new platform GSM-R (GSM-Railways) will replace the various, mutually incompatible, national railways radio networks, which are based on analogue transmission techniques. In several countries outside of Europe, railways are also preparing the introduction of GSM-R. A Memorandum of Understanding (MoU) has been signed in most European countries (see Figure 5.3.1) and implementation is progressing well.

The main applications for GSM-R are at present:

- Voice Group Call Service and Voice Broadcast service for voice communication of the operational staff
- Data services for ETCS, the European Train Control System.

In future there will be a wide range of railway applications based on GSM-R.

5.3.1 GSM-R Features

UIC (Union Internationale des Chemins de Fer), the international railway body, started to investigate the introduction of a Pan-European mobile railways communication system in 1987/1888.

The identification of features for mobile railways communication and their characteristics was an ongoing task for several years, due to the differences in railways operation in the

---

[1] The views expressed in this section are those of the author and do not necessarily reflect the views of his affiliation entity.

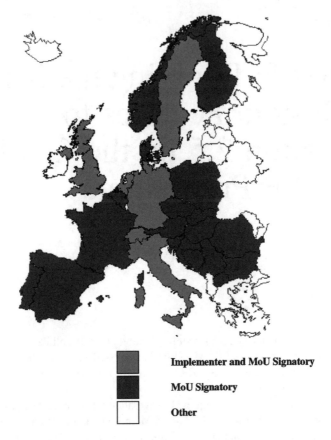

Figure 5.3.1 Implementation of GSM-R in Europe

different countries. It resulted in the FRS (Functional Requirements Specification) and SRS (System Requirements Specification).[2] Essential applications include:

- ATC (Automatic Train Control)
- Voice communication for shunting groups (the people arranging coaches for passenger and freight trains) and between train drivers and shunting staff
- Communication between train drivers and ground staff
- Multiple driver communications within the same train
- Railways emergency calls: this feature enables the set-up of a group call in case of dangerous situations; the addressed parties have to receive the call with high priority; the call set-up is performed in less than 2 s.

SRS and FRS also contain requirements for specific features, for operation, performance and quality of service, including:

- Communication for train speeds of up to 500 km/h;
- Fast and guaranteed call set-up;

[2] See www.eirene-uic.org.

- Usage of functional addressing (e.g. a train number or coach number) for setting up a call and for display of calling/called party;
- Link assurance, an indication that the connection is maintained;
- Push-to-talk button for the group communication;
- Set-up of urgent or frequent calls through a single keystroke or similar;
- Automatic and manual test modes with fault indications;
- Automatic mobile network management;
- Control over system configuration;
- Location dependent functions, e.g. connection to the local ground.

These requirements were not on the table when work started. As it happens often, they were elaborated in parallel with the realisation, always ahead and leading the development, but subject to adaptation and improvement due to feedback.

The study initiated by UIC at the end of the 1980s included:

- the identification of a suitable radio band in Europe;
- the choice of an appropriate radio communication system.

5.3.2 Identification of a Radio Band

The identification of a common frequency band was a key prerequisite for international operation as well as for economies of scale.

As a result of studies within the UIC and of discussions with CEPT, co-ordinating the management of radio frequencies in Europe, the 900 band was shown to be most suitable for several reasons such as favourable radio propagation conditions and ease of applying GSM technologies.[3]

CEPT agreed in 1990 to reserve the bands 870–876 MHz (for *uplink*, i.e. mobile station transmit) and 915–919 MHz (for *downlink*, i.e. base station transmit). Later, severe problems were discovered because the railways downlink band would have been to close to the GSM uplink band. After lengthy investigations, CEPT changed the recommendation in 1995 and assigned 876–880 MHz as the uplink band and 921–925 MHz as the downlink band.

5.3.3 Choice of a Radio Communication System

The European railways had to overcome the situation of a multitude of railways communication networks, using different analogue techniques and different frequency bands. The solution had to be a common standard. Development of an own standard was not an option.

The major goals for a Pan-European railways communication systems were identified:[4]

- digital technology;
- integration of services;
- usage of a proved standard system;
- European wide roaming and mobility management;
- suitability for railway specific services and performance;

[3] cf. Evald, J., Frequencies. In: EIRENE newsletter No. 1 (http://www.eirene-uic.org/eirene/).
[4] cf. Grethe, W., Güner, T., Münning, D., GSM-R – The New Mobile Radio Standard For Railways, Vehicular Technology Conference, IEEE, Boston, MA, Fall 2000.

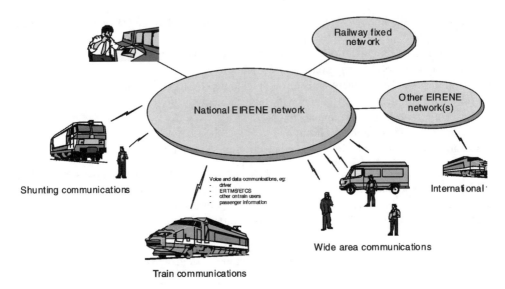

Figure 5.3.2   GSM-R Network Structure

- interworking with railway and public fixed networks;
- the possibility to use public mobile networks.

In 1993, Commission 7 of UIC approved a recommendation to choose GSM as the basis for the new standard. This was the result of a thorough study; there were technical and economical reasons for the selection. This recommendation was re-examined and re-confirmed in 1995.

The final choice had had to be made between

- GSM, already in 1993 the *de facto* standard for public cellular mobile communication (end of 1993: 1.4 million subscribers), and
- emerging standards for Private Mobile Radio (PMR), in particular TETRA, only existing as draft paper work at that point in time.

5.3.4 Standardisation in SMG

In 1993, the project EIRENE[5] was set up by UIC to co-ordinate the development of the planned Pan-European mobile railways communication and to elaborate the framework for interoperability (Figure 5.3.2).

Based on the UIC requirements, work started in SMG to add the GSM features necessary for railways operation. The development was driven by interested companies and by PT 61V and later STF 139, sub-projects of PT12/PT SMG funded by UIC and the EC.

In the first phase of GSM railways standardisation, the main features to be added were:

- voice broadcast service

---

[5] The acronym stands for European Integrated Railway radio Enhanced Network (and the word means "peace" in Greek).

Table 5.3.1   GSM 900 frequency bands

Band	Uplink (MHz)	Downlink (MHz)
GSM-R	876–915	921–960
GSM-E	880–915	925–960
GSM-P	890–915	9935–960

- voice group call services
- call priority
- fast call set-up

All these features have been specified so that they can be introduced into public GSM networks as well.

In addition, the railways frequency band had to be added to the GSM standard. This had been done in such a way that a GSM mobile station supporting the railways band has also to support the GSM 900 primary band and the GSM 900 extension band (see Table 5.3.1).

As a consequence, roaming into public GSM networks is possible where GSM-R coverage is not feasible or not sufficient.

Another important issue in the development was the radio channel performance. The main aspects are:

- High speed
- Railways specific radio environment

High-speed trains are intended to reach velocities of up to 500 km/h, whereas GSM had been designed for velocities of up to 250 km/h. The main consequences of high speed relate to Doppler effects and handover performance. The railways specific radio environment is determined by longish cells along the railway line, antennas located close to the tracks, embankments, cuttings and tunnels.

These aspects have been analysed in detail in Research and Development projects like:

- DIBMOF (Dienste Integrierender Bahn MObilFunk, Integrated Services Railway Mobile Radio), conducted in 1990–1997 by the German railways together with partners from the industry and funded by the German Ministry of Research and Technology.
- DEUFRAKO-M, a co-operation project of French and German railways from 1990 to 1993, an important step towards a joint European approach.

GSM usage by fast moving mobile stations has also been an SMG work item. Radio channel measurements have been carried out on railways including high velocity lines. Based on the radio channel analysis, simulations have been performed for various GSM channel types.

As a result, it has been shown that the channel performance required by railways applications can be achieved with a suitable planning of the radio coverage along the railway line. Enhanced receiver concepts have been shown to be unnecessary.[6]

---

[6] Göller, M. Application of GSM in High Speed Trains: Measurements and Simulations. In: Proceedings of the IEE Colloquium and Communications in Transportations, London, May 1995.

5.3.5 Completion of the work

Work went on to cover further railways requirements in the GSM standard. Among the new features were

- train registration procedures, realised based on supplementary service FOLLOW ME;
- AT commands for the interface between the mobile station and computer;
- availability of user information for the dispatchers.

The development in SMG benefited from the feedback from MORANE (MObile RAdio for railway Networks in Europe), a project of six European railways and various industrial partners launched in 1996, responsible for the specification, development, testing and validation of GSM-R. MORANE had trial sites in France, Germany and Italy. It successfully completed work in 2000.

The intention of GSM-R had been to develop a mobile communication system fulfilling the railways requirements and using the GSM standard without deviation. This has been achieved.

Chapter 6: GSM Goes to North America

Don Zelmer[1]

6.1 US Regulatory Status in the 1990s

During the first part of the 1990s, the regulatory climate in the US cellular environment was fashioned by the components of the time. The migration from analogue to digital was underway. The TDMA/FDMA debate had finished and a decision had been made (with TDMA winning) and the TDMA/CDMA public relations wars were in full swing. CDMA was thought by some to be the saviour of the world, capacity wise, and others were firmly committed to TDMA and some even to enhancements of TDMA.

A new US President took office in 1993 and one of the concepts that was brought forward, by his administration, was the idea of auctioning spectrum to the highest bidder within certain bounds. The concept was further elaborated in that a portion of spectrum, in the 1900 MHz band, that had historically been used for point to point microwave, and had recently been designated as Future Public Land Mobile Telecommunications System (FPLMTS) spectrum by the ITU, would be auctioned by the US government. The idea was to relieve increasing capacity problems in the 850 MHz wireless band and to increase revenue to help to balance the US budget.

With so many different options apparently available, the US Federal Communications Commission (FCC), under the leadership of Reed Hundt, came up with some novel concepts for licensing a portion of the FPLMTS band, which eventually became known as Personal Communication Services (PCS). A plan was put into place that would grant companies, that proposed new and innovative technological concepts, commercial licenses to try, build, and perfect their concepts, which were called "Pioneer Preference Licenses". The licenses were to be sold at a discount compared to the auctioned commercial licenses. In January 1994, three of the 50 candidates were successful and received licenses at a discounted cost based on a "technology beauty contest": Omnipoint, American Personal Communications and Cox Enterprises.

[1] Editor: Don Zelmer, Cingular Wireless (don.zelmer@cingular.com); Contributors: Chris Wallace, Nokia (chris.wallace@nokia.com), Ed Ehrlich, Nokia (ed.ehrlich@nokia.com), Gary Jones, VoiceStream (gary.jones@voicestream.com), Mark Younge, VoiceStream (mark.younge@voicestream.com), Quent Cassen, Conexant (quent.cassen@conexant.com), Alain Ohana, Cingular Wireless (alain.ohana@cingular.com), Mel Woinsky, Telecom Standards Consulting (mel@woinsky.com), Asok Chatterjee, Ericsson (asok.chatterjee@ericsson.com), The views expressed in this module are those of the author and do not necessarily reflect the views of his affiliation entity.

During this time, the FCC continued its historical approach to licensing by letting the market place decide what technology to implement into any particular licensed band. This helped to open the floodgates for multiple technologies ultimately being implemented in the wireless marketplace in the US by not creating a selection criteria for the standards process.

As part of the vision supplied by the FCC that shared many FPLMTS concepts, PCS was to enable wireless communications for the masses. At the time, cellular was primarily a business tool, still quite expensive and cumbersome to use. PCS was to allow robust and ubiquitous service to small, inexpensive portable devises with wireline voice quality supporting many advanced features. The FCC wanted to use this commercial potential of PCS to stimulate technical creativity to surpass the current analogue and digital technologies. Unfortunately, there were also aggressive PCS auction dates associated with this activity. These dates eventually compromised the technical revolution aspects of PCS.

In September 1993, the FCC released their Report & Order, formally authorising PCS in the 2 GHz "emerging technologies" band. The Commission allocated 120 MHz for licensed PCS services and 20 MHz for unlicensed PCS devices in the 1850–1990 MHz band.

One aspect of the auction process, which was to have significant consequences for GSM, was the requirement that the technology choice for a prospective bidder had to be declared prior to the beginning of the auction process. As the various potential bidders were building their business cases, there was initially great reluctance to declare a technology choice. As GSM represented the greatest viable commercial threat to the incumbent digital technologies, it was important to have a commercial entity supporting the GSM technology through the standards process. The first company to declare its technology choice was MCI, Inc. This November 1993 decision by MCI proved to be crucial in helping GSM survive the early days of the standards process as it withstood assaults on all fronts from US incumbents.

The first broadband PCS auctions for the PCS A and B bands began in December 1994. And by March 1995, bids for this 60 MHz of spectrum totalled $7.7 billion.

6.2 Wireless Standards in the US

Standardisation in the US is generally limited to an industry agreement to a particular architecture and related interfaces to which conformance is voluntary. ANSI[2] accreditation of an organisation means that the standards developed by that organisation adhere to ANSI approved guidelines and procedures for standards development.

However, there was a perception in the American market place that if a technology was standardised, it would meet end user expectations. Further, that it would interoperate seamlessly with another product that also meets the same standard. Neither of these perceptions was necessarily true. American standards are usually written to allow for significant flexibility in implementation and proprietary feature development.

In the US, standards development for wireline products and services has traditionally been done in Committee T1,[3] an ANSI accredited standards body sponsored by the Alliance for Telecommunications Industry Solutions (ATIS). Committee T1 was formed shortly after the

---

[2] The American National Standards Institute (ANSI) is a private, non-profit organization that administers and coordinates the US voluntary standardization and conformity assessment system. ANSI has served in its capacity as administrator and coordinator of the US private sector voluntary standardization system for more than 80 years.

[3] Accredited by ANSI, Committee T1 develops standards and other deliverables that facilitate the deployment of interoperable telecommunications systems and services in North America and throughout the world.

divestiture of AT&T in 1984 in order to allow broad industry participation in the development of standards and to plan the development of standards going forward to meet industry needs. Traditional cellular standards were being developed in the Telecommunications Industry Association, a manufacturers trade association. This group was responsible for all cellular and mobile radio standards development in North America.

6.3 The Formation of the Joint Technical Committee (JTC)

The development of standards to support the PCS market was significantly appealing to both US standards bodies and each was able to argue the right to claim them as their own. The architecture and network interfaces required to support full service PCS was thought to probably more closely emulate those of the wireline network. However, wireless air inter-faces and mobility management issues were clearly the domain of the cellular group. Rather than perform redundant work in each organisation along with creating unnecessarily devia-tions, both groups agreed to a T1 proposal to work together in the development of air interfaces for PCS. The resulting entity, the Joint Technical Committee on Wireless Access (JTC), was formed between TIA[4] TR45 and Technical Subcommittee (TSC) T1E1 under Committee T1. The JTC was later moved to TSC T1P1 and TR46 as PCS continued to evolve in TIA and Committee T1.

Initially, the objective of the JTC was to develop a unified Common Air Interface (CAI) for the provision of PCS. While this was an admirable goal, it may have been a little naive based on the potential size of the PCS market and the number of different commercial interests involved. In addition, the industry was polarised between those who had significant cellular interests and those that were new entrants. Also, the available time was compressed by the auction schedule.

In November 1993, 16 proposals for air interfaces were submitted to the JTC. Within 6 weeks, the JTC successfully consolidated the number of air interface proposals to eight (the total number was later reduced to seven). The final air interface candidates included:

- Composite CDMA/TDMA which was the technology proposed by Omnipoint
- CDMA (Upbanded IS-95) which was proposed by Qualcomm and Lucent
- Personal Access Communications System (PACS), a combination of PHPS and WACS which was proposed by Bellcore and Hughes
- North American TDMA (Upbanded IS-54) proposed by Ericsson
- PCS 1900 (Modified DCS 1800 or Upbanded GSM) proposed by Ericsson, Nokia, Nortel, and Siemens
- DECT based proposed by Ericsson
- Wide Band CDMA proposed by Oki

Seven air interfaces was still a long way from the original goal of a single CAI solution expected from the JTC by the industry. Many of the proposed technologies were thought to be more appropriate for low tier (low mobility) operation than for high tier operation. This combined with the aforementioned cellular/non-cellular view of the PCS market formed a market view in which no single technology appeared to be able to serve. In addition, each

---

[4] Also accredited by ANSI, the Telecommunications Industry Association (TIA) is a US based communications and information technology trade association industry that plays an important role in the development of commu-nications standards.

technology was at a different level of maturity with respect to commercial viability at 2 GHz and no proponent was interested in withdrawing their proposal in the interest of unifying the industry around a common solution.

The multiplicity and the inability to reduce the number of air interfaces was an issue which plagued the JTC throughout its existence. This was especially irritating to several commercial interests who had hoped that PCS would follow an air interface evolution similar to cellular. The JTC repeatedly attempted to seek guidance from Committee T1 and TIA on further minimising the number of air interfaces but neither committee was able to provide criteria for further reducing the number of proposed standard air interfaces. In addition, multiple technical standards were consistent with the FCC policy of letting a competitive market place determine the winners.

In an effort to gain characterisation information and to possibly reduce the number of proposed technologies for standardisation, the JTC considered requiring RF propagation simulations and the subsequent testing at a common environment be completed by each technology proposed for standardisation. This was opposed by manufacturers of cellular based technologies who felt that it was a waste of resources because their technologies had been previously proven in other trials and deployments. Further, they argued that the JTC did not have the authority to require the simulations and testing as a condition for standardisation.

However, there was significant support for the simulations and testing especially from the operator community who wanted as much information about each technology as possible in order to help with deployment decisions. US West volunteered their Boulder (Colorado) Industry Test Bed (BITB) for the field-testing. Subsequently, the JTC decided to require each technology to perform simulations and to commit to testing at BITB as a condition to go to letter ballot unless it presented an "onerous burden" on that technology.

Despite the discord, individual air interface work progressed rapidly as some technologies were already standardised at 900 MHz and only needed to be modified to operate at PCS frequencies. This was especially true for PCS 1900 that had a close cousin; DCS 1800 already deployed in the 2 GHz region.

6.4 Americanisation of GSM

The ETSI GSM standard is a complete architectural specification for a wireless mobile telecommunications network. All that was required to satisfy the JTC requirements was an air interface specification that was a subset of the complete GSM works. As a result, PCS 1900 set the pace for standards development in the JTC based largely on Phase 2 specifications. PCS 1900 was the first technology to complete its simulation work and to be tested at BITB. However, it was difficult to describe the air interface without hincluding some additional supporting information. Some of that information included certain regulatory requirements, removal of references which were relevant only to Europe and the addition of references that were relevant to the US market. One area of considerable concern by the US editors was the use of a UK dictionary for spell check purposes. It was agreed that a US dictionary would be used to facilitate the translation process.[5] As a result, the complete ANSI standard for the PCS1900 radio interface, J-STD-007, is approximately 2400 pages. By March 1995, PCS 1900 had received approval for publication as an ANSI standard. And

---

[5] This chapter, written by Americans, was translated into the King's English by the editor of this book.

in November 1995, Vice President Al Gore inaugurated the nation's first PCS system – the PCS1900 network from American Personal Communications (APC) – with a phone call from the White House.

Work to develop American versions of the GSM MAP and SS7 A interface was initiated in TR46 as it was outside of the scope of the JTC. In addition, Stage 1 service description and A-interface projects were initiated as well as an Interworking & Interoperability (I&I) project between GSM and IS 41 MAPs in TR46. These three standards were sent out for letter ballot approval in late 1994.

6.5 The Conclusion of the JTC

The simulation and testing decision turned out to be the beginning of the end for the JTC. The opponents of the simulation and testing program escalated their objections within the TIA. This coupled with baseless accusations of TIA procedural violations within the JTC formed the basis for the TIA decision to stop its participation in the JTC.

As a result of this decision, it was necessary to find an appropriate standards fora from which the maintenance and evolution could be continued. The TIA had assumed that all the air interface standards proponents would elect to perform the document maintenance and future work in TIA. However, this was not necessarily the case for many of air interface groups especially among potential operators. The PCS 1900 group was one of those air interfaces that did not support a move to TIA.

The Committee T1 reaction to the TIA resolution was to pass a resolution of their own in support of the JTC. Several discussions between the TIA and the Committee T1 yielded little progress and the matter was escalated within the TIA to the Technical Standards Sub Committee (TSSC) and a special meeting was called to hear arguments.

The special meeting of the TSSC heard the arguments from the interested parties and then deliberated privately. The result was a recommendation that after completing its work a TAG would transition to a Joint Project Committee (JPC) managed between Committee T1 and TIA. The JPC was a new concept developed by the TSSC for future joint work. Unlike the JTC these JPCs would be unique entities supporting a single project. Further, a JPC would be led by either TIA or Committee T1 which was not the case with the JTC. The TSSC proposed that the upbanded cellular technologies (e.g. NA TDMA, CDMA) and Wideband CDMA and DECT based technologies form JPCs under TIA. Conversely, PCS 1900, Composite CDMA/TDMA, and PACS would be under Committee T1.

The Committee T1 response to the TSSC proposal was generally positive. The T1 Advisory (T1AG) group held several conference calls to discuss the TSSC proposal and a response suggesting some minor modifications to the plan was sent to the TSSC. Meanwhile, the cellular based technologies grew impatient and initiated a move back to TR45 where their technologies were initially developed. The DECT based technology had long been doing its work in TR41, a CPE (Customer Premise Equipment) development group. Based upon participant preference, this left PCS 1900 and Composite CDMA/TDMA (both use GSM network architectures), Wideband CDMA and PACS in a T1P1 led JPC.

The movement of the cellular technologies from TR46 to TR45 prompted the TIA to host a joint meeting between the groups to eliminate duplication of effort. The resulting decisions left the PCS 1900 services and network and air interface work along with the Composite CDMA/TDMA, PACS and Wideband CDMA air interface work alone in TR

46 jointly with TSC T1P1 as the lead group. TSC T1P1 assumed a leadership role in the development of these standards with TR46 concentrating on issues of common interest between CDMA/TDMA based and GSM systems such as PCS interference and intersystem interworking.

The PCS 1900 standards were successfully developed in an environment where some parties did everything possible to delay or block the development of PCS 1900 standards.

6.6 The North American PCS 1900 Action Group (NPAG)

About half way through the standards development cycle market pressures necessitated the development of more features for PCS 1900 for the North American market. Many of these features could not be developed in the standards bodies in time for service rollout. This and the need for PCS 1900 proponents to be able to work out common issues led to the formation of the North American PCS 1900 Action Group (NPAG) which was a logical step in the evolution of PCS 1900 as a North American technology.

The NPAG membership was composed of operators who were actively promoting PCS 1900 for deployment in North America. The intent was to identify and resolve issues that affected the deployment of PCS 1900 in North America. The NPAG formed Technical Advisory Groups (TAGs) composed of experts from the manufacturers and operator communities. The TAGs worked under well-defined scope and charter and reported results to the plenary.

The NPAG initiated TAGs to address Vocoder, Data, Services, Handset, Billing, Standards, EMC, and Roaming issues as well as ad-hoc groups to investigate E911, Lawful Intercept, and Network Management. The vocoder TAG was the steward of the US-1 EFR, which became commercially available in 3Q96.

The turmoil in the standards fora had concerned NPAG for quite some time. In addition, GSM related issues that required intervention were continually surfacing. The standards TAG was initiated to monitor these for the PCS 1900 community. The TAG became very active in the JTC issue by authoring several contributions to TIA and Committee T1 that presented the view of the PCS1900 community. In addition, the group successfully co-ordinated the allocation of an E.164 global title translation from T1S1 after two previous attempts had failed.

6.7 The Enhanced Full Rate Vocoder (EFR)

North American PCS operators had generally recognised that wireline quality speech would be a critical element in the success of PCS in their markets. The NPAG believed that is was feasible to develop this capability in PCS 1900 systems in time for the North American PCS rollout. Traditional environments and processes for vocoder development had yielded satisfactory results in the past. It was recognised that the development process for a new vocoder would have to be expedited significantly in order to meet the aggressive time to market requirements. NPAG felt that a market driven, task oriented forum was required for the vocoder development in order to increase the probability of success.

NPAG initiated a TAG to work with equipment manufacturers and develop an improved quality speech vocoder that would be commercially available in time for service rollout. The participation in the TAG consisted of those companies who had committed to market GSM based PCS 1900 systems in North America. There were five proposals presented in the first TAG meeting.

NPAG shared the proposed development schedule for the vocoder with the manufacturers. There was significant concern about being able to meet the schedule without compromising the quality of the vocoder. Further, it became evident that there were critical areas in the development process that, if not managed aggressively, would delay the availability of the vocoder. Eventually the process was slowed significantly while trying to reach consensus on the evaluation criteria for vocoder selection.

The resulting delay postponed the vocoder selection and jeopardised the commercial availability of the vocoder. In order to break the deadlock, the operators authored a joint letter to each of the manufacturers reiterating the need for the earliest possible commercial availability of the new vocoder. The letter directed the manufacturers to collectively decide which of them had the best solution.

The manufacturers met and individually evaluated each others design in their laboratories. At the completion of their tests and they collectively selected the Nokia Sherbrook design for the EFR vocoder, whose specification had been completed in August 1995. The EFR was later tested at an independent laboratory. The results indicated that the EFR had achieved the desired quality improvement within the complexity objectives. The EFR was then proposed to the JTC for standardisation. The standard was forwarded on for letter ballot and was approved as the "US-1" EFR.

The EFR was also proposed to the ETSI SMG for standardisation. However, the Speech Experts Group (SEG) had been working on several proposals and were considering conducting an extensive testing program to select an EFR vocoder. Proponents of adopting the US-1 EFR pointed out that it had been thoroughly tested and was in the final balloting stages in TSC T1P1 in the US and could quickly be adopted and deployed. The GSM MoU Association sent a liaison to ETSI requesting that the US-1 be adopted as the EFR as well. In October 1995 at ETSI SMG plenary in Vienna the US-1 EFR was accepted as the ETSI SMG EFR. The selection of US-1 was a significant milestone as it represented the first time that ETSI SMG had accepted technical work from another standards development organisation. This agreement set the stage for closer co-operation between TSC T1P1 and ETSI SMG.

6.8 AMR Codec

In late 1995 some of the NPAG companies were having problems with their business cases for future expansion. The issue was that with the explosion of users of wireless, it was apparent that system capacity would be a problem in the near future. Studies in the US had shown that an additional 160 MHz of new spectrum would be required to satisfy the demand for wireless services but at that point in time, no spectrum relief from the regulatory bodies was on the horizon. Additionally, as a result of the PCS license auction process, some GSM operators ended up with only 10 MHz of spectrum (5 MHz uplink and 5 MHz downlink). One of the solutions to compensate for the small amount of spectrum available was to introduce a half-rate speech codec. However, the competitive picture in North America and the necessity to offer a high speech quality was making the existing GSM half-rate speech codec (VSELP at 5.6 kbit/s) simply unacceptable. Consequently, the US marketplace requested that an Enhanced Half Rate (EHR) speech codec be standardised as soon as possible and immediately after the approval of the EFR standard.

In response to that request and others coming from Europe, SMG set up a Speech Quality Strategy Group (SQSG) to evaluate and refine the future strategy for the development of new speech codecs (ex: narrowband or wideband) and new advanced speech related features (Tandem Free Operation (TFO)). The SQSG released its report in mid-1996 and recommended to start a feasibility study on a new multi-rate and adaptive[6] speech codec, later called Adaptive Multi-Rate Speech Codec or AMR. The key objective of AMR was to provide at the same time a high quality half-rate mode in good to very good propagation conditions and a very robust full-rate mode so that the same high quality could be offered down to the cell edge. Both modes had the potential to significantly increase the GSM capacity without an impact on the offered quality.

When the feasibility study was completed in late 1997, the final approval of the AMR as a new work item was still not considered as critical in Europe because most operators there were barely starting to implement EFR and were not ready to support a new speech codec and its possible impact on the installed base. At the same time, SMG was defining the basis for the selection of the UMTS standard and it was clear that a speech codec with the same flexibility as that offered by AMR would also be required for UMTS. It was then natural to consider AMR as a strong candidate for the future GSM speech codec as well as UMTS. The work item was approved under this assumption.

Because it was a major driver behind the new speech codec, the GSM Alliance[7] was particularly interested and consequently highly involved in the development of AMR.

The development of AMR was completed as part of the GSM Release 98. The codec specifications were approved in February 1999. The development was very successful, all requirements being met by the selected candidate. In half-rate mode, AMR provides a very high quality down to 13–16 dB C/I. In full-rate mode, AMR will provide the same high quality level down to 4 dB C/I.

In the following months, on the basis of comparative tests performed in Europe and Japan, the codec group of 3GPP recommended that AMR be also selected as the default speech codec for the new UMTS standard. This recommendation was approved by TSG-SA in early 1999.

AMR is now the centre piece for the development of the speech service in all future evolutions of the GSM and UMTS standard. This includes voiceover 8-PSK EDGE in GSM or voiceover IP for both GSM and UMTS.

6.9 The GSM Alliance

With the PCS auctions complete, the executives of companies that had committed to deploy PCS1900 turned their attention to the task of knitting together a North American GSM network that could effectively compete against the nation-wide TDMA and CDMA carriers.

---

[6] The concept of an adaptive speech codec is one that changes voice coding (or source coding) rates depending on some external constraint like the RF environment or the required capacity. For example, in a poor RF environment more bits are directed to the protection of the encoded voice channel, while in a good RF environment fewer bits are directed to RF protection and the highest possible speech quality can be provided to the subscriber. Similarly, by reducing the coding rate, it is also possible to use less of the available radio bandwidth (half-rate channels in GSM or higher spreading factor in W-CDMA) and increase the system capacity accordingly.

[7] The GSM Alliance consisted of seven major companies from North America which were: Aerial, Omnipoint, BellSouth PCS, Microcell 1-2-1, Pacific Bell Wireless, Powertel, and Western Wireless. Note: Aerial, Omnipoint, Powertel, and the PCS portion of Western Wireless have all been absorbed by Voicestream Wireless.

The NPAG had been very successful in advancing the technical capabilities of GSM; however, it lacked the administrative horsepower required to make the business decisions and commitments required to forge a GSM footprint over North America. The GSM Alliance LLP was formed so that the companies that declared to deploy PCS1900 could form a business partnership to facilitate roaming and other interconnectivity agreements. In addition, the Alliance served as GSM public relations by promoting the technology and recruiting prospective GSM operators.

At about the same time the GSM operators took the opportunity to formalise the NPAG structure by establishing it as a regional interest group under the GSM Association (at that time the GSM MoU Association). The NPAG TAGs were reorganised into Sub Working Groups similar to the structure of the GSM Association. The re-organised entity was called GSM North America.

6.10 US Committee T1 and ETSI SMG Co-operation

Official representatives from the US operator community began attending ETSI SMG at SMG#9. These early meetings were attended on an observer basis by kind permission of the Chair of ETSI SMG. As the standardisation process continued in the US, it became apparent to the US representatives that a more formal arrangement was needed between TSC T1P1 and ETSI SMG. This arrangement was negotiated over a number of SMG plenary meetings and eventually resulted in a working procedures agreement. These procedures became the basis of an excellent and practical working partnership between these two technical groups.

In an effort to contribute to the GSM specification development, TSC T1P1 proposed and agreed within the SMG community, to develop standards for areas of US need such as 14.4 kbs per time slot, terminal location, calling name identification, and emergency services which were to be standardised in the US and submitted to SMG for inclusion into the ETSI GSM standardisation process. These services would be accepted by SMG and SMG would have the ultimate responsibility for ensuring that the TSC T1P1 developed services would be properly incorporated into the base GSM specifications. One of the most important additions to the base GSM specifications was the inclusion of the 1900 MHz radio specific aspects. This addition formally integrated the US PCS1900 system as part of the GSM family.

Calling name identification, 14.4 kbs per time slot, and EFR were completed in North America and submitted and accepted by SMG. Location and emergency services were worked and eventually transferred to 3GPP where they are being standardised as this is being written (mid-2001). Some issues, such as location, are not trivial and have required a significant amount of development with some "false starts". Completion of the standardisation of location services will probably be completed by the end of 2001.

6.11 A Change to the US Process

As a result of the standards efforts by TSC T1P1, it also became apparent that the differing processes between the two standards bodies were making maintenance of the US standards an impractical burden. The speed and scope of change at each SMG plenary made the US maintenance process impractical. Another method was needed to maintain the relevancy of the US documents and keep within the guidelines of the ANSI process. After much discussion, the concept of using ETSI SMG documents as normative reference in the US standards documents was finally agreed. This allowed the maintenance and evolution to continue unimpeded without unnecessary duplication of resources and effort. In fact, this concept would form the basis of the SDO relationship to the 3GPP.

6.12 Closing

The success of the co-operation between ETSI SMG and TSC T1P1 was a brilliant counter-example to the sometimes intense inter-regional standards competition that has marred many international standards efforts. In fact, the experience gained, during the GSM standards process, provided the basis for key elements of the co-operational working methods for 3GPP.

Chapter 7: The UMTS Related Work of the European Commission, UMTS Task Force, UMTS Forum and GSM Association

Section 1: The European Research

João Schwarz da Silva[1]

7.1.1 Introduction

Compared to today's reality, the mobile and wireless communications evolution perspective of the world, at the start of 1985, looks retrospectively rather conservative. While the promise of an accelerated development of mobile communications was sensed as a likely possibility, due notably to the anticipated success of GSM, the most optimistic scenarios for market deployment called for a few million subscribers at the turn of the century. Some 10 years later, there were about 13 million users of mobile communications in Europe and estimates suggested that the market potential was of the order of 50 million at the turn of the century and will be 100 million by 2005. Back than it was also felt that while the fixed telephones penetration rate would not exceed 50%, personal mobile communications, in all forms, would perhaps reach nearly 80% of Europe's population.

7.1.2 Research into Advanced Communications in Europe (RACE)

7.1.2.1 RACE Definition Phase

In 1985, the telecommunications, computing and broadcasting sectors accounted for an annual turnover of over 500 billion Euros world-wide. Telecommunications were central to the performance of the services sector and crucial to the business competitiveness. It

[1] The views expressed in this section are those of the author and do not necessarily reflect the views of his affiliation entity.

was clearly felt that the prosperity of Europe was critically dependent of good communications. On the demand side, the call for new and more sophisticated services and applications was expected to change rather rapidly, with businesses needing more flexible services, higher transmission capacities for fast data and image transmission at more competitive tariffs. On the supply side, the lowering of the internal barriers to trade was going to present new opportunities for network operators and service providers that would be looking to distinguish themselves from their competitors.

Recognising these trends, European industry ministers[2], launched a "Definition Phase" of the RACE programme. This definition phase established that there was scope and need for a European framework for collaboration in R&D in telecommunications and led to a decision adopted by the European Council in December 1987[3] calling for the first phase of RACE within the Second Framework for Research and Development.

7.1.2.2 RACE Phase I (1988–1992)

While the main objective of RACE was to contribute to the introduction of Integrated Broadband Communications (IBC) progressing to Community-wide services by 1995, specific objectives of phase I included inter alia:

- To promote the Community's telecommunications industry
- To enable European network operators to compete under the best possible conditions
- To enable a critical number of member states to introduce commercially viable IBC services
- To support the formation of a single European market for telecommunications equipment and services

In the RACE community it was recognised at the time, well before the commercial introduction of GSM, that a new generation of mobile technology would be necessary to cater for the perceived challenges of the 21st Century. Such a new generation was seen as comprising not only novel radio techniques, but also an open and flexible fixed infrastructure based on state-of-the-art technology. A work plan was hence developed in June 1987, a Call for Proposals on mobile communications was launched and as a result the RACE's Mobile Project (R1043) was launched. Bringing together some 20 partners (comprising industrial organisations, operators and academic partners), the project identified two main classes of mobile communication services, namely; UMTS (targeted to provide speech and low to medium data rate services, with virtually complete geographical coverage) and MBS (targeted to provide mobile units with very high bit rate services in hot spot areas). The primary aims of the project included the specification of air interface parameters, the specification of signalling system functions and the supporting networking infrastructure, the identification of the required frequency spectrum (later on made available at the WARC 92), the submission of contributions towards ETSI, and the ITU (both CCIR and CCITT) and the identification of evolution scenarios from second generation systems.

---

[2] Council decision of 25 July 1985 on a definition phase for a Community action in the field of telecommunications technologies.

[3] Council decision of 14 December 1987 on a Community programme in the field of telecommunications technologies – R&D in advanced communications technologies in Europe (RACE programme).

On its completion the project presented its key achievements[4] including the main concepts of UMTS. The driving forces for UMTS were seen to be the requirement for better quality, universal coverage, additional services and higher capacity. The UMTS concept included a standardised system, supporting mobile access in almost any location, indoors or outdoors, city or rural areas, in the home, office or street, a wide range of terminals and services, a low cost pocket size personal communicator for the mass market. It was then estimated that the UMTS penetration rate would reach 50% of the European population in 2005, corresponding to a subscriber base of 100 million.

It is no doubt clear that, in setting up an ambitious objective, namely the development of a Community-wide market for telecommunications services and equipment, the RACE I programme, did stimulate the commitment of the major European telecommunication operators, equipment manufacturers and leading-edge users, to pursue mutually beneficial goals. The quality and cost/effectiveness of traditional services was enhanced and a new generation of innovative services was introduced. In the area of mobile communications, Project R1043 laid the ground for the second phase of the RACE programme having made seminal contributions to the development of UMTS.

7.1.2.3 RACE Phase II (1990–1994)

Following the success of the first phase of RACE, its second phase was launched[5] in the context of the Third European Framework Programme (1990–1994) of research and technological developments. RACE phase II envisaged that a number of actions would be launched, particularly in the area of communication technologies, with the principal objective of enabling the broadband network to take on the emerging new services, constructed on 'open standards', and to make use of flexible and cheaper integrated services. Such actions included a community research effort of a pre-normative and pre-competitive type in order to ensure the inter-operability of systems on the basis of common standards and protocols. Very much at the core of the RACE II programme were the research activities on UMTS and MBS.

Further to a Call for Proposals, a number of UMTS related projects were retained (projects PLATON, MONET, CODIT, MAVT, ATDMA) with one project (MBS) specifically addressing the concept of a mobile broadband system operating in the 60 GHz frequency band. Close collaboration was established with ETSI, and the mobile projects were requested to devote a significant effort to standardisation issues. It was also felt essential to ensure that different projects carried out collaborative work, particularly in joint systems engineering and in the preparation of Common Functional Specifications. Issues for which collaboration was required included propagation prediction and channel modelling, air interface definition, source and channel coding, error correction and modulation, cell design, architecture and coverage, handover, channel and resource management, network and mobility management, security and authentication, performance assessment scenarios and quality measures. Other crucial aspects also dealt with were the ones dealing with marketing studies, service requirements, evolution and implementation strategies, operational and functional requirements.

In light of their achievements, three particular projects will be briefly described, namely CODIT, ATDMA and MONET.

---

[4] RACE Mobile Telecommunications Workshop, 5–6 May 1992, Nurnberg, Germany.

[5] Council decision of 7 June 1991, adopting a specific research and technological development programme in the field of communications technologies (1990–1994) (91/352/EEC; OJ L192.8, 16.07.91).

7.1.2.3.1 CODIT

The overall objective of UMTS Code Division Testbed (CODIT) was to explore the potential of Code Division Multiple Access (CDMA) for the Universal Mobile Telecommunication System (UMTS). An advanced system concept based on CDMA was sought including advanced radio technologies (radio interface, radio transceivers, etc.) and advanced subsystem architectures (micro- and picocells, macro-diversity, fast and soft handover, frequency management, radio network planning methods, etc). The first European CDMA system demonstrator (testbed) comprising test mobile stations, radio base stations and a radio network controller was designed and built, and the CODIT system concept was validated in laboratory and field trials.

The CODIT project also succeeded in the development of a system concept aimed at meeting the major requirements of a third generation UMTS which were:

- handling of pico-, micro- and macrocells with a simple deployment of spectrum resources allowing multiple operators as well as private networks;
- indoor and outdoor operation with a high grade of service;
- support of low power pocket terminals for high quality speech services;
- easy access to (known) data networks;
- variable bit rate bearer for advanced data services;
- ability to support a large number of users (50% penetration rate).

7.1.2.3.2 ATDMA

The overall objective of the Advanced TDMA Mobile Access (ATDMA) project was to contribute to the identification of the most appropriate radio access system for provision of mobile narrowband service connections to IBCN. The project concentrated upon advanced TDMA techniques such as the concept of an adaptive TDMA air interface which automatically adapts to suit different operating environments and service needs for UMTS. A simulation testbed comprising a radio network control and signalling traffic model and a radio system capacity model was built to evaluate different aspects of the radio access concepts for the ATDMA system concept. Key issues to which the project ATDMA system concept contributed included:

- Support of the required services under different operating environments and fulfilment of the needs of multiple operators with low infrastructure and low terminal costs;
- Flexible air interface capable of recognising and supporting different cell types and allowing high bit rate services;
- Different base station interconnection interfaces were required for various operators;
- An adaptive air interface with ability to accommodate the varying needs of different environments, services and quality;
- Transmission bandwidth for various cell types;
- Use of linear or non-linear modulation schemes;
- Radio control issues such as DCA or central frequency planning, packet or call reservation, inband or dedicated signalling channels, degree and response time of power control;
- Balance between allowable interference and noise levels, forward error correction codes, robustness of the speech codec and ARQ;

- Demonstration of a fair basis of comparison between the CDMA and TDMA radio access techniques.

7.1.2.3.3 MONET

While the previous two projects sought to develop the systems concepts for UMTS and devoted significant resources towards the definition of the air interface, project Mobile Network (MONET) aimed to develop network standards for UMTS. Two important goals in UMTS were (1) to integrate the infrastructure for mobile and fixed communications, and (2) to offer the same range of services as provided by fixed communication networks. An additional benefit of mobile networks was the possibility to offer unique services such as navigation, vehicle location, and road traffic information. Given that it was expected that within the next two decades, UMTS pocket telephones would become a mass market consumer item, the basic challenge was, therefore, to define a fixed infrastructure capable of supporting a huge volume of mobile connected traffic. Project MONET achieved the following results:

- Development and specification of new concepts for handover, call handling, location management, security, telecommunications management, databases and base station interconnection, to limit the signalling load due to the mobility of the user and to allow UMTS terminals to be used anywhere (public, home, business and vehicle environment);
- Definition of a UMTS network architecture as part of IBCN, permitting maximum exploitation of the IN and commonalties with UPT;
- Validation of performance and signalling load of the proposed network architecture and protocols by means of simulation.

7.1.2.4 Consensus work in RACE

To reach a common understanding regarding the key issues confronting the various projects, and speed-up consensus development in anticipation of the discussions that were taking place within ETSI, special interest groups were established in some key areas including:

Radio network modelling: this area of activity comprised the reference scenarios defining the environment in which UMTS was expected to operate, the mobility models, the traffic models, the radio channel characteristics, the interference models and radio resource management procedures.

Base station system functions: this area comprised issues such as synchronisation of base stations, backward-forward handover, multichannel handover for multimedia services, adjustable parameters of the radio interfaces, flexible assignment of capacity, flexible power transmission, minimisation of signalling load, etc.

Propagation and channel characterisation: the basic objective of this activity was to achieve common models for propagation and channel characterisation based on measurements and analysis. This entailed the harmonisation of the measurements campaigns, the exchange of the measured data, and the comparison of the analysis carried out in different propagation environments for macro, micro and pico cells.

Common testing requirements: since for UMTS two distinct air interface schemes were considered, work was initiated towards the establishment of a common platform for testing requirements and a common basis for system comparison taking into account the limited total spectrum availability, the cost of implementation of terminals, infrastructures and services, the network evolution and the network complexity.

7.1.2.5 The RACE Vision of UMTS

Throughout the R&D work in RACE phase II, it was realised that in order to be successful, UMTS should offer significant added value compared to its predecessors. The approach adopted was therefore not to start from the current or near telecommunications scene, but to position UMTS in a situation around the year 2000 when significant advances in both fixed and mobile communications would have been achieved. A key driver for the work in UMTS was the concept of integration. Whereas mobile systems such as GSM and DECT had been designed as stand-alone systems, the UMTS network was defined as an integrated part of networks for fixed telecommunications. This way, services made available for fixed users were accessible by mobile users as well, and infrastructure costs could be reduced by sharing expensive network resources. Integration with broadband ISDN was defined as the target scenario. The principles of the Intelligent Network (IN) concept were to be used to provide the necessary flexibility in the network, to provide rapidly the services and applications the various user segments were demanding. This integration philosophy was backed by the following observations: firstly, requirements for prospective fixed and mobile networks were very similar. Many important trends in telecommunications, like customisation, interactive control, high quality, tailor made service offerings, etc. applied to both fixed and mobile environments. Secondly, a development towards an integrated personal communication environment was envisaged, in which users would be able to have access to telecommunication services, irrespective of whether the means of access were fixed or mobile. Users would expect that services available on the fixed networks would also be available on mobile networks and vice versa. Moreover, there should be no noticeable difference in user-interface and control procedures. For network operators and/or service providers, integration implied that they did not have to install and maintain duplicate platforms for service creation, service management and service control.

The vision of UMTS as it emerged from the work undertaken within RACE,[6] called for UMTS to support all those services, facilities and applications which customers already enjoyed while having the potential to accommodate, yet undefined, broadband multimedia services and applications with quality levels commensurate to those of fixed IBC networks. A thorough discussion on the above issues took place in the context of the RACE mobile projects, which concluded on the need to offer to the wider European community of mobile sector actors, a vision of UMTS. A document was hence drafted which was the subject of agreement by all projects. Given the significant impact that such a document was expected to have, the RACE projects advocated a gradual evolutionary path which called for the establishment of a UMTS Task Force, to be followed, if required by the creation of a UMTS Forum.

---

[6] RACE Vision of UMTS, Workshop on Third Generation Mobile Systems, DGXIII-B, European Commission, Brussels, January 1995.

7.1.3 Advanced Communications Technologies and Services ACTS (1994–1998)

The RACE mobile programme concluded its activities in 1995 but by then the European Fourth Framework Programme was supporting the next phase of collaborative mobile R&D into ACTS. This new programme while capitalising on the RACE experience, was conceived as a demand-driven R&D programme of demonstration trials that would prepare the ground for a European-wide, internationally competitive, broadband telecommunications infrastructure that naturally had a mobile communication dimension. The ACTS mission was to articulate potential solutions to those remaining technological issues seen to be impeding evolution to the wide-scale use of advanced digital broadband communications throughout Europe. Within the ACTS programme, the mobile and wireless communication dimension related to the need to provide seamless service across various radio-environments and operational conditions for a range of user-defined and customised advanced multimedia services. Key issues included system and service integration with the relevant fixed network to ensure continuity of multimedia service provision. Three particular aspects namely, services, platforms and technologies were considered as prime objectives of the ACTS programme in the area of mobile and satellite communications.

Services: the demonstration and proving of new novel services and applications taking into account the full implications of user environment, system characteristics and service provision and control.

Platforms: demonstration of the viability of a major system or service-provision architecture using technology demonstrators that would consider issues such as: feasibility, acceptability, quality of service, general fit-for-purpose, interworking and integration capability of the demonstrated radio system or integrated network.

Technologies: proving the validity of new, or novel, components or sub-system technologies, including multi-mode transceivers, tools for network planning, methods to achieve secure communications and system concepts.

7.1.3.1 UMTS System Platform

Further to the development of the UMTS vision elaborated in the RACE programme, UMTS was seen as needing to support all those services, facilities and applications which customers already enjoyed (e.g. GSM) but also had to accommodate, yet undefined, broadband multimedia services and applications with quality levels commensurate to those of fixed IBC networks. In this context the leading questions were:

- What were the cardinal services that UMTS should support?
- What were the "future-proofing" UMTS bearer requirements in macro-, micro- and pico-cell environments?
- What were the applications likely to be supported from UMTS?
- How would second generation technologies evolve towards UMTS?

UMTS was seen as an opportunity to exploit the 2 GHz band with a unified and universal personal mobile telecommunications system for multi-operator environments. It was a multi-function, multi-service, multi-application digital system that would use end-of-the-century

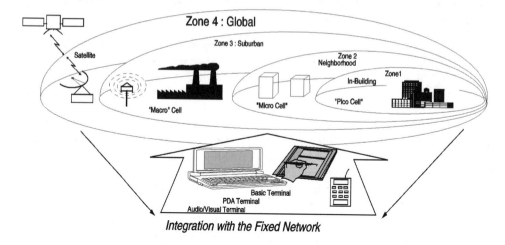

Figure 7.1.1   System environments

technology to support universal roaming and offer broadband multimedia services with up to 2 Mb/s throughput. Figure 7.1.1 illustrates the range of service environments, from in-building to global, in which UMTS was to be deployed while Figure 7.1.2 portrays the technological capabilities of UMTS, measured in terms of terminal mobility and required bit rates as compared to those of second generation platforms such as GSM.

7.1.3.2 Evolving from Second to Third Generation

Finding the solution to the issue of evolution and migration path from second to third generation (see Figure 7.1.3), particularly from a service provision point of view, was also

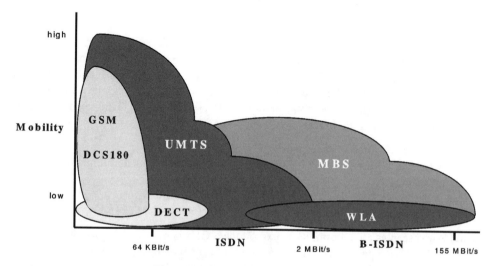

Figure 7.1.2   Mobility versus bit rates

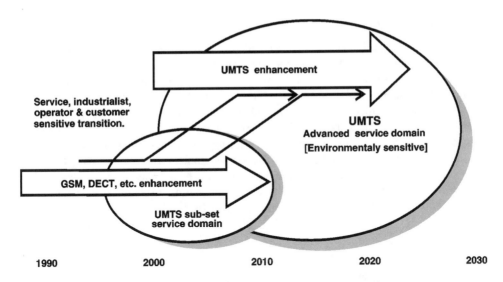

Figure 7.1.3   Evolution from second to third generation

the subject of intense debate within the ACTS community. Agreement was reached on the need to ensure a smooth market-led transition between second and third generation systems. Indeed these standards needed to ensure a smooth customer, operator and industry-sensitive transition at the appropriate time with multi-mode, hybrid, transceiver technology used to provide multi-standard terminal equipment. Thus enhanced second generation technology could offer some of the lower bit-rate user and service requirements but UMTS would also satisfy the advanced-service, broadband, multimedia demand (Figure 7.1.3).

7.1.3.3 Key UMTS Projects in ACTS

As for the RACE programme, a number of key UMTS related projects are briefly described below.

7.1.3.3.1 FIRST

It was the primary objective of project FIRST to develop and deploy Intelligent Multimode Terminals (IMT) capable of operation with UMTS as well as with multiple standards and with the ability to deliver multimedia services to mobile users. Secondary objectives arising from the primary objective included the feasibility of delivering multimedia services to the mobile user and the investigation of requirements for such services. The project undertook activities aimed at the innovation of key IMT sub-systems (e.g. adaptive transceivers) capable of operation with multiple second generation standards and with those air interface standards expected to emerge for UMTS. IMT technology demonstrators were produced and tested as a result of these activities and the groundwork for the development of reconfigurable radio in Europe was established.

7.1.3.3.2 OnTheMove

The project aimed at the development of a standardised mobile application program interface (Mobile API) to facilitate and promote the development of a wide spectrum of mobile multimedia applications. An architecture, the Mobile Applications Support Environment (MASE), to support both mobile-aware and legacy (i.e. non-mobile-aware) applications was developed. Some of the project challenges included adaptation to varying quality of service, robustness in the face of disconnected links, roaming between different operators and network types, reconfigurable real-time multi-party connections, flexible coding, personalised information filtering, and support for heterogeneous user equipment. A sequence of field trials and demonstrations of an interactive mobile business system offering time critical financial information services to business users were carried out, in order to capture user requirements, to validate results on the interface specification and to develop inputs to standardisation bodies.

7.1.3.3.3 FRAMES

This project comprised two technical objectives namely UMTS system specification and demonstration. System specification included requirements and synthesis, by means of liaison with the GSM MoU group (and later with the anticipated UMTS MoU group). The project defined a number of hybrid multiple access based adaptive air interfaces (which were the basis for the ETSI decision on UTRA in January 1998), the radio network functions and their implications upon the access networks. The project developed an implementation platform which was capable of validating the feasibility of the UMTS system definition, capable of supporting and substantiating the technical specification towards third generation of mobile communication as a basis for UMTS definition, validation as well as standardisation. Validation tests of the UMTS air interface using the demonstrator and demonstrations of the selected applications were carried out and lead to contributions to standardisation bodies.

7.1.3.3.4 STORMS

The objective of this project was to define, implement and validate a software tool to be used in the phases of designing and planning of the UMTS network. The set of reference environments considered, ranged from indoor coverage (picocells) to regional coverage as supplied by satellite systems integrated with the terrestrial cellular structure. The set of spatial traffic distributions and traffic intensities ranged from typical office/business to residential in rural and low-density areas. The work also involved a comprehensive definition of the categories of UMTS environments with the corresponding identification of the most appropriate propagation models, so that the environment and radio channel model were logically mapped. The combined electromagnetic and traffic description was the starting point of the dimensioning process for the cellular structure which also took into account those system features (e.g. handover strategies, macrodiversity) that imply a particular usage of radio and network resources.

7.1.3.3.5 RAINBOW

In the perspective of an heterogeneous and mass market for mobile communications, several architectural and service issues remained still to be solved in the framework of UMTS. This project hence addressed some of these crucial topics, representing significant turning points in the migration to the third generation mobile network and services. Specifically, the project demonstrated the feasibility of a "generic" radio independent UMTS network infrastructure, able to supply mobility transport and control functions for different radio access techniques, defined the boundary between radio dependent and radio independent parts of UMTS, studied the integration/interworking configurations of such an infrastructure within the IBC context and the expected IN based functionality located in the core network, identified solutions for the migration from second generation mobile systems to UMTS through the above infrastructure, and verified the impact and suitability of multimedia services, including variable bit rate services on the fixed UMTS network architecture.

7.1.4 Conclusions

As we are about to approach the 1 January 2002, from which date onwards UMTS networks are expected to be commercially introduced in Europe, it is important to reflect on the last 10 years of R&D in the context of the European Union research programmes. Thanks to the efforts of the RACE and ACTS research community, what was considered in 1988 as a wild wireless dream, is about to be realised. The major milestones met over these years are depicted in Figure 7.1.4. Starting from the decisions on frequency spectrum taken by the WARC 92 in which the EU R&D effort was instrumental, following with the development of the seminal UMTS vision document that raised the key issues to be tackled and defined the

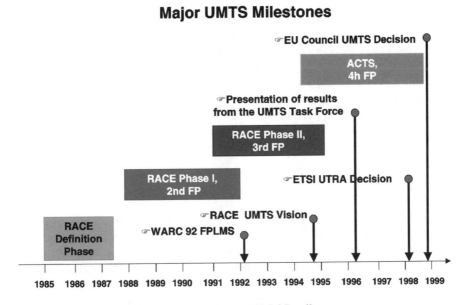

Figure 7.1.4   Major UMTS R&D milestones

challenges ahead, culminating with the ETSI decision on UTRA in January 1998 to which key contributions were made by the ACTS community. All projects have contributed significantly towards the definition and development of UMTS and numerous major contributions were made on key enabling technologies, service platforms, network aspects, planning tools, some of which have been introduced in second generation systems. These significant achievements made in these years illustrate the profound impact that collaborative precompetitve and prenormative research and development work can have.

However, research and development alone cannot be considered in isolation from the regulatory and policy framework, particularly when the issues under consideration have Pan-European and global implications. Very early on the RACE and ACTS community did recognise the need to strongly articulate R&D with the regulatory and policy initiatives and has hence actively contributed to the setting up of the required framework. In a companion chapter a description is given of the work on the regulatory and policy framework that has culminated in the UMTS decision of December 1998.

Today Europe has an undisputed expertise in mobile and wireless communications, its mobile communication operators and equipment manufacturers benefit from a clear competitive edge when compared to other world regions, and it is very well positioned to shape the evolution and development towards the next wireless generation. Both the number of cellular subscribers and penetration rate at the end of the year 2000, exceed by orders of magnitude the expectations of 10 years ago and it is expected that world-wide the overall number of cellular subscribers will reach 1 billion in 2003, far exceeding the number of fixed Internet users. It is widely believed that, in the near term, the preferred means of access to the Web and its resources will be wireless, with mobile e-commerce and e-working gradually replacing today's business and work models.

Much like in 1988, and in light of the technological and market developments that have taken place since then, the issues on today's research agenda, include inter alia; (1) the consideration of the requirements for additional spectrum to support the emergence of information society services and applications in both licensed and unlicensed frequency bands, (2) the requirement to investigate how a number of competing and complimentary heterogeneous wireless systems and networks can be alternatively used to convey services in optimised cost/ effective ways, (3) the pressing need to develop new wireless architectures particularly for ad-hoc networks for personal and home environments, (4) the need to maintain the momentum on R&D in reconfigurable and cognitive radio systems and networks, (5) the increasing emphasis that needs to be placed on service development and middleware issues, and finally (6) the continuous search for innovative ways to maximise transmission capacity, optimise air interface schemes and develop the underpinning enabling wireless technologies.

Acknowledgements

Thousands of researchers and engineers have over the years contributed to the design and development of UMTS. Credit to this superb team effort is not only due to my colleagues in the European Commission who helped steer and integrate the work, but is mainly due to the RACE and ACTS project managers who spared no efforts to reach the stated objectives of their projects and who understood the need to collaborate and share their research results. In so doing they contributed to the creation and dissemination of a wealth of knowledge, that is today fully recognised world-wide. Amongst these researchers, particular tribute should be paid to Morrison Sellar, Rodney Gibson, Ed Candy, Bob Swain, Per Goran Andermo, Alfred Baier, Allistair Urie, Cengiz Evci, Frank Reichert, Bosco Fernandes, Giovanni Colombo, Ermano Berruto, Heiki Huomo, Werner Mohr, Evert Buitenwerf, and Hans de Boer. The reader is invited to consult the following references for further detailed information on the EU R&D approach to UMTS.

Further Reading

1 RACE Mobile Telecommunications Workshop, 5–6 May 1992, Nurnberg, Germany.
2 RACE Mobile Telecommunications Workshop, 16–18 June 1993, Metz, France.
3 RACE Mobile Telecommunications Workshop, 17–19 May 1994, Amsterdam, Germany.
4 IEEE Personal Communications Magazine, Vol. 2, No. 1, February 1995.
5 RACE Mobile Telecommunications Summit, 22–24 November 1995, Lisbon, Portugal.
6 IEEE Communications Magazine, Vol. 34, No. 2, February 1996.
7 ACTS Mobile Telecommunications Summit, 27–29 November 1996, Granada, Spain.
8 ACTS Mobile Telecommunications Summit, 7–10 October 1997, Aalborg, Denmark.
9 IEEE Communications Magazine, Vol. 36, No. 2, February 1998.
10 ACTS Mobile Telecommunications Summit, 8–11 June 1998, Rhodes, Greece.
11 IEEE Personal Communications Magazine, Vol. 6, No. 2, April 1999.
12 ACTS Mobile Telecommunications Summit, 8–11 June 1999, Sorrento, Italy.
13 The European Co-ordinated Approach to 3G, 9–10 September 1999, Beijing, China.
14 Mobile Communications in ACTS, Infowin Report, October 1999.
15 IEEE Communications Magazine, Vol. 37, No. 12, December 1999.
16 IST Mobile Telecommunications Summit, 1–4 October 2000, Galway, Ireland.

Chapter 7: The UMTS Related Work of the European Commission, UMTS Task Force, UMTS Forum and GSM Association

Section 2: UMTS from a European Community Regulatory Perspective

Ruprecht Niepold[1]

7.2.1 The Significance of UMTS from a European Community Policy Perspective

7.2.1.1 Mobile Communication – an Important Society Phenomenon and an Economical Success

With nearly 250 million GSM users in Europe and an average penetration of over 60% of the population at the end of 2000, mobile communications has become a true mass market phenomenon which has a double significance for the European economy:

- It represents by itself a remarkably dynamic sector. Telecommunications service in western Europe in 2000 accounted for a market of some €200 billion with about 12.5% growth rate, 38% of which is accountable to mobile communications. The GSM Association estimates that since 1996 some 445 000 jobs have been created by the GSM sector and that the cumulative investment amounts to some €70 billion.
- Mobile communications are becoming pervasive to all activities of society. Changing private and professional lifestyles have created a surging demand for communications "on the move" and "reachability". Mobile communications in that sense have therefore already today – although essentially limited to voice communication – contributed to an

---

[1] The views expressed in this paper are those of the author and do not necessarily reflect the views of the European Commission.

increased efficiency of economic and private activities, well beyond the mobile communication sector itself.

7.2.1.2 UMTS – a Truly New Generation of Services

Third generation mobile communication (3G) has the potential to allow higher transmission bandwidth and allowing for a better efficiency in radio spectrum usage. Enabled by UMTS but not limited to this specific technology platform a new quality of wireless services is about to emerge. Although the exact profile of the new applications remains largely untested, the following characteristics are relevant for the new service generation:

- The possibility to convey data with a large bandwidth enables the *wireless transmission of a vast range of content forms* such as high quality audio, still and moving pictures, large data streams including access to the Internet. A vast range of range of multimedia services as opposed to a simple voice communications is both technically and economically at reach.
- 3G communications aim at realising for the first time *a truly global access capability*. *Roaming*, which was offered for the first time in the case of GSM, was an essential ingredient of its success as it gradually became *a pervasive feature*, is aimed at from the outset in the case of 3G.
- As an innovative feature compared to GSM, the large service offer with ubiquitous access will allow for a *personalised service profile*.
- *Location determination* of 3G terminals will become a key feature of the new services. The position information is expected to add significant value to services requested by an individual user, as information can be targeted with respect to the physical location of the terminal.
- Integrating *smart cards* in 3G terminals will effectively support transaction type operations, based on personal identification and security features embedded in the future mobile terminals.

These new dimensions clearly qualify 3G as a key element in realising the Information Society.

7.2.1.3 The Policy Objective Associated with UMTS

The *overall policy objective for the European Community*[2] is to create the environment, which enables the development of the new networks and services in accordance with a demand by users, and thereby to contribute to developing this sector and the added value

---

[2] The European Community consists presently of 15 member states (Austria, Belgium, Denmark, Finland, France, Germany, Greece, Ireland, Italy, Luxembourg, The Netherlands, Portugal, Spain, Sweden, and United Kingdom). In many policy areas, among them activities related to information society, the European Community co-operates closely with non-Community neighbouring countries through the European Economic Area (EEA, including Switzerland, Norway, Iceland, Liechtenstein); these countries incorporate in certain cases Community legislation and participate in Community research programmes. In the context of new countries in the process of becoming new members of the Community (Tcheky, Hungary, Poland, Cyprus, Lithuania, Slovenia, Romania, Bulgaria, Malta, Turkey, Slovakia), the existing Community legislation will become applicable to these countries upon accession. Many of them already today participate in Community R&D programmes.

it generates in many societal activities. As a side effect, it helps to maintain a high level of industrial competitiveness and fosters technological progress. Building on the success of the second-generation mobile communication to enter the age of 3G has therefore been recognised at an early stage as an important Community interest.

The challenge for the European Commission was to accompany the preparation of 3G with regulatory measures applicable in the European Community in order to support the *development of an interoperable service platform at Pan-European level* via *minimum regulatory intervention*, thereby *creating a market at the European scale*.

7.2.1.4 Layers of Political Action to Prepare for and Realise UMTS

To achieve this objective, political action at Community level is essentially anchored around four layers:

- *Support of technology development at pre-competitive level*: research and development help to prepare new technologies and to develop and validate the innovative equipment and methods. For the Community, this has been undertaken under the successive Community Framework Programmes.[3]
- *Encourage a common technical platform*: although the Community institutions are not directly engaged in establishing standards, the policy objective is to encourage all relevant players to identify, if possible, an open platform based on a common set of specifications, which would allow for roaming capability and interoperable networks, and thereby open the way for a competitive multi-vendor environment while potentially reducing the cost of equipment through economies-of-scale.
- *Harmonisation of spectrum usage*: sufficient spectrum bands must be made available for 3G services; moreover, the use of such spectrum needs to be harmonised, in order to facilitate roaming and to support an efficient use of such scarce resources. Overcoming fragmentation of spectrum simplifies the design of terminal equipment, as multi-band technology can be avoided.
- *Offering an appropriate regulatory environment*: it determines the legal conditions under which services can be offered and spectrum can be used. This implies also to organising the relations with other service providers and networks operators (e.g. interconnection modalities) as well as the rules for certifying equipment to allow free movement and placing on the market. The regulatory environment significantly influences the conditions under which the industry invests in the new market opportunities and therefore is critical for a sound development of the new sector.

7.2.2 The Background of the Community Telecommunications Regulatory Environment

In order to understand the involvement of the Community in 3G, it is useful to recall the milestones of the Community telecommunication regulatory environment, which set the boundaries for UMTS to become a reality today.

---

[3] Refer to Chapter 7, Section 1.

7.2.2.1 Regulatory Environment at Community Level: Harmonisation and Liberalisation

At the outset of the 1980s, the telecommunications sector was characterised by state monopolies in the member states. The first decisive step to deregulate the sector was the Green Paper on Telecommunications (1987) [1],[4] which proposed a step-wise implementation of a series of principles aimed at liberalising and harmonising the sector throughout the Community.

A concrete roadmap was established with the "Telecommunications Review" in 1992 [1]. It set out the pivotal date of 1998 to achieve a full liberalisation of all telecommunications services (including voice) as well as of infrastructures.

The subsequent period of 6 years (1992–1998) was one of the most intensive ones in terms of regulatory reform, but it also coincided with new forms of services reaching maturity at the market level, in particular the launch of the second generation mobile services (GSM) in 1992, the rapid deployment of cable networks and the availability of satellite communication services.

From the regulatory perspective, the Community reacted to these developments by setting an advanced agenda for liberalisation of the new services. A series of Green Papers initiated public consultations and the subsequent adoption of specific measures, such as for the satellite sector [2], the mobile sector [3], and the cable sector [4].

In parallel, a first attempt was made to establish the Community legislation, which would provide for the harmonised regulatory frame of the sector. In particular, principles were established governing the licensing telecommunications services (Licensing Directive 1997 [5], the Directive on interconnection of networks 1996 [6], and the provision of universal service (ONP Directive applied to voice telephony and universal service, 1996 [7]). Legally speaking, all Directives require a transposition into national law, which can take different forms taking into account the specific national legal environment. The Commission verifies regularly the transposition and publishes results in the form of Implementation Reports since 1998 [8].

In parallel with the harmonisation measures, the application of the competition rules to the telecommunications sector was set out. In this context, the Mobile Directive [9], which eliminated all historical exclusive and special rights, only allows a limitation in the number of licences based on essential requirements and on the lack of frequency available. These limitations have to be justified and based on objective, proportionate and non-discriminatory criteria.

As a consequence of these undertakings, the mobile sector was already fully liberalised in 1996, i.e. 2 years in advance to the fixed voice services. By encouraging competition from the outset, this early liberalisation certainly contributed to the success of the deployment of GSM services and networks.

7.2.2.2 International Environment: WTO General Agreement on Trade and Services (GATS)

In parallel with the European liberalisation process, the WTO GATS negotiations (Uruguay Round) resulted in the conclusion of the Agreement on Basic Telecommunications Services

---

[4] All superscript indications in square brackets correspond to the reference documents listed in the table at the end of the chapter.

in February 1997. Its entry into force in 1998 coincided with the full liberalisation of tele-communications in the Community, which corresponds to the Community commitment under WTO GATS.

According to the international obligations enshrined in the WTO GATS agreement (partly also in the WTO Technical Barriers to Trade Agreement (TBT)), market access may not be limited and national treatment has to be ensured for mobile communications services from other trading partners. Since frequencies are an indispensable condition for access to a mobile market, disciplines have been laid down. GATS allows some freedom to allocate and assign frequencies subject to the obligation to do so in a reasonable, objective and impartial manner; restrictions may not be more burdensome that necessary to ensure the quality of the services and cannot be used as a disguised barrier to trade.

7.2.2.3 Origins of the Community Policy on Information Society

Politically, the reform effort in the telecommunication sector was embedded in and supported by the overall movement towards implementing the single market, which included services. The White Paper on growth, competitiveness and employment of 1993 [10] made this link particularly visible as it contained a whole section on the Information society. As a follow-up, the so-called Bangemann Report [11] to the Council launched an Action Plan, which contributed to paving the way to realising the Information Society. These early efforts have been pursued since and are today conducted inter alia under the eEurope Initiative [12] and the eContent Programme [13].

7.2.2.4 Community Co-ordination on Standardisation

An important transition towards a harmonised environment at Community level was the creation of the European Telecommunication Standardisation Institute (ETSI) at the initiative of the Commission in 1988. Until then, standardisation in telecommunications at Community level was undertaken by the Conférence Européenne des Postes et Télécommunications (CEPT). The main novelty introduced with the creation of ETSI as an independent body was the possibility of all actors (and not only administrations) to contribute to standardisation efforts.[5] ETSI closely co-operates with the standardisation efforts undertaken by the ITU. ETSI's mandate includes not only the development of standards as open and common platforms where innovative systems and services are emerging, but also producing so-called harmonised standards where legislation of the European Community requires legally binding specifications (e.g. security specifications, tolerable interference levels, etc.).

One of the first and most successful undertaking by ETSI was the elaboration of the GSM platform which allowed the launch of GSM in 1992. The standardisation work undertaken by ETSI since then which has resulted in continuous revisions and gradual evolution towards upgraded functionalities (speech codecs, SMS messaging, GPRS, application development tools, etc.) has been the key to achieving today's world-wide acceptance of GSM.[6]

---

[5] ETSI has acquired a global membership. Its CEPT origin comprised from the outset the CEPT "footprint" (CEPT presently has 43 members, i.e. includes the 15 members of the European Community, the EEA countries as well as the accession countries). However, ETSI is also open to members from outside this geographical region.

[6] Refer in particular to Chapter 2, Section 2 and Chapters 3–5.

7.2.2.5 Community Approach to Radio Spectrum Management

With respect to *spectrum management and grants of authorisations to use spectrum*, the responsibility lies with national administrations of the Community. Until the end of the 1980s, harmonisation of spectrum usage was achieved on a case by case basis through Community legislative acts, where Community co-ordination was deemed necessary. This was notably the case for achieving the harmonisation of radio spectrum for cellular digital land-based mobile communications in 1987 [14]. After WARC 92, the member states entrusted the CEPT with this task, in close co-operation with the Community institutions such as the European Commission, where Community interests are at stake. A Memorandum of Understanding (MoU) between the European Commission and CEPT was concluded, as well a Framework Agreement covering studies to be conducted to identify harmonisation needs in the radio spectrum area. Legally speaking decisions taken by the CEPT are not binding to its members. This has facilitated in many cases compromises since individual CETP members can opt out, but may also result in a lack of legal certainty as for the implementation of harmonisation measures advocated by CEPT.

7.2.2.6 Community Approach on Equipment Certification

The legal basis necessary to ensure the *free movement, placing on the market and putting into service of IMT2000 equipment*, is laid down in the Radio and Telecommunication Terminal Equipment Directive (*R&TTE Directive* [15]). It entered into force in April 1999 and introduced a new approach facilitating the complex procedures of telecommunications terminal type approval developed during the 1980s. It included for the first time radio equipment. This R&TTE Directive allows the manufacturer to proceed by self-certification of equipment against harmonised standards, which describe certain essential requirements to be met in order to avoid harmful interference and or to ensure the protection of health. Once complying with the essential requirements, equipment can be circulated freely through the Community, can be placed on the market and operated.

7.2.3 Preparing and Implementing the Regulatory Environment for UMTS

From the outset, the European Commission considered that the introduction of 3G was an industry and market-led process.

Before deciding on new regulation, the European Commission supported the creation of an independent industry-led association that gathered all interested players and competitors (manufacturers, operators, content providers, and administration representatives). The *UMTS Forum*[7] was launched in 1996 and did pioneering work to determine the basic factors, which were to impact on the development of 3G services. The Forum elaborated market and communications traffic predictions and formulated requests concerning regulation.

In 1997, the European Commission launched *a public discussion via a Communication to the Council and the European Parliament* [16] and summarised its *policy proposal* [17], based on comments received and on input from the UMTS Forum. The generally advocated approach was to rely on existing legislation for licensing and competition issues, while

---

[7] Refer to Chapter &, Sections 3 and 4.

adopting specific harmonisation measures in view of a co-ordinated introduction of 3G in the Community. In 1998, a *Harmonisation Decision* ("UMTS Decision" [18]) was proposed to Council and the European Parliament. This was subsequently discussed and adopted in a record time of 9 months. The UMTS Decision entered into force in January 1999. This Decision, together with the *Licensing Directive* [5] and the *Mobile Directive* [9], constitutes today the basic legal reference for the adoption by member states of their national licensing modalities.

7.2.3.1 Key General Provisions Governing the Licensing of Operators of Third Generation Networks and Services

- The Licensing Directive requires member states to issue licenses on request.
- In the case of scarce resources such as spectrum, *the number of licenses can be limited*, but national administrations must ensure a transparent, non-discriminatory and pro-competition decision process in doing so and in attributing licences.
- Spectrum assignment decisions must ensure an *efficient usage of the spectrum* and promote *competition*.
- The Licensing Directive further specifies the conditions under which member states can impose *fees* (to cover administrative costs of spectrum management) or *charges* (for the utilisation of spectrum in case of scarcity).

7.2.3.2 Harmonisation Conditions Provided for by the UMTS Decision

- It *sets out a time frame* within which member states must prepare their licensing conditions (by 1 January 2000) so that applicants who wish to do so can start 3G services (by 1 January 2002 at the latest). This time frame was based on the results of public consultation. The UMTS Decision does not impose the start of 3G services by a certain date.
- It identifies ETSI as the body delivering an open common 3G-platform proposal, UMTS, which needs to be compatible with the standardisation concept, which was under preparation within ITU and called IMT2000.
- In order to benefit from the advantages of a common platform in terms of roaming, the UMTS Decision asks *member states to license at least one network according to ETSI's UMTS proposal*. Otherwise, license applicants can adopt any platform within IMT2000 recommendations.
- The UMTS Decision incorporates a mechanism aimed at reinforcing the legal certainty of CEPT Decisions to be adopted by the Community member states. The UMTS Decision foresees the possibility for the Commission to issue *mandates to CEPT in view of harmonising the spectrum utilisation throughout the Community*. The provisions of the UMTS Decision make CEPT deliverables pursuant to these mandates applicable within the Community, since CEPT Decisions by themselves do not have a mandatory effect on CEPT members.
- *Trans-border roaming is encouraged*, but is not made mandatory as it is considered to be in the economic interest of operators and users.
- Member states can take measures in cases where it is necessary to ensure *coverage of less populated areas* with UMTS services.

7.2.3.3 Spectrum for UMTS[8]

After the UMTS Decision entered into force in January 1999, the Commission *issued four mandates to CEPT in relation to spectrum harmonisation.*

- A first mandate addressed the amount of spectrum needed for 3G and has resulted in member states agreeing to make available 155 MHz of spectrum (nearly identical with the IMT2000 core band) as of 1 January 2002.
- A second mandate covers the spectrum scheme subsequently worked out by CEPT.
- The third mandate asked CEPT to prepare a harmonised position in view of the decisions which were to be taken by WRC2000 on identifying IMT2000 extension bands to be made available beyond the IMT2000 core band. CEPT delivered in the form of a European Common Proposal which was in substance adopted by WRC2000.

The Decisions adopted by CEPT pursuant to these three mandates can be found under [19].

- A more recent fourth mandate [20] launched the process of CEPT to identify the spectrum bands which the Community intends to make available in addition to the present IMT2000 core band, subject to market demand. CEPT was asked to deliver in iterative steps, to allow for a suitable adjustment of the envisaged time frame and choice of bands according to the 3G market development.

7.2.3.4 UMTS Equipment Certification: a Pre-Condition to Free Movement and Placing on the Market within the Community

Following the adoption of R&TTE Directive in 1999 ETSI was requested to establish the technical specifications against which manufacturers could certify their equipment. UMTS equipment was listed among the product categories requiring specifications. At that time, ITU was just about to finalise its IMT2000 recommendations.

In December 1999, the European Commission invited ETSI to work out a Harmonised Standard for IMT2000 equipment. A formal request (mandate from the Commission to ETSI in July 2000) confirmed this request. The technical work is currently under way, and a specialised Technical Working Group has submitted a draft set of recommendations to ETSI. While these specifications are, once formally adopted, the legal basis for applying the provisions of the R&TTE Directive in the Community, they will at the same time be proposed for consideration to the ITU (Working Party 8F) as a reference basis for free movement of 3G equipment world-wide.

7.2.3.5 Intellectual Property Rights (IPRs)

The debate on standards for 3G both in ETSI and in the ITU revealed two potential problems resulting from the conditions of access to patents covering technology which is necessary to implement 3G networks of terminals.

The possible blockage of a standard where the rights in connection with used technical solutions are not made available by holders of essential patents is, under the present IPR policy of ETSI or the ITU, addressed by obligations for parties to declare essential patents

---

[8] Refer also to Chapter 7, Section 5.

during the standard development process. However, these rules do not provide for a mechanism to assess whether patents claimed to be essential are effectively so. This absence of procedures can result in lengthy and costly disputes.

A further issue became visible in the case of IMT2000, which is a complex technical platform. Some feared that the cumulative royalty payments for all essential patents involved would become prohibitively high and could even jeopardise the business case as a whole.

National regulators as well as the European Commission have not proposed specific measures to meet these two challenges, considering that the matter needs to be solved among industrial actors themselves. Ultimate safeguard measures are under the current Community competition rules, provided evidence is given that a patent holder abuses a dominant position.

However, industry itself has since 1998 worked on finding a suitable scheme for IMT2000. A group of industrial players (manufacturers, operators) has proposed a *"3G Patent Platform"* [21] which attempts to ensure stable and reasonable royalty conditions to all players adhering to the platform and proposes a method to assess the essentiality of patents. Given the competition aspects underlying this proposal, the Patent Platform is currently subject to clearance by regional or national competition authorities (Japan Fair Trade Commission, US Department of Justice, European Commission).

7.2.4 The Present Situation

At the time of writing (March 2001), 11 out of 15 member states have issued 3G licenses. Table 7.2.1 gives an overview of the outcome of these licensing processes. These member states represent more than 90% of the present GSM user population in the Community.

7.2.4.1 Results of the Licensing Process

The *number of licences* differs in the member states. In most cases the regulators proposed four to six 3G licences, to exceed (or at least equal) the number of existing 2G operators. However, over the period of licensing (1999–2001) the interest of potential applicants decreased, leaving for the time being two member states, France and Belgium with less 3G licences issued than offered.

As for the *technical platform of IMT2000*, so far all licensed operators in the EU have opted for UMTS.

Closely, but not exclusively linked to the number of licensed operators, the *amount of radio frequencies to be assigned per operator* is not homogeneous among member states. The choices result from the influence of several factors, weighed differently by national administrations: the overall available spectrum in each member state, the number of licences retained by each member state (e.g. including provisions to "float" the number of licenses within certain limits and to let the auction process determine the degree of competition in the case of Germany and Austria), specific asymmetries of spectrum assignment foreseen to promote the market entry of newcomers, etc. From a user point of view, the variations in spectrum allocation should not affect the 3G operation or impose constraints to the terminal design. All member states have adhered to the CEPT Decisions on the availability of spectrum and on the spectrum scheme taken subsequent to mandates 1 and 2 issued by the

Table 7.2.1 Current situation of 3G licensing in member states (as of 31 March 2001)

MS	App[a]	Status[b]	Date	Number of licenses[c]	Total price (€billion)	Duration of licenses (years)	Population coverage obligation	Frequencies allocation per license[d]	Roaming 2G.3G
AUS	A	Done	11.00	6 (4)	0.83	20 from licence award	25% by 31.12.2003 50% by 31.12.2005	12 packages of 2 × 5 MHz, and five packages of 1 × 5 MHz	Yes
B	A	Done	3.01	3 (3) (four licences offered)	0.45 (three licences)	20	30% > 3 years 40% > 4 years 50% > 5 years 85% > 6 years	All licences: 2 × 15 MHz +1 × 5 MHz	Yes
DK	A	Pending	10.01	4 (4)	t.b.d	t.b.d	t.b.d	t.b.d	t.b.d
FIN	CB	Done	3.99	4 (3)	0	Network license: 20 freq. Lic. 10 renewable	No specific obligation, but Ministry to ensure implementation of licenses	All licences: 2 × 15 MHz + 1 × 5 MHz	Yes
F	CB + P	Under way	7.01	4–2 (3) (four licences offered)	9.8 (two licences)	15	Voice: 25% > 2 years; 80% > 8 years Data: 20% > 2 years; 60% > 8 years	All licences: 2002: 2 × 10 MHz 2004: 2 × 15 MHz + 1 × 5 MHz	Yes

Table 7.2.1 (continued)

MS	App[a]	Status[b]	Date	Number of licenses[c]	Total price (€billion)	Duration of licenses (years)	Population coverage obligation	Frequencies allocation per license[d]	Roaming 2G.3G
DE	A	Done	8.00	6 (4)	50.8	20	25% by end 2003 50% by end 2005	Number was determined by auction: five licences: 2×10 MHz $+ 1 \times 5$ MHz; 1 license: 2×10 MHz p.	Possible but no obligation
GR	t.b.d	Pending	t.b.d	t.b.d (3)	t.b.d	t.b.d	t.b.d	t.b.d	t.b.d
IRL	CB + P	Pending	7.01	4 (3)	t.b.d	t.b.d	Class A: June 2004: 53%; 2005: 80% Class B: end 2004: 33%; end 2006: 53%	All licences: 2×15 MHz $+ 1 \times 5$ MHz	Yes
I	A	Done	10.00	5 (4)	13.82	15	7.2004: regional capitals 1.2007: main provincial towns	Two licences: 2×10 MHz; two licences (new entrants): 2×15 MHz	Yes

Table 7.2.1 (continued)

MS	App[a]	Status[b]	Date	Number of licenses[c]	Total price (€billion)	Duration of licenses (years)	Population coverage obligation	Frequencies allocation per license[d]	Roaming 2G.3G
LUX	CB	Pending	< 6.01	4 (2)	–	t.b.d	Subject to market development	t.b.d	t.b.d
NL	A	Done	7.00	5 (5)	2.7	Until end 2016	1.1.2007: cities > 25 000 inh. + main communication points	Two licenses: 2 × 15 MHZ + 1 × 5 MHz Three licenses: 2 × 10 MHz + 1 × 5 MHz	In principle yes (w.o object)
P	CB + P	Done	11.00	4 (3)	0.4 + yearly fee	15	20% > 1 year 40% < 3 years 60% > 5 years	All licences: 2 × 15 MHz + 1 × 5 MHz	Yes
SP	CB + P	Done	3.00	4 (3)	0.5 + yearly fee (14.1)[e]	Until 8.2020; 10 extendable	1.8.2001: cities > 250 000 inhabitants	All licences: 2 × 15 MHz + 1 × 5 MHz (progressive freeing of spectrum)	Yes

Table 7.2.1 (continued)

MS	App[a]	Status[b]	Date	Number of licenses[c]	Total price (€billion)	Duration of licenses (years)	Population coverage obligation	Selection criterion	Frequencies allocation per license[d]	Roaming 2G.3G
SUE	CB + P	Under way	11.00	4 (3)	0.05 + 0.15% turnover annual fee	15 (network license)		Selection criterion	All licences: 2 × 15 MHz + 1 × 5 MHz; max of two new entrants receive GSM frequency (900 and 1800 MHz)	Access to GSM spectrum
UK	A	Done	4.00	5 (4)	38.5	Until 31.12.2021	80% population by end 2007		A: 2 × 15 MHz + 1 × 5 MHz; B: 2 × 15 MHz; C,D,E: 2 × 10 MHz + 5 MHz	Yes

[a] Approach for licensing decision: A, auction; CB, comparative bidding; P, payment of fees.
[b] Done, licensees have been designated; under way, formal process of licensing engaged; pending, licensing conditions not yet determined.
[c] Number of 3G licenses issued; in brackets: number of current 2G operators.
[d] 2 × ...MHz, paired bands (FDD); 1 × ... MHz, unpaired bands (TDD).
[e] Estimate based on the level of yearly fee for 2001 extended over an assumed licence duration of 20 years.

European Commission to CEPT. There are, however, some differences on the date of effective availability of spectrum between member states.

Specific licensing conditions vary among member states. Most member states favour *"national roaming" as a means to re-equilibrate competition* between existing 2G operators applying for 3G licenses and 3G new entrants. In most cases, the rights for a new 3G entrant to access the 2G network of an incumbent 2G operator who has obtained a 3G licence are conditioned ("sunset clause" terminating the access rights under specific conditions or after a certain period, minimum deployment required to enjoy the access rights, etc).

Licence duration (including the effective start of the licensing period) as well as *deployment obligations* also vary significantly. These conditions reflect specific policy objectives pursued by member states, (e.g. ensuring a rapid infrastructure deployment in all regions). The options for 3G licensees to share part of their network infrastructure are not regulated in a uniform way. While *network infrastructure sharing* appears feasible in some member states, it seems de facto ruled out in others who apply very strict rules to preserve competition in the new market.

Member states have made different choices for the *licensing method*. About half chose *auctions* whereas the other used *comparative bidding* (with or without payments). The licensing method is not prescribed by Community legislation, and both methods are a priori compatible, as long as the detailed implementation conditions remain in line with the principles set out in the Licensing Directive and the UMTS Decision.

Consequently, and depending on which approach was adopted, the *price for spectrum to be paid by successful bidders varied significantly* (€0–650 on a per capita basis). Whereas auctions immediately make the value of spectrum visible, this is not the case for those comparative bidding exercises where no payment is required. In these cases, the value of spectrum remains undetermined and is transferred as such to the future licensees. In certain member states where administrations used comparative bidding procedures and fixed a price for spectrum, the objective was to establish comparatively moderate "entry" thresholds. In some member states calculations of prices were based on the results from auctions in other member states. In doing so, administrations had to consider specific national conditions including various factors such as anticipated size of the 3G market, number of licences impacting on the future competition situation, duration of license, investments needed for establishing the network depending on geography/demography. In the case of auctions, such modelling is undertaken by the future market players themselves (often together with their investment partners), who are closer to the market reality than administrations.

7.2.4.2 First Assessment of the Licensing Results

The process of 3G licences has attracted considerable attention for several reasons. The comparatively high price paid in certain auction cases was a new experience for players in the Community. Furthermore, the different 3G licensing exercises took place at a moment when – at the outset independently from the 3G market perspective – the financial environment of the communications and the Internet sector experienced a phase of readjusting following a period of sharp market evaluation. Finally, operators acquired 3G licences while the 3G market remains largely untested. While it is too early to draw final conclusions, three aspects have already been highlighted:

- *Competition*: a debate has started on whether the licensing procedures ensured neutral market access conditions, in terms of competition, when considering the variety of licensing conditions imposed by member states and the differences in licensing decision approaches. With three exceptions (France, Sweden and Italy), *all incumbent GSM operators applied successfully for a 3G licence*. The pattern of licence applicants for 3G shows *a clear Pan-European strategy* by a few incumbent 2G operators attempting to secure a Pan-European footprint. Their number approximately corresponds to the number of networks which can be accommodated in the available spectrum (IMT2000 core band). However, they do not cover all member states and are in some of them only present through minority participation in 3G consortia. This leaves room for competition with "local" 3G licensees. All licensees have a record of telecommunications activities, although there are frequent cases of "new entrants", i.e. operators not present so far on the market of a specific member state. It is too early to assess the *degree of competition* which will eventually emerge. While it is clear that the unexpected high costs of the licences in some cases and high investments required to deploy 3G networks put pressure on 3G licensees, some observers indicate that these difficulties are temporary and will be by far compensated by the market potential and the resulting promising revenue streams. Others predict consolidations as inevitable and put forward their fear of oligopolistic structures developing in the mobile sector.
- *Fragmentation of licensing conditions*: the present situation *shows that licensing conditions and approaches to the decision process vary widely*. This is rooted in the legal framework in place, which does not attempt to regulate in detail the 3G licensing modalities, and widely leaves it to the member states to determine the conditions applicable to 3G licenses. The regulator was generally decided on the basis of public consultations launched at national level. This fragmentation will inevitably create distortions in the way 3G networks and services will be deployed in Europe and could affect the way a Pan-European market for 3G develops. Not only will 3G markets in individual member states pick up at different speeds. There may also be trans-border effects in that costs and other conditions of licences in one country may influence Pan-European operators in their decision to enter another market.
- *Market evolution*: the business case of 3G cannot be predicted with certainty, as future 3G operators lack a feedback on the market acceptance of the services which 3G promises to deliver. 3G offers through technological innovation the potential of a new service generation, but it has become visible that the current 2G networks – for instance by introducing high bit-rate and packet-switched modes by upgrading GSM networks with GPRS technology – can provide for services which partly overlap with the scope of 3G services. The higher than expected costs and difficulties in launching 3G networks and services have contributed to raising expectations on these intermediate technical solutions. The boundaries of 2G and 3G markets will become blurred.

7.2.5 Open Challenges

The Commission issued a Communication on the introduction of third generation mobile communications in March 2001 [22]. While taking stock of the situation and identifying critical layers possibly affecting the deployment of 3G, the Commission confirmed the confidence in the 3G market perspectives and proposed some action lines to help ease the transition from 2G to 3G.

Taking into account the lessons learned so far and anticipating the future evolution of wireless communications, two main undertakings proposed by the Commission and currently under the institutional decision process are particularly relevant: the reform of the existing communication legislation for communication and the proposal to establish the operational foundation for a spectrum policy at Community level.

7.2.5.1 Review '99

Following the telecommunications legislation put in place in view of the full liberalisation of the telecommunications sector in 1998, the European Commission last July proposed a *new regulatory framework*, after intensive consultation with the concerned sector. This process labelled "Review '99" simplifies the existing legislation and aims at promoting a competitive European market for telecommunications services. Several of its provisions will have an impact on the 3G development, depending on the date of implementation (2002–2003 as proposals are under the co-decision procedure at the time of writing).

The following *scope and general principles* characterise the Review '99:

- the new legislation is to cover communications and access services as well as communication infrastructure; services content is not regulated by the Review, but all forms of networks are covered (e.g. telecommunications, broadcasting);
- EU legislation is to set out principles, not details;
- increased autonomy of the National Regulatory Authorities, in combination with safeguard measures to ensure Community consistency of regulation;
- simplification and clarification of legal framework;
- minimum ex ante rules, reliance on competition rules otherwise: as competition increases, regulation decreases.

Five Directives have been proposed [23]: a Framework Directive as well as Directives on Authorisation, Access and Interconnection, Data Protection and Universal Service. In addition a Regulation has been proposed (adopted in December 2000 [24]) to make the unbundling of the local loop mandatory.

The Review proposes important new rules that are relevant in the context of 3G. In order to simplify the *licensing*, a separation of service licenses from granting rights to use radio spectrum is advocated. Only for the latter will member states continue to use individual licences, whereas otherwise use of general authorisation should be generalised. This separation will allow as an optional measure member states to adopt the implementing frequency trading (*secondary spectrum market*). This will open the way for a more flexible and efficient usage of spectrum, and allow 3G operators to manage the spectrum asset to maximise the economic advantage of spectrum pursuant to their evolving needs and the market reality. For cases where the market does not function, it allows the imposition of *asymmetric measures on players with a significant market power (SMP)*. However, the threshold for significant market power has been raised compared to the present regulation. In addition, relevant markets will have to be defined by the regulators when scrutinising significant market power. This flexible regulatory approach seems to be particularly indicated for 3G markets, the definition of which may evolve. *Number portability* will be extended to the mobile sector, including 3G, which should increase competition among wireless operators. *Location-based data* are referred to in the proposed new Data Protection Directive where basic principles on the use of these data are outlined. The Framework Directive encourages *agreements for co-location and facility sharing*. It is expected that the provision will contribute to easing the problems encountered by the mobile network operators in finding base station locations.

7.2.5.2 Spectrum Decision

The experience with 3G in Europe has shown that where a *political* consensus exists, spectrum harmonisation can be achieved within the framework of spectrum management by CEPT. Legal certainty of the results of such a harmonisation process could be ensured through the mandate mechanism already in place in the UMTS Decision. The conditions of access to 3G spectrum however show the need to reflect generic issues such as spectrum pricing or licensing conditions.

In order to set up a Community frame to discuss spectrum policy aspects relevant to Community policies, the European Commission started a reflection process on a radio spectrum policy for the European Community. Following the publication of a Spectrum Green Paper in 1998 [25] and a subsequent public consultation [26], a proposal for a Decision was submitted to Council and the European Parliament. This so-called *Spectrum Decision* [27] lays the *foundation for a spectrum policy in the Community*:

- putting into place a *policy platform* (i.e. Senior Official Spectrum Policy Group) which is responsive to technological, market and regulatory developments in the area of radio communications and which appropriately consults all relevant radio spectrum user communities. The group will provide guidance to the Commission on the need for harmonisation of the use of radio spectrum in the general context of Community policy and on regulatory and other issues related to the use of radio spectrum which impact on Community policies;
- establishing a *legal framework for the harmonisation of the use of radio spectrum*, where necessary to implement Community policies in the areas of communications, broadcasting, transport and R&D. In accordance with the provisions of the Spectrum Decision and on the basis of the advice from the group referred to above, the Commission will grant mandates to the CEPT to achieve this aim. Where required, the implementation of the solutions worked out by the CEPT in response to the Commission mandates shall be legally ensured;
- ensuring *co-ordinated and timely provision of information on radio spectrum use and availability* in the EC. This is necessary for investments and policy making. Existing Community and international (e.g. WTO) requirements for publication of information on the use of radio spectrum need to be complemented and the information needs to be aggregated at a European level in a user-friendly manner;
- ensuring that appropriate Community and European positions are developed in view of *international negotiations relating to radio spectrum* (e.g. ITU/WRC, WTO) where issues at stake are covered by Community policies. The Community should make sure to adopt common positions in advance of such negotiations with regard to the objectives to be achieved.

The Spectrum Decision applicability is not limited to the communication sector but covers all areas where Community interests depend on the availability of spectrum. For 3G, the Decision will allow the continuation of the practice of mandating ERC on spectrum issues in the same manner as so far undertaken pursuant to the UMTS Decision. However, the High Level Group is expected, besides helping *to identify spectrum requirements for specific Community policies*, to provide for a political frame *establishing consensual recommendation on more generic spectrum policy aspects*. The 3G case has revealed that there is a need to attempt to find common answers at Community level to approach with respect to best practices of pricing of spectrum, of relocation of spectrum, of efficient usage of spectrum, etc. The proposal to increase transparency of the usage of spectrum by gathering and making available relevant information has been recognised as crucial when attempting to price spectrum and is a necessary complement to a possible future secondary market for spectrum.

References

[1] Towards a dynamic European economy: Green Paper on the development of the common market for tele-communications services and equipment, COM (1987) 290, 30.06.87; Commission Directive of 16 May 1998 on competition in the markets in telecommunications equipment, 88/301/EEC, OJ L131, 25.5.1988; Commission Directive on 28 June 1990 on competition in the markets for telecommunications services, 90/388/EEC, OJ L192/10, 24.7.1990.

[2] Towards Europe-wide systems and services: Green Paper on a common approach in the field of satellite communications in the European Community, COM (1990) 490, 20.11.1990.

[3] Towards the Personal Communications Environment: Green Paper on a common approach to mobile and personal communications in the European Union, COM (1994) 145, 27.4.94.

[4] Green Paper on the liberalisation of telecommunications infrastructure and cable TV networks: Part I: Principle and timetable, COM (1994) 440, 25.10.1994; Part II: A common approach to the provision of infrastructure in the European Union, COM (1994) 682, 25.1.1995.

[5] European Parliament and Council Directive 97/13/EC of 10 April 1997 on a common framework for general authorisations and individual licenses in the field of telecommunications services; OJ L 117, 07.05.1997.

[6] European Parliament and Council Directive 97/33/EC of 30 June 1997 on Interconnection in Telecommunications with regard to ensuring universal service and interoperability through application of the principles of Open Network Provision (ONP), OJ L199/32, 26.7.97.

[7] Proposal for a European Parliament and Council Directive on the application of open network provision (ONP) to voice telephony and on universal service for telecommunications in a competitive environment, COM (1996) 419, 11.9.1996.

[8] Communication from the Commission to the European Parliament, the Council, the Economic and Social Committee and the Committee of the Regions: Sixth Report on the Implementation of the Telecommunications Regulatory Package, COM (2000) 814, 7.12.2000.

[9] Commission Directive 96/2/EC of 16 January 1996 amending Directive 90/388/EEC with regard to mobile and personal communications; OJ L 20, 26.01.1996, p. 59.

[10] White Paper on growth, competitiveness, employment – the challenges and ways forward into the 21st Century, COM (1993) 700, 5.12.1993.

[11] Europe and the global information society – recommendations to the European Council, 26.5.1994.

[12] "eEurope Action Plan", prepared by the Council and the European Commission for the Feira European Council of 19–20 June 2000 (Brussels, 14.6.2000). An "eEurope+" plan is about to be established under which the accession countries will promote in parallel with eEurope comparable actions.

[13] Council Decision adopting a multi-annual Community programme to stimulate the development and use of European digital content on the global networks and to promote the linguistic diversity of the information society, 2001/48/EC, 18.1.2001.

[14] Council Directive 87/372/EEC on the frequency bands to be reserved for the co-ordinated introduction of public Pan-European cellular digital land-based mobile communications in the European Community, OJ L196/85, 17.07.87. This Directive was adopted in parallel with: Council Recommendation 87/371/EEC of 25 June 1987

on the co-ordinated introduction of public Pan-European cellular digital land-based mobile communications in the European Community, OJ L196/81, 17.07.87.

[15] European Parliament and Council Directive 1999/5/EC of 9 March 1999 on radio equipment and telecommunications terminal equipment and the mutual recognition of their conformity; OJ L 91, 07.04.1999.

[16] Communication from the Commission to the European Parliament, the Council, the Economic and Social Committee and the Committee of the Regions of 29 May 1997 on the further development of mobile and wireless communications; OJ C 131, 29.04.1998, p. 9 and OJ C 276, 04.09.1998.

[17] Communication from the Commission on strategy and policy orientations with regard to the further development of mobile and wireless communications (UMTS); OJ C 214, 10.07.1998.

[18] European Parliament and Council Decision 128/1999/EC of 14 December 1998 on the co-ordinated introduction of a third generation mobile and wireless communications system (UMTS) in the Community; OJ L 17, 22.01.1999.

[19] CEPT/ERC/DEC/(97)07 on the frequency bands for the introduction of terrestrial Universal Mobile Telecommunications System (UMTS); CEPT/ERC/DEC(99)25 on the harmonised utilisation of spectrum for terrestrial Universal Mobile Telecommunications System (UMTS) operating within the bands 1900–1980 MHz, 2010–2025 MHz and 2110–2170 MHz; CEPT/ERC/DEC(00)01 extending ERC/DEC/(97)07 on the frequency bands for the introduction of terrestrial Universal Mobile Telecommunications System (UMTS).

[20] Mandate to CEPT to harmonise frequency usage in order to facilitate a co-ordinated implementation in the community of third generation mobile and wireless communication systems operating in additional frequency bands as identified by the WRC-2000 for IMT-2000 systems, LC 01 /02, 9.3.2001.

[21] European Parliament and Council Directive 1999/5/EC of 9 March 1999 on radio equipment and telecommunications terminal equipment and the mutual recognition of their conformity; OJ L 91, 07.04.1999.

[22] Communication from the Commission to the Council, the European Parliament, the Economic and Social Committee and the Committee of Regions: the introduction of third generation mobile communications in the European Union: state of play and the way forward, COM (2001) 141, 20.3.2001.

[23] "Review 99" Directive proposals: Proposal for a Directive of the European Parliament and the Council on a common regulatory framework for electronic communications and services, COM (2000) 393, 12.07.2000; Proposal for a Directive of the European Parliament and of the Council on the authorisation of electronic communications networks and services, COM (2000) 386, 12.07.2000; Proposal for a Directive of the European Parliament and of the Council on access to, and interconnection of, electronic communications networks and associated facilities, COM (2000) 384, 12.07.2000; Proposal for a Directive of the European Parliament and of the Council on universal service and users' rights relating to electronic communications networks and services, COM (2000)392, 12.07.2000; Proposal for a Directive of the European Parliament and of the Council concerning the processing of personal data and the protection of privacy in the electronic communications sector, COM (2000) 385, 12.07.2000.

[24] Regulation of the European Parliament and of the Council on unbundled access to the local loop (adopted by Council on 5.12.2000).

[25] Green Paper on radio spectrum policy in the context of European Community policies such as telecommunications, broadcasting, transport, and R&D, COM (1998) 596, 9.12.1998.

[26] Communication from the Commission to the European Parliament, the Council the Economic and Social Committee and the Committee of the Regions: next steps in radio spectrum policy results of the public consultation on the Green Paper, COM (1999) 538, 10.11.1999.

[27] Proposal for a Decision of the European Parliament and the Council on a regulatory framework for radio spectrum policy in the European Community, COM (2000) 407, 12.7.2000.

Chapter 7: The UMTS Related Work of the European Commission, UMTS Task Force, UMTS Forum and GSM Association

Section 3: The UMTS Task Force

Bosco Eduardo Fernandes[1]

7.3.1 The Building-up of an International Momentum

By 1994 the close collaboration between the European R&D projects and ETSI's Special Mobile Group (SMG) was well established with substantial contributions in all areas of mobile communications work but especially in UMTS Radio aspects. ETSI was considered to be the main forum to address telecommunications standards issues arising from the research projects. Links between the European R&D mobile project line and ETSI, in particular ETSI's SMG5, were maintained on the basis of an agreement between the European Commission (EC), ETSI and European Telecommunications Industrial Consortium (ETIC). Close collaboration with CEPT took place on a number of issues such as frequency allocation, satellite operation and network evolution. There was also a close link to the ITU-R Task Group 8/1 (TG 8/1) originally called CCIR Interim Working Party 8/13, this group was formed at the end of 1985 to define the Future Public Land Mobile Telecommunication Systems (FPLMTS). Over 30 countries and most relevant major international organisations participated in the work on FPLMTS. The World Administrative Radio Conference (WARC) of the International Telecommunication Union (ITU), in March 1992, identified global bands 1885–2025 and 2110–2200 MHz for FPLMTS, including 1980–2100 and 2170–2200 MHz for the mobile satellite component.

The major challenge facing TG 8/1, however, was the relationship between FPLMTS and the emerging Personal Communications Networks (PCNs), the European implementation of 1800 GSM, Digital European Cordless Telecommunication System (DECT) and the as yet

[1] The views expressed in this section are those of the author and do not necessarily reflect the views of his affiliation entity.

undefined third generation of the European system (UMTS) and the events in the rapidly changing Japanese and North American arenas.

7.3.2 The Trilateral FAMOUS Initiative

In Europe, the EC's Green Paper on Mobile and Personal Communication[2], a major step for a common vision was open to public consultation. This document called for continuing support for the evolution towards UMTS as a common future basis for personal communications.

Since 1991, a group of administration representatives and experts from Japan, the US and the EU had pursued yearly exchanges on issues related to third generation mobile communication systems, this group adopted the name FAMOUS (Future Advanced Mobile Universal System). The main objective of this group was to provide a forum for discussion on matters related to system development activities, how standardisation and frequency allocation in the different regions are concerned and how to foster international co-operation regarding in particular interoperability on a global basis of third generation mobile communications systems. However, barriers were obviously expected due to regional industrial interest and questions of system revolution, evolution or migration were not technical problems and could only be seen as those of a long-term European strategic telecommunications policy.

At the Fifth Trilateral FAMOUS Meeting in Lisbon in May 1995, a recommendation was made for region and nation standards bodies to promote a consensual standardisation process which was to be market-led and which encouraged open interfaces. Due to the fact that markets in different countries/regions differed, global standards should address the kernel of services and interfaces necessary for international compatibility. Regional extensions could specify more detail. And all standards should be flexible (technology agile) to allow for evolution/migration and to reflect user needs as expected by the market place in developing as well as developed countries. This would lead to an open and formal process in order to adopt technical standards considering efficient use of spectrum and there would be no formal policy restrictions to prevent the consideration for introduction of any technologies into regions, including those originated in the US.[3]

This was the beginning of what was later to be known as the IMT-2000 family concept.

7.3.3 The MoU Personal Mobile Telecommunications

Other national groupings were beginning to appear around the same time. This could have developed into a complete fragmented opinion and divergence in an overall European strategy introducing third generation mobile systems and its requirements.

In December 1993, a Memorandum of Understanding (MoU) for Personal Mobile Telecommunications (PMT) was signed in Germany, a new body! It was an open forum with 31 members from different national industry sectors or subsidiaries of international industry located in Germany. The MoU PMT developed short- and long-term aggressive objectives with a phased roadmap indicating goals up to 2005. The European presence were mostly

---

[2] Towards the Personal Communications Environment: Green Paper on a common approach to mobile and personal communications in the European Union, COM(94) 145 Final. European Commission.

[3] A report from this meeting and copies of the recommendations agreed can be found in the CD ROM folder E.

sceptical observers who would have registered should the activity have caught momentum. It did not gain European support and was closed later.

7.3.4 Empirical Findings

The EC research community felt that a lack of a common vision and the ultimate transition strategy from GSM to UMTS was missing and had yet to be developed. The industry and ETSI were totally involved in second generation systems and did not take third generation seriously. The RACE vision of UMTS that advocated a gradual evolutionary path from second towards third generation mobile broadband networks was still a consensus of the European Research Programme and although presented to ETSI and other standard bodies as well at many international conferences, it was not taken seriously enough. A lack of acceptance of the key messages of the European research community could be sensed. This was further aggravated by lack of firm industrial commitment in the Standards bodies and the regional European representation in the ITU TG 8/1 who were from administrations but not necessarily participants at ETSI. The development of UMTS standards within ETSI SMG5 was beginning to stagnate in the technical definition of UMTS. Some operators and also participants in ETSI SMG5 expressed concerns at the EC about the slow progress ((e.g. no clear understanding of UMTS radio system)) and avenues for making effective use of information were not being opened ((e.g. establishing a full European influence in ITU-TG8/1)) due to lack of a clear supporting policy and process.

European UMTS work paralleled ITU's FPLMTS activities and consequently the primary standards decision forum laid outside Europe's direct control. Europe, therefore, had a choice

- to stand back from the ITU process except for fundamental networking decisions, or
- to use its full experience to influence, or produce, global standards.

The latter option was the sensible course and yet ETSI-SMG5, the centre of European expertise, had not the power to defend Europe's position in the ITU. Indeed ETSI was unable to express its voice on behalf of ETSI in the ITU even though it had a standardisation group for UMTS. No other European body was able or willing to pick up the reins to play this role of expressing a single European voice either. This was a major weakness that needed a quick solution.

7.3.5 The RACE UMTS Vision

At this stage the RACE community decided to refine its "RACE UMTS Vision" paper and to present it in a workshop called "Towards Third Generation Mobile Communication Systems" in Brussels on 31 January 1995. The main objective was to achieve European industrial consensus and support on how to proceed. The real need, however, was to find a quick solution to the above problem and harmonious development of GSM's, DCS1800's, DECT's, etc. full exploitation in commercial or technological terms, with seamless service transition and enhancement beyond the year 2000, coupled with the intention of maintaining Europe's commanding position of personal mobile communications in third generation deep into the 21st Century.

7.3.6 Doom's day!

A day's workshop was organised at the invitation of the EC with distinguished speakers from different industry sectors and standards bodies to present their views on the evolution towards UMTS and also the research vision paper on 31 January 1995. The audience were project managers of the RACE community and some 180 high ranked registered guests from all over Europe. The clear message was:

UMTS's development needed a clear European strategic direction on a number of issues. Figure 7.3.1 illustrates the requirement and core issues.

Figure 7.3.1 shows an Advisory Board with the role to guide, co-ordinate, and focus on disseminating the UMTS strategy for the four issues. In some cases, e.g. migration, it might have had the responsibility for forming strategy. On the other hand the ERC/ERO would play a lead role concerning frequency allocation matters.

The Advisory Board would guide (and lead?) UMTS contributions to the ITU and generally act as a quick-responding focal point for such matters. This activity needed to be co-ordinated with national administrations since it was highly desirable for Europe to balance the single-minded effectiveness demonstrated by the US, Japan, Canada, etc. towards FPLMTS. In the US the FCC co-ordinates participation and there was an "International Co-ordination Group" within their standards bodies chaired by Qualcom Inc.

After setting the scene, a drop out of the electrical power supply in the middle of the workshop led to security measures of everyone having to immediately leave the room and resuming the workshop a couple of hours later. Of course, it was undoubtedly important to accomplish the mission of the workshop. Was this a bad omen? A quick decision to reduce the speakers time to 10 min with key messages made it possible for a good end discussion to derive a conclusion.

The outlook was bright and in spite of uncertainties, reserved opinions and objections the majority of the people present, from both the manufacturers and the operators agreed

- that it was important to pursue the matter;

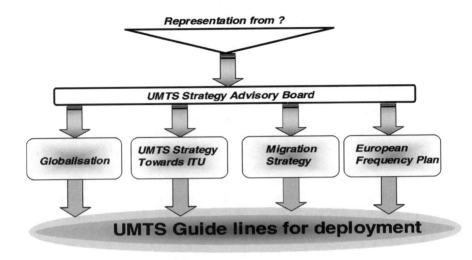

Figure 7.3.1   UMTS requirement and core issues

- that UMTS should be retained as its brand name;
- to create a UMTS Task Force as an advisory group and task them to propose a European strategy for UMTS was the right way forward.

This was a clear signal with the mission statement and objectives of the Task Force yet to be developed.

7.3.7 The Launch of the UMTS Task Force

The Task Force was established in February 1995 as a consequential conclusion of the "Towards Third Generation Mobile Communications System Workshop (Brussels, 31 January 1995)" which involved all parties relevant to UMTS. Its mandate issued by the EC was later confirmed by the European Parliament and Council. It was charged to carry out its mission and objectives as an advisory group who should propose a UMTS strategy for Europe. The details should be defined by the participants and it should produce a report within 6 months. At the first meeting on 24 March 1995 the chairman and the secretary were installed in a common agreement of the participants and EC as observers.

The rest of its members were people with a collective background and who had the following qualifications:

- expertise in terms of industrial issues (manufacturers, mobile and fixed network operators);
- standards and R&D;
- policy and regulatory issues.

7.3.7.1 Establishing Common Grounds

The Task Force first developed and agreed on a common *Mission Statement*:

To propose a European migration strategy and policy towards the implementation of UMTS that includes consideration of:

- user/service requirements of UMTS
- the path to UMTS
- European ITU representation
- frequency allocation philosophy
- regulatory and licensing issues

Then based on the outcome of the workshop it agreed on its *Prime Objectives*:

The initial study objectives were:

- To characterise UMTS's service, application and networking capabilities, requirements and potential.
- To map a European strategy for the introduction of UMTS in a customer, operator, industrial, technology and time sensitive manner.
- To propose procedural structure(s), with broad representation, for steering UMTS towards service introduction.

The secondary Objectives being *Publicity Objectives*:

- To make public its final consensus views by means of a workshop and/or report, and to consider targeted reports.
- To distribute progress reports from time-to-time with the Task Force's agreement."

It was also essential to develop a *Position Statement* that could be followed during coarse of the work:

'The Task Force saw UMTS as the next distinct mobile development that will support convergence of wired and wireless services, including personal communications services; it aims to match wired capability (recognising radio environment limitations), significantly extend user communication capabilities beyond that of a fully enhanced second generation mobile technology.

The Task Force saw mass commercial introduction of UMTS in 2005 with a life extending to 2025. Some services associated with UMTS may be emulated by enhanced second generation technology but some broadband needs may demand UMTS presence by 2002.

The Task Force sees UMTS as having considerable service enhancement potential during its life.

7.3.7.2 The First Challenge

The Task Force was confronted all of a sudden with a new situation that arose, i.e. European Radio Office (ERO) were about to launch a study of the "transition towards UMTS". Critical inputs to this exercise were needed by November 1995 and April 1996. ERO needed to know what the frequency requirement was, when and how. UMTS timing was important from a frequency planning point-of-view to clear the spectrum in a way that does not leave empty spectrum. Indeed, administrations would well insist on an operator's commitment to use the spectrum before giving the instruction to present users to clear the band. ERC was preparing for WRC 95. ETSI SMG5 had received a number of proposals but not everyone agreed to the figures.

It was vital to secure the common harmonised spectrum bands allocated for FPLMTS in WRC 92, however, the assignment of the 180 MHz band was not known and when it would to be made available for the use of third generation. The Task force was confronted with a very difficult situation at its very first meeting.

7.3.7.3 Highlights of the Path to Completion

The following meetings were very progressive in a way but also became very political. Of course, there was a lot of criticism by those who were not participants of the Task Force but to some extent also from the members of the Task Force, e.g. there was no way Europe was going to destroy the world-wide successful footprint of GSM and of course forget DECT that felt a lack of standards support and the threat of PHS down its neck. At same time, the ETSI

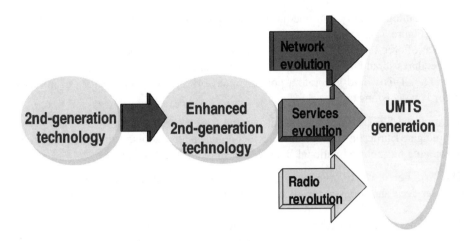

Figure 7.3.2   The evolution of UMTS

Programme Advisory Committee[4] announced the establishment of the "Global Multimedia Mobility" (GMM) project.

What the Task Force lacked was a clear market study forecast of the UMTS development and revenue predictions. What role was the Internet to play in data delivery services and were the spectrum estimates submitted to SMG5 correct? What were the services and applications the market was excepting to see in the time frame of 2005? As a result, the EC decided to exude a market study and an RFQ with clear requirements generated within the Task Force was sent out to potential analysts/consultants.

Was ETSI going to complete a stable standards release in time for deployment? This discussion led to the development of a very realistic phased approach to the UMTS schedule.

It was clear that those who provided mobile services, their equipment providers and their customers must see a continually improving mobile marketplace without major dislocations as radio technologies and regulatory scenarios evolve in the future.

A common 2 GHz RF front end arrangement, e.g. based on basic 5 MHz FDMA (FDD/TDD) channels, should be standardised as part of a general UMTS spectrum core organisational structure, with specific base band processing schemes as required in various IMT200 operational environments besides in Europe. However, none of the, at that time used, multiple-access schemes, i.e. FDMA, TDMA or CDMA on their own appeared ideal for all IMT2000 operating environments; high rate services in particular, even in very small cell coverage areas, were likely to require hybrid access schemes which also made full use of the directional properties of each user's signals.

Radio interface had to be revolutionary and the Future Radio Wideband Multiple Access Systems (FRAMES) R&D project was established to deliver the necessary input to standards.

There were also a number of other reasons why and how FRAMES was created.

In order to retain the huge investments in the network infrastructure the consensus was to maintain the evolutionary/migration of the second generation system components. Furthermore, the evolution of services with the need to develop good high quality multimedia services was the way forward. The investigations within the R&D projects of RACE clearly

---

[4] Chapter 8, Section 1, Paragraph 4.

indicated that high quality multimedia services of at least 20 MHz per network operator was required (Figure 7.3.2).

The Task Force felt that a much lighter regulatory policy, licensing policy and co-ordination procedures should be laid down in a document named "European Regulatory Framework for UMTS". This would generate a policy framework to give sector actors an information base for key decisions on:

- Harmonised SPECTRUM for global roaming;
- Promulgate European concepts into a wider arena and harmonise with global requirements
- Achieving harmonised national licences, or single European licences.

The Task Force was to urge the regulatory authorities to avoid auctioning of licences for UMTS services and/or spectrum to ensure the highest possible growth of mobile communications.

At this time, a fear of a erecting a new regulatory body at the EC level was sensed and led to major discussions. It was alleged to be a territory of national interest and legislation and not an issue for the EC.

The Task Force paved the way to completing its work within the allocated extended time frame of 12 months and ignored all political interferences and smartly overcame the huddles. By 1 March 1996 the one time historical event for presentation of the UMTS Task Force Report[5] – the future of wireless as envisioned by a consensus building process was completed and its goal met.

7.3.8 Key Findings

The Universal Mobile Telecommunication System (UMTS), will take the personal communications user into the new information society. It will deliver information, pictures and graphics direct to people and provide them with access to the next generation of information-based services. It will move mobile and personal communications forward from second generation systems that are delivering mass-market low-cost digital telecommunication services.

The principal recommendations were:

- The development and specification of the UMTS such that it offers true third generation services and systems.
- UMTS standards must be open to global network operators and manufacturers.
- UMTS will offer a path from existing second generation digital systems, GSM900, DCS1800 and DECT.
- Basic UMTS, for broadband needs up to 2 Mb/s, should be available in 2002.
- Full UMTS services and systems for mass-market services in 2005.
- GSM900, DCS1800 and DECT should be enhanced to achieve their full individual and combinational commercial potential.
- UMTS regulatory framework (services and spectrum) was to be defined by the end of 1997 to reduce the risks and uncertainties for the telecommunications industry and thereby stimulate the required investment.

---

[5] The road to UMTS "in contact anytime, anytime, anywhere, with any one", Brussels, 1 March 1996.

- Additional spectrum (estimated at 2×180 MHz) must be made available by 2008 to allow the UMTS vision to prosper in the mass market.

According to the mandate given by the European Parliament and Council, the Task Force[6] recommended the establishment of an open UMTS Forum as the central body charged with the elaboration of European policy towards the implementation of UMTS and based on industry-wide consensus. It was not foreseen what would happen to the Forum after it completed its task.

The road to UMTS "in contact anytime, anywhere, with any one" had been built. What once was only speculation of future technology was becoming a reality. Of course there is a risk of the availability of the standard and regulatory framework being put in place to achieve the agreed milestones in a reasonable time schedule. Nevertheless, it was also now up to the EC to table a "Communications" paper, followed by a "Directive" for its member states to accept. At this time, a call for tender for a "Market Study" was also out and a potential winner in sight. A convenor was appointed to establish the articles of association and create the UMTS Forum. It was now time to create a Forum and to build on the recommendations above and move on.

The Report was well received by a number of organisations including the ITU-TG8/1 group.

[6] The following are acknowledged: the members of the European Parliament and Council in 1995/1996. Without their trust and support the Task Force would never have been created nor could it have survived; the EC who inspired the RACE and ACTS project managers to share their research results and support this activity and to the project managers of the RACE and ACTS Mobile project line who gave me their strong support; the members of the UMTS TF: J. da Silva, R.S. Swain, E. Candy, E. Buitenwerf, D. Court, J. Desplanque, A. Geiss, F. Hillebrand, L. Koolen, V. Kumar, B. Eylert, J.Y. Montfort, M. Nilsson, P. Olanders, J. Rapeli, G. Rolle, E. Vallström, M.T. Hewitt and P. Dupuis.

Chapter 7: The UMTS Related Work of the European Commission, UMTS Task Force, UMTS Forum and GSM Association

Section 4: The UMTS Forum

Thomas Beijer[1]

7.4.1 The Creation of the UMTS Forum

The telecommunications industry has seen numerous organisations and groups come and go during the last 30 years. It is impossible to remember them all. In spite of this, I am not aware of any other organisation before the UMTS Forum with the same kind of mission, membership or working style. The concept of "fora" did exist long before the UMTS Forum, but those organisations were mostly dealing with technical matters, often with the purpose to create de facto standards outside the recognised standardisation bodies. The UMTS Forum had a totally different mission and at the time it was started it did not resemble anything else.

The story – as far as it was visible to a wider community – starts in late January 1995. A workshop was organised in Brussels with the purpose of discussing the evolution towards third generation mobile systems. I attended that workshop being (as I guess were most other participants) rather puzzled about the purpose of the meeting. Listening to the discussion did not help, in fact it was the other way around. At the very end of the meeting, Bosco Fernandes who was leading the workshop drew the conclusion that there was a need to create a "UMTS Task Force", referring to discussions he had had with a few senior people in the telecom business. In particular he referred to Jan Uddenfeldt, head of Ericsson technical development, who according to Bosco had recommended this move. Of course I asked Jan about this, but it turned out that he was as surprised as me, saying that he was not aware of having ever expressed an opinion on this proposal.

[1] The views expressed in this section are those of the author and do not necessarily reflect the views of his affiliation entity.

Anyway, the UMTS Task Force was set up. Bosco Fernandes describes this part of the UMTS Forum history in Chapter 7, Section 3. His description will most probably look different (and better informed) than my more anecdotal version above. But the truth has many facets. One of them is the way it looks when looking at the events from the outside.

The UMTS Task Force was a precursor to the UMTS Forum. It was a closed group consisting of 19 handpicked persons. It worked for about 1 year, without leaking much information about what was going on to the wider communications community. At the end, in March 1996, the so-called "UMTS Task Force Report" was published. This is where the UMTS Forum was born.

For completeness it is worth noting that besides the UMTS Task Force, there was also another precursor of the UMTS Forum. In the middle of the 1990s an initiative was taken by some German players to form the "Personal Mobile Telecommunications MoU" (PMT MoU). This organisation went on until late 1996. It tried to build a vision for the future that emphasised the personal nature and the convergence between fixed and mobile services. Its membership never spread much internationally.

Going back to the UMTS Task Force, the report they produced became an important turning point in the development of third generation systems. It painted a new picture of the future of mobile communications, different from those being discussed in standardisation. To materialise that vision it was necessary to change the mindset of people and companies; to learn to think more in terms of multimedia rather than ordinary telephony. There was also a need to back up this vision by other means, in terms of spectrum, regulatory support and to achieve a commercial kick-start. This is why the Task Force suggested the establishment of a UMTS Forum to help the process.

Recent experiences from the GSM development and from the GSM MoU had also taught the industry that having a technical standard is far from all that is needed for a commercial success. There is also a need to take care of the "everything else", whatever it may be. A UMTS Forum could do that. A totally open forum with the mandate to discuss whatever it found appropriate was an excellent idea, since it allowed participation by *everybody* in *all* sorts of debates provided he or she supported the UMTS vision. No more spectrum discussions behind closed CEPT doors! No more service and business discussions behind closed GSM MoU doors!

The UMTS Forum was initially established as a fairly loosely organised body. A group of people, recruited mainly from the remains of the UMTS Task Force, formed the initial UMTS Forum Steering Group. This group prepared the first meetings of the Forum. It was chaired by Ed Candy of Orange, who was also the first UMTS Forum chairman.

Starting an organisation without having any membership, mandate, procedures or rules is certainly a challenge. It is not surprising if the first meetings were rather confusing and sometimes even chaotic. I believe that the people attending the first 2–3 meetings had rather vague ideas about what the whole thing was really about and they probably had different expectations. Both organisational matters and matters related to the mission had to be taken care of. One of the more interesting discussions during those first meetings was the one about the establishment of working groups. The Steering Group proposed four different groups, dealing with regulation, spectrum, market and technology, respectively. There was no problem in agreeing on the first two of those. The market group was accepted with some hesitation. Concerning the technology group there were a lot of objections, mainly based on the fact that there was already enough technology oriented bodies around to take care of standardisation and research, much more than the members had resources to attend. The purpose and mission of such a group was unclear. Still, at the end it was decided to have it. My recollection of those decisions is that they were not brilliant examples of a democratic procedure. But again, it is difficult to apply rules that do still not exist. The decisions were accepted, leaving to the working groups themselves to find out what to do.

It was a lucky move that those working groups were formed at an early date, without getting stuck on formalities. Most of them managed to get started immediately, producing useful output without being too concerned about the fact that they reported to a body that – strictly speaking – did not exist.

In the Forum plenary much of the time during that first year was spent on organisational matters; legal status, articles of association, finances, document handling, etc. Necessary things but not very exciting. Assuming that the reader is less interested in these questions I will not discuss them here. It remains to be noted that at the end of the day – on 16 January 1996 – the Forum was established as a legal entity, with all the administrative requisites requested by Swiss law. It had a set of officers appointed for various purposes and shortly after the start it also had a secretariat set up. Ed Candy changed employer at that time and had to terminate his office as chairman. I was appointed chairman of the Forum. Belonging to Telia, Sweden, I was a representative of the operator community at that time, something that the membership (and in particular the manufacturers?) apparently found important. My feeling is that they were anxious to avoid every suspicion that the UMTS Forum was a show run by any particular manufacturer. It must be kept in mind that the purpose of the Forum was still not widely understood.

Who were the members? On the day the Forum was formally established it had around 60 members. As mentioned earlier, it was open to any company or organisation that supported the UMTS vision, as described in the Articles of Association. No geographical restriction applied. However, in reality the membership was unbalanced: it had a very strong dominance of European players; almost all of us belonged to the telecom sector; regulators and to some extent also operators were underrepresented.

During the next few years there were several attempts to rebalance the membership of the Forum and to live up to its ambitions of a global organisation. This was not entirely success-ful. Improvements were certainly made with respect to participation from operators and regulators but when it comes to the "globalisation" and the participation of the Information and Communications Technology (ICT) things developed very slowly. Interest from the ICT industry was low. There are several reasons for this: first, the awareness of UMTS in the ICT industry was still relatively low in those days. Secondly, the Forum followed very much the legacy working methods from the telecom world and – it must be admitted – the Forum's discussions were extremely telecom oriented, all the members being obsessed by the problems surrounding licensing and spectrum. On top of it all came the strong pressure to produce the first report matching the schedule of the European Parliaments Decision on UMTS, an activity that by definition is both Eurocentric and telecom regulations driven. The few ICT members that participated at that time must have found those discussions very uninteresting from their point of view.

The globalisation of the Forum's membership started to happen only during 1998, and slowly. The ICT industry did not join the Forum during my chairmanship. Nowadays this has improved.

7.4.2 The Forum's Mission and Work

The Articles of Association outline the mission of the Forum. In essence, it all boils down to the following concrete points:

- Create momentum and commitment from all players involved
- Ensure the awareness and endorsement of UMTS at the highest political level as a key building block for the future information society
- Promote a favourable regulatory environment for UMTS
- Speed up the licensing process and ensure early spectrum availability
- Promote the long-term spectrum supply
- Ensure, promote and globalise a common vision of UMTS
- Co-ordinate views towards international fora such as ITU

Looking at those different tasks it is apparent to everyone that one of the most important target audiences of the Forum were regulators and various political instances. Also, the text of the Articles of Associations contains numerous references to European regulatory institutions and activities. To avoid any misunderstanding it must therefore be emphasised that the UMTS Forum is an independent body. It does not report to any other organisation and it writes its own mandate.

Notwithstanding this, the Forum was in the beginning largely driven by events on the European arena. The European Commission and Council were preparing for a Council Decision on the next generation mobile service and it was seen as a top priority for the Forum to state the view of the industry on this subject. The Forum decided to produce a report "A Regulatory Framework for UMTS". During the first half of 1997 this report consumed almost all the energy of the Forum and all the working groups contributed to it. The report covers a wide range of topics: vision, market, technology, business structure, economy, spectrum and regulation.

The work was carried out under great pressure and also some nervousness. There was a strong desire to meet the expectations of the European Commission, but those expectations were not fully understood and the schedule of the EU kept changing all the time. Sometimes the Commission pushed for an interim report to be produced in a couple of days, only to come back the day after, stating that the schedule had been changed. The situation was sometimes rather frustrating. In any case, at the end of the day, in June 1997 the report was finalised and published.

Browsing trough Report 1 today it must be admitted that it is rather general. All the content is trivial knowledge today, but it was not at the time it was written. It must also be borne in mind that the strength of this document – rather than the degree of inventive thinking – was the fact that it did represent the consensus view of the entire industry. Of course, to achieve this the Forum sometimes had to make compromises. To understand how this worked it must also be recalled that the Forum unlike most other industry constellations did not have any area where it had a decisive power of its own. It was a pressure group trying to play the role of a glue between different kinds of activities. It touched the domains of other bodies, but without having the final say on anything. As a consequence of this, differences of opinion between the members never became very critical, but they could always be resolved (after some smaller fights) by diluting the texts of the reports. Taking the discussions about national roaming as an example (this issue was extremely controversial between operators and regulators at that time and maybe still), there was no reason to shed much blood for the sake of a few words in a piece of text, when the final decision would be taken elsewhere anyway. Hence, the really tough fights never happened in the UMTS Forum.

Report 1 was certainly produced ahead of the Council Decision (No. 128/1999/EC) so there was full opportunity to take its advice into account. However, when reading the Council Decision it is not obvious on which points this report has influenced it. The Council Decision itself is also very general. However, as many times before the Taoist philosophy applies: the Road is more important than the Goal, or – if you prefer – the process of writing this report was as important as the report itself. The work of the Forum during this period did attract strong attention and there were numerous occasions to present and discuss this work in all sorts of constellations. Just to mention one example: the Forum was the first organisation in the history of the European Union to be invited to speak at a Council working group, being offered the opportunity to present its case. This was a nice indication of the importance the highest European level attributed to UMTS.

Another pleasant example of this kind of attention came from Japan. The first steps to build the 3GPP had started in spring 1997, and the Forum was active in that process. Of course Report 1 was sent to the Japanese colleagues as soon as it was finished. At all meetings during the next 6 months following that, Japanese delegates were seen carrying around a small booklet in A5 format that they consulted from time to time during discussions with Europeans. Report 1! This was by the way what inspired the UMTS Forum to print all its reports in that format afterwards.

Report 1 was produced under extreme time pressure and time had not allowed full penetration of the more difficult issues. Now that the Forum was no longer being chased by the time schedule of an external body it was felt that there was time to dig deeper into those issues and to produce some more specialised reports. Several items for deeper study were identified, such as the technical vision, spectrum pricing, licence conditions, business case, minimum spectrum per operator at the start-up phase, spectrum requirements in general and market research.

All these studies started at the same time in early autumn 1997. It took more than 1 year before any output of that work was ready and then all the reports came more or less at the same time.

7.4.2.1 Technical vision

Going back to the UMTS Task Force Report for a moment, this was as mentioned earlier as an important turning point in the development of third generation systems. At the time it was written the development of third generation had already been going on for about 10 years, without making much progress from a standardisation point of view. (The achievements with regard to spectrum must be recognised however!) The reason for that is to be found in a lack of business incentive. The earlier development focussed on plain old mobile telephony, offering hardly anything more than what could already be achieved by second generation systems. The Internet explosion had still not influenced the standardisation. Hence, the business motivation for going to third generation was low. One of the most important contributions of the UMTS Task Force Report was that it brought the concept of mobile multimedia into the picture. A new business idea became visible.

Hence, one of the more urgent tasks of the Forum after Report 1 was to launch the new ideas about mobile Internet on the global scene in a powerful and convincing way. To do this the Forum produced a second report "The Path Toward UMTS – Technologies for the Information Society". Picking a few sentences from the scope statement, the report aimed to crystallise what the critical technologies required for UMTS were and how they would be integrated to achieve the goals. The report focused on the introductory years from 2002, but the longer-term potential beyond 2005 was also addressed, seeking to position UMTS not only against today's mobile and fixed systems but also against the backdrop of future developments in communications. Among things the report covered were things like the IMT 2000 family, relations to 2G technologies, service creation, business structure, key technologies, terminal aspects and support of IP traffic and IP addressing.

The most striking difference between the vision of UMTS as painted in Report 2 and the perception we have today 21/2 years later is that the all-IP network – now widely accepted – is not reflected there. Neither is the horizontally layered network architecture, and the separate service layer. Those developments were not generally accepted at the time Report 2 was written. The layered architecture had been discussed and recommended in the ETSI GMM Report in the context of Global Information Infrastructure (GII) but it had still not gained full acceptance in the sense that it had influenced the standardisation in practice. The all-IP approach was at an even more embryonic stage, so far only considered within a few individual companies.

Report 2 was produced within the Technical Aspects Group of the Forum. That group had since the beginning been struggling to find its role in the Forum. With the work on Report 2 those problems were brought to a happy end.

7.4.2.2 Spectrum pricing

One of the aspects that had been swept under the carpet in Report 1 was the issue of spectrum auctions. It was well known that it would never be possible to reach a unanimous agreement on any sort of strong statement in favour or against this concept. The position was locked. For operators the spectrum auction idea was nothing less than the Devil's invention. Manufacturers were also against this, while regulators – whatever their opinion was officially or unofficially – would never subscribe to any definitive statement on this subject. The general impression was that most telecom regulators were not very enthusiastic about these new directions, but other higher order governmental powers were driving the issue and telecom regulators did not control the decisions themselves. Anyway, after some discussion the Forum decided to address also spectrum pricing. The regulators resolved their problem by doing as Pontius Pilatus. They stated already from the start that they were not part of this activity and not bound by any statement whatsoever.

Report 3 "The Impact of Licence Cost Levels on the UMTS Business Case" was written. To avoid stepping on anyone's sensitive toes the words "spectrum auction" were carefully avoided. Instead, the concept of "high up-front licence charges" was invented and used as a fig leaf. The recommendation was of course to avoid such charges and a large number of arguments in that direction were presented.

7.4.2.3 UMTS Business Case

The idea of modelling the operator business case for UMTS was extensively discussed already during the production of Report 1. The discussion came up in various contexts, such as the time of validity of spectrum licences, the impact of excessive spectrum prices and the need for new service pricing schemes. It was also driven by a concern that the roll-out of UMTS may become costly. Some estimations were presented already in Report 1. They indicated that UMTS would render significantly longer pay-off times than second generation system. An attempt was made to assess this with more accuracy by having a group looking specifically at this aspect. This was never successful. To make it short, this work trespassed far too deep into the confidential area of the members.

7.4.2.4 Licensing Conditions

The work on licensing conditions started in the Regulatory Aspects Group, and it resulted in Report 4 "Considerations of Licensing Conditions for UMTS Network Operations". A portal statement of that report said that licensing must be market driven within a regulatory environment that takes account of the two overriding principles of encouraging competition at both the infrastructure and service level along with the wider picture of the legitimate interests of consumers, business and society. It also raised the general principle that any qualification process prior to the main selection should be open and non-discriminatory to remove only obviously unsuitable candidates. The latter statement goes back to a concern that existing operators of second generation systems would not be entitled to even apply for UMTS licences. There were rumours around at that time that this was the position of the European Competition Authority. Those rumours were probably true, but such a regulation was never officially suggested. The objections came before the suggestion itself.

Among the topics of Report 4 were eligible candidates, roll-out and coverage obligations, roaming (and in particular national roaming), infrastructure sharing, experimental licences and regional licences.

The report caused a lot of animated discussions in the Forum. Many of the issues addressed in Report 4 go back to the rather widespread concern about the cost of the UMTS infrastructure. In Europe there was a continuous debate going on for years about licence conditions for second generation mobile systems through initiatives such as the Mobile Green Book, Code of Conduct for mobile operators and the 1999 Regulatory Review. Ideas concerning Universal Service Obligation for mobile services, national roaming obligations, mandatory split of network operations and service provision, etc. were floating around in that debate. On top of it all came the spectrum auction proposal and the concerns about the cost for the UMTS roll-out, due to the high frequency band. Hence, the industry and in particular the operators were somewhat nervous about the costs and did whatever they could to avoid any extra, non-business driven burden that may occur through excessive licence conditions.

7.4.2.5 Spectrum

The spectrum issues were – and still are – of outstanding importance for the Forum. There were three main tasks in this area: to define a strategy for the use of the spectrum that was already allocated to UMTS; to estimate the need for additional spectrum and to find a strategy to obtain this, and finally; to define a strategy for spectrum assignments to operators, taking account of the need to reconcile two conflicting requirements, the need of regulators to ensure adequate competition and the need of operators to have enough spectrum to work with during the first years.

I will not talk more about the Forums deliberations on spectrum. This is covered in Chapter 7, Section 5 by Josef Huber of Siemens who lead this work and who is clearly the best person to describe what happened in this area.

7.4.3 Has the Forum Been Successful?

Let's admit that it is far too early to write the history of UMTS and also a bit early to write the history of the UMTS Forum. If everything develops the way the industry expects, the UMTS history has in fact hardly even started. Concerning the UMTS Forum it must be kept in mind that it is still a living, active organisation. Hence, what is said below concerning the Forum, refers only to the events up to the beginning of 1999 when I handed over the chairmanship to my successor. In spite of the short period, I think that there is a possibility to apply a "historical perspective" on the work of the Forum up to that date. 1999 was a turning point for the Forum in the sense that it brought a change of the nature of the Forum's work and a shift of focus from regulation to ICT industry matters.

Having said that, it is already admitted that one area where the Forum did not fully reach its goals in the 1996–1999 time frame, was the development of relations with the ICT industry. I have already explained the problems we had with that. It remains to be noted that fortunately, considerable progress was made in that area later.

Looking again at the list of tasks describing the mission of the Forum, I believe that the Forum has been successful in most of them. It did contribute very strongly to the globalisation of the vision of UMTS. To be precise and fair, the Forum did not create the vision on its own, but we did a major job in communicating it. The formation of 3GPP was no doubt the most important step to ensure the global vision of UMTS and to ensure that at least one of the IMT 2000 family members would have sufficient strength to become a global standard. The credit for this should go to ARIB and ETSI in the first place, but the UMTS Forum became an important catalyst in the process. Discussions on co-operative arrangements in standardisation across regions had taken place before the Forum entered the scene, but due to uncertainty and some disagreements in the ETSI camp they were more or less put on hold, resulting in extremely valuable time being lost. The Forum's involvement in this made the ball roll again.

But that was not all that helped to launch the UMTS vision. I have already mentioned the reports. The Forums newsletter "Mobilennium" was a best seller. In addition to that, there was an endless stream of conferences and other types of events where the Forum carried out its "missionary" tasks. These included customised visits to China and South Africa, presentations to individual ministries and national events, presentations at research fora, presentations at GSM Association plenary meetings, discussions with CEPT and participation at several kinds of ITU events, including the WRC 97 and TG 8/1 meetings. There was always a great interest in what the UMTS Forum was doing and many doors were opened.

I wish I could say that everything the Forum did was a success, but I am afraid I cannot. The sad thing is that on one of our more important missions, I think we failed. Not because we did a bad job, but because the political institutions did not understand or they did not listen to us. Or was there something wrong with our strategy?

As the reader has most probably noticed, much of the efforts of the Forum were spent on regulation and spectrum licensing. The ambition was – it was repeated like a mantra – "to establish a common regulatory framework, bringing a favourable regulatory environment for the successful introduction and growth of UMTS". To do this, it was felt necessary to raise the political awareness of UMTS and the awareness of the importance UMTS will play in the development of the information society. Was this successful? Well, we certainly managed to raise the political awareness of mobile communications. But political attention is something double-edged, and the result of that political attention was not always what we were really looking for.

The common regulatory framework did not happen. Now that the UMTS licensing has happened in most countries (or at least the selection rules and licence conditions are known) we can observe that the approaches adopted in different countries have never before been so different from each other as they are today with UMTS. And most regrettably we have to note that in many countries the use of that double-edged weapon of political attention, only led to the mobile industry hitting itself in the head. Regulation has imposed extremely expensive arrangements on the industry in some countries, arrangements that will no doubt damage the business for many years. Not all countries have adopted such regimes, but there is a considerable risk that the entire industry – operators as well as manufacturers, wherever they are located – will suffer from it. Today, in mid-February 2001 at the time of writing, the financial papers are full of concerns about the cost of the UMTS role-out and full of bad news about the finances of the communications industry. Figures from the stock market speak loud and clear.

A lot of this could have been avoided.

But the game is not over. I am sure the industry will recover. UMTS is just too strong a concept to succumb to this kind of backlash. Besides, we have seen this pessimism and nervousness before. It was there in Scandinavia in the early 1980s, the year before the start of the NMT system and it was there in Europe in the early 1990s before the GSM system took off.

Let's finally shift to the happiest part of the UMTS Forum story. Spectrum! The most important and successful contribution of the Forum to the development of the mobile industry was – in my opinion – the spectrum area. The Forum laid the foundation for the estimation of the future spectrum needs. That work was almost entirely accepted by European regulators. The results and the proposals went further to the ITU TG8/1 where they were accepted. They went further again to the WRC 2000 where the spectrum was allocated as proposed. This was a great success for the entire community of the mobile industry and users.

The UMTS Forum was a team with numerous persons involved, contributing in different ways. It is therefore with great hesitation I specifically point to the contribution of an individual. However, when it comes to the extremely hard and skilled work of Josef Huber, I think that is something really special, which must not disappear into oblivion.

Josef worked for many years as the chairman of the Spectrum Aspects Group. Using his breathtaking energy and personal charm he managed to find a way into the future for the mobile business by ensuring supply of spectrum, the raw material on which all radio communication is built. This was also part of the globalisation activities, many of which happened due to Josef's initiatives. Hats off!

Visit the Forums web site at www.umts-forum.org

Chapter 7: The UMTS Related Work of the European Commission, UMTS Task Force, UMTS Forum and GSM Association

Section 5: Spectrum Aspects

Josef Huber[1]

One of the most difficult tasks in spectrum management of the 1990s was and still is dealing with the forecasts: user penetration and user demand, globalisation, service and technology innovation are some main criteria impacting the planning of this basic resource, which has to be long-term due to its global relevance. Up to now, we realise the mobile market has always been underestimated. It is exploding and turning into the fastest growing and largest field in communications. The present forecasts estimate that between 2003 and 2004 the number of mobile terminals will surpass the number of fixed line subscriptions. The year 2001 will add a new market stimulating technology to the mobile networks, the third generation component UMTS [6,8]. At this very significant moment of the development of mobile communications world-wide, it is important to have a look at the frequency situation. Frequencies are a limited resource, like pieces of land, they are available just once. It is difficult not to exaggerate the importance of frequency planning and management, but probably the most important lesson that the industry learned so far in the past is to seriously consider identifying more frequency spectrum than appears necessary at the planning stage. Mobile communications in general – and GSM in particular – have been more successful than anyone foresaw and they are going to be extended coming initially from pure voice, then data and then to Internet/Intranet related services. Data traffic including the Internet/Intranet will have a tremendous impact around the world. Various forecasts signal higher subscription numbers than for voice, and also far higher data traffic. Voice and data traffic have different characteristics and will therefore be handled differently. Such developments change the traditional way of calculating frequency spectrum demand.

---

[1] The views expressed in this section are those of the author and do not necessarily reflect the views of his affiliation entity.

With the foundation of the UMTS Forum the industry work on spectrum aspects was concentrated in the Forum's Spectrum Aspects Group. The UMTS Forum took up its investigations on spectrum issues in 1996 and concentrated on three main questions:

1. How much frequency spectrum will be needed in the timeframe of 10–15 years for mobile cellular services?
2. What is the UMTS-specific spectrum demand for the public operator in the initial phase up to the year 2005/06?
3. To what extent is global harmonisation of frequency spectrum necessary and feasible?

In the light of various individual estimates on frequency demand [1–4,7] it deemed necessary to develop an appropriate methodology and traffic models for spectrum calculations based on an industry-wide consensus. This methodology should be common in order to make it applicable for terrestrial as well as for satellite spectrum investigations; the traffic models and the data base should also be useable for regional adaptations considering a country's needs.

7.5.1 The Global Frequency Situation for Cellular Mobile Services Towards Third Generation

Already in 1996 the world-wide growth of GSM in the 900/1800/1900 MHz bands was giving a clear vote for allocating new frequency spectrum for the upcoming new services. This opened the door for a global UMTS frequency band world-wide identified as IMT 2000 Core Band. It avoids impacts on existing services networks except for countries which have fully adopted the PCS spectrum plan including the US. The introduction of third generation services in these countries needs a series study which takes into consideration the national demand and global roaming as an essential IMT 2000 requirement. The IMT 2000 band plans are shown in Figure 7.5.1.

The diagram shows the ITU identification from the Worlds Radio Conference in 1992 and

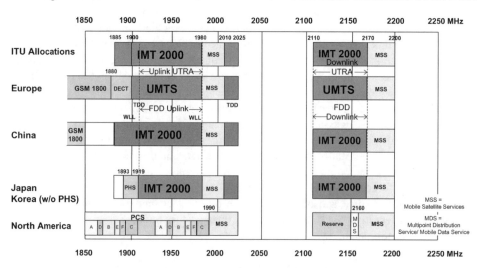

Figure 7.5.1   Global spectrum situation in the 2 GHz range before WRC-2000

1995 and the planning/allocation in some main countries which will go forward from GSM or other standards to IMT 2000/UMTS. It has to be kept in mind that the IMT 2000 Core Band (earlier FPLMTS band) was harmonised world-wide in 1992 through WARC 92, and that in 1993 the US adopted the 1.9 GHz PCS frequency plan which overlaps with the IMT 2000 Core Band. The consequences of this development are:

- CEPT in Europe can make available nearly all the ITU spectrum (except 15 MHz, already allocated for DECT). This results in 155 MHz of spectrum for terrestrial services with additional 60 MHz set aside for UMTS satellite services. A number of European countries have granted UMTS terrestrial licenses.
- China considers the ITU allocations as being quite similar to those in Japan and Korea. It may be assured that the major part of the ITU bands could be made available. Some segments are designated for wireless access systems (PHS, DECT). No final plan or decision has been made yet.
- South Korea already indicated full IMT 2000 spectrum allocations for paired and unpaired use: 1885–1920 MHz, 1920–1980 MHz, 2010–2025 MHz, 2110–2170 MHz. The license process has already been finalised.
- The Japanese Ministry of Post and Telecommunications (MPT) did its planning for third generation systems in nearly the same way as the Europeans with the difference, that the frequency band 1895–1918.1 MHz is already allocated to PHS services. Three IMT 2000 licenses were granted on 12 July 2000.
- Thailand, Hong Kong, Singapore, Australia, New Zealand and Indonesia have started to prepare the license process for IMT 2000/UMTS, some licenses are meanwhile already decided.
- North America: the introduction of PCS services and the auctioning led to a split into licenses of 2×15 MHz and 2×5 MHz up to 1980 MHz blocking the up-link part of the IMT 2000 Core Band.

The PCS 1.9 GHz plan was adopted mainly in North America and partly in South American states. It allows the operators to make their own choices of standards and technologies. A number of different systems (IS 95, IS 136, GSM, etc.) can be deployed, along with second and third generation systems. In order to avoid market fragmentation, several countries in Latin America kept both IMT 2000 Core Band and the PCS 1.9 GHz band reserved for future mobile systems, while waiting for a solution. One solution towards harmonisation with the world-wide IMT 2000 developments was decided in the year 2000 by the largest state in South America, Brazil and by Venezuela, some others followed.

As shown, the GSM 1.8 GHz band (1710–1850 MHz) leaves the 2 GHz IMT 2000/UMTS band unblocked for the deployment of third generation systems. It is a fact that due to the GSM market success more and more countries are granting licenses in this band world-wide, it will be the dominant second generation band after the 900 MHz band allocations.

7.5.2 How Much Frequency Spectrum will be Needed in Future?

Since the beginning of the successful growth in mobile cellular networks, repeated discussions took place on the amount of spectrum which will be needed to satisfy the users' demand. In parallel, research and development work increasingly dealt with better spectral efficient technologies.

The UMTS Forum's Working Group on spectrum issues recognised the discrepancies in both areas, in the spectrum demand estimations as well as in the area of spectral efficiency [1–5]. It realised, that a world-wide common approach and industry consensus will be the only means in order to achieve acceptable world-wide and regional spectrum requirements. It was obvious, that the priority consideration has to be in the context of spectrum utilisation based mainly on three inputs:

- Market (public, private) figures
- Service and traffic database
- Standards and technology spectral efficiency values

It was quite clear from the first analysis of various scenarios, that the spectral efficiency could be impacted by the number of operators, either by sharing a common spectrum or being owners of public licensed spectrum. The realistic assumption was to follow the conventional model where spectrum would be allocated on a per-operator basis ("public licensed frequencies") and for private/corporate use spectrum would be shared amongst operators ("self-provided applications").

7.5.2.1 Spectrum Investigations

In order to come to a stepwise industry consensus, it was agreed to develop the market model on the European scenario and then bring it onto the global level with the appropriate modifications. Thus, co-operation with various organisations, the GSMA, CEPT-ERC, ETSI, 3GPP was required.

7.5.2.1.1 Market Model

The market model was built using the first substantial study on the global mobile market with a European focus, which was done in the UMTS Forum [12]. Several market scenarios were considered and finally one scenario was selected as a basis for spectrum investigations. This scenario considers exclusive spectrum per operator and includes the license exempt networks for corporate networks and private homes.

7.5.2.1.2 Services and Traffic

The market study has shown in contrast to the past, that in future there will be a multiplicity of applications/services in addition to a single service "mobile telephony". In order to come to a simplified services and traffic model for a global consensus, a limited number of service categories was chosen representing the variety of future services with typical traffic bitrates (see Table 7.5.1) and expected penetration rates for 2005 and 2010.

7.5.2.1.3 Standards and Technology

The third step was to define an appropriate methodology for calculating spectrum. All different environments from rural to urban outdoor and indoor were taken into account. The methodology is shown in Figure 7.5.2. It was accepted with some minor additions and modifications on a world-wide discussion level in the ITU-R. As shown, it takes into account

Table 7.5.1 Penetration rate urban environment year 2005/year 2010

Service categories	User net bit rate	Penetration urban year 2005 (%)	Rate environment year 2010 (%)	Applications
High interactive multimedia	128 kbps	0.5	5	Videophone, real time services
High multimedia	≤2 Mbps	5	18	WWW type services
Medium multimedia	<384 kbps	8	18	WWW documents, video streaming, Internet/ Intranet access
Switched data	14.4 kbps	10	10	e-mail access, data
Simple messaging	14.4 kbps	25	40	SMS, enhanced messaging
Speech	16 kbps	60	75	Voice

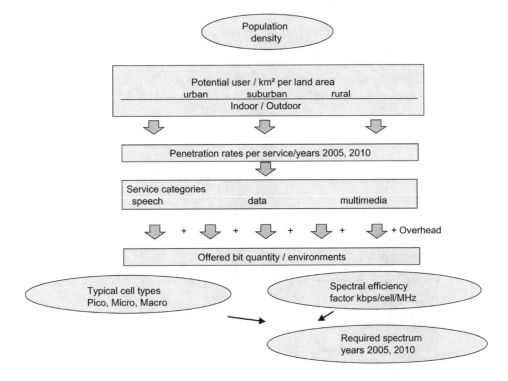

Figure 7.5.2 Calculation methodology spectrum

the population density distribution, market assumptions like penetration rate and busy hour calls, etc. It uses a traffic model with the described service categories for different applications. Finally, the technology values came in. Here typical cell sizes for pico, micro and macro cells and spectral efficiency figures are estimated. The technology related spectral efficiency values are assumed on a network level. GSM is used as a basis, added with a spectral efficiency factor dealing with the technology improvements expected in the third generation. The spectral efficiency factor expresses an average spectral efficiency due to the fact, that the total mobile market will be served by a mix of 2G/3G technology.

As already mentioned, the total spectrum calculation included all services as presently covered by second generation systems (mainly speech) and future multimedia services covered by third generation systems. Also, terrestrial as well as satellite components were considered.

7.5.2.2 Spectrum Calculation Results

The calculation results for the European terrestrial and satellite spectrum demand are shown for the years 2005 and 2010 in Figures 7.5.3 and 7.5.4.

If already available spectrum is deducted from the total values, an *additional spectrum demand* of

- 185 MHz for terrestrial services
- 30 MHz for satellite services

can be calculated in the year 2010 [10]. In the year 2005, it appears that the total available terrestrial spectrum for GSM 900 (2 × 35 MHz), GSM 1800 (2 × 75 MHz), DECT (20 MHz) and UMTS (155 MHz), which is 395 MHz, may satisfy the upcoming demand. More or less the same can be said for the satellite side. Thus, the main conclusion of this investigation is, that additional spectrum will be needed – at least for terrestrial mobile services – in the years between 2005 and 2010.

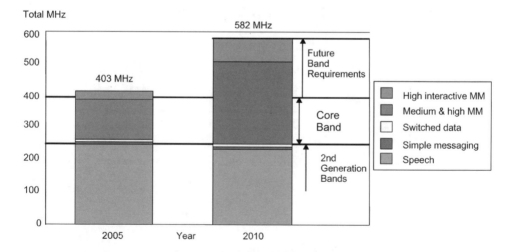

Figure 7.5.3   Terrestrial spectrum estimates for the years 2005/2010 (EU 15)

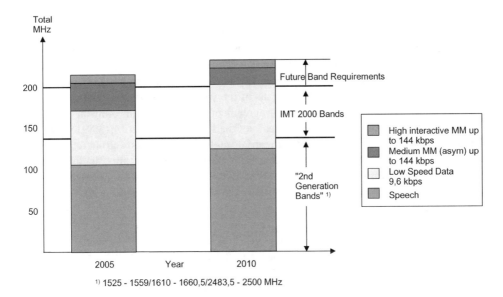

Figure 7.5.4 Satellite spectrum estimates for the years 2005/2010 (EU 15)

7.5.2.2.1 World-wide Achievements

As the UMTS Forum approved its report [10] on spectrum calculations a way of achieving world-wide consensus between industry and administrations had to be found. One of the contributory factors to the global spectrum discussions has been that the ITU has approached IMT 2000 in a focussed manner. A Task Group (8/1) was formed in the ITU-R. Maintaining

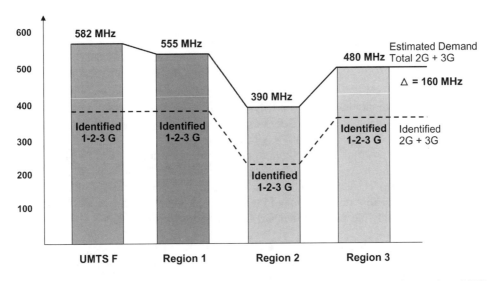

Figure 7.5.5 ITU-R agreement (March 1999, Fortalezza) on terrestrial spectrum estimates (year 2010)

this focus was considerably assisted by the fact that WARC 92[2] identified specific frequency bands for IMT 2000. The existence of a footnote identifying IMT 2000 spectrum, the timing of the availability of the spectrum, from the year 2000 onwards, was also a factor in developing this focus. Thus, the UMTS Forum submitted inputs into ITU-R TG 8/1 and started to promote its recommendations on how to proceed in order to come to a global agreement. IMT 2000 Seminars were organised in the regions of the world and supported with the focus on spectrum issues. The outcome was a final agreement in the ITU-R meeting in Fortalezza, Brazil in March 1999. It resulted in the release of the ITU-R document M.1390/1391 [13,14] providing the methodology to calculate terrestrial/satellite IMT 2000 spectrum requirements. Further, a general consensus was achieved using parameters for spectrum calculations in the regions, and the additional spectrum has been calculated. These results are contained in Report ITU-R [15], which was approved and adopted by Study Group 8 (SG 8). They are shown in Figure 7.5.5.

It can be seen, that the total mobile spectrum demand differs from region to region. Nearly the same relation exists for the already identified spectrum. The UMTS Forum's total estimate is the highest and there is still the belief in the industry, that the Forum's figures are reasonable. The world-wide achievement on the ITU-R level was not on the figures for total spectrum demand, as it was understood, that the regions have to reflect their demographic situation and their needs for mobile spectrum. On the basis of this agreement, the regional organisations developed their positions on the identification of future extension bands.

7.5.2.2.2 Region 1

CEPT[3] represents 43 member states in Europe. It supported the UMTS Forum's methodology and views with minor modifications in the database. Priority was given to the band 2500–2690 MHz as additional spectrum for IMT 2000/UMTS with the band 2520–2670 MHz specifically for the terrestrial component. In addition, the GSM 900/1800 bands should be secured for migration to IMT 2000/UMTS in the longer term. Arab states and African states in principle supported the position of CEPT. However, the extension band 2520–2670 MHz allocation seems to be difficult in African states even for later allocations.

7.5.2.2.3 Region 2

The calculation methodology was supported, however, with different parameter assumptions – especially for data services and user densities in the various environments. Thus, the outcome of the spectrum calculations was considerably lower than for the other regions. CITEL[4] developed positions mainly supporting the extension band 1710–1885 MHz. There was no common position because of the difficulties in the harmonisation of existing frequency band use. The US recognised the urgent need for spectrum and that spectrum discussions are vital for administrations in order to meet their own needs as they develop and supported considering multiple standards in the context of IMT 2000.

---

[2] WARC 92 – World Administration Radio Conference in 1992.

[3] CEPT – European Conference of Postal and Telecommunication Administrations. Its task is to develop radio communication policies and to co-ordinate frequencies.

[4] CITEL – Inter-American Telecommunication Commission belongs to the Organization of American States (OAS), represents the needs and interests of the 34 member states of the region.

7.5.2.2.4 Region 3

APT[5] represents 30 member states, four associate members and 40 affiliates. APT accepted the methodology and supported the UMTS Forum's view with modifications on the parameters in the traffic model resulting in a total spectrum demand of 480 MHz. Also, the 2520–2670 MHz extension band was supported. The bands from 806–960 MHz and 1710–1885 MHz were considered for later migration to IMT 2000.

7.5.2.2.5 The World Radio Conference 2000 (WRC 2000)

In May 2000, the World Radio Administrations met to discuss the overall spectrum situation with a main focus on IMT 2000/UMTS. The agenda item 1.6.1 was already created in the WRC 97, where the Spectrum Aspects Group together with the ERC Task Group 1 (ERC TG1)[6] took the initiative. Comprehensive regional preparations started long before the actual convention. The UMTS Forum's contributions continued with another Report [11] on Candidate Extension Bands for UMTS/IMT 2000, Terrestrial Component. A number of promotion activities took place in co-operation with the ITU, with the GSM Association, the Global Mobile Suppliers Association (GSA), with CEPT, etc. In its Report [11], the UMTS Forum examined all candidate bands for IMT 2000 proposed by ITU-R TG8/1 in the Draft CPM Report. This review was made from an industry point of view giving particular attention to the interests of end-users world-wide. The final result of WRC 2000 was a success for the mobile community:

- It is a clear and stable identification of 160 MHz of additional spectrum, on a global basis, for the terrestrial component of IMT 2000/UMTS. It should be made available in the time frame up to 2010, subject to market demand;
- It identifies certain second generation mobile bands for IMT 2000 to support future evolution to third generation systems;
- It identifies existing MSS bands <3 GHz for the IMT 2000 satellite component; and
- It allows access to the "Core Bands" for High Altitude Platform Stations (HAPS) acting as a platform for base stations of the IMT 2000 terrestrial component (Figure 7.5.6).

An important element of the WRC 2000 result was the flexibility, set out in the WRC 2000 resolutions:

- The administrations remain free to decide, whether they wish to implement IMT 2000.
- If they do, then the choice of band(s), the amount of spectrum and the timing of its availability – all remain under the administration's control.
- If not, identification for IMT 2000 does not prevent the bands from being used for other applications instead.

WRC 2000 agreed on the following for the terrestrial spectrum:

- *Protection of existing IMT 2000 "Core Bands"*: the status of these bands remains unchanged.
- *Identification of 2500–2690 MHz extension band in all three ITU regions:* this will be the

---

[5] APT – Asia-Pacific Telecommunity, founded in 1979 by intergovernmental agreement, responsible for regulatory recommendations in telecommunications.

[6] ERC – European Radio Committee is part of CEPT and is responsible for radiocommunication matters.

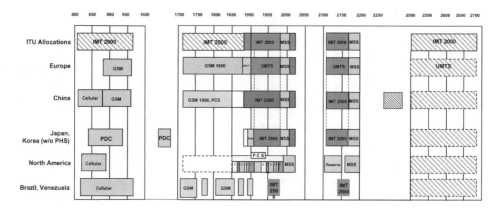

Figure 7.5.6   IMT 2000/UMTS frequency spectrum after WRC 2000

main extension band for Europe and Asia for the terrestrial IMT 2000 (initially only 2520–2670 MHz).

- *Identification of 1710–1885 MHz in all three ITU regions:* this is the preferred IMT 2000 band for the CITEL countries, it will allow eventual evolution of GSM 1800 to IMT 2000 at a later stage.
- *First and second generation mobile bands <1 GHz:* specifically 806–960 MHz are identified for global IMT 2000 for migration in the longer term.

The WRC 2000 achievements opened the discussions on the review of the use for third generation services. In this discussion, care needs to be taken especially for the GSM bands and the PCS bands for two reasons:

- world-wide roaming plays an increasing role, thus a harmonised migration of currently used bands towards UMTS seems obligatory;
- the potential overlap of the PCS plan with the IMT 2000/UMTS Core Band and the GSM 1800 band leads to compatibility problems.

The US has started already with two interim reports (NTIA, FCC from 15 November 2000) following President Clinton's memorandum from 13 October 2000 requesting a strategy for 3G telecommunications. The first results are expected in the year 2001.

7.5.3 What is the UMTS-Specific Spectrum Demand per Operator in the Initial Phase?

One of the most important issues in the UMTS licensing process is the amount of frequency spectrum per operator. Of course, this question has to be investigated under the assumption, that second generation spectrum will still be in use by second generation systems including their services, mainly voice. Thus, it was necessary to make a market split between 2G and 3G radio networks in order to come to reasonable calculation results. The UMTS Forum market forecasts for Western Europe from the Market Report [12] were used in the calculations for minimum spectrum per operator. First of all, it was necessary to define traffic for UMTS out of the forecasts in total. For speech and low speech data it has been assumed that 90% of speech services as well as low speed data services will go via GSM radio networks in the respective GSM spectrum. This results in a UMTS scenario with 20 million medium/high multimedia and 17 million speech users and low speed data for the year 2005 (EU15).

The work of the Spectrum Aspects Group used the methodology which was developed for the total spectrum calculations and assumed further, that in the specific UMTS case the Erlang B resource calculation would not be relevant. However, instead of using Erlang B a "Quality of Service" factor was introduced in order to secure the expected QoS level for packet transmission. All this information is laid down in Report No. 5 [9] on "Minimum spectrum demand per public terrestrial UMTS operator in the initial phase"! The report studies the minimum frequency bandwidth per operator from the following points of view:

- *Spectrum point of view*: there are limited paired and unpaired bands available for UMTS. They have to be subdivided by the expected number of UMTS licenses per country. Some reserves for operators with fast market development may be taken into account.
- *Market point of view*: subscriber and traffic forecasts made by the UMTS Forum lead to the spectrum requirements as calculated. This determines the amount of spectrum an operator will need to satisfy an equal share of the market. It is assumed that the full scope of service classes will be offered by UMTS operators.
- *Technical point of view:* the preferred frequency raster is 5 MHz, 2×5 MHz will be the basis for the Frequency Duplex Mode (FDD) considerations. It is assumed that channel spacing can vary by around 4.4–5 MHz depending on the deployment scenario. Thus the figure of 5 MHz includes an allowance for guard bands.

The UMTS Forum studied eight different deployment scenarios. Traffic asymmetry, which appears with Internet-based multimedia services was included. Parameter assumptions were made according to the standardisation of UMTS including higher spectral efficiency figures than used in the earlier studies. The outcome of these calculations is shown in Table 7.5.2.

Table 7.5.2   Outcome of studies on minimum spectrum per UMTS operator

Scenario	Operator frequencies [MHz]		Radio cell layers	Factor for traffic capacity[a]	UMTS service capability
	Paired	Unpaired			
1	2×5	–	1		
2	2×5	5	2	1–2	Limited
3	$2 \times$	–	2		
4	2×10	5	3	2–3	Some restrictions
5	2×15	–	3		
6	2×15	5	4		
7	2×20	–	4	3–4	Full
8	2×20	5^b	5		

[a] Traffic capacity factor = 1 corresponds to a calculated traffic requirement of 6 Mbps/km^2 during the busy hour. The overall radio capacity is higher, it goes up with the MHz per operator. It is used to fulfill QoS requirements and to cover traffic peaks.

[b] Optional 10 MHz for more capacity.

The deployment scenarios provide different service capabilities and traffic capacities. All these scenarios are feasible from a technical viewpoint. The scenarios with limited capabilities will impact the business plan of the operator. As the purpose of the report is to examine the minimum spectrum required to start up a UMTS service and cater for demand up to the year 2005, some judgement must be made on the suitability of these scenarios. The choice of which figure will be used cannot be the same for all countries and markets. There will be a trade off between the cost of rolling out a network and the benefits of having extra operators, since extra spectrum (all other things remaining equal) should lead to a less costly roll-out. Scenarios 1, 2 and 3 do not cater for the projected demand for *all* UMTS services. However, if an operator should go into a specific service segment, the scenarios cannot be excluded as not being possible. Scenarios 4 and 5 allow the deployment of a complete hierarchical cell structure with a full range of services. Pico cells might only be deployed in a small percentage of a city. The remaining scenarios 6–8 allow the operator full service offerings and capacity if it captures a larger market share than the forecasted figures equally shared by the assumed number of operators per country.

7.5.4 Global Harmonisation of Frequency Bands for Mobile Services

With the introduction of IMT 2000/UMTS from the year 2001 onwards, the rapid globalisation of the mobile telecommunications industry is anticipated to continue. Global standards and spectrum, which is common globally, are critical elements in the realisation of a global system providing the highest benefits to the users through globally competitive sources of equipment (due to large-scale product implementation, minimum cost and complexity, etc.) and rapid deployment. This in turn is critical to realising the ITU's vision and efforts, in collaboration with the global industry, to developing a world-wide mobile communications system and service. It is therefore suggested that the additional global spectrum should be contiguous and common in all three regions.

It is important to note that "identifying" spectrum for IMT 2000 through footnotes does not preclude the use of this spectrum for other services in accordance with the radio regulations and other regional and national regulations. Thus the systems currently deployed and their associated investments should be protected. Based on market demand and national requirements, national administrations can make available for IMT 2000 part or all of the spectrum identified. So for example, some countries may not require all of the band, some may make the spectrum available in "hot spot" city centre areas and some may allow operators to evolve or migrate their current systems to IMT 2000 as required by market developments.

7.5.5 Outlook for the Future

It is certain that spectrum-related work will continue. It is forecasted for the year 2010, that there will be more than 2 billion mobile users on the globe and probably 1 billion subscriptions for Internet-based services. These services will produce a high amount of traffic far beyond the traffic volume coming from voice communications. There will also be computer to computer communications driving the mobile penetration rates even higher than expected. Also, the technology side has to make its contributions on improving spectral efficiency and providing flexibility regarding the multiplicity of UMTS services. QoS will be an essential impact on spectrum demand as it determines the allocation of spectrum resources considerably. The UMTS Forum's spectrum investigations in the past have shown, that the "packet transport" of information over the air-interface brings better spectrum utilisation – as long as the additional overhead is not dominating. The latter has to be proven in the future.

The UMTS Forum's work has shown, that industry consensus is an important factor in stabilising the frequency planning process in order to achieve successful results. It cannot take the uncertainties fully away from the market forecasts and it is a comprehensive job to synchronise the views in the industry – but it is worth it for the high responsibility frequency management implies. Thus, the UMTS Forum will continue to work in the spectrum field making contributions to public consultations and to the ITU-R. ITU-R itself has started follow-up activities, which are concentrated in the ITU-R Working Party 8 F. Here, the implementation of IMT 2000 in the newly identified bands, the further evolution of IMT 2000, frequency sharing, traffic asymmetry issues, etc. will be studied over the next years.

References

[1] Ramsdale P., One2One Spreadsheet, number of operators, total GSM 900/1800, DECT UMTS spectrum requirements for the universal mobile telecommunications system, 24 May 1996.
[2] Harris J., Davis J., Spreadsheet model for TGMS spectrum analysis, Cellular Systems Unit, BT Labs, BT Laboratories, 16 November 1995.
[3] Telia Research AB, Technical Report UMTS Frequency Estimation, 24 May 1996.
[4] DTR/RES-03077, Vers. 0.4.6 ETSI Technical Report, Traffic capacity and spectrum requirements for multi-system and multi-service applications co-existing in a common frequency band (DECT), 03 June 1996.
[5] Satellite Spectrum Requirements, ICO Global Communications, 18 July 1996.
[6] UMTS Task Force Report, Brussels, 1 March 1996.
[7] ETSI/SMG Task Force on UMTS Frequencies, Vers. 6, 06–07 May 1996, Malmö, Sweden.
[8] UMTS Forum Report No. 1, A regulatory framework for UMTS, June 1997.
[9] UMTS Forum Report No. 5, Minimum spectrum demand per public terrestrial UMTS operator in the initial phase, September 1998.
[10] UMTS Forum Report No. 6, UMTS/IMT 2000 spectrum, December 1998.
[11] UMTS Forum Report No. 7, Candidate extension bands for UMTS/IMT 2000 terrestrial component, March 1999.
[12] UMTS Forum Report No. 8, The future mobile market, global trends and developments with a focus on Western Europe, March 1999.
[13] ITU-R Document M. 1390 (2000) 1999.
[14] ITU-R Document M. 1391. Methodology for the calculation of IMT satellite spectrum requirements, March (2000) 1999.
[15] ITU-R [IMT-SPEC] approved by SG 8.

Chapter 8: The UMTS Standardisation Work in ETSI

Section 1: The Initial Work (up to Spring 1996)

Philippe Dupuis[1]

8.1.1 Introduction

Some experts started working on, or maybe we should say dreaming about, third generation mobile communications in the mid-1980s, even before second generation mobile communications took shape in GSM. UMTS was invented then. It was initially just a vague concept, something which had to one day take over from GSM and therefore had to be superior to GSM. There was also a view that the capacity of GSM would be exhausted just after a few years and that UMTS should thus follow very quickly. This was not a workable proposal as the industry could not throw away GSM developments and adopt a new system so rapidly. This initial UMTS concept had to evolve into a workable proposal. The main purpose of this section is to show how this happened.

8.1.2 The Genesis of the UMTS Concept

The UMTS concept has emerged from the R&D work funded by the CEC under the RACE program. In parallel at the ITU and particularly in the Comité Consultatif International Radio (CCIR) interest emerged for the elaboration of a single world standard for public mobile communications. This rapidly focused on next generation, i.e. digital solutions. It is interesting to recall the various activities which took place in these two areas.

8.1.2.1 The RACE Projects

The RACE programme was initiated by the CEC to fund R&D in the area of telecommunications. It consisted of a collection of precompetitive co-operative R&D projects associating companies, laboratories and universities belonging to several countries of the European communities. It generally addressed broadband communications. Initially the goal was called B-ISDN but later it became Integrated Broadband Communications (IBC). There was a

---

[1] The views expressed in this section are those of the author and do not necessarily reflect the views of his affiliation entity.

mobile component which included Mobile Broadband System (MBS) and an advanced digital cellular system called UMTS, limited to about 2 Mb/s.

A first RACE project on future mobile systems had been awarded around 1987. The project leader was Philips Research in Cambridge.[2]

It seems strange a posteriori that, at the very time that the GSM group was working on the specification of a digital cellular system, work could start independently. Many reasons can explain or justify this. First the RACE project was led by manufacturers while GSM was an initiative of the traditional telecom operators. Some probably thought that RACE would limit itself to supplementing and supporting the work on GSM with some basic research. But others had certainly in mind the design of a system which would compete with GSM.

In 1989 RACE 1043, a full phase 1 RACE project, was launched to continue the work undertaken by the first project. It covered the same area with a more important budget and in total 26 partners from 13 countries. It is probably within RACE 1043 that the expression UMTS was used for the first time. In 1990 I chaired the Mobile Expert Group of ETSI whose task was to review activities in the sector of mobile telecommunications. I remember that we had a tense meeting with a representative of RACE 1043, Ed Candy of Philips Research. The GSM and UMTS visions obviously belonged to different worlds. On ETSI's side the current assumptions were that UMTS was to be considered as third generation, emerge around 1996 and that standardisation activities had to be carried out within ETSI, or within CCIR if one was aiming at a world standard.

In 1992 RACE phase 2 was launched to cover the 1992–1996 time frame, including a "mobile project line", a set of projects in the area of mobile communications. The budget of this mobile project line was quite impressive totalling thousands of man × months of effort. At this time I was involved in a small group whose task was to establish the overall relevance of RACE 2. To this end I made sure that there was a clear understanding of the features by which UMTS would differentiate from GSM. Among these features were an enhanced speech quality, the integration of IN concepts, multimedia services, increased capacity to allow the emergence of a mass market, etc. Nobody was expecting at this time that GSM phase 2 + would eventually meet some of these requirements. We also clearly stated that the objective of the RACE projects was not to produce a full specification of UMTS, only possibly to provide an input to the standardisation group in ETSI.

Despite this the RACE office in the CEC, headed by Roland Hueber, had in mind to control or at least play a major role in the process that would bring UMTS to the market. This was indeed their last opportunity to influence the future of the telecommunication industry as the Integrated Broadband Communications were then being eclipsed by the dominance of the Internet. In 1995 Roland Hueber invited representatives of the mobile industry, operators, and manufacturers to a meeting in Brussels, the purpose of which was to propose to set up an UMTS Memorandum of Understanding (MoU) quite similar to the GSM MoU. This MoU would have defined the conditions of the introduction of UMTS in Europe. In the opening speech Roland Hueber, the head of the RACE office made several agressive statements. In particular he said that the time taken to bring GSM to the market had been unduly long. The proposal to set up a MoU with the support the CEC was not well received by the audience. At that time it was generally understood that UMTS could only come as an evolution of GSM. The GSM MoU had also started activities on third generation and one tended to assume that a

---

[2] At the European Seminar on Mobile Radio Communications (Brussels, 7–8 April 1987), already mentioned in Chapter 2, a presentation on the project was made by R. Gibson of Philips Research.

possible UMTS MoU would have to be brought under the same roof as the GSM MoU. Finally, as a compromise, the meeting eventually agreed to create an UMTS Task Force to investigate the matter further. The work of this Task Force is reported in Chapter 7 section 3.

In the meantime CEC funded R&D projects continued within the framework of the ACTS programme.[3]

8.1.2.2 ITU, FPLMTS and IMT200

In the mid-1980s, even before producing results, the work of the GSM had attracted the attention of observers outside of Europe. That the Europeans had undertaken to develop a common cellular standard was indeed something unusual. At the same time the interest of international roaming had been demonstrated in the NMT system. Emerging countries were also in favour of a unique standard which would stimulate competition between the international equipment suppliers. I remember a high level panel discussion during the opening session of the ITU's Asia Telecom 1985 conference in Singapore during which the CEO of Telecom Singapore made a plea in this direction. Richard Butler, the Secretary General of ITU, said his organisation would be ready to work in this area but the representatives of Bell Labs and NTT both stated that the cellular market was growing so quickly that it was not possible to lose time on developing an international standard. Initiatives in ITU indeed came from the working level in CCIR. A working party was set up which was successively called CCIR TG 8/1, CCIR IWP 8/13, and eventually ITU-R TG 8/1. It was chaired by Mike Callendar of BC Tel in Canada. For many years Mike Callendar was the apostle of a single world-wide next generation mobile standard. The working party invented an unpronounceable acronym FPLMTS (Future Public Land Mobile Telecommunication System) which later was replaced by IMT 2000.[4] In the mean time a World Administrative Radio Conference in 1992 had designated frequencies around 2000 MHz to be used by a future world mobile telecommunication system. I remember several discussions with Mike Callendar in the early 1990s. He had been impressed by the work done in the GSM Permanent Nucleus, or later in PT12, and would have liked to see the same arrangements in ITU. But at the same time people expressed doubts on the capability of ITU to agree on a single standard and produce a full specification. One started then to mention a "family" of standards, the IMT 2000 family, of which UMTS would be the European component.

8.1.3 The Early Work in SMG

As explained earlier, ETSI's Mobile Expert Group in 1990 had reviewed the whole area of mobile telecommunications. On UMTS the report recommended that ETSI set up a technical committee in charge of UMTS standardisation. It took some time before ETSI acted on this recommendation[5] and it was eventually decided in 1991 that the terms of reference of technical committee GSM would be extended to include UMTS standardisation. As reported earlier the name of the committee was then changed and became SMG. Not everybody in

---

[3] TDoc SMG 487/94: preliminary information on ACTS.

[4] TDoc SMG 306/94: developing a roadmap to FPLMTS (ITU).

[5] The main reason why the Mobile Expert Group study had been undertaken was to decide whether ETSI should work on the development of the DCS 1800 specifications although at this time only the UK had the intention of building 1800 MHz. All the other issues addressed in the report were considered less urgent.

ETSI was happy about this decision. Some would have preferred responsibility for UMTS to be given to technical committee Radio Equipment and Systems (RES) who dealt with radio specifications in general, or to a new committee.

However, putting GSM and UMTS under the same roof made sense. First it ensured that the experience acquired on GSM would more surely benefit UMTS. But more important even was the fact that SMG would be in a position to determine the relative positioning of GSM and UMTS both in terms of features and timing.

But in 1992 SMG had still a lot to do on GSM which was just starting commercial operation. Not much time could be spared in plenary meetings to discuss UMTS. An easy solution was adopted which was to delegate UMTS matters to a new subcommittee, SMG 5.

SMG 5 was chaired by Stein Hansen of Norway in 1992–1993 and Juha Rapeli from Finland in 1993–1996. SMG5 meetings attracted representatives of the RACE projects, as well as from the research departments of operators or manufacturers. Members of this UMTS community, as we may call it, were quite different from those of the GSM community who were closer to operational matters. Often companies sent junior staff or even beginners to these meetings. As the chairman of SMG I regret that we did not succeed in achieving the unification of these two populations.

As a consequence, the presentation of the SMG 5 report, usually on the last day of the SMG plenary meeting, did not generate a huge interest from the SMG participants. In such conditions SMG 5 tended to consider itself as an independent body. Quite often we had to remind them that some liaisons with ETSI had to be channelled via SMG. They also would have liked to have an independent project team rather than using the services of PT12. Another argument revolved around the fact that they had created subgroups to deal with services, radio aspects, etc. while some felt that in most cases the expertise should be sought in the other subcommittees particularly SMG 1–3.

SMG 5 had also the task of attempting to unify the views of European participants in the ITU meetings on FPLMTS/IMT 2000 or in some cases to elaborate a European input to such meetings.

SMG 5 undertook to issue framework documents defining the objectives or requirements for the various aspects of UMTS.[6] The first versions of such documents were approved by SMG in 1993. Through these documents the same vision of UMTS still emerged: an entirely new system based on the GSM model but different which would one day replace GSM.

No one of those who had an experience in operational matters in SMG could imagine how such a scenario could work. As a result in 1995 SMG had not yet a clear vision of the introduction of third generation mobile services. Nor had the GSM MoU Association who had undertaken to study the matter (see Chapter 6, paragraph 2.6). This was going to come from another ETSI initiative.

8.1.4 The Global Multimedia Mobility Concept

In 1995 ETSI created a Programme Advisory Committee (PAC). At the November 1995 Technical Assembly of ETSI, PAC suggested that a small expert group make proposals on the migration from the second to the third generation of personal communication systems. This group was chaired by Bernard Depouilly of Alcatel and worked very actively. SMG was represented by Gunnar Sandegren of Ericsson.

[6] TDoc SMG 477/93: UMTS work programme.

At the beginning of 1996 a draft report was circulated which was formulating a concept, or vision, called Global Multimedia Mobility (GMM).[7] It was based on the postulate that different access networks and core networks can be associated in a flexible manner, provided a suitable interface specification exists. Among "core networks" the report identified ISDN, GSM, B-ISDN, ATM, TCP/IP. SMG was well prepared to accept this vision having specified the A-interface at the junction of the GSM radio access network and the GSM core network and undertaken to develop for GPRS an interface between the GSM radio access network and TCP/IP based networks.

Quite concretely this concept enabled the visualisation of the following step by step approach to UMTS.

In a first phase the UMTS Radio Access Network (UTRAN) would be developed and associated with an evolution of the GSM core network, including a TCP/IP component. This approach could be implemented in both new and pre-existing networks, thus allowing the building of new UMTS networks as well upgrading pre-existing GSM networks where UTRAN could be introduced progressively to boost capacity and/or functionality. UTRAN would of course interwork directly with Internet Protocol (IP) networks, a capability that GSM networks were to acquire already with the introduction of the General Packet Radio Service (GPRS).

In a second phase a UMTS core network could be specified to focus mostly on broadband multimedia services and applications.

With this approach the introduction of UMTS became a viable proposal. It indeed preserved the investments made by manufacturers and operators in the GSM technology while moving forward in the area of both radio technology and networking concepts, as well as eventually making multimedia services available in the most cost effective way.

The GMM concept thus played an important role in the re-orientation of the work towards realistic objectives and scenarios even if it was not favourably received in the GA (see Chapter 8, Section 2).

---

[7] TDoc SMG 194/96: global multimedia mobility (ETSI/TA24(96)45).

Chapter 8: The UMTS Standardisation Work in ETSI

Section 2: The Creation of the UMTS Foundations in ETSI from April 1996 to February 1999

Friedhelm Hillebrand[1]

The foundations of UMTS were created in ETSI, mainly in the Technical Committee SMG, during the period from 1996 to 1998. This standardisation work had four major phases, which are treated in the first four paragraphs of this section:

- *UMTS priorities and work distribution*: in 1996 (see paragraph 8.2.1)
- *The UMTS strategy consensus:* from April 1996 to February 1997 (see paragraph 8.2.2)
- *Basic concepts of the UMTS standard*: from March 1997 to March 1998 (see paragraph 8.2.3)
- *UMTS reports and raw specifications*: from February 1998 to February 1999 (see paragraph 8.2.4)

Then the UMTS work was transferred to the Third Generation Partnership Project in early 1999.

The rest of this section treats legal and organisational issues and conclusions:

- *IPR issues* (see paragraph 8.2.5)
- *The initiation of 3GPP, the new global organisation* (see paragraph 8.2.6)
- *The work in ETSI complementing the 3GPP work* (see paragraph 8.2.7)
- *Conclusions* (see paragraph 8.2.8)

This whole section treats the development from April 1996 to the end of 1998 from an SMG plenary perspective. I was chairman of SMG during this period. The report is focussed on the SMG related aspects: strategy and decisions. The report follows closely the events visible at the SMG plenary and points frequently to reference documents, which can be found on the attached CD-ROM. More information about the technical work can be found in Chapters 10–20.

---

[1] The views expressed in this section are those of the author and do not necessarily reflect the views of his affiliation entity.

8.2.1 The Agreement on the UMTS Priorities and UMTS Work Distribution in ETSI in 1996

8.2.1.1 The Global Multimedia Mobility Report (GMM Report)

The "Global Multimedia Mobility" report[2] was produced by ETSI's Programme Advisory Committee as described in Chapter 8, Section 1, paragraph 4. It describes "a standardisation framework for multimedia mobility in the information society". It reviewed medium- to long-term trends in the telecommunication environment and technology. It recognised the variety of technologies and competition also in network operation. It proposed a framework architecture and described four domains: terminal equipment, access network, core transport network and application services. The report states explicitly that there is and will be a multiplicity of core networks and it recognises competition as a key element. The essence for the ongoing standardisation work is in the part "Conclusions and Recommendations".

8.2.1.2 Reactions by ETSI Members and Responses of the ETSI Technical Bodies

The elaboration of the GMM report triggered a controversial debate in ETSI about a reorganisation of UMTS work. One manufacturer launched an initiative to take UMTS out of Technical Committee SMG and to create a new organisation in ETSI in June 1996.[3]

All technical committees in ETSI were invited to comment on the draft report. SMG#19 (June 1996) received a report and proposal by Gunnar Sandegren, the SMG vice-chairman, who had represented SMG in the Program Advisory Committee.[4] SMG elaborated and agreed on a full set of comments on the conclusions and recommendations for the ETSI General Assembly.[5] SMG endorsed the conclusions and recommendations in principle and offered to bear the responsibility for specifying the radio access network for UMTS. SMG saw no need for a (superior) co-ordination committee. Instead SMG proposed to implement technical co-ordination by bilateral mechanisms between the technical bodies involved.

In a second General Assembly contribution[6] SMG offered to keep the responsibility for the standardisation of all UMTS services and particularly that of the UMTS radio access network. SMG offered to serve in these areas the interest of the entire ETSI community. The main reasons given were:

- To maintain Europe's leading position in mobile communication by building UMTS on the footprint of GSM.
- Competence for carrying out such work can be found in SMG only.
- GSM is the system best aligned to the GMM.
- Technical Committee SMG has developed a working methodology for the handling of evolving complexity in a changing environment

It should be noted that Technical Committee SMG was the only technical body, who fully considered the GMM report and produced a written response to the General Assembly.

[2] GMM Report Part A: Executive Summary, Conclusions and Recommendations in SMG 194/96 and ETSI/GA26(96)7.

[3] Meeting of the Interim Board.

[4] SMG 517/96.

[5] SMG 560/96 (revision of 554 and 539) (General Assembly #26 Temporary Document 16).

[6] SMG 549/96 (Annex 2 in General Assembly #26 Temporary Document 15).

8.2.1.3 The Decisions on the GMM Report in the ETSI General Assembly 26 in June 1996

The General Assembly received several input documents and had an intensive debate. The report was noted. The range of subjects was very wide, therefore an approval was not possible. Regarding the conclusions and recommendations the following agreements were reached:

1. The modular architecture framework for all services and a multiplicity of networks was endorsed in principle (Conclusions 1 and 2).
2. Recommendation 1 to refine the GMM architecture and to establish or identify a group for it was referred back to the ETSI Board.
3. Conclusion 3 and Recommendation 2 regarding the provision of harmonised spectrum were seen as not relevant in ETSI.
4. The promotion of the GMM concept to other standardisation organisations was agreed (Conclusion 4 and Recommendation 3).
5. Conclusion 5, that the UMTS core network should be based on the evolving existing core network standards (e.g. GSM or ISDN) was referred back to the ETSI Board. Also Recommendation 4B to make sure that the evolving core network standards comply with the GMM framework was referred back to the ETSI Board.
6. It was endorsed, that it is an ETSI priority to develop the radio access network for UMTS (Recommendation 4).

This means that the General Assembly saw the framework architecture principles as useful and confirmed the priority to develop the UMTS radio access network. All other issues were referred back to the ETSI Board. This outcome showed that no ETSI consensus in key areas had been reached during the elaboration of the report.

8.2.1.4 The ETSI Board Meeting #1 in August 1996

The Board agreed additional guidance mainly regarding the UMTS network aspects, which was added to the Executive Summary of the GMM Report:

> When continuing the work on the 'GMM standardisation framework', it must be taken into consideration that UMTS services require standards for new terminals, a new access network *and*enhanced or new network capabilities in existing core transport networks, e.g. GSM and narrowband ISDN. There will be a planned evolution path from existing GSM and narrowband ISDN networks to support UMTS and other future mobile services (Annex E to the Meeting Report).

In addition the creation of a "GMM co-ordination group" was initiated. This group should at least:

> ensure an overview of the relevant work of Technical Committees NA and SMG....and that the relevant bodies were proceeding in the same direction (Meeting Report 5.2).

Based on an SMG input document,[7] which I presented, the ETSI Board confirmed the Technical Committee SMG proposals and invited SMG to produce revised terms of reference to cover the existing GSM and UMTS work and make clear provision for the necessary liaison with other ETSI technical committees and projects concerning UMTS. The revised

---

[7] SMG 564/96, 558/96 and 549/96, identical with ETSI/B1(96)9.

SMG terms of reference were elaborated and agreed in due course by the SMG#20 in October 1996 and approved by the Board#2 in October 1996.[8]

8.2.1.5 Bilateral Agreements between Technical Committees NA and SMG

In bilateral talks between delegations of Technical Committee NA (network aspects) and Technical Committee SMG, Technical Committee NA confirmed that they saw the UMTS core network evolution based on ISDN as their responsibility. Technical Committee SMG invited NA1 (services) to participate in the SMG1 work on UMTS services. NA1 sent regularly a liaison person. SMG offered NA a subcontract in order to fulfil their additional requirements on the UMTS radio access network. This subcontract was not elaborated further. SMG invited NA regularly to provide their additional requirements. But it turned out that there were no additional needs for the use of the UMTS radio access in a ISDN based core network.

8.2.1.6 Conclusions on a Core Network Evolution and Work Distribution

At the end of a cumbersome process an agreement within ETSI on strategic targets was reached in October 1996, which is fully reflected in the revised Technical Committee SMG terms of reference:

1. Specification of all UMTS services aspects in Technical Committee SMG with the widest possible participation;
2. Specification of one new UMTS radio access network in Technical Committee SMG;
3. Specification of the GSM core network evolution for UMTS in Technical Committee SMG;
4. Specification of the core ISDN core network evolution for UMTS in Technical Committee NA;
5. A loose co-ordination by the GMM Co-ordination Group with participation from all involved groups

Technical Committee NA and other forces coming from the "fixed network world" tried to change this agreement in 1997 and to establish a completely new organisation for UMTS, which had meant a disbanding of SMG and the distribution of all GSM and UMTS work on several other new bodies (see e.g. ETSI Board#7 and #8 Reports). But these initiatives did not lead to a success since they did not gain sufficient support in ETSI.

In addition even the activities on the ISDN based core network evolution did not lead to specifications for a services' opening in 2002, since the community interested in this matter did not find sufficient momentum and support from their membership.

8.2.2 The Strategy Consensus for UMTS Achieved from April 1996 to February 1997 (SMG#19–#21)

The original UMTS concept was developed in the framework of the RACE program in 1986. It described UMTS as "mobile access to broadband ISDN". This concept was transformed into a more market-oriented "New UMTS" strategy based on the agreements reached in the

[8] SMG revised terms of reference in SMG 662/96.

UMTS Task Force Report, and the results of the UMTS Forum in close co-operation with the GSM Association in 1996/7. The main elements of the "New UMTS" are:

- Services' innovation (e.g. VHE concept)
- Continuity for GSM services and evolution from GSM
- High performance interworking with the Internet

The process to take back the lead and to build an initial consensus on the UMTS strategy needed three SMG plenary meetings: SMG#19 in June 1996, SMG#20 in October 1996 and SMG#21 in February 1997. Sections 2.1–2.3 describe this process in chronological order. They try to highlight the main aspects of this difficult transition and re-orientation process. Section 2.4 summarises the results.

8.2.2.1 The Initiative on UMTS taken back by SMG at SMG#19 in June 1996 at Kista, Sweden

The UMTS work had been dealt with in the meetings before June 1996 by the sub-group SMG5, a community working relatively independently from the rest of SMG as described in Chapter 8, Section 1, paragraph 3. The SMG plenary was so loaded with GSM work that little time was left for UMTS. The biggest problem was the lack of a clear vision and strategy for third generation mobile services as identified in Chapter 8, Section 1, paragraph 3.

In spring 1996 many SMG members realised that it was of strategic importance to deal properly at the SMG plenary with UMTS. When I came into office as chairman I changed the format of the SMG plenary, in order to gain sufficient time for UMTS. The time for GSM was shortened by enabling a new change request procedure which delegated decision power to the subgroups and saved plenary time.[9] In addition the treatment of controversial items was reorganised.[10] This allowed a new plenary format:[11] GSM was treated from Monday morning to Wednesday noon, UMTS from Wednesday noon to Thursday evening. Friday was reserved for postponed controversial issues, which could not be resolved earlier in the week. Small teams were charged to seek a solution for these postponed controversial items.

SMG received a presentation of the UMTS Task Force report.[12] A full working relationship was established with the UMTS Forum and its working groups. I invited the chairmen to present their ideas to the SMG#19 and to all following plenaries. I was offered a seat in the Forum's Steering Group. A co-operation agreement[13] between ETSI and the UMTS Forum was endorsed by SMG and approved by the ETSI Board. The overlap in work was removed. The activities of SMG on spectrum issues (in SMG5 and in the SMG spectrum task force) were closed and a close liaison to the Forum's Spectrum Aspects Group established.[14] I became the official ETI representative in the Steering Group of the UMTS Forum, in order to ensure a close co-operation between the UMTS Forum and ETSI SMG.

The chairman of SMG5 responsible for UMTS covered a wide range of subjects which had been treated by SMG5 in his status report. He presented several ETSI telecommunication reports and ETSI telecommunication specifications for approval: vocabulary, introduction,

[9] See Chapter 5, Section 2, paragraph 5.2.4.5
[10] See Chapter 5, Section 2, paragraph 5.2.4.1
[11] SMG 352/96 rev.1.
[12] See Chapter 7. Section 3 and text of the report in the CD ROM folder E.
[13] SMG 547/96.
[14] See Chapter 7, Section 5.

security principles, USIM. The plenary felt uneasy in approving these documents, since the plenary had not reached an earlier agreement on strategic questions, in addition some documents were seen as too detailed and oriented too much towards the old UMTS philosophy which considered UMTS as an access network providing access to broadband ISDN, completely separated from GSM. On the other hand there was the intention to honour the work of SMG5. Therefore these documents were approved for publication.

SMG5 requested guidance on the relationship between UMTS and FPLMTS. A list of questions[15] was asked by SMG5. The SMG5 chairman had asked SMG#18 (April19 96)[16] on positions regarding ITU without receiving an answer. It was stated and agreed that SMG should influence the ITU FPLMTS[17] work. But SMG had no relationship with ITU groups. It was also not possible to agree on a way forward at SMG#19. But it was agreed to come back to these issues in the following plenary and also to discuss relevant inputs to the ITU in that meeting.[18]

SMG had shifted the work of the working groups reporting to SMG5 on services, radio and network aspects to the "normal" SMG subgroups responsible for GSM work: SMG1, 2 and 3 and re-focussed SMG5 into a co-ordinating role.[19] SMG#19 received the first SMG1, 2 and 3 reports on UMTS. Some other subgroups expressed interest in UMTS work (security, data, testing).

In addition SMG#19 had a full debate on the GMM report and reached conclusions on the strategic orientation of UMTS as described in paragraph 1.2.

During SMG#19 the SMG plenary managed to win back the initiative on UMTS strategy. Initial results were reached and a way forward became visible. In the parallel activities in the ETSI General Assembly and Board (see paragraph 1) the responsibility of SMG for the leading role in UMTS was defended. But it was clear to me that this lead role needed to be justified by rapid progress with demonstrable results.

8.2.2.2 The UMTS Strategy Debate at SMG#20 in October 199 96 at Sophia Antipolis

8.2.2.2.1 Strategic Agreements

The UMTS key features[20] and proposals to seek a focus on wideband and multimedia services[21] were discussed. The initial conclusions were that the *UMTS services*support the full range of applications from narrowband to wideband (2 Mbit/s as a target) services and that multimedia services need a high bitrate and a high degree of flexibility (see Meeting Report 7.6.3). This was the first strategic agreement in SMG on the orientation of the UMTS services. It is a "soft" conclusion, but it provides direction. Standardisation is all about consensus building! Here is the beginning of the UMTS services consensus.

---

[14] See Chapter 7, Section 5.

[15] SMG 416/96 and 524/96.

[16] SMG 349/96.

[17] FPLMTS = Future Public Land Mobile Telecommunication System, a term replaced later by IMT-2000.

[18] The situation regarding ETSI ITU contributions was difficult. Previously the Technical Assembly would have approved ETSI contributions. The ETSI reform had delegated all technical matters to the technical bodies. This would implicitly mean that technical committees could approve ETSI contributions to the ITU. But several delegates were reluctant to do this since no explicit rules existed for this case.

[19] See new SMG5 terms of reference approved by SMG#18 in April 1996 in SMG 231/96.

[20] SMG 725/96.

[21] SMG 718/96, 729/96, 727/96.

A meeting on the focus of UMTS work had been held and results were noted by SMG#20.[22] The first firm agreements on *UMTS phases and milestones* were triggered by a contribution of Mannesmann, Orange, E Plus, Vodafone and DeTeMobil. It contained principles, a phases definition, and a UMTS development schedule on key requirements for the radio interface. This document had been prepared to provide strategic guidance to the UMTS standardisation work.[23] It was aligned with the results of the UMTS Task Force and the Third Generation Interest Group (3GIG) of the GSM MoU Group. It was presented to the GSM MoU plenary in due course. The paper presents a UMTS concept which is oriented to market needs as seen by network operators. It proposes a modular approach and a phases and milestones concept. SMG drew some firm conclusions from it. The major elements were:[24]

- The UMTS development should follow a phased approach
- UMTS should have a modular approach (separation of access and core network)
- UMTS phase 1 uses a new BSS and an evolved GSM core network and should be operational by 2002
- UMTS phase 2 (with a new core network) around 2005
- UMTS concept (basic studies including selection of the radio interface) during 1996–1997
- UMTS detailed standardisation in 1998-1999

These key milestones became the polar star of the future UMTS work program. What is even more remarkable, they were achieved!

8.2.2.2.2 Results of the SMG Subgroups

A number of documents elaborated by SMG5 were presented for information. They covered areas like work program, framework of network requirements, overall radio requirements, framework for the telecommunication management, satellite integration. SMG5 provided information about the ITU work and contributions sent to the ITU by SMG5. Several other contributions to the ITU were endorsed.

The first specification on services principles presented by SMG1 was approved.

The work plan for the access technology selection[25] proposed by SMG2 was endorsed. A paper on radio selection procedures was presented for information.

It was decided to charge the SMG3 System Architecture Group (the later SMG12) with UMTS network architecture aspects in addition to GSM architecture. The task was previously taken care of by SMG5. This transfer, allowed the exploitation of the GSM know-how for UMTS.

8.2.2.3 The Agreement on the Strategy for the Relationship between FPLMTS and UMTS and the UTRA Definition Procedure Reached at SMG#21 in February 1997 in Paris

SMG#21 received reports from the chairmen of the UMTS Forum and the Third Generation Interest Group (3GIG) of the GSM MoU Group.

---

[22] SMG 641/96.
[23] SMG 682/96. I had produced the first draft and circulated it to the mentioned operators.
[24] See Meeting Report of SMG#20 in the CD ROM folder B1.
[25] SMG 601/96, Annex II.

8.2.2.3.1 Private Cordless Systems

A discussion on the impact of private and cordless systems took place based on an input of the UMTS Forum. It was referred to SMG1 and 2 for further study. This was a difficult subject for a technical committee focussed on a public cellular system.

8.2.2.3.2 Strategy for the Relationship between FPLMTS and UMTS and Process for Contributions to the ITU

The relationship between UMTS and FPLMTS and also the roles of ETSI and the ITU in these programs were hot potatoes. The role of the ITU was indispensable for several fundamental aspects such as free circulation, spectrum harmonisation, etc. It was however clear to most participants that the ITU would not be able to reach a decision on one radio solution due to the diverging interest of the members. There were also serious doubts, whether ITU would be able to deliver the necessary detailed specification for services opening and system evolution on time. On the other hand a recognition by the ITU was necessary for access to the market and spectrum in many countries.

After the ETSI reform 1995/1996 no explicit procedures for ETSI contributions to the ITU existed any more. Before the reform they had to be endorsed by the Technical Assembly. This body had been disbanded and the decisions on all technical matters had been allocated to the technical bodies. But the technical committees had not been explicitly authorised to submit contributions on behalf of ETSI. The ETSI Board had no decision competence in these matters. But there were people who were reluctant to endorse technical contributions to the ITU in a technical committee and to submit them with the source "ETSI".

Therefore, both was needed: a strategy for the relationship between FPLMTS and UMTS and the roles of the ITU and ETSI and a process for contributions to the ITU.

The Strategy for the Relationship between FPLMTS and UMTS

Based on the trilateral discussions of the European Union, US and Japan on "Future Advanced Mobile Universal Systems (FAMOUS)",[26] I drafted a strategy paper on "Third Generation Standardisation Policy: The Role of the ITU and ETSI, the Relationship between FPLMTS and UMTS" and agreed it with key persons in the Technical Committee SMG.[27] The paper was submitted in their names and my name. It described the interest of the US, Japan and Europe. It summarised the FAMOUS agreements of May 1995 on the standardisation of the third generation, which had not been implemented but instead had fallen into a "sleep mode". The FAMOUS recommendations of May 1995 recognised the different market needs in different regions. They called for:

- Evolution paths from all second generation systems
- Global standards should address the kernel of services
- Regional standardisation organisations should specify the details

The contribution proposes to recognise that there will be several different FPLMTS systems. Efforts should be made to enable roaming. These systems should be seen as an FPLMTS family of systems. The paper identifies the first family members.

---

[26] Report on FAMOUS, see Chapter 7, Section 3.

[27] G. Sandegren (SMG vice-chairman), J. Rapeli (SMG5 chairman), A. Bergmann (PT SMG leader) and A. Samukic (UMTS program manager), see SMG 63/97.

A new focus for the ITU work was proposed. It included the following: recognise family members, facilitate roaming between family members, framework standardisation, seek commonalties between members, free circulation, spectrum issues.

A proposal for the work focus of a "regional" standardisation organisation like ETSI was made. It included the following: standardisation of UMTS, contributions on framework standardisation and roaming between family members to the ITU, submission of an air-interface candidate, which meets or exceeds the essential ITU requirements as a member of the FPLMTS family.

Regarding the relationship between FPLMTS and UMTS, it was proposed that ETSI standardises UMTS as a true third generation system with evolution paths from GSM and ISDN.[28] UMTS should be designed and planned as an FPLMTS family member and the UMTS radio interface should be submitted as a candidate to the ITU for approval as one member of the FPLMTS family.

The document was discussed intensively. There were questions whether ETSI could "prescribe" what the ITU should do. The response was that the document "proposes" elements for discussion. But a consensus could be reached on the principles. The document was updated and approved as an SMG strategy document.[29] It was agreed to distribute the document widely to the ETSI Board, UMTS Forum, GSM MoU Group, ANSI T1P1, ITU low level groups, etc. for information. It was slightly revised at SMG#22. The substance was submitted to the ITU.[30] After an intensive discussion with other regional standardisation organisations, the ITU accepted an FPLMTS family of systems (later called the IMT-2000 family of systems) as a realistic and implementable approach.[31] A new orientation was given to the international co-operation between ITU, ANSI T1, TIA, ARIB and TTC by the IMT-2000 family of systems concept. It proposes a viable way forward:

- Acknowledgement that one standard for IMT-2000 is not acceptable for the different market needs and business interests
- Modular framework as a target
- Search for as much commonalties as possible in functions, modules, etc.

This concept ended the strong confrontation between the different communities of interest and opened the way to a reasonable and successful co-operation between the previously hostile parties. It enabled the recognition of several radio technologies by the ITU. A substantial harmonisation of third generation systems was achieved.

Process for the Approval of Contributions to the ITU

It was not possible to agree on a process to submit contributions to the ITU which were approved by SMG with the source "ETSI". Since it was urgent to have a process available, SMG avoided the discussion on principles in ETSI by seeking a pragmatic procedure. In 1996 SMG approved contributions that had been submitted in the name of one or several ITU member companies. This was felt to be inadequate, since it did not show the level of support reached in ETSI. Therefore, a new form was developed: the document was submitted by one or several ITU member companies, e.g. "member xyz in the name of ETSI SMG". This form was used frequently in the following years, e.g. for the submission of the UMTS radio access.

---

[28] ISDN was mentioned in line with the ETSI agreements on UMTS priorities, see paragraph 8.2.1 above.

[29] SMG 290/97.

[30] These formal contributions were elaborated in due course and endorsed by SMG#22bis in August 1997.

[31] The first agreement in the ITU was reached in the SG11 Rapporteur's meeting in June 1997.

It showed clearly to the ITU, that a consensus of the technical committee responsible for UMTS backed such a document.

8.2.2.3.3 Relations with Japanese Standardisation Organisations

G. Sandegren of Ericsson proposed a co-operation with the Japan FPLMTS activities including an opening of the ETSI SMG work for Japanese and ANSI T1P1 participation and contributions.[32] An ad hoc group with participation from France Telecom, Alcatel, Nortel, Siemens and others drafted a mandate for a mission of the SMG chairman to Japan in March 1997 with the following tasks:

- to present and discuss the ETSI vision based on the SMG strategy paper FPLMTS/UMTS (see paragraph 2.3.2);
- to invite ARIB/TTC to present their views in ETSI;
- to report back on the perceived willingness of Japanese representatives.

The proposed mandate was approved by SMG#21.[33] This mission was approved by the ETSI Board#5 in February 1997, since this initiative opened a new external relationship.

This mission lead to a close co-operation with the Japanese standardisation organisations (for details see paragraph 3.6). For the UMTS strategy this initiative meant that a solution would be sought, which included Japan. This was an important opening towards a fully globally applicable solution.

8.2.2.3.4 UMTS Radio Aspects

SMG#21 in February 1997 provided guidance on strategic aspects of the UMTS Terrestrial Radio Access (UTRA):

- SMG2 shall focus on the UMTS UTRA, this means not considering satellite or aeronautical requirements;
- SMG2 shall consider wireless fixed access only in so far as this does not add additional complexity or performance degradation;
- UTRA shall meet the essential ITU requirements;
- SMG confirmed 384 kbit/s as a requirement for the pedestrian environment.

SMG approved the "UTRA definition procedure and time schedule".[34] The main elements were:

- UTRA *definition* procedure instead of a *selection* procedure, i.e. co-operation from the beginning instead of competition, which leads to large confrontations.
- Every company can contribute to every aspect of every proposed solution.
- Planned agreement on groupings at SMG#22 in June 1997.
- Planned selection of one UTRA concept at SMG#24 in December 1997.
- Submission of UTRA to the ITU as an FPLMTS candidate.

---

[32] SMG 222/97.

[33] SMG 240/97.

[34] SMG 163/97 and 164/97. Early versions of these documents had been prepared in two meetings as an input to SMG by a subset of the SMG steering group: N.P.S. Andersen the SMG2 chairman and me as SMG chairman, G.Sandegren and A. Maloberti the SMG vice-chairmen and supported by A. Bergmann the PT SMG co-ordinator.

- Further elaboration of the selected UTRA concept as a joint activity with contributions from all interested parties in 1998.

It was further agreed not to perform comprehensive hardware validation tests in the public sphere as it had been done for GSM in 1986 since many aspects can be clarified by simulations. In addition it was assumed that each contributing company would perform the necessary validation tests and report results.[35]

8.2.2.3.5 The Transfer of UMTS Work from SMG5 to the SMG Plenary and the other Subgroups

It was agreed to transfer the work on a role model from SMG5 to SMG1 (services) and the work on an architecture reference model from SMG5 to SMG3 (network aspects).

In order to streamline the UMTS work further, to progress the UMTS document structure and work program and to examine the SMG5 mandate an extraordinary meeting of the "SMG Steering Group"[36] was held at the request of SMG in March 1997. It agreed on a revised draft work plan and came to the conclusion to close SMG5 since the work had been successfully integrated into the other subgroups. SMG endorsed this conclusion by correspondence.

8.2.2.3.6 The Initiation of the UMTS Baseline Document

In order not to "loose" strategic agreements, which were documented in meeting reports, SMG charged me as SMG chairman, A. Bergmann the PT SMG leader and A. Samukic the UMTS program manager, to produce the "UMTS Baseline Document UMTS 30.01". The first version was submitted to SMG#22 in June 1997.[37]

It was updated regularly and served as an orientation document during the UMTS work in SMG. It was also a valuable reference, to show external groups the agreements reached. The further strategy decisions were documented in the meeting reports and saved in the UMTS Baseline Document. It evolved substantially over time. Every SMG plenary endorsed a new version.[38]

8.2.2.3.7 The New UMTS Work Plan and the First Structure of the Permanent Output Documents

The new UMTS focus needed a new work plan focussed on the essential work items and a new structure of the permanent output documents. It had been drafted by the SMG Steering Group (see paragraph 8.2.3.6) and was endorsed by correspondence.[39]

---

[35] SMG 165/97.
[36] SMG chairman, vice-chairmen, subgroup chairmen, PT SMG co-ordinator.
[37] SMG 384/97 shows the first version submitted to SMG#22.
[38] An early version is in SMG 384/97, the last version is in SMG 112/99.
[39] SMG 385/97 shows the version submitted to SMG#22.

8.2.2.4 Conclusion: UMTS Strategy Consensus Reached in the Period from April 1996 to February 1997 bySMG#19–21

8.2.2.4.1 Strategy Consensus on a "New" UMTS Vision

The following *agreements in substance* were reached:

- Key UMTS characteristics
- Key services aspects (including the evolution from GSM services)
- Key requirements for the UMTS Terrestrial Radio Access
- Modular system concept:

  - Separation of core and access network
  - New UMTS radio system
  - GSM core network evolution

- Strategy on the relation between FPLMTS and UMTS

The following work and organisation agreements were reached:

- Creation of the UMTS baseline document to save strategic decisions
- UMTS phases and milestones
- UTRA definition process
- New work plan with new streamlined work item and document structure
- Streamlined organisation: all work in the proven structure for GSM in order to exploit the know-how and resources and to secure cross-fertilisation
- Work distribution between ETSI and ITU
- Opening towards a co-operation with Japan

These cornerstones formed a firm base for the following work phases.

8.2.2.4.2 The Evolution of the UMTS Vision

Table 8.2.1   The evolution of the UMTS vision

	UMTS vision in the late 1980s	"New" UMTS vision in 1997
Core Ideas	Integration of all existing and new services into one new universal network	Focus on innovative services *and* Support of GSM services
Partner networks	Broadband-ISDN	Intranets and Internet
Introduction	Migration from existing networks	Evolution from GSM services and networks
Roaming	New development, INAP based	Evolution of GSM roaming, MAP based
Standardisation	FPLMTS as one monolithic standard in the ITU	IMT-2000 family of systems as framework standard in the ITU
	UMTS in ETSI for filling of gaps and option selection	Detailed specifications in ANSI, ETSI, ARIB/TTC

8.2.3 The Elaboration of the Basic Concepts of the UMTS Standard from March 1997 to March 1998 (SMG#22–25)

This work phase achieved – starting from the strategy consensus described in paragraph 8.2.4 – the agreement on key services' concepts, the decision on one UTRA concept and the agreement on several network concepts. They formed the basis for the elaboration of the UMTS reports and raw specifications in the following phase (see paragraph 8.2.4).

This paragraph describes selected issues and is structured according to the major subject areas: 8.2.3.1, work management, 8.2.3.2, services, 8.2.3.3, system architecture, 8.2.3.4, radio, 8.2.3.5, ITU, 8.2.3.6, relations with Japan. All other technical subjects not treated here are dealt with in Chapters 10–20.

During this work phase SMG received regular progress reports and many contributions from the UMTS Forum and the GSM MoU Group's Third Generation Interest Group (3GIG). In autumn 1997 the GSM MoU Group agreed to extend the co-operation agreement with ETSI to third generation and to provide some funding for the UMTS work of PT SMG.

8.2.3.1 UMTS Work Management

The UMTS work plan contained a framework planning. Since standardisation is contribution driven, a work plan of a standardisation group provides visibility to members for areas and timing of sensible contributions and an invitation to contribute.

The second important content of the UMTS work plan is the identification of work items as well as the definition of the structure of the output documents (technical reports and technical specifications).

The UMTS work plan evolved during the period. It was updated at every plenary meeting[40].

Another important tool for the work management was the ITU work program. This was driven by needs of the ITU (see paragraph 8.2.3.5).

8.2.3.2 The Clarification of the UMTS Services Concept

Further clarification on services was reached at SMG#22 in June 1997 by an agreement on dual mode of operation between GSM and UMTS and handover between GSM and UMTS as mandatory features.

- GSM only, UMTS only and GSM/UMTS terminals should be allowed.
- GSM-UMTS handover is needed in both directions, i.e. from and to UMTS or from and to GSM.

A *UMTS role model* was approved by SMG#22.[41] But it became clear that more study was needed, to clarify the model's validity for private environments, etc. Questions were asked about the role of the UMTS role model. After many controversial discussions the deletion of the UMTS role model was agreed at SMG#26 in June 1998, since it was seen as being outside the scope of standardisation.

---

[40] An early version is in SMG 385/97, the last version is in SMG 111/99.
[41] SMG 302/97.

In order to make a sensible use of the GSM/UMTS handovers, the services must "continue" between GSM and UMTS. In order to secure this, the work item "*Service Continuity and Provision* of VHE (Virtual Home Environment) via GSM/UMTS" was approved by SMG#24 in December 1997.[42]

Another important agreement was the approval of the *basic UMTS services philosophy* at SMG#24:[43] UMTS specifies service capabilities, not complete services in order to offer more flexibility. This text was inserted into the UMTS baseline document in order to serve as a "polar star" for the further UMTS services work.

In order to progress the UMTS services work a program of *technical reports in key innovation areas* had been agreed. But it was very difficult to find the necessary resources and contributions for this work. This was discussed with concern at SMG#23 in October 1997. This mobilised members' efforts. SMG1 was then able to present some first drafts for information to SMG#24 in December 1997. It was possible to complete key documents for approval by SMG#25 in March 1998:

- Services and services capabilities (UMTS 22.05)
- Terminals and smart card concepts (UMTS 22.07)
- New charging and accounting mechanism (UMTS 22.24)
- Mobile multimedia services including mobile Internet and intranet services (UMTS 22.60)
- Virtual home environment (UMTS 22.70)
- Automatic establishment of roaming relationships (UMTS 22.71)
- Advanced addressing (UMTS 22.75)

8.2.3.3 UMTS System Architecture

8.2.3.3.1 UMTS System Architecture Aspects at SMG#22 and 23

The progress in architecture and network aspects was so slow that SMG#23 in October 1997 charged the SMG chairman to call for an extraordinary SMG Co-ordination Group[44] meeting in order to plan the work on UMTS network aspects. This activity lead to the proposals to define an evolving GSM-UMTS core network and to create a subgroup SMG12 System Architecture.

8.2.3.3.2 The Agreement on Evolution in the GSM-UMTS Core Network at SMG#24 in December 1997

SMG approved a contribution of the PT SMG co-ordinator[45] which established the formal requirement that the UMTS core network is an evolution of the GSM core network and that therefore the GSM core network should be treated as a subset of the UMTS core network from Release 99 onwards.

Another contribution[46] called for evolution as a formal requirement, the definition of guiding principles and a clearer responsibility for the definition of these.

---

[42] SMG 1044/97.
[43] SMG 1041/97.
[44] SMG chairman, subgroup chairmen, PT SMG leader.
[45] SMG 882/97 elaborated by Ansgar Bergmann, endorsed by me, submitted with source "SMG Chairman".
[46] SMG 1043/97.

Several delegates proposed to establish the System Architecture Subgroup of SMG3 as an autonomous subgroup.

Some progress in substance was reached by the approval of a work item description for a new interface between the network subsystem and the base station system.[47]

NTT DoCoMo described their plans for migration to the third generation core network at the request of SMG.[48] The basic idea is to attach the IMT-2000 radio access network to the GSM-UMTS core network and to provide an interworking function between the PDC core network and the GSM-UMTS core network. The project plan foresaw the elaboration of additional specifications to the GSM specifications in TTC, the Japanese standardisation organisation for network aspects.

8.2.3.3.3 The Creation of the SMG Subgroup SMG12 Responsible for System Architecture at SMG#25 in March 1998

SMG3 reported that the progress was very slow. The work moved towards a common Release 99 for UMTS and GSM.

After thorough discussions between the SMG management and key contributors a proposal was made to SMG#25 to "promote" the SMG3 subgroup on system architecture into a SMG subgroup SMG12 with a lead responsibility on system architecture matters.[49] SMG12 would work under services requirements coming from SMG1 and would make the basic architectural decisions which would serve as guidance to the other subgroups. SMG endorsed this proposal. This re-organisation was crucial for the future system architecture work.

8.2.3.4 UMTS Terrestrial Radio Access (UTRA)

8.2.3.4.1 The Agreement on the UTRA Requirements and on Five UTRA Concepts Reached atSMG#22 in June 1997

As planned the elaboration and approval of the *requirements for UTRA* (high level and detailed) *and* the *selection procedure* for the choice of radio transmission technology could be achieved at SMG#22 in June 1997. It was possible to approve the relevant SMG2 input documents[50] in this meeting unanimously.

It was also planned to agree on a limited number of concepts in this meeting. SMG2 had studied a wide variety of possible solutions since December 1996 and proposed *five concept groups* for UTRA.[51]

- Alpha: Wideband CDMA (Code Division Multiple Access)
- Beta: OFDMA (Orthogonal Frequency Division Multiple Access)
- Gamma: Wideband TDMA
- Delta: Wideband TDMA/CDMA
- Epsilon: ODMA (Opportunity Driven Multiple Access)

---

[47] SMG 1112/97.
[48] SMG 1118/97. (not available in the CD-ROM, can be retrieved from the ETSI archive).
[49] SMG 247/98.
[50] SMG 478/97, 431/97 and 432/97.
[51] SMG 480/97.

These five concept groups were approved by SMG. It was stressed that they are open to contributions by all companies.

During the summer and early autumn the initial technical proposals which showed great variations were analysed and integrated into one common technical concept per concept group.

8.2.3.4.2 The Agreement on the UTRA Decision Procedure at SMG#23 in October 1997

It was expected that the UTRA decision planned at SMG#24 in December 1997 would be highly controversial, since there was so much interest from different manufacturers involved, and that a consensus would be very difficult to achieve. I proposed therefore a procedure to SMG#23 in October 1997 with the following main elements:

- SMG#24 should make all efforts to reach a consensus solution
- SMG#24 may hold an indicative voting using the same procedures as in the decisive voting
- If no consensus can be achieved atSMG#24, a SMG#24bis meeting should be held on 28–29 January 1998 in order to vote
- The a.m. voting processes should be supervised by ETSI's legal adviser

This proposal[52] was approved by SMG.

The procedure provided planning security to all interested parties, since it was announced and agreed well in advance of the decision. In addition the process was designed as a process of "slowly drying cement", which showed the picture clearer over time. Moreover, this process allowed all parties to communicate intensively within the agreed schedule.[53]

The process needed to be waterproof against formal complaints to the General Assembly.[54] Therefore, in order to ensure a high level of neutrality and formal/legal correctness, I had asked the ETSI legal adviser to supervise it.

The indicative voting is performed normally as a show of hands, i.e. one vote per ETSI member. This may lead to other results than the formal voting procedure, which uses weighted voting.[55] Therefore, I proposed that the indicative voting at SMG#24 should use the same procedure as the weighted voting. Technical Committee GSM and SMG had used such a voting process for the first time in the election of the second SMG vice-chairman during the summer of 1997.[56]

In order to make the voting on UMTS waterproof, I reviewed the rules in the ETSI Directives. It turned out that there were a number of unclear items and missing rules. Therefore, I produced in co-operation with the SMG Co-ordination Group and ETSI's legal advisor a document "Procedure of voting for indication of intent of UTRA in SMG#24". This document explained also the existing rules for the SMG delegates, who were not familiar

---

[52] SMG 858/97.

[53] G. Sandegren called this process later ""the extended coffee break". Thomas Haug had used coffee breaks intensively to resolve controversial matters.

[54] Many colleagues mentioned to me that all such efforts would be useless, since somebody would complain to the General Assembly anyhow.

[55] The number of votes depends on the telecommunication turn-over of an ETSI member. The financial contribution to ETSI is proportional to the number of votes a member has.

[56] Alistair Urie commented later, that this has been done in order to exercise the weighted voting procedure.

with formal voting. The document was approved after a lively question and answer session at SMG#24.[57] An update approved at SMG#24[58] was used for the vote at SMG#24bis.

8.2.3.4.3 The Alternatives for the UTRA Decision

SMG2 had dealt with UTRA at many SMG2 plenaries, ad-hoc meetings and concept group meetings. The requirements on UTRA and a grouping into five concepts had been endorsed by SMG#22 in June. Since then a detailed evaluation of the proposals has been performed and the different original proposals have been combined into one single proposal for UTRA per concept group. SMG2 had not been able to agree on a single proposal or to merge the different proposals into one.

It was claimed that all concepts fulfil the high level UTRA requirements. However, SMG2 remarked that the area of private and residential operation and the use of unpaired spectrum were "not areas on which the concept groups have placed the highest attention" and "that it may be necessary to consider modification of any adapted UTRA concept to improve these aspects...".[59]

In a first step SMG#24 qualified the Epsilon concept (ODMA) as an advanced relay protocol applicable to all other concepts. Therefore, the selection needed to be made between the four concepts: Alpha, Beta, Gamma and Delta.

8.2.3.4.4 The Debate at SMG#24

The meeting was attended by 270 delegates including delegations from ANSI T1P1 (US), RITT (China), ARIB (Japan), GSM MoU Association, UMTS Forum and five ETSI board members.

Contribution from other Organisations

Thomas Beijer, the UMTS Forum chairman stressed the need to select only one radio interface for UTRA. Adriana Nugter, the GSM MoU Association chairperson stressed the requirement to select only one global radio interface and that all GSM MoU members should have an influence on this decision.[60] The European Commission expressed their deep concern about the IPR situation and called for transparency[61]. The DECT Forum stressed the need that fixed and private services need to be provided by UTRA with at least the flexibility DECT offers. The chairman of ANSI T1P1 said that they were evaluating the different alternatives. A group of 18 operators co-operating in SMG2 expressed their concern on open technical issues, e.g. guard bands.

Akio Sasaki (ARIB, Japan) spoke about the situation and principle attitude in ARIB.

ARIB prepared a WCDMA solution for the IMT-2000 radio access. A principal decision for Direct Sequence CDMA (WCDMA) had been made in January 1997. This is the only solution studied in detail. ARIB has succeeded in having nearly the same parameters as the Alpha concept. There is also work on other solutions in Japan (e.g. IS95 evolution, OFDMA).

---

[57] SMG 995/97.

[58] SMG 1157/97.

[59] Meeting Report of SMG#24 (CD ROM folder B1), Section 4.1.1.

[60] The last ETSI General Assembly had decided not to grant (non-European) associate members voting rights in technical matters.

[61] The IPR issue is treated systematically in paragraph 8.2.5.

NTT DoCoMo presented a letter to SMG confirming its support of WCDMA and the GSM core network evolution. They have an MoU for co-operation on third generation issues with Telecom Italia and Telecom Finland and with several parties in Asia and the Pacific.

The Contributions by SMG Members

Ericsson, Nokia and NEC submitted documents stating the superiority of the *Alpha concept* over all other concepts in respect of capacity, coverage, maturity, etc. Siemens and Motorola objected. No additional documents were submitted regarding the Beta and Gamma concepts.

The *Delta concept* was supported by Alcatel, Bosch, Italtel, Motorola, Nortel, Siemens and Sony. The GSM compatible evolution, the inherent TDD solution and the much less critical IPR situation were stressed[62].

Fujitsu objected to some aspects. Nortel and Motorola objected to Fujitsu. France Telecom presented results of their technical work which favours the Delta concept. Philips saw a superiority of the Delta concept for private application and terminal cost.

The Indicative Voting at SMG#24

Since no consensus could be reached an indicative voting took place as agreed (see paragraph 8.2.3.3.2). One hundred and forty-one ETSI members cast their vote as indicated in Table 8.2.2.

Table 8.2.2   The indicative voting at SMG#24

Concept	Solution	Votes	%
Alpha	WCDMA	716	58.45
Beta	OFDMA	0	0
Gamma	WTDMA	0	0
Delta	TD-CDMA	509	41.55

There were 19 abstentions of with a total weighting of 167 votes. According to the ETSI rules the abstentions are not considered for the decision. The necessary majority of 71% of the votes cast was not reached by any concept. The spread of votes clearly showed a polarisation and deep division of the ETSI membership. Therefore I called the SMG#24bis meeting to resolve the issue. The agenda and an update of the voting procedure document[63] was already approved at the end of SMG#24.

8.2.3.4.5 The UTRA Decision at SMG#24bis on 28–29 January 1998 in Paris

The period between SMG#24 and SMG#24bis had seen intensive lobby campaigns by both camps.[64] Six manufacturers of the TD-CDMA camp formed the UMTS Alliance. But the

---

[62] The IPR issue is treated systematically in paragraph 5.

[63] SMG 1157/97.

[64] The heat of the controversy and the firm grouping of supporters justifies to my mind the term "camp".

WCDMA camp convinced more operators using the argument that only WCDMA would allow a global solution.

There were also efforts by several parties to act as mediators to reach a consensus solution without a result. I was also involved in numerous discussions. But it was not possible to come to a consensus. Therefore the SMG#24bis meeting needed to take place. It is interesting to note that nobody requested a postponement. Obviously a decision was expected by the voting at the meeting.

The SMG#24bis meeting attracted an all-time high number of 316 delegates including delegations from North America and Japan as well as the chairpersons of the GSM MoU Association and the UMTS Forum. In terms of the number of delegates, but also the number of weighted votes represented, this was equal to a well attended General Assembly of ETSI.

Concept Enhancements

Both camps presented enhancements to their solutions. The TD-CDMA camp described their relevant results. The WCDMA camp described enhancements of the WCDMA concept by the introduction of key aspects from the TD-CDMA concept in order to improve the capabilities for uncoordinated operation. Motorola reviewed this proposal and objected to several findings.

Position Papers

BT supported the selection of the Alpha concept reasoning it to have the highest potential to become a global standard. BT also proposed that SMG should transfer the issue to the ETSI General Assembly in March in case SMG could not find a solution. Several speakers stated that this would also not help since SMG#24bis had already the voting participation of a General Assembly.

A document with 16 GSM operators in Asia-Pacific supported WCDMA reasoning it would offer the potential of a common standard with Japan. It was commented that these were 16 out of 84 GSM operators in the region and therefore a minority.

Several operators pleaded for a delay of 8 weeks in order to give sufficient time to the ongoing efforts to find a compromise solution and to resolve IPR issues. This proposal did not find the consensus of the meeting.

Alcatel stressed the advantages of TD-CDMA for cordless applications. NTT DoCoMo presented their view that WCDMA is best suited for mobile multimedia services. Motorola stressed the advantages of TD-CDMA for (dual-mode) terminals.

Weighted Voting at SMG#24bis in Paris

Since no consensus was reached a weighted voting took place[65] with the following results: 198 ETSI members voted. One thousand, five hundred and twenty-three weighted votes expressed an opinion (i.e. they did not abstain) (Table 8.2.3).

One hundred and sixty-three votes abstained. This result mean that no decision was reached since the necessary 71% majority was not achieved.

Building of the Consensus between Major Manufacturers Outside the Plenary

The meeting was adjourned in the late afternoon of 28 January until the next morning in order provide time for discussions between interested parties. The major companies involved had their SMG delegations and one or several high level executives in Paris. The companies

---

[65] SMG 1157/97: agreed voting procedure.

Table 8.2.3   Weighted voting at SMG#24bis

Concept	Solution	Votes	%
Alpha	WCDMA	931	61.1
Beta	OFDMA	0	0.0
Gamma	WTDMA	3	0.2
Delta	TD-CDMA	589	38.7

resided in hotels in the neighbourhood of the meeting place. They had rented meeting rooms and worked there like control centres of armed forces in a battle. I had met most senior representatives before the SMG meeting or at coffee breaks. So I was aware that they closely followed the proceedings.

Since the share of votes cast was nearly unchanged between the two meetings (58.5 and 61.1%) and since this was far from the necessary 71%, both camps came to the conclusion that a solution could not be reached by another voting, since there was no possibility of bringing in more voters or convincing supporters of the other side. However, the vote was not unnecessary because its result did set the starting points for final negotiations.

Such a situation can easily lead to a blocking or a falling apart of a whole standardisation program. But the GSM community shares large successes and saw a huge market ahead. Therefore, soon after the voting both Alfa and Delta groups met in their own meeting rooms to prepare compromise proposals to the competing camp. After the Alfa group had drafted their proposal on a flip-chart in a bullet-point format they called to Dr. Konhäuser of Siemens and asked him to come for discussions. Dr. Konhäuser said that they also had a proposal and that the whole Delta group wanted to join in the discussions. When both groups met at 8 p.m. the discussion was based on the bullet-point presentation prepared by the Alfa group. Some small improvements were made and the agreement was ready in less than 30 minutes. The agreement identified a way forward by merging WCDMA as an FDD solution with the TDD part of TD-CDMA. This removed a weakness of WCDMA with regard to UMTS require-ments and integrated a key module of the competing TD-CDMA solution. Due to the extra-ordinary importance of the issue some high level executives like Mr. Å. Persson and Dr. Uddenfeldt of Ericsson, Dr. Neuvo of Nokia and Dr. Konhäuser of Siemens were present to confirm that the agreement also reflected the true position of these key companies.

I was invited to participate in these talks at 10 p.m. when the agreement in principle had been reached and asked to advise on the formulation of the proposal, the presentation and treatment in SMG[66].

Then the key players lobbied for the proposal and found the support of a number of operators: Cellnet, Deutsche Telekom, France Telecom, Mannesmann Mobilfunk, NTT DoCoMo, Telia, Telecom Italia Mobile, T-Mobil and Vodafone. In addition it was supported by the Japanese manufacturers Matsushita (Panasonic), Mitsubishi Electric and NEC.

Building the Consensus at SMG#24bis on 29 January 1998

The consensus proposal was presented to the meeting on 29 January in the morning.[67]

---

[66] Photos of this historic event can be found in the CD ROM folder F.
[67] SMG 29/98.

Since this was a far reaching proposal, first an intensive round of discussions was held with questions and comments for clarifications. Several interventions showed support already. Then the meeting was interrupted for about 2 hours to allow all delegates to discuss the proposal with their colleagues present in Paris or at home.

When the meeting was continued it was clarified that an approval of the consensus proposal meant that it would "set the basis for the future development of UTRA performed by SMG2 together with the already approved SMG documents, which are under change control: high level requirements, procedure for the UTRA definition, UMTS 21.01, UMTS 30.03 (Evolution)".[68]

The technical content of the consensus proposal was presented in a separate document:[69]

"This consensus agreement contains the key elements and advantages of WCDMA and TD/CDMA, and contains the following elements:

1. In the paired band we adopt the radio access technique proposed by the ETSI Alpha group, that is WCDMA (Wideband Code Division Multiple Access), TDoc SMG 903/97.[70]
2. In the unpaired band, we adopted the radio access technique proposed by the ETSI Delta group, that is TD/CDMA (Time Division/Code Division Multiple Access), TDoc SMG 897/97.[71]
3. In the process of selecting the technical parameters the following shall be the objectives:

   – Low cost terminal
   – Harmonisation with GSM
   – FDD/TDD dual mode operation
   – Fit into 2 × 5 MHz spectrum allocation"

This proposal was approved by all ETSI members present.[72] There were no abstentions. NTT DoCoMo stated support for this decision as the basis for the development of IMT-2000.

It was noted that the companies who had proposed the consensus agreement had committed themselves to co-operating to resolve the IPR issues and providing a progress report to SMG#25 in March 1998.

The Relationship with the Press

Since the press followed the activities around the UTRA decision very closely since late 1997, the SMG Co-ordination Group[73] discussed the proceeding and proposed a process, which was endorsed by SMG#24bis at the beginning of the meeting:

- Only ETSI should inform about the results
- A press release would be elaborated by the ETSI press officer and me as SMG chairman and be presented to SMG for approval
- The press will be invited on 29 January at 5 p.m.
- No intern results should be communicated to the press

---

[68] SMG#24bis Meeting Report (CD ROM folder B1), Section 2.5.
[69] SMG 39/98 (also Annex 5 of the Meeting Report) is an extract of the technical proposal from SMG 29/98.
[70] SMG 903/97 copied on the attached CD-ROM in folder B2.
[71] SMG 897/97 copied on the attached CD-ROM.
[72] This included also Qualcomm.
[73] Chairs of all SMG groups and liaison persons to network operators 'and manufacturers' associations.

There was no information or leakage of information to the press during the SMG#24bis meeting. At the end of the meeting a press release was elaborated and endorsed.[74]

A press conference was held immediately after the meeting. I was charged by SMG with informing the press. The UTRA decision generated a lot of attention in the press. It was designated as a timely landmark decision.

8.2.3.4.6 Observations

It is remarkable that, except for France Telecom, no GSM operator presented a contribution or declared a position at SMG#24. There were very few position papers by operators using political arguments only at SMG#24bis. No technical, operational or planning aspects were addressed by other operators other than France Telecom. This was a complete change compared to the situation in the GSM radio decision in 1987.

8.2.3.5 Contributions to the IMT-2000 Programme of the ITU

8.2.3.5.1 Introduction

The ITU had a program to develop world-wide standards for FPLMTS (Future Public Land Mobile Telecommunication System) since the late 1980s. This program was later called IMT-2000 (International Mobile Telecommunications 2000). It was the intention to develop one global system with one radio interface, which would allow global roaming. Spectrum in the 2 GHz band had been designated for this purpose (see Chapter 7, Section 5). But the progress was slow and no way was visible to come to a decision on one radio interface given the great divergence of interest of the participants.

The solution for a way forward for the ITU program was the concept of a FPLMTS/IMT-2000 family of systems (see 8.2.2.3). On the basis of this concept agreed in the ITU Technical Committee SMG was very active in providing contributions to the ITU.

8.2.3.5.2 Urgency to Contribute, Work Program, Co-ordinators

Nokia, Siemens, Ericsson and Alcatel stressed the need for a sufficient priority for ITU contributions at SMG#22 in June 1997.[75] The meeting established a work program for contributions to ITU-R and ITU-T[76] and appointed two full-time ITU co-ordinators.[77]

8.2.3.5.3 SMG2 as ITU Radio Transmission Technology Evaluation Group

There was also an agreement in principle at SMG#22 to offer to ITU SMG2 as an ITU RTT (Radio Transmission Technology) evaluation group. Due to lack of resources SMG2 would however focus their work on the evaluation of the RTT candidate proposed by SMG.[78]

---

[74] Press release at SMG#24bis Meeting Report (CD ROM folder B1), Annex 4.
[75] SMG 519/97.
[76] SMG 570/97.
[77] Makis Kokos and David Williams, sponsored by Nokia resp. Lucent.
[78] SMG 753/97.

8.2.3.5.4 The Approval of Contributions to the ITU September 1997 Meetings

To discuss and approve the contributions to the ITU-T and ITU-R meetings in September it became necessary to call an extraordinary SMG Plenary SMG#22bis on 18 August in London. It was attended by around 50 delegates. The chairman of the SPS ITU co-ordination group participated at the meeting.

The meeting studied 33 potential input papers to the ITU-T SG11 WP3 meeting in September 1997 and to the ITU-R TG8/1 meeting and workshop in September 1997. It compared the documents with agreed SMG positions and identified 12 papers to fully present the agreed positions. The meeting approved these papers to be presented to the mentioned ITU meetings by supporting and originating companies with the indication of these companies as source on behalf of SMG. The main focus of the endorsed papers is

- Reporting on the UTRA definition process; and
- Contributions to the IMT-2000 family of systems concept

The meeting also identified that whereas all studied European and American contributions featured the IMT-2000 family of systems concept, several papers contained

- Controversial material;
- Contradictions to agreed ETSI/SMG positions; and
- Technical proposals at a much higher level of detail than the status of work in SMG, where SMG will possibly or likely come to different conclusions during its future work.

It became clear that a closer link of all direct ITU inputs of the different companies to ETSI work was needed, in order to secure the commonality between IMT-2000 and UMTS.

8.2.3.5.5 The Approval of the Contributions to the ITU December 1997 Meetings

The co-ordinators reported that the IMT-2000 family concept was anchored in the ITU-R recommendations.[79] The contributions to the December 1997 meetings of ITU-R and T could be approved during SMG#23 in October 1997.

The following SMG plenaries approved regularly ITU contributions mainly regarding the IMT-2000 family concepts both in ITU-R and T. The contributions in the next work phase were mainly related to UTRA (see paragraph 8.2.4.4.1).

8.2.3.5.6 IMT-2000 Results in the ITU

The ITU developed framework standards for IMT-2000, which were applicable globally. "A key objective of the ITU efforts to promote 3G standards is to allow for the incorporation of a wide variety of systems while also promoting a high degree of commonalty of design beneficial to fostering world-wide use and achieving economies of scale. Within this framework the ITU has defined five radio interface standards for the terrestrial component of IMT-2000. These radio interfaces have been designed to provide compatibility with existing services and also to provide broadband services at high data rates, up to 2 Mbps. It is envisioned that these data rates will permit operators to provide a wide range of services...".[80]

---

[79] SMG 833/97.

[80] Text quoted from Federal Communication Commission's (US) Notice of Proposed Rule Making FCC 00-455, no. 17, the consultation document for the US 3G spectrum hearing process.

ETSI SMG submitted UTRA FDD and UTRA TDD. ETSI Project DECT submitted DECT. All three were accepted and endorsed as IMT-2000 family members.

8.2.3.6 Co-operation with ARIB and TTC in Japan

8.2.3.6.1 The Exploratory Mission of May 1997 showed a Strong Japanese Interest in co-operating with ETSI

The exploratory mission as mandated by SMG#21 in February 1997 had been performed by the SMG chairman supported by the PT SMG co-ordinator. It was carried out together with a mission of the UMTS Forum's chairman in May 1997.[81] Very open and constructive discussions were held with the Ministry of Post and Telecommunication, the standardisation organisations ARIB and TTC and NTT DoCoMo, the leading mobile network operator in Japan. There was a strong interest on the Japanese side to co-operate with the UMTS Forum and ETSI SMG.[82] The next steps in the dialogue were agreed: radio workshop in Europe, experts meeting on network aspects, information exchange on spectrum issues and a next "general meeting" in August.[83] The ongoing activities were covered by an extended mandate approved by SMG.[84]

8.2.3.6.2 The First Co-operation Steps Lead to Firm Agreements and a Close Co-operation

A joint radio workshop was held by SMG2 and ARIB on 30 June and 1 July 1997. The experts meeting between TTC and SMG on network aspects took place on 22–23 July. The general meeting between the two sides[85] took place in London on 25–26 August 1997.[86] As a result a proposal for an "Agreement on Common Working Procedures between ARIB and SMG"[87] was agreed. This foresaw mutual participation in the relevant ARIB and SMG groups as active observers with the right to speak and submit documents.

Furthermore, it was agreed to organise multilateral meetings between ARIB, SMG and relevant North American standards organisations starting in February 1998 (i.e. after the ETSI decision on UTRA). This process was endorsed by ARIB in October and by SMG#23 in October 1997 and the ETSI Board.

Regarding network aspects it was agreed to continue the "SMG-TTC network aspects working group" in order to identify modules which would ease global roaming, study a universal A-interface, examine the potential of the SIM. The proposal[88] was approved by SMG and TTC.

[81] No other SMG official joined me. The whole issue was seen as a ""hot potato". It attracted heavy attacks by some ETSI delegates, who did not participate in SMG, during the summer and autumn.

[82] It was visible during the talks that European companies with business in Japan had prepared the ground. Otherwise the rapid progress would not have been possible. Gunnar Sandegren informed SMG#23 in June 1997 that the FRAMES project and NTT DoCoMo had agreed to harmonise their W-CDMA proposals for Japan and SMG.

[83] SMG 398/97 contains the Executive Summary of the report. The full report is contained on the CD-ROM in B2 as file "Exploratory Mission Japan 97".

[84] SMG 571/97.

[85] The "Japanese side" formed by ARIB and TTC delegates and the "UMTS side" formed by delegates from SMG, UMTS Forum, ECTEL TMS and GSM MoU Association.

[86] SMG 757/97 Meeting Report.

[87] SMG 757/97 Annex 2.

[88] SMG 757/97 Annex 3.

In practical terms this allowed NTT DoCoMo to participate in SMG and SMG2 meetings and European companies to participate in ARIB meetings, both as active observers.

8.2.4 The Production of the UMTS Reports and Raw Specifications in ETSI from February 1998 to February 1999 (from SMG#25 to SMG#28)

8.2.4.1 Strategy Evolution

8.2.4.1.1 DECT Evolution as an IMT-2000 Candidate

ETSI Project DECT[89] had the intention of submitting a solution based on the DECT evolution as an IMT-2000 candidate. DECT has had a remarkable succesds based in the world market. It is the leading cordless technology. The interested companies wanted to ensure a long-term evolution by integrating 3G features into DECT. In order to achieve this, the recognition of the ITU as an IMT-2000 family member, the ongoing access to the existing spectrum and get access to a part of the 3G unpaired spectrum was needed. Since SMG held the prime responsibility in ETSI with regard to contributions on radio aspects to the ITU for IMT-2000, ETSI Project DECT needed to involve SMG.

In the discussions before SMG#26 (June 1998) I sensed substantial opposition in SMG to present another ETSI IMT-2000 candidate to the ITU, especially after the difficult agreement on one UTRA candidate.

On the other side I learnt from Günter Kleindl, the ETSI Project DECT chairman, that they intended to address the cordless/private market with a TDD solution and that they needed a way forward and an evolution for DECT, which had gained a substantial acceptance in the world market.

The DECT chairman wrote a letter to SMG.[90] At my proposal SMG#26 (June 1998) noted the intention of ETSI Project DECT to propose DECT as an IMT-2000 family member. It was clarified, that DECT had not been submitted to the UTRA definition procedure and could therefore not be a part of UMTS, but needed to be an independent IMT-2000 family member. This common understanding was also confirmed by an additional contribution from ETSI Project DECT.[91] SMG sent a letter to ETSI Project DECT confirming the common understanding and endorsing the plans of ETSI Project DECT.[92]

This was a fine example of the peace making potential of the IMT-2000 family of systems concept.

8.2.4.1.2 The Agreement in Principle on the Default Speech Coding Algorithm for UMTS

The subgroup SMG11 proposed to use the AMR speech coding algorithm as the default solution in UMTS. SMG#26 (June 1998) saw AMR as a good candidate and agreed to support it as a working assumption. More information was requested for a firm decision. It was agreed that work with ARIB (Japan) on the default speed coding algorithm should proceed.

---

[89] Digital Enhanced Cordless Telecommunication.
[90] SMG 374/98.
[91] SMG 296/98.
[92] SMG 509/98.

8.2.4.1.3 Use of SIM and Smart Cards in UMTS

The subgroup SMG9 proposed a smart cards strategy to SMG#27 (October 1998).[93] Derived from the requirement that UMTS evolves from GSM it was requested that UMTS terminals support existing GSM SIMs. Furthermore, it was confirmed that the UMTS Integrated Circuit Card (UICC) is a mandatory element of UMTS. This was endorsed by SMG.

The existing GSM SIM technology was confirmed as the basis of the UMTS SIM technology. This includes also the SIM tool kit and the associated security mechanisms. It was estimated that the necessary modifications for UMTS would not be difficult.

A revised smart cards strategy paper was approved by SMG#28 in February 1999.[94]

Security review

All security issues were reviewed and problems were listed by SMG10. This list became the basis for security requirements and lead to a complete review of the existing report on UMTS security (UMTS 33.20). This document was endorsed in SMG#25 in March 1997.

8.2.4.2 The Services Work for UMTS Release 99

The debate on the precise content of UMTS Release 99 started at SMG#25 in March 1998 and continued at the following SMG#26 plenary in June 1998. A first draft of the specification UMTS 22.00 "UMTS phase 1" was presented for information and discussion to SMG#27 (October 1998).[95] SMG#27 approved the basic lines and main direction in principle. SMG#28 in February 1999 approved the final version.[96] SMG1 had also updated all work items in order to support the definition of the content of UMTS phase 1.[97]

The two reports on Advanced Addressing and Automatic Establishment of Roaming Relations were approved by SMG#27 in October 1998.

The specification UMTS 22.15 "Charging and Billing" was completed by the end of 1998 and approved by SMG#28 in February 1999.

After SMG#28 in February 1999 all UMTS services work was transferred to 3GPP.

8.2.4.3 UMTS System Architecture Work

8.2.4.3.1 UMTS System and Network Aspects at SMG#26 in June 1998

SMG12 had joint meetings with other subgroups and a meeting with officials of Technical Committee NA responsible for the ISDN core network evolution for UMTS, and discussions with TTC, the Japanese standardisation organisation for network aspects.

8.2.4.3.2 UMTS System and Network Aspects at SMG#27 in October 1998

Finally a first stable draft of UMTS 23.01 "General UMTS Architecture" was presented to SMG#27 with the intention of getting it approved in the next meeting. Also the specification

---

[93] SMG 624/98 (revised version of the original input in SMG 668/98).
[94] SMG 178/99, Annex B.
[95] SMG 555/98.
[96] SMG 38/99.
[97] SMG 681/98.

UMTS 23.10 "UMTS Access Stratum: Services and Functions" as well as UMTS 23.20 "Evolution of the GSM Platform towards UMTS" were presented for information.

8.2.4.3.3 UMTS System and Network Aspects at SMG#28 in February 1999

In addition the specification UMTS 23.30 "Iu Principles" was presented for information.

8.2.4.3.4 Transfer of the Work to 3GPP

Th work items and specifications relevant to UMTS system architecture were transferred to 3GPP in the beginning of 1999. The work relevant to GSM and UMTS was transferred in mid-1999. Much work was left to 3GPP, since the progress in SMG was not overwhelming.

8.2.4.4 The UMTS Radio Work from the UTRA Decision in January 1998 to the End of 1998

8.2.4.4.1 The Contributions to the ITU and the SMG Specification Work

SMG2 finalised the description of UTRA as a Radio Transmission Technology (RTT) proposal for IMT-2000 in June 1998. It was unanimously approved by SMG#26 (in June 1998). SMG declared their willingness to participate actively in the future ITU consensus building process.[98] The evaluation report needed for the ITU was completed in September 1998.

During the 1998 the selected UTRA concept was refined and the FDD and TDD modes were harmonised. The names used were no longer WCDMA and TD-CDMA, but instead UTRA FDD and UTRA TDD. This lead to detailed descriptions of the UMTS Terrestrial Radio Access Network (UTRAN) including the mobile station. This included all radio protocols terminated in UTRAN and the interfaces as well as the description of the functionality requirements of the network nodes and the terminal. The results were endorsed by SMG#28 in February 1999 and transferred to 3GPP as the basis to write actual technical specifications to be completed by the end of 1999.

8.2.4.4.2 An Initiative Towards Harmonisation at SMG#27 in October 1998

Qualcomm repeated their proposal from June 1998 to create a Convergence Working Group to work on the convergence of the W-CDMA and cdma2000 radio access standards.[99] In the attachments a letter to a US Senator was copied for information and the IPR position was repeated saying that Qualcomm would license their IPR only for standards meeting a set of technical criteria based on three "fairness" principles:

- A single converged world-wide CDMA standard should be selected;
- The converged standard must accommodate IS41 and GSM core networks equally;
- Disputes on technological issues should be resolved on demonstrably better performance, features, cost; or else on compatibility with existing technology.[100]

---

[98] The three actions proposed at the end of SMG 488/98 were endorsed.
[99] SMG 564/98.
[100] i.e. ANSI95, formerly named IS95.

In the discussions many members strongly commented in favour of convergence, in particular to resolve outstanding IPR issues. Different views were expressed regarding the right forum: ITU-R, SMG2, outside ETSI.

Joint Motorola and Nokia results were presented concluding that for dual-mode terminal implementations, complexity due to two chip rates, etc. were insignificant. The real issues were the added complexity and cost in the radio part due to different frequency allocations and bandwidths. Based on these results new topics for harmonisation were proposed, that would have greater impact than issues discussed earlier like chip rate, etc.

ARIB (Japan) reported on their efforts and achievements to harmonise the two CDMA solutions.[101]

It was also stated that already a high degree of harmonisation had been achieved and that SMG is not the right forum for the negotiation of IPR issues.

Niels P. S. Andersen, the SMG2 chairman commented that from the discussion it was clear that SMG wanted to continue harmonisation and convergence of radio transmission technologies and that the detailed work since June had not shown the optimum way forward. He proposed to allocate at the next SMG2 plenary an agenda item and sufficient time to deal with the issue in a top down approach. He stressed that this would be a contribution driven process.

This proposal was endorsed by SMG. It was accepted by Qualcomm under certain caveats, which were agreed by SMG.

8.2.4.4.3 The Initiative of the Operator's Harmonisation Group

During the course of 1998, some major mobile operators (Airtouch, Bell Atlantic, Bell Mobility, BT, France Télécom, NTT DoCoMo, Sprint PCS, TIM, T-Mobil, Vodafone) shared concerns about the still unresolved IPR situation for W-CDMA and the potential threats to the entire mobile community in case they weren't cleared. Some informal talks on how to overcome all this were started. Quite naturally, it took some time to establish an entirely open discussion between members of different camps, but in a series of meetings (called *Operators Harmonisation Group (OHG)* meetings,) resulting from these contacts, mainly in Beijing (14/15 January 1999) and in London (22/23 March 1999), a framework for a world-wide harmonised CDMA standard was achieved.[102] The main goal was to agree on unique basic parameters for each of the three CDMA modes:

- the *Direct Spread* (DS) mode, to be standardised by 3GPP;
- the *Multi-Carrier* (MC) mode, to be standardised by 3GPP2; and
- the *Time Division Duplex* (TDD) mode, to be standardised by 3GPP.

In the 25/26 May 1999 meeting at Toronto, the agreement was joined by the major manufacturers. In addition to the basic parameters of the air-interface modes, the agreement paved the way for all Access Network/Core Network interoperability scenarios

- Multi-carrier and evolved TIA/EIA-41;
- Multi-carrier and evolved GSM MAP;
- Direct spread and evolved TIA/EIA-41;
- Direct spread and evolved GSM MAP.

[101] SMG 641/98.

[102] Main elements of these results were integrated immediately in the SMG2 specifications (see e.g. SMG2 status report of February 1999.

by mutually including suitable "hooks" and "extensions" in both the 3GPP and 3GPP2 specifications.

8.2.5 Intellectual Property Right (IPR) Issues

8.2.5.1 The IPR Situation in ETSI and in ETSI Technical Committee SMG

ETSI established an IPR policy.[103] This provided a clear framework for the activities of Technical Committee SMG. ETSI members have obligations to inform ETSI about essential IPR they hold in a standardisation work item. I reminded all SMG participants regularly to fulfil their duties (e.g. at SMG#23 and #24) by quoting the ETSI Directive: "Each member shall use its reasonable endeavours to timely inform ETSI of essential IPRs it becomes aware of. In particular, a member submitting a technical proposal for a standard shall, on a bona fide basis, draw the attention of ETSI to any of that member's IPR which might be essential if that proposal is adopted." The ETSI legal adviser produced a document on a regular basis containing the information received from members.

An ETSI Technical Committee has no leverage to resolve IPR issues. It can contribute to ease a solution by organising an open system specification process, where every interested company can register IPR so that they get negotiation power in licensing negotiations. This has been done by the whole UMTS specification process. A special feature of the UTRA definition process was, that every company was able to contribute to every concept and to secure IPR which could be used in cross-licensing agreements.

8.2.5.2 The Lack of Declarations to ETSI and the Lack of an Initiative to Deal with the Issue (IPR at SMG#24 in December 1997)

A "Report on Essential IPRs Declared in Relation to the Work of SMG#24" was given by ETSI's legal advisor to SMG#24.[104] She informed the meeting that the ETSI Secretariat had been given on a confidential basis a list of companies possibly having UTRA relevant IPR.

I reminded the ETSI members present to fulfil their duties with regard to the ETSI IPR policy to inform ETSI of any essential IPR they possess in relation to UMTS.

For several UTRA concepts no IPR had been declared. It was noted that the list presented "is far from being complete". Concerns were expressed by some network operators.[105] It was recognised by the meeting that the concerns expressed by the network operator cannot be resolved by the ETSI rules. As chairman I remembered at the meeting that the ice breaking activity on IPR matters for GSM had been conducted by the GSM MoU Group in the late 1980s. But no reaction occurred to this.

8.2.5.3 The Emerging Will to Co-operate on a Solution in Order to Stabilise the Reached UTRA Decision (IPR at SMG#24bis in January 1998)

As chairman I reminded the delegates to fulfil the duties of ETSI members to inform ETSI

---

[103] See ETSI Rules of Procedure, Annex 6 (available on the ETSI server). The background is explained in the book by Stephen Temple, ETSI, a Revolution in European Telecommunication Standards Making, 1991, p. 44.

[104] SMG 1066/97.

[105] SMG 1061/97 statement of the three German operators.

about essential IPR they hold in respect of UMTS. About 60 companies were estimated to hold IPR on UTRA, but only 20 had informed ETSI.

The chairperson of the GSM MoU Group declared that due to the uncertainty about the situation "IPR questions are not considered in GSM MoU as a decisive point at this time".[106]

Motorola stated that the IPR situation was strongly in favour of TD-CDMA. Major manufacturers of the WCDMA camp informed the meeting of their activities to resolve IPR issues. Qualcomm claimed to have a strong IPR position in WCDMA and TD-CDMA, but needed to investigate this and to prepare an IPR position. Siemens answered that their investigations had not identified any essential IPR of Qualcomm relevant for the Delta concept (TD-CDMA).

The European Commission presented a "Declaration on UMTS"[107] declaring that the Commission intends to combat misuses of a dominant UMTS IPR holder by means of the competition law, should they happen.

The meeting "concluded that, on the IPR question, ETSI/SMG cannot go further than the ETSI IPR policy defines; further IPR arrangements have to be found outside of SMG, e.g. between the big commercial players".[108]

Later in the meeting a commitment of 23 major manufacturers and network operators was given to co-operate on reasonable guidelines for IPR problems and to report progress at the next meeting.[109]

8.2.5.4 The Creation of the UMTS IPR Working Group and the Announcement of Qualcomm's IPR Position (IPR at SMG#25 in March 1998)

A large number of manufacturers and network operators informed the meeting that they had created an IPR Working Group.[110] This lead later to the creation of the UMTS Intellectual Property Association (UIPA).

Qualcomm stated their position[111] that they would be willing to enter into licensing arrangements provided that convergence would be achieved between the Wideband (CDMA) proposals that have been developed in the US, Japan and Europe. The convergence should be achieved in such a way that operators who so choose can deploy CDMA radio access in an evolutionary manner (i.e. the IS95 operators).

8.2.5.5 Qualcomm's Request to Change Five Points in UTRA and SMG's Response (IPR at SMG#26 in June 1998)

Qualcomm had denied to grant licenses for essential IPR they own for WCDMA and confirmed their request to change WCDMA before SMG#26.[112] In addition Qualcomm had launched a campaign in the US Government and Parliament. Even though ETSI was

---

[106] SMG#24bis report (CD ROM folder B1), p. 3.
[107] SMG 22/98. The European Commission is a counsellor in ETSI and has the right to participate in ETSI meetings as an observer with the right to speak and submit contributions.
[108] SMG#26bis Meeting Report (CD ROM folder B1), end of Section 2.2.
[109] SMG#24bis Meeting Report (CD ROM folder B1), Annex 6.
[110] SMG 241/98.
[111] SMG 240/98.
[112] SMG 493/98.

accused heavily, ETSI was not heard. Therefore an open letter was produced responding to several accusations.[113]

At the request of Qualcomm I offered an agenda item to present their position at SMG#26.[114] They requested to change five points in UTRA.[115] A presentation and a question and answer session took place. The first conclusions of it were:

- The proposal for a single chip rate was transferred to SMG2 for a technical analysis. Qualcomm was invited to provide a proper technical documentation of the resulting improvements.
- Regarding the openness for other core networks it was commented that this is more an IMT-2000 question to be treated in the ITU.
- Regarding the request for synchronised base stations it was clarified that UTRA allows both synchronised and non-synchronised operation.
- Regarding the optimal speech codec Qualcomm was invited to participate in the work of SMG11.
- The request on emissions in the audio spectrum was already resolved.

It should also be noted that the North American GSM Alliance objected to several Qualcomm proposals to change WCDMA. The Alliance claimed that the changes would reduce system capacity and limit system capabilities and performance.[116]

SMG declared their general openness to contribute to the further harmonisation of the IMT-2000 CDMA proposals.

8.2.5.6 The Re-enforcement of Qualcomm's IPR Position at SMG#27 in October 1998

A statement of Qualcomm was presented which re-enforced their IPR position.[117] They pointed to a proposal for a resolution.

The UMTS IPR Working Group has identified three alternative approaches for the resolution of IPR issues related to third generation.[118] Comments from the industry were expected in December 1998.

8.2.6 The Global Co-operation was Intensified and Completely Restructured to Secure the *Integrity and Consistency of GSM and UMTS Specifications*

8.2.6.1 Agreement on the Cornerstones of UMTS between ETSI, ARIB and TTC (Japan) as well as ANSI T1P1 (US)

During discussions in bilateral contacts with ARIB/TTC and ANSI T1P1 it became visible at the end of 1997, that a common UMTS concept could become possible between ETSI territory and the leading grouping in Japan and the GSM community in North America.

---

[113] SMG 446/98.
[114] SMG 471/98.
[115] SMG 482/98 in combination with 471/98.
[116] SMG 450/98.
[117] SMG 601/98.
[118] SMG 608/98.

This common concept would be based on UMTS service innovation, UTRA and the GSM core network evolution. This result had been prepared and enabled by a network of interested companies active at the global level.

8.2.6.2 *The Efficient Global Open Organisation of the UMTS and GSM Specification Work in the Third Generation Partnership Project (3GPP)*

8.2.6.2.1 Problem Situation

The implementation of the agreement on the UMTS cornerstones within the existing organisations would have been unmanageable. Three committees in different continents would have worked on the UMTS radio specifications (SMG2, T1P1.5 radio sub-working group, ARIB), several other committees would have worked on network aspects (SMG12 and SMG3, T1P1.5 network sub-working group, TTC). The situation in other key areas like services, SIM, O&M would have been comparably difficult. There would have been no overall decision-making body for conflict resolution. Therefore, the global strategic agreement on the UMTS cornerstones called for a new more efficient global organisational solution, in order to lead the agreement on the cornerstones to a complete and consistent UMTS system specification available in time for the market.

8.2.6.2.2 Proposal to Initiate a Partnership Project for GSM and UMTS

In order to secure the integrity of GSM and UMTS, the cohesion between GSM and UMTS, the ongoing cross-fertilisation between UMTS and GSM and an efficient specification work,

I proposed in the fourth quarter of 1997 to create an ETSI Partnership Project for UMTS and GSM to several network operators and manufacturers. This would provide a single lean working structure and would be open to all committed parties world-wide. This Partnership Project model had been developed in the ETSI reform in 1996, but had never been used.

This Partnership Project for GSM and UMTS was proposed by several GSM network operators at SMG#24 in December 1997.[119] The document "Future Organisation for GSM and UMTS Standardisation", source T-Mobil, Mannesmann Mobilfunk, E-Plus Mobilfunk, aimed at a smooth and efficient standardisation process for the evolution of GSM and towards UMTS. The GSM community is now a global community of operators and manufacturers but has experienced difficulties in opening up for a wider participation in ETSI/SMG. Organisations from outside Europe still cannot become full ETSI members. Even voting rights for associate members in Technical Bodies were not endorsed by ETSI's General Assembly in November 1997. Present working methods with ANSI T1P1 on common GSM specifications are proven as a best possible solution for co-operation with other standard bodies, but they are very complex. This situation calls for a closer and more efficient overall co-operation. The GSM MoU Association and especially the Asian Pacific Interest Group (APIG) of GSM MoU have expressed their desire to participate fully in GSM work and in third generation standardisation and to ensure roaming with Japan.

For these reasons, these three companies proposed the establishment of SMG as the joint working structure among the interested bodies to produce GSM and UMTS standards for ETSI (as an ETSI partnership project) and for the other interested bodies, avoiding parallel work and overhead co-ordination; current budget allocations for ETSI/SMG to be considered

---

[119] SMG P-97-1062 submitted by T-Mobil, Mannesmann and Eplus.

as an asset for this possible joint working structure; the SMG chairman to carry out an exploratory mission in this sense."[120]

8.2.6.2.3 Endorsement of the Partnership Project by Technical Committee SMG and Mandate for Exploratory Missions

After an intensive discussion and some revisions of the document it was approved by SMG#24 in December 1997.[121] This included a mandate for exploratory missions to Japan and the US. This mission mandate was endorsed by the chairman of the ETSI General Assembly and Board as well as the ETSI director general in a meeting on 13 January 1997 in Sophia Antipolis.

I gave a first progress report to SMG#24bis in January 1999.[122] Then I led – as mandated by SMG – an exploratory mission to Japan on 3–11 February 1998. We found a strong interest in such an intensified co-operation[123]. The following summary was agreed in the meeting on 5-6 February 1998 between ETSI SMG, UMTS Forum, GSMA, ARIB, TTC and ANSI T1P1:

> "1. There is interest to create common specifications for IMT -2000 in the areas of terminals, radio access networks and core networks
>
> 2. It was recognised that the development of these standards in parallel organisations would be slow and could lead to unnecessary differences
>
> 3. It was agreed that the best way to proceed would be to further explore the creation of a common working structure (a "Project") to produce common specifications.
>
> 4. Such a Project, built on agreed common interests, would need appropriate recognition by as well as relationship / membership to the standardisation bodies.
>
> 5. A procedure for a transition phase into the full implementation of the Project would need to be worked out."[124]

This was the basic agreement for the creation of the Third Generation Partnership Project.

8.2.6.2.4 Lead Taken by the ETSI Board

Creation of the UMTS Globalisation Group

Due to the fundamental importance the creation of such a Partnership Project had for ETSI, the ETSI Board created the UMTS Globalisation Group with a strong SMG participation, who undertook the negotiations for the implementation (see Chapter 9, Section 1). This group was chaired by Karl Heinz Rosenbrock, the ETSI director general.

The interest of the GSM and UMTS community was especially actively supported by three members of the ETSI Board in the UMTS Globalisation Group and in ETSI Board meetings: Wolf Haas of Mannesmann Mobilfunk, Kari Lang of Nokia and Tom Lindström of Ericsson. Their contribution was decisive for the ultimate success, the creation of 3GPP.

Great help for the ultimate resolution came also from the UMTS Forum and the chairman

---

[120] Extract from the SMG#24 Meeting Report (CD ROM folder B1), Section 3.3

[121] SMG P-97-1154.

[122] SMG P-98-0009.

[123] Report on the Japan mission in SMG P-98-0112.

[124] This summary is contained in SMG P-98-0112 as appendix B13. It was also presented to SMG and the ETSI Board in February 1998.

Thomas Beijer. His diplomacy, bridge building ability and the ability not to give up strategic targets was an indispensable key to the ultimate success.

SMG was represented in the UMTS Globalisation Group by Gunnar Sandegren (SMG vice chairman), Francois Grassot (ECTEL TMS chairman) and me as SMG chairman.

Factions in the UMTS Globalisation Group

There was a very strong polarisation between three factions in the UMTS Globalisation Group. They were not officially organised. Therefore, I will name and describe them briefly:

"Greater ETSI faction": they wanted an even stronger and greater role for ETSI in the future in UMTS. They feared a big loss, if SMG the greatest producer of deliverables in ETSI were to "emigrate" to the Partnership Project. They tried to bring non-European partners into ETSI. But after the decision of the ETSI General Assembly to grant these parties only associate membership without voting rights, this was not attractive to the non-Europeans. In the UMTS Globalisation Group they tried in the beginning to block the Partnership Project and later to minimise the scope of work to be transferred to the Partnership Project.

"Fixed-mobile convergence faction": this community came from a fixed network background. They hoped to reach fixed-mobile convergence by bringing the relatively independent GSM/UMTS work into an organisation which would be created by melting the existing fixed network committees with SMG under the leadership of the fixed side. Their UMTS vision was dominated by fixed-mobile convergence as a high priority. They lacked in their groups dealing with the fixed network evolution towards third generation a sufficient support and momentum. They wanted to keep GSM and UMTS in ETSI in order to reach their targets and to exploit the SMG momentum and know-how. They tried in the beginning to block the creation of the Partnership Project. Later they tried to limit the scope of the transferred work as much as possible.

"GSM-UMTS faction": due to the global acceptance of GSM and the UMTS cornerstones they needed an efficient globally open work structure which dealt with all system aspects. Their prime concern was the progress of their GSM-based UMTS vision. This was complemented with work on the "mobile-fixed convergence". They saw that the number of mobile users would very soon be much bigger than the number of fixed users. They wanted to transfer all GSM and UMTS work into the Partnership Project.

The Decision of the ETSI General Assembly Which Freed the Way for 3GPP

Due to the strong polarisation in ETSI a decision of the General Assembly on principles was needed in September 1998. During this General Assembly I was charged with negotiating a compromise, which was acceptable to the whole ETSI membership. The compromise proposal foresaw:

- to create 3GPP for an initial phase of UMTS;
- to keep GSM in ETSI; and
- to create the ETSI Project UMTS for long-term UMTS aspects.

This proposal was endorsed with a very high majority. It cleared the way to the 3GPP agreement signed in December 1998.

The Success of 3GPP

The first 3GPP Technical Meeting in December 1998 attracted 350 delegates and the level of participation and contributions remained high. 3GPP produced a common set of Technical Specifications for UMTS based on service innovation, UTRA and the GSM core network evolution. The work was started in December 1998. The Technical Specifications of UMTS Release 99 were completed in December 1999. Some smaller issues were resolved by March 2000. For this purpose all pure UMTS work was transferred from SMG to 3GPP during the first quarter of 1999. The responsibility for the common GSM and UMTS specifications was transferred in the third quarter of 1999. 3GPP was supported by a large number of SMG contributors and SMG leaders. The full-time program managers of the SMG technical support were made available to 3GPP. All proven SMG working methods were made available to 3GPP.

In autumn 1999 ANSI T1P1 and TIA with UWCC proposed the transfer of the remaining GSM work (mainly EDGE, SIM and mobile station testing) to 3GPP in order to ensure the cohesion between the classic GSM and UMTS. The ETSI Board endorsed this proposal based on a review and recommendation of SMG. A Board ad-hoc group was installed with SMG representation. The negotiations between the partners led to an acceptance in principle in May 2000. Therefore, the remaining GSM activities were transferred to 3GPP in mid-2000. ETSI Project UMTS attracted 30–50 delegates and did not have the momentum to produce UMTS specifications. It was closed in 2000.

So finally the SMG vision of 3GPP was realised. The creation of 3GPP ensures the integrity of GSM and UMTS, the cohesion between GSM and UMTS and the cross-fertilisation of GSM and UMTS. 3GPP allows all interested and committed organisations, e.g. regulators, network operators and manufacturers world-wide to participate in the work with equal rights.

8.2.7 Complementary Work to 3GPP in ETSI

8.2.7.1 The Transposition of 3GPP Documents in ETSI Documents

After the creation of 3GPP the question arose, how to "transpose" the 3GPP documents into ETSI documents and whether there is a need for additional documents. 3GPP elaborates and approves common Technical Specifications and Technical Reports, which should be transposed into ETSI documents. I developed the following concept, which was endorsed by SMG and the ETSI Board.[125]

3GPP is acknowledged by the ETSI internal rules as an ETSI Technical Body. Therefore Technical Specifications and Reports approved by 3GPP are to be recognised directly as ETSI Technical Specifications and Reports without another "ETSI internal approval". They can be published directly by the ETSI Secretariat.

Besides Technical Specifications and Reports there are in ETSI European Standards (ENs). They are approved by an ETSI Technical Body and then in a second step by the whole ETSI membership with the assistance of the National Standardisation Organisations.

A broad demand survey[126] regarding the demand for ENs in autumn 1999 showed that a demand for ENs exists for the purposes of the R&TTE-Directive only (access of terminals to the market). All other demand can be covered by Technical Specifications.

These ENs should be elaborated and approved by a "pure" European Committee (e.g.

---

[125] P-99-751.
[126] P-99-736.

Technical Committee SMG). In this process 3GPP results should be referred to as much as possible.

8.2.7.2 The Elaboration of European UMTS/IMT-2000 Harmonised Standards for Terminals Pursuant to the R&TTE Directive

In order to avoid barriers to international trade the European Commission requested ETSI in a letter in December 1999 to produce European harmonised standards pursuant to the R&TTE Directive, which "would typically describe emission masks ensuring proper coexistence of the different members of the IMT-2000 family and that it would be aligned with similar standards outside the Community".

I was charged by SMG with forming a small delegation and to talk to the different parties involved to explore a way forward. A strategic framework and several technical documents were elaborated and endorsed by SMG. The principles were endorsed by the ETSI Board. A joint ERM/SMG Task Force was formed in May 2000 to do the technical work.

The strategic framework document developed by me and endorsed by Technical Committee SMG[127] identifies the regulatory requirements and contains the following key targets for the standardisation work:

- ETSI needs to produce harmonised standards for all IMT-2000 systems.
- The work can reference ITU, 3GPP2 and TIA specifications directly. There is no need to transpose these into ETSI documents.
- The harmonised standards will be produced by a joint ERM/SMG Task Force and EP DECT.
- The first release of the harmonised standard needs to be completed ideally in October 2000.

8.2.7.3 Technical Committee MSG, the New Body for ENs

The work, which remains in ETSI, is the elaboration and approval of ENs needed for regulatory purposes. For this task I proposed to create a new body Technical Committee MSG (Mobile Standards Group) and its terms of reference. This was endorsed by SMG[128] and approved by the ETSI Board. It started in June 2000.

8.2.7.3.1 EP SCP (ETSI Project Smart Card Platform)

The generic smart-card work and the work on common lower layer functions for smart-cards of all 3G systems was transferred to EP SCP which was created in March 2000.

8.2.8 Conclusions

During the period from April 1996 to February 1999 ETSI Technical Committee SMG created a UMTS strategy consensus on a vision which was based on services' innovation, GSM evolution and Internet orientation. All basic concepts for the UMTS standard were

---

[127] SMG P-00-194.
[128] SMG P-00-183.

elaborated and agreed. Very difficult decisions like the UTRA decision were taken. On this basis a set of reports and raw specifications were produced. The creation of a globally open efficient new working structure, the 3GPP, was initiated and brought to life. All UMTS and GSM work was transferred to 3GPP. The necessary changes in ETSI were initiated. Then the Technical Committee SMG was closed at the end of July 2000 since its mission was fulfilled.

Chapter 9: The Third Generation Partnership Project (3GPP)

Section 1: The Creation of 3GPP

Karl Heinz Rosenbrock[1]

Having read the title, it should not surprise you that this section deals with the creation of the Partnership Project for the standardisation of a Third Generation Mobile Communications System (3GPP).

Why, you may ask, in a history book about the GSM and UMTS development, do I want to talk about the establishment of a partnership project? Isn't it the most natural thing to do? This is, of course, a stance an insider can take today – after nearly 30 months of 3GPP's creation and the smooth and successful running of this project.

As this section will eventually show, it took quite some time, filled with tough and even passionate discussions, before the goal was achieved. Approaching this idea from a rather philosophical point of view, one should not be too surprised about the big efforts needed, because already the old Greek ancestors knew that "prior to being successful the Gods will demand some sweat"…[2]

This section starts with some general considerations leading the European Telecommunications Standards Institute (ETSI) membership towards a global approach in standardisation and then deals with the establishment of an ad hoc group of the ETSI Board (UGG = UMTS Globalisation Group) to address the matter of global standardisation in this context and the related meetings and discussions. Afterwards, the 3GPP will be described in a rather general manner, highlighting how it works, who the stakeholders are and dealing with the results achieved so far. The section is rounded up with the relationship towards the International Telecommunication Union (ITU) and other initiatives as well as a few concluding remarks.

9.1.1 First Approaches to Globalisation

The re-engineering process ETSI, the "Excellent" Telecommunications Standards Institute in Europe, undertook in the years 1995/1996 – only 7 years after its creation – resulted in among others a kind of mission statement for the Institute: "Making international standards happen first in Europe".

---

[1] The views expressed in this section are those of the author and do not necessarily reflect the views of his affiliation entity.

[2] One of my cruel translations of a German idiom "Vor den Erfolg haben die Götter den Schweiß gesetzt"…

Classical examples of ETSI success stories that witness this slogan are among others: the Global System for Mobile Communication (GSM); Digital Enhanced Cordless Telecommunications (DECT); Digital Audio Broadcasting (DAB); Digital Video Broadcasting (DVB); Terrestrial Trunked Radio (TETRA), just to name a few of them.

In positioning ETSI in the standardisation landscape, it became clear that the trends and changes towards globalisation, convergence and new value chains would lead to the creation of a huge volume of standards making space. Furthermore, it was not tenable for ETSI to try to fill the entire space. Choices had to be made. In addition, the investigations revealed that collaboration by means of appropriate partnerships could be a promising formula.

ETSI consciously withstood the temptation to become a global standards body. But it has always undertaken great efforts to ensure that all of its products, i.e. deliverables, such as European Norms (ENs), ETSI Standards (ESs), ETSI Technical Specifications (TSs), etc. satisfy real market needs and have the potential to become global standards. The ITU remains ETSI's global partner of choice. But the fast moving markets were expected to require ETSI to supplement this with various international partnerships on a case-by-case basis.

The high level task force that undertook the ETSI review in 1995/1996 advised the Institute that it had to sustain its core competence of making high quality standards for large and complex telecommunications systems. But, if neat demarcation lines are going to cease to exist, then ETSI must inevitably move more into the IT, audio-visual and other fields. It should do this in good partnerships, where other Standards Developing Organisations (SDOs) or appropriate fora and consortia are willing to co-operate with ETSI. Retrenchment by ETSI was not considered to be in Europe's interest.

Other results of interest here, of ETSI's re-engineering process after 7 years of existence were:

- to reduce the hierarchical structure in the Technical Organisation to a minimum;
- to delegate power (of approval, etc.) to the Technical Bodies where the main work is being done;
- to focus on semi-autonomous projects;
- to aim at proper project management;
- to allow the creation of ETSI Partnerships Projects (EPPs);
- to streamline and rationalise the ETSI Working Procedures;
- to improve the use of electronic tools for further rationalisation and innovation;
- to use audio and video conferencing;
- to broadcast inter-active meetings;
- to increase standards promotion activities;
- to facilitate and to promote direct electronic access to ETSI documents and deliverables, free of charge.

Regarding ETSI's external relations the advice given was: that ETSI should add to its strength through partnerships in complementing areas, ceding some sovereignty on a case-by-case basis to achieve common purposes.

ETSI should continue its dialogue (in GSC/RAST) with its major regional/national counterparts, with the objective of strengthening arrangements for effective co-operation and be prepared to enter into bilateral co-operation on a case-by-case basis.

With this short excerpt of some basic results stemming from the ETSI re-engineering process undertaken in the middle of the 1990s, we have the fertile soil, i.e. the driving forces, motivations, basic elements needed in order to establish a partnership project.

The tremendous success story of GSM may have even led some ETSI members to believe that it would have been the most natural choice to repeat this with the third generation mobile communication system within ETSI, too. But for insiders it became clear that such a success could not be guaranteed another time.

After having dealt with the ETSI internal change in orientation from Europe-centric to international and global, let's have a short look at the first attempts at getting into closer contact with our partners. Let's start with our American friends.

In Sections 5.3 and 5.4 the standardisation work on PCS 1900 in ANSI T1P1 as well as the new co-operation between ANSI T1P1 and ETSI Technical Committee SMG are described.

The relationship between ANSI T1P1 and ETSI TC SMG started in 1996/1997. Both Technical Committees were working on independent sets of Technical Specifications, i.e. GSM 1900 in the US and GSM 900/1800 in Europe, etc. Parallel working with different speeds includes the risk of differences that may result in incompatibilities. In order to avoid these difficulties both SDOs agreed to merge the two independent sets of specifications into a common one and to further develop it commonly using a co-ordinated approach, i.e. each work item and the results were approved in both committees and incorporated into the common specifications.

Despite the fact that the co-operation between T1P1 and SMG was excellent, it suffered a little from the fact that the double approval process and the difficult co-ordination process at several levels, e.g. first in T1P1 and then within SMG, were not very efficient and too time consuming. One of the lessons learnt from this exercise was: Why don't we really co-operate, i.e. work together, from the very beginning – then avoiding any type of "approval ping-pong"? This was another good reason to consider what and how to improve the standardisation work for the third generation...

Now let's have a look at the relationships with our friends from Asia.

In the People's Republic of China, network operators had implemented large GSM networks based on existing ETSI standards. In order to avoid divergence, Chinese authorities (RITT) joined ETSI as an associate member and participated fully in the work of ETSI TC SMG since 1997. Thus, it was possible to fully integrate the Chinese requirements into the standardisation process within TC SMG in order to secure the integrity of GSM between China and the "rest of the world".

In the meantime – with the Universal Terrestrial Radio Access (UTRA) decision in January 1998 – contacts with Japanese ARIB/TTC colleagues had been established within ETSI TC SMG. These contacts had started in spring 1997, 1 year earlier

Discussions on the Technical Committee working level at the end of 1997 and the beginning of 1998 between ETSI, ARIB/TTC, and T1P1[3] led to the hope that the creation of a common UMTS concept applicable in all territories – and thus de facto globally – was possible. Such a concept could be based on UMTS service innovation, UTRA and the GSM core network evolution.

In other words, there was from the beginning a lot of goodwill available from all sides to do/undertake something in common. But how to do it? It became clear that to perform the UMTS standardisation within the three (or more) existing organisations would have been nearly unmanageable. Three committees in different continents could have developed the

---

[3] More information is in Chapter 8, Section 8.2.6.2.3

UMTS radio specifications whilst several other committees would have worked on network aspects. The situation in other key areas, such as services, Subscriber Identification Module (SIM), and Operation and Maintenance would have been equally difficult. Furthermore, there would have been no overall decision-making body for the resolution of possible conflicts. All these difficulties sketched out here simply called for a new and much more effective global organisational solution.

As the friends from ARIB were quite interested in developing a common radio interface whose key parameters had been agreed in Japan and in the UTRA radio interface decision in January 1997, a delegation from the ETSI TC SMG undertook an exploratory mission to Japan on 5 and 6 February 1998 in order to find out whether or not a kind of co-operation with them would be possible. Further information is given in Chapter 8, Section 8.2.6.2.3.

These informal contacts on the working level revealed that there was a good resonance on the Japanese side. In an association like ETSI, there is not only the working level, there are other levels as well, e.g. the General Assembly (GA), the highest ETSI authority, and the Board, a body with some 25 clearly identified powers delegated by the GA. In 1997/1998 the first ETSI Board could have looked back at about 18 months of existence and had, of course, to play its role... Without acting like a "donkey who eats up the grass that has grown over (above) an old and nasty story,[4] one has to admit that the communication/co-operation between the ETSI Board and the ETSI TC SMG suffered a little bit from irritations, mis-understandings, mal-perceptions, etc. In other words, it was far from optimum at that time... Nevertheless, there was a role to play from a more political strategic point of view. And now the question was what to do in order to make something useful happen?

What do you do, when you do not exactly know how to proceed? You create a committee. At least the politicians are supposed to do so. Well, within ETSI it was the Board that – after an interesting extraordinary meeting at the Frankfurt Airport on 27 February 1998 – decided to create an ETSI UMTS Globalisation Group (UGG), i.e. not a committee but an ad-hoc group, but what is the difference? The next section will tell you more.

Coming back to the extraordinary ETSI Board meeting on 27 February 1998, to simply mention "it was an interesting one" is, of course, correct, but an understatement. We had already quite an emotional ETSI Board#11 meeting dealing, with among other items, the question of how to standardise UMTS in the most useful manner. As no consensus could be achieved, a specially convened Board meeting was required.

At the beginning, a recall of the ETSI Board#11 results concerning the third generation mobile standardisation was made. In addition, reports from ETSI TC SMG were given, especially about their exploratory contacts with potential partners.

Regarding the ETSI strategy and policy for the standardisation of a third generation mobile communications system, there were quite differing opinions and fears expressed. The two extreme positions were something like:

- create a new and independent 3G forum; and
- keep all 3G standardisation within ETSI.

With all shades of compromise in between – among others why not use an ETSI Partner-ship Project (EPP)? By the way, at that time it was not very clear what an EPP was. There existed a general description in the ETSI Rules of Procedure, but it dated from 1995/1996 and covered some basic characteristics only, allowing for a great variety of different implementa-

---

[4] Wenn endlich Gras über eine (traurige) Angelegenheit gewachsen ist, kommt ein Esel, der das Gras auffrißt.

tions... Thus, asking three people about their interpretations, one could well be confronted with four descriptions...

The result of this heated discussion was the idea to create a kind of starter group, UGG. And the Board agreed that the group should consider the requirements for the globalisation of GSM-based UMTS and make recommendations as to how this may be achieved.

9.1.2 The ETSI UMTS Globalisation Group

The UMTS Globalisation Group (UGG) was an ad-hoc group established by the ETSI Board from where it got its first draft Terms of Reference. They have been refined since based on the experience gained in the meantime.

9.1.2.1 Terms of Reference of UGG

In the following you will find the UGG Terms of Reference as revised during the first UGG meeting and approved by correspondence by the ETSI Board.

9.1.2.1.1 Objectives of the Group

The objective of this Group was to consider the actions, which are required to enable UTRA and "GSM-based" UMTS specifications to be prepared and promoted in a manner, which makes them attractive to global partners such that they will be implemented world-wide.

To achieve this objective the Group should:

- provide strategic management of those activities which fall within the scope of this Group;
- investigate the development of relationships with external partners and identify their expectations for UMTS;
- propose an organisational structure which meets the expectations of the Institute and external partners, taking into account the recommendations of the ETSI GA ad-hoc Group on fixed/mobile convergence (after their approval by the ETSI GA);
- consider what transition arrangements are necessary to move towards a new organisational structure;
- propose a mechanism, which enables all active partners to take part in the approval of related specifications.

In their work the Group may need to take into account the following factors:

1. Management characteristics (*How do stakeholders define and approve strategic direction? How is the "work-programme" defined and approved to carry out the strategic objectives?*)
2. Business model (*How are "regional" priorities, based on their business model, defined such that the standards are truly global?*)
3. Financial model (*How are "overhead" costs assigned?*)
4. Operating principles (*How is actual standardisation work carried out? How are the standards approved in different regions/countries?*)
5. Maintenance work (*How are improvements, maintenance of standards performed?*)
6. European fall-back (*What is the fall-back solution for Europe if there are disagreements at*

strategic/operational level? How can European interests be safeguarded if other regions do not want "European solutions" and if they retain the right to develop their own solution since they have a "multi-standards" market?)

7. IPR policy (*Is the ETSI IPR policy the one to use?*)
8. Relationship with other ETSI activities
9. Relationship with ITU (How to maximise the effectiveness of ETSI's input into the IMT-2000 activities of the ITU?)

9.1.2.1.2 Composition of the Group

The Group will have the following composition:[5]

Chairman	ETSI director-general
Vice-chairmen	Two to be elected by, and from, the group.
Eight board members	Mr Davidson, Mr Etesse, Dr Haas, Mr Kaiser, Mr Lang, Mr Lathia, Mr Lindström, Mr Salles
SMG	Two representatives (Mr Hillebrand and one to be advised)
GSM MoU Association	One representative (Dr Nugter)[6]
UMTS Forum	One representative (Mr Beijer)
FMC Group	FMC Group chairman (Mr Hearnden)[7]

9.1.2.1.3 Reporting Arrangements

The Group will report to the ETSI Board.

9.1.2.1.4 Duration of the Group

Since the detailed GSM-based UMTS phase 1 specifications will be completed by the end of 1999, the Group should aim to complete its work in the fastest timeframe, with the objective of producing their final recommendations no later than the September 1998 Board meeting. This should enable relationships with external partners to be established, and organisational changes to be implemented by autumn 1998.

Commenting a little on these Terms of Reference, the following needs to be said.

The additional text in italics and brackets is explanatory and had been added by the UGG secretary in order to facilitate understanding.

Regarding the duration, to finish the work by September 1998 was extremely ambitious. This time pressure came from the fact that the Japanese partners were seeking to have an operational 3G system by the year 2001!

Despite an awkward UGG meeting schedule (see paragraph 9.1.2.3), that prevented the key

---

[5] The chairman may co-opt additional representatives as required to assist in the completion of the tasks of the group.

[6] The representatives of the GSM MoU Association and the UMTS Forum are guests, invited to assist the Group, but which are not bound by the governing rules of ETSI.

[7] The FMC Group chairman will participate for as long as the FMC Group exists. It is expected that the FMC Group will be closed during ETSI GA#30.

people from taking any summer vacation, it was not possible to fully meet that ambitious goal. Nevertheless, as you will read in the following, 3GPP was able to have their first meetings at the beginning of December 1998 at Sophia Antipolis.

9.1.2.2 Composition of UGG

The UGG ad-hoc group met very often (13 times) in the period between March and November 1998! This was the reason that the UGG members had some difficulties in attending all the meetings. This situation again led to repetitive discussions owing to the fact that some results of meeting X were challenged by members in meeting X + 1 who were unable to participate in the previous one.

The UGG meetings were chaired by myself, the ETSI director-general.

The Vice-chairmen were Mr. Phil Davidson and Mr. Kirit Lathia. Mr. Adrian Scrase acted as secretary.

The eight ETSI Board members were already mentioned in paragraph 9.1.2.1. It happened that at some UGG meetings additional Board members participated.

The ETSI TC SMG was basically represented by Messrs. Friedhelm Hillebrand, the SMG chairman, by the SMG vice-chairmen, Messrs. Alan Cox and Gunnar Sandegren and by François Grassot, the chairman of a manufacturers' co-ordination group for the SMG work.

When the constitution of the Group was discussed for the first time, it was immediately agreed that the GSM MoU Association and the UMTS Forum were to be invited to participate in the ETSI Board ad-hoc group UGG. Under this arrangement, the benefit of the GSM MoU Association and UMTS Forum expertise could be fully realised without them being bound by the governing rules of ETSI. The GSM MoU Association was represented by Ms Adriana Nugter and the UMTS Forum by Mr. Thomas Beijer, its chairman.

Mr. Chris Roberts acted on behalf of the European Commission as ETSI counsellor.

As the Fixed Mobile Convergence Group (FMC) was closed by the ETSI GA #30 in Spring 1998, there was no FMC delegate participating in UGG. But in the later meetings there was some representation from the ETSI TC NA (Network Aspects), e.g. Messrs. François Lucas and Hans van der Veer.

In general, there were between 10 and 17 delegates present, except for UGG meeting #12 where we reached 29 delegates including the SMG STC chairmen and vice-chairmen.

9.1.2.3 UGG Meeting Schedule

UGG met 13 times in 1998 as regular UGG meetings as follows:

UGG#1	On 13 March in Brussels, hosted by Airtouch
UGG#2	On 24 March in Sophia Antipolis, hosted by ETSI
UGG#3	On 08 April in Brussels, hosted by Airtouch
UGG#4	On 20 April in Sophia Antipolis, hosted by ETSI
UGG#5	On 12 May in Brussels, hosted by Airtouch
UGG#6	On 08 June in Sophia Antipolis, hosted by ETSI
UGG#7	On 29 June in Frankfurt, hosted by ETSI
UGG#8	On 28 July in London, hosted by DTI
UGG#9	On 19 and 20 August in Brussels, hosted by Ericsson
UGG#10	On 14 September in Munich, hosted by Siemens

UGG#11	On 30 September to 02 October in Sophia Antipolis, hosted by ETSI
UGG#12	On 28–29 October in Amsterdam, hosted by Lucent
UGG#13	On 23 November in Frankfurt, hosted by Nokia

In addition, it was basically UGG that conducted exploratory and negotiation meetings with potential partners as described in the following.

9.1.2.3.1 GSC/RAST

GSC stands for Global Standards Collaboration and RAST for Global Radio Standardisation. These are loose co-operations between recognised Standards Developing Organisations (SDOs) from different regions in the world: ACIF (Australia), ARIB (Japan), ETSI (Europe), ITU (global, with its two sectors radiocommunications & telecommunications), TIA (US), TSACC (Canada), TTA (Korea), and TTC (Japan). The GSC/RAST activities date back to February 1990 to a meeting of the former ITSC (Interregional Telecommunications Standards Conference) at Fredericksburg, VA, where ITU-T, T1, TTC and ETSI delegates met for the first time.

In the meantime the activities within GSC/RAST have resulted in a number of bilateral and multilateral actions/collaborations and a considerable creation of trust between the participating organisations.

One can even state that without the mutual exchange of information and co-operation within ITSC/GSC/RAST and the creation of trust between the bodies from different regions, the establishment of the third generation partnership project would have been more difficult and complicated if not impossible...

During the GSC/RAST meetings that took place from 30 March to 1 April 1998 at the ETSI premises in Sophia Antipolis, a lot of discussion focussed on the question about how best to standardise the next generation of mobile communication. That meeting offered additional occasions for exploratory discussions with potential partners.

9.1.2.3.2 Meeting the Japanese Partners

In Japan standardisation in the area of telecommunications is undertaken by two SDOs:

- ARIB (Association of Radio Industry and Businesses) for radio matters, and
- TTC (Telecommunication Technology Committee) for the fixed network part.

After the ETSI TC SMG delegation that visited Japan on 5 and 6 February 1998, UGG met as follows with the Japanese partners:

- on 2 and 3 April 1998 in connection with the GSC/RAST meeting at Sophia Antipolis
- on 28 and 29 May 1998 in Tokyo
- on 29–31 July 1998 in London
- on 7-9 October 1998 in Tokyo together with T1, TTA and
- on 2–4 December 1998 in Copenhagen together with T1, TTA and CWTS.

9.1.2.3.3 Meeting the American Partners

In the US, there is the American National Standards Institute (ANSI), responsible for standardisation. The actual standardisation work is performed by more than 300 SDOs that got the ANSI accreditation.

In the telecommunications area, the SDOs with which ETSI has the closest links are:

- The T1 Committee; and
- The Telecommunications Industry Association (TIA).

On 19 March 1998, in conjunction with an ETSI TC SMG Plenary that took place at Sophia Antipolis, a discussion on the working level with representatives of T1P1 and TC SMG took place in order to explore how to standardise UMTS in the future. The GSC/RAST meeting in 1998 at Sophia Antipolis allowed for further exchange of information with both T1 and TIA representatives. At the end of May/beginning of June, Mr. François Grassot was authorised by UGG to represent ETSI at a meeting with T1P1 and inform them about the UGG work as well as the negotiations that took place on 28 and 29 May with the Japanese partners.

Afterwards, UGG undertook the following negotiations with the Americans:

- 25 and 26 June 1998 in Seattle. The T1 Committee was officially informed about the UGG work and was invited to join the UMTS standardisation initiative.
- 10 July 1998 in London. An ANSI delegation met UGG and welcomed the UMTS initiative.
- 25 September 1998 in Washington. UGG met the ANSI delegation again, which was enriched by a few T1 and TIA representatives. The idea of the UMTS initiative to be treated by means of a common partnership project was addressed as "paradigm shift" by the Americans.
- 26 and 27 August in Paris. This meeting between UGG and T1 (P1) Committee representatives resulted in a breakthrough regarding the engagement of T1 delegates in the further preparation of the 3GPP. The T1 delegates then took part in the meetings with all other potential 3GPP partners, i.e.
- 7–9 October 1998 in Tokyo; and
- 2–4 December 1998 in Copenhagen.

9.1.2.3.4 Meeting the Korean Partners

In Korea the SDO responsible for standardisation in the fields of telecommunications is the Telecommunications Technology Association (TTA).

After the ARIB/ETSI/TTC meeting on 28/29 May 1998 in Tokyo, Mr Kirit Lathia was authorised by UGG to inform the Korean colleagues about the 3GPP preparations. Further contacts were maintained by the ARIB/TTC colleagues that resulted in inviting TTA to the next common meetings:

- 7–9 October in Tokyo; and
- 2–4 December in Copenhagen.

9.1.2.3.5 Meeting the Chinese Partners

The Research Institute of Telecommunications Technology (RITT) of China via the Chinese Academy of Telecommunications Research (CATR) joined ETSI as an associate member and represented the de facto SDO of the People's Republic of China.

In the meantime, since 1999, standardisation in the UMTS related area is being performed by the China Wireless Telecommunication Standard (CWTS) group.

The Chinese authorities are currently considering the creation of a Chinese Telecommunications Standards Institute (CTSI)) and have consulted ETSI for support...

In addition to the contacts mentioned already in paragraph 9.1.1, the following was done in order to involve the Chinese partners:

26 May 1998 in Beijing. In conjunction with a UMTS seminar organised by the UMTS Forum, some unofficial meetings took place with RITT and CATR representatives in order to inform them about the UMTS standardisation initiative and the plans to create a 3GPP.

Afterwards, some contacts were established between the Japanese and the Chinese partners. With assistance from ARIB/TTC the Chinese partners were invited to the next common meetings:

- 7–9 October in Tokyo; and
- 2–4 December in Copenhagen.

9.1.2.4 Discussions within UGG

9.1.2.4.1 General

The discussions within UGG have been tough and passionate. None of the UGG members held the monopoly to be nice or nasty. In cases where it was impossible to achieve consensus within UGG, the ETSI Board and even the ETSI GA had to play their decisive role, and that has been necessary a few times...

The ETSI internal discussions within UGG and the negotiations with the different partners – as described in paragraph 9.1.2.3 – as well as the feedback/decisions stemming from the ETSI Board and GA meetings resulted in the establishment of the documents that were approved during the final preparatory 3GPP meeting in Copenhagen on 2–4 December 1998:

- The 3GPP Description;
- The 3GPP Agreement; and
- The 3GPP Working Procedures.

The exact wording of these basic papers for the creation of the 3GPP can be found on the 3GPP website at http://www.3gpp.org.

In paragraph 9.1.3 and its sub-paragraphs a rather general description of 3GPP will be given without going in to too much detail.

It is not the intention to record here all discussions that took place because what counts at the end of the day are the results aren't they? Furthermore, as history very often proves the heated discussions of yesterday may even be regarded as "water under the bridge", especially when the development of events have proven that fears that may have existed at the beginning, either have not materialised of are of no more relevance today...

9.1.2.4.2 The Start of UGG Discussions

During the second UGG meeting Mr Dieter Kaiser prepared schematic diagrams to show a possible structure for the proposed project as viewed from the ETSI side, and from the side of potential partners. These diagrams were discussed in depth and the following points were noted:

- The potential partners would need to be willing to adjust their own internal organisational structures in order to contribute to the new project (to avoid duplication of effort within the partner organisations).
- Some parts of the existing ETSI TC SMG would need to be moved into the new project.
- The remaining work would be undertaken within ETSI TC SMG as normal.
- The ETSI TC SMG structure would remain largely unchanged, since the output of the new project would be passed to SMG for further regional treatment, particularly for regulatory deliverables.
- The membership of the existing ETSI TC SMG could be widened to include external partners, rather than creating a new project. This option was not supported by the UGG, particularly since there would be no fallback position.
- Participation in ETSI TC SMG, from potential partners for the new project, could be increased on an interim basis pending the creation of the project.
- The participation within the new project should be on an individual members basis, and the creation of regional views should be avoided (i.e. there should not be regional blocking).
- The new project could be structured in a similar manner to the EP TIPHON.
- The new project must be made attractive to potential partners.

From the points listed above, the diagrams were updated and served as a basis for the development of the 3GPP structure.

In addition, Mr Friedhelm Hillebrand prepared a document, whose principles had been endorsed by TC SMG, giving the proposed responsibilities of the new project. The text was edited during the meeting and a revised draft agreed as follows:

The UMTS Project elaborates, approves and maintains the necessary set of common specifications for:

- UMTS terrestrial radio access (UTRA): W-CDMA in Frequency Division Duplex (FDD) mode and TD-CDMA in Time Division Duplex (TDD) mode.
- GSM platform evolution towards UMTS (including mobility management and global roaming).

The common specifications will be developed in view of global roaming and global circulation of terminals.The common specifications aim at forming the technical basis of an IMT-2000 family member. The UMTS project elaborates contributions to the ITU on relevant aspects of the IMT-2000 family.In the framework of agreed relationships the UMTS project elaborates common specifications for approval and publication as standards, or parts of standards, by ETSI and/or other standards bodies (such as ANSI, ARIB/TTC and RITT).

This can be considered a starting point for the 3GPP scope and project description.

Furthermore, Mr Hillebrand also prepared a document giving the proposed responsibilities of SMG following the creation of the new project. It was agreed that this document should be further discussed during the next meetings of the Group.

Mr Tom Lindström and Mr Gunnar Sandegren prepared a draft Project Requirements Definition for the new project. Some preliminary work was undertaken to edit the draft but this was not completed due to time constraints. Members were requested to study the draft and to make proposals for amendments.

UGG noted that in accordance with an action requested during its first meeting, a press release had been made immediately prior to ETSI GA meeting #30 in Spring 1998 announcing ETSI's plans to globalise its work on UMTS.

9.1.2.4.3 First Contacts with the Partners

The recent GSC and RAST meetings in March/April 1998 at Sophia Antipolis made good progress and experienced a good collaborative atmosphere that ran through both meetings. The meetings had been well attended (approximately 80 participants) from Europe, Canada, US, Australia, Korea, and Japan.

The informal meetings that had taken place with T1 representatives during the SMG plenary (19 March) and with T1 and TIA representatives on 31 March had been positive, with the US representatives expressing an interest in participating in the proposed ETSI Partnership Project. There had also been good support for the offer made by ETSI to provide the administrative umbrella under which the new activity could be launched.

A more extensive discussion had taken place with delegates from ARIB and TTC on 2/3 April, immediately after the GSC/RAST meeting. Views had been exchanged on how such an ETSI Partnership Project could be structured and operated. The concerns of most importance to the Japanese delegates appeared to be the following:

1. The name of the ETSI Partnership Project. "UMTS" was seen as a European term, which would be inappropriate in a global environment.
2. ETSI dominance. The Japanese wished to participate as equal partners, with equal rights, and equal responsibilities.
3. Use existing structures. The benefits of using existing ETSI structures (rather than creating a new forum for example) were supported by the Japanese representatives.
4. IPR issues.

9.1.2.4.4 Continuation of UGG discussions

Mr Tom Lindström and Mr Gunnar Sandegren had prepared a revised Project Requirements Definition which took into account the comments made during the last meeting and those received subsequently.

The Group focused on the matter of partnerships and memberships of the proposed ETSI Partnership Project. There were differing views expressed concerning the relationship between partners and members and whether each member should be required to be associated with a partner. After some discussion a clear view emerged that the ETSI Partnership Project should be established with one or more partners who must be a recognised standards body to be eligible. It was understood that the term "recognised" meant different things in different regions but the intention was clear. Members could be any entity that wished to contribute to the work of the ETSI Partnership Project but must be members of a partner organisation, with the fall back position that they would in any case be eligible for ETSI associate membership and could through that route obtain membership of the new activity. It was agreed that the principles of the two terms "member" and "partner" should be included within the Project Requirements Definition. It was understood that the formation of an ETSI Partnership Project would not require the creation of a new legal entity since the existing ETSI structure provided a suitable framework to operate within.

It was understood that a literal reading of the terms "member" and "partner" could exclude bodies such as the GSM MoU Association and the UMTS Forum. Dr Adriana Nugter and Mr Thomas Beijer agreed to consider how best their organisations could be included within the proposed organisational scenario.

A brief discussion took place on how voting could be conducted within the new activity and a clear preference for one member one vote was expressed. It was agreed that consensus should be the normal method of decision-making.

At UGG meeting #4, Mr Tom Lindström presented a revised Project Requirements Definition for the ETSI Partnership Project which took into account the comments received from UGG members. Mr Kari Lang proposed a complementary pictorial representation of the proposed ETSI Partnership Project which was noted.

Also proposed were methods for funding the ETSI Partnership Project. A long discussion followed during which the advantages of member or partner funding were considered. At this stage it was not clear what the funding was required for, nor what magnitude of funding was required, but some difficulties were envisaged with the concept of individual member funding. It was believed that the funding requirements were low and were probably restricted to additional secretariat assistance and a small STF. After full discussion it was agreed that the starting point should be that ETSI is prepared to absorb the cost of running this Partnership Project for a duration of say 2 years. This starting point could be re-discussed on the basis of unexpectedly high costs, or the wish of potential partners to take part in funding the activity.

It is interesting in this context to note – looking back with hindsight and without anger – that the costs for the Mobile Competence Centre within ETSI reached the order of magnitude of 5 million EUR already in the year 2000 and will exceed 6 million EUR in the year 2001…

A short discussion took place on the voting principles to be applied within the ETSI Partnership Project. It was basically agreed that one member one vote should be applied, but some precaution should be included to stop decisions being taken which were unlikely to be acceptable to specific regions. It was not proposed that regional voting should be applied, but some care be taken to ensure that decisions were generally acceptable to all participating regions.

By the way, the topic of voting has been discussed with changing focus during nearly all UGG meetings as well as during the meetings with the 3GPP partners.

9.1.2.4.5 Further Discussions with the Japanese Partners

During the meeting with ARIB and TTC on 28 and 29 May 1998, the talks had centred around two proposals; one prepared by ARIB/TTC and the other prepared by ETSI based on the results of UGG#5. After a long discussion, all three parties agreed to base their further discussions on the ETSI proposal which had been modified during the meeting to take account of concerns raised by the ARIB/TTC. The proposal was endorsed in principle by the three parties subject to final approval.

UGG discussed whether a task force should be created to accelerate the technical work pending the creation of the partnership project. However, it was felt that the existing SMG structure provided a good basis for the technical work to proceed until overtaken by the partnership project.

UGG noted that whilst the ETSI proposal had been endorsed by all three parties there remained a significant amount of work to be done in defining clearly the work to be undertaken and the working rules under which that would be done.

Nevertheless, this meeting with the Japanese partners was a kind of first breakthrough in the direction of 3GPP and led to unofficial and official contacts with the SDOs of the US, Korea and China.

9.1.2.4.6 Back to UGG discussions

Within UGG meeting #6 it was agreed that a collective letter should be sent to the membership informing them of the progress made so far and the expectations for the future. This should include a request for support from the membership and an invitation to provide constructive comments.

Furthermore, it was concluded that the SMG representatives to the UGG would keep ETSI TC SMG fully informed of the progress being made, i.e. something self-evident.

The UGG chairman agreed that the ETSI secretariat would prepare a draft 3GPP partnership agreement, but sought guidance on the format that should be used. It was thought that either a common agreement signed by all partners could be used or a declaration could be established which each joining partner could sign. Finally, the chairman volunteered to prepare a draft based on ideas presented by UGG members. Well, whether we talk here about the ETSI secretariat or later about the chairman preparing some drafts, in plane English it basically meant that the UGG secretary, Mr Adrian Scrase, did an excellent job in performing all these and other tasks!

Furthermore, it was agreed that Mr Adrian Scrase would undertake to prepare draft working procedures based on the ETSI Directives. In addition, it was concluded that voting procedures would be contained within the draft working procedures and that options would be provided, where necessary, for decision at a later stage.

9.1.2.4.7 Discussions with ARIB and TTC in London

During the meeting with the Japanese partners in London on 29 and 30 July, the draft 3GPP project description and the remainder of the draft 3GPP documentation was discussed, reviewed and further refined.

Regarding the meetings of ETSI's UMTS Globalisation Group representatives with the future 3GPP partners, one has to state with great satisfaction and relief that they have always been – in quite some difference to the tough ETSI internal discussions within UGG – guided by great openness, a lot of trust and a very good spirit of co-operation!

A big compliment and expression of gratitude is extended to all who participated in these fruitful discussions!

9.1.2.4.8 Looking Back at UGG Discussions – Without Anger

Back to reality, in UGG meeting #9, four contributions were presented proposing different 3GPP scopes and different work areas to be included in 3GPP.

During the presentation of these documents, UGG considered whether the 3GPP scope should include all of the work currently within ETSI TC SMG. The inclusion or exclusion of VHE was also considered. This led to the development of two different basic opinions:

1. that 3GPP should contain only those work items falling within the "three bullet points" as per document ETSI/BOARD-Global#9(98)08:

   - UTRA – wideband CDMA in FDD mode and TD-CDMA in TDD mode
   - GSM core network evolution
   - Terminals for the above, and

2. that 3GPP should include the majority of the work currently being performed in SMG but excluding those items of a European regulatory nature.

It was not possible to reach consensus on either of these views. UGG also considered keeping ETSI TC SMG in its current form but opening up the membership to include other players by means of associate memberships and co-operation agreements.

UGG was unable to reach consensus on any of the views expressed but agreed to raise the issue to the Board for resolution.

Commenting on this today with hindsight and the knowledge of the further evolution within 3GPP, we can only smile… But at that time, i.e. August/September 1998, the matter was discussed like the question within *Macbeth*: "…to be or not to be?"

9.1.2.4.9 An Extraordinary ETSI GA was Needed

Not only UGG, but the ETSI Board, too were unable to solve these problems. This situation necessitated calling for an extraordinary meeting of the ETSI GA to decide on the structure and scope of the Partnership Project. That meeting took place at the ETSI headquarters on 29 September 1998.

After explanations of the current situation on third generation mobile systems by the ETSI Board chairmen, Mr David Hendon, and a number of documents related to the proposed 3GPP by the ETSI director-general, an intensive discussion on the proposed Partnership Project followed.

Afterwards, the ETSI GA chairman, Dr Antonio Castillo, put the following three questions to a vote:

Vote #1: Are you in favour of the principle of the creation of an ETSI Partnership Project for a Third Generation Mobile System (3GPP)?

Result: 92.62% of the votes cast were in favour.

Vote #2: Do you agree with the creation of an ETSI Partnership Project (3GPP) that includes: GSM core network evolution towards a third generation mobile system?

Whilst 63.4% of the votes cast were in favour, vote #2 failed to reach the required 71%.

Vote #3: Do you agree with the creation of an ETSI Partnership Project (3GPP) that includes UTRA (W-CDMA in FDD mode and TD-CDMA in TDD mode)?

Result: 95.9% of the votes cast were in favour.

The GA chairman then stated that in view of the fact that Vote #2 had failed to reach the necessary 71%, the Assembly had to return to that question for re-discussion and to try and reach an agreement on that point.

An interesting discussion ensued and the GA chairman then asked the ETSI TC SMG chairman, Mr Friedhelm Hillebrand, to convene an ad-hoc group with those ETSI members concerned to reach a compromise proposal, which might achieve a successful vote result.

Mr Hillebrand reported back that the ad-hoc group had reached an agreement by consensus, based on BT and Ericsson contributions and combining the ideas of those who looked for a short-term solution with those who looked towards the longer term. He stated that they had achieved a strategy statement and that the final (detailed) decisions were to be taken at future normal General Assemblies with the full 3GPP documentation available.

The GA chairman then put the following question for a vote:

Vote #4: Do you agree that the GA endorses the following principles for the forthcoming negotiations:

1. to recommend ETSI to initiate an ETSI Partnership Project (EPP) to be known as "3GPP" for third generation mobile system specification work.

   – The 3GPP will be a partnership between ETSI and recognised Standards Developing Organisations and other partners; industrial entities are members of 3GPP and provide the technical input.
   – The 3GPP will develop specifications for the initial phase of a complete third generation mobile system based on UTRAN and evolved GSM core network (a "G-UMTS" system).

2. To initiate the setting up of an ETSI Project (EP) for UMTS:

   – the new ETSI Project UMTS will collect current and future ETSI activities relevant to UMTS outside those G-UMTS areas to be handled in the 3GPP.

3. Furthermore it is agreed that:

   – Work on GSM standardisation needs to continue, for example, for those network operators who do not get a UMTS licence. It is proposed that this work should remain in ETSI itself for the time being at least. For this and other reasons, it is proposed that ETSI SMG should continue to exist.
   – No duplication of work between 3GPP and ETSI EP UMTS, new SMG.

Result: 94.5% of the votes were cast in favour. Thus, a solution for the scope and further pursuit of 3GPP was, eventually, found… What a heavy birth!

9.1.2.4.10 Back to UGG

Despite these ETSI GA decisions, the next UGG meeting #12, immediately following was not very effective – owing to some personal polarisation…

9.1.2.4.11 Discussions with Nearly all the Partners in Tokyo

For the period from 7 to 9 October 1998 all six 3GPP Organisational Partners (OPs), ARIB,

CWTS, ETSI, T1, TTA and TTC, had been invited to a preparatory OP meeting in Tokyo. Unfortunately, the Chinese partners were unable to join. The other OPs met for the first time in this 3GPP context. A TIA delegate and a delegate from the European Commission acted as observers.

After a detailed review of 3G related activities of each SDO, an in-depth discussion and further detailed elaboration of the recent draft 3GPP documentation took place.

The following results were achieved:

- The formation of the 3GPP was approved by ARIB, ETSI, T1, TTA and TTC.
- The draft Partnership Project Description was written and was agreed upon by the OPs as a framework document subject to formal confirmation.
- The principles for a Draft Partnership Project Working Procedures were agreed upon by the OPs.
- The project start date was set to 7 December 1998.

For the final preparatory meeting in Copenhagen from 2 to 4 December 1998, the following steps were agreed:

- The head of SDO delegations (HoDs) to get together to produce the Terms of Reference for Technical Specification Groups (TSGs) prior to the next meeting and this proposal was approved.
- Project Co-ordination Group (PCG) meetings will be rotated among different regions to minimise the burden on one region.
- When created, the 3GPP website will be open to all members.
- PCG will be formed on 2–4 December and TSG meetings will start on 7–8 December 1998.

9.1.2.4.12 UGG Fine-tuning

During UGG meeting #12, some further fine-tuning of the basic 3GPP documentation was done. In addition, in the presence of SMG Technical Subcommittee (STC) chairmen and vice-chairmen, an exchange of views on the 3GPP start-up took place, such as:

- Responsibilities of 3GPP and ETSI TC SMG
- Co-operation between 3GPP TSGs and ETSI SMG STCs
- Achieving a smooth work transfer
- Future TSG chairmanships

9.1.2.4.13 Approval by the ETSI GA

During ETSI GA meeting #31 on 19 and 20 November 1998, the following decisions related to 3GPP were taken as a briefing for the ETSI delegation:

The ETSI GA:

- unanimously approved the creation of an ETSI Partnership Project for the third generation mobile system (to be known as 3GPP);
- unanimously endorsed the draft Partnership Project Description;

- unanimously requested the ETSI Board (through its UMTS Globalisation Group) to complete the negotiations with potential partners and to oversee the creation of the Partnership Project;
- unanimously requested the ETSI Board to agree and maintain on behalf of ETSI the final versions of the Partnership Project Description, the Partnership Project Working Procedures, and the Partnership Project Agreement;
- unanimously authorised the director-general to sign the Partnership Project Agreement.

9.1.2.4.14 Happy End in Copenhagen

It was in Copenhagen where the last 3GPP preparatory meeting with all six OPs took place on 2–4 December 1998.

Here, the final fine-tuning of the 3GPP documentation was achieved. In addition, the 3GPP agreement[8] was signed (in a nice framework provided by the host, TeleDenmark) by the following OPs: ARIB, ETSI, T1, TTA, and TTC.

Unfortunately, the partners from CWTS (China) were not authorised to sign the 3GPP agreement yet. Furthermore, owing to the fact that the UMTS Forum was unable to participate in Copenhagen, they were prevented from co-signing the 3GPP agreement as a first Market Representative Partner (MRP).

During that Copenhagen meeting, another discussion ensued about the role the MRPs should play. Finally, it was concluded that the high competence of MRPs should be used in order to identify market requirements, thus enabling 3GPP standardisation to meet the needs of the market. An MRP is an organisation invited by the OPs to participate in 3GPP with the objective of offering market advice to 3GPP and to bring into 3GPP a consensus view of market requirements.

9.1.3 What is 3GPP?

The 3GPP is a global standardisation initiative created in December 1998. Its task was to develop a complete set of globally applicable Technical Specifications for a third generation (3G) mobile telecommunications system based on the evolved GSM core network and an innovative radio interface known as UTRA. The Project is based on a concept devised by ETSI aimed at facilitating better co-operation between regional standards organisations, fora and other industry groupings. 3GPP is a collaborative activity between officially recognised SDOs, with the participation of other industry groups and individual members.

Partnership in 3GPP is open to all national, regional or other SDOs, irrespective of their geographical location – within the project the participating SDOs are referred to as OPs. The OPs may invite MRPs to participate: these may be any organisation from anywhere in the world that can offer market advice to 3GPP and bring a consensus view of market requirements that fall within the project's scope. Individual membership is open to companies and organisations within the communications industry that are active members of one of the OPs. The truly global nature and the breadth of the market interest in the task of specifying this 3G system is evident from the identity of the 3GPP partners (see further in sub-paragraph 9.13.2) and all agree that 3GPP is proving a highly successful initiative.

---

[8] The 3GPP Agreement and the 3GPP Project Description can be found on the attached CD-ROM in folder C1.

The 3GPP has no legal status. Ownership (including copyright) of the specifications and reports it produces is shared between the partners. The 3GPP process includes a conversion (transposition) of the project's output into official standards and reports by one or more of the OPs.

The specifications being prepared by 3GPP are evolved in part from the enormously successful GSM standard, which is currently (February 2001) serving over 400 million subscribers in more than 140 countries. Building on this massive installed base, the system being specified by 3GPP will be an attractive upgrade path for existing operators and users. It also has an assured compatibility with GSM – good news for both operators and users who are unable, or unwilling, to upgrade to 3G.

Paramount among the 3GPP specifications is the definition of UTRA, the innovative radio access technology that is the key to the new system's high data rates and dramatically improved performance. UTRA is spectrum-efficient and supports FDD and TDD modes. This interface has been accepted by ITU as a member of the IMT-2000 family – or more correctly, as *two* family members: IMT-DS, the FDD mode; and IMT-TC, the TDD mode. IMT-2000 family membership requires the ability for users to roam globally and seamlessly, which implies interoperability with other family members: 3GPP thus co-operates closely with the 3GPP2 project which is specifying another family member, a 3G CDMA system based on an evolution of the ANSI-41 architecture.

9.1.3.1 How Does 3GPP Work?

3GPP has been designed to minimise delays and inefficiencies. As a result, it has a "flat" organisational structure and a large degree of distributed autonomy. Overall project planning and co-ordination is the responsibility of the PCG, with input primarily from the OPs, guided by the MRPs. It is mainly at this level that regulatory requirements, provided by the telecommunications administrations and governments around the world, are taken into account.

The development of the specifications is performed by TSGs and their subordinate working groups. Here, the main participation is by technical experts from the individual members of the OPs. Individual members in their capacity as ITU members are also responsible for carrying the results of the 3GPP work to the ITU.

At the time of the creation, 3GPP had structured around four principal aspects of the 3G system being defined as TSGs:

- TSG CN: core network
- TSG RAN: radio access network
- TSG SA: services and systems aspects
- TSG T: terminals

Each TSG is authorised to develop and approve specifications and reports within its scope, and TSG SA also has a role of co-ordinating the work of the TSGs at a more detailed level than the PCG. The result is a process that is able to rapidly produce and approve specifications and reports in response to the needs of the market, although it is important to note that the deliverables do not have a formal status until they have been transposed by one or more of the OPs.

The formal status is necessary for regulatory and other purposes in the various regions, and all the OPs have committed themselves to complete this process rapidly. Each OP will apply its own procedures, appropriate to their respective regions. As the official European SDO within 3GPP, ETSI recognises the 3GPP output as ETSI Technical Specifications and ETSI Technical Reports without a need for any further endorsement within the Institute. This means that the 3GPP documents are published – within a matter of a few weeks – as identical text directly as ETSI deliverables. In addition, ETSI is transposing, i.e. adapting, a few of the initial 3GPP specifications into ENs for specific European regulatory requirements. This is happening in parallel with the publication of the initial Technical Specifications and will not impede the implementation of 3G in Europe.

9.1.3.1.1 Electronic Working

3GPP has taken a leading role in changing the traditional ways of standards making. A very heavy dependence is now placed on electronic working, both outside and within meetings, advancing a trend that started in ETSI a year or two ago. This means that paper copies of draft documents have been almost entirely eliminated, saving time and expense, and making a significant contribution to the environment.

Given that 3GPP has participants from all over the world, the use of the Internet, e-mail exploders and other such facilities have proved invaluable for distributing and sharing information, working drafts and so on. Delegates to meetings had already become used to downloading working documents from the Internet and having updates to the documents distributed in meetings on CD-ROM. But in recent months, in meetings around the world, participants have experienced the benefits of a local area network (LAN) solution as the latest step in improving working methods. ETSI's headquarters premises already have LANs in all its meeting rooms, but most other venues currently rely on temporary LANs (wired or radio), using equipment and support kindly donated by individual members.

Such facilities permit delegates to access all the meeting documents electronically from their laptop computers. As a result, the huge burden of producing paper copies (as many as 10 000 pages per delegate for some meetings) can be eliminated. Delegates can access new documents as soon as they are available, rather than having to wait for paper copies to be made or for the documents to be distributed by other means, such as diskette or CD-ROM.

9.1.3.1.2 Project Support

For administrative and support purposes the 3GPP Partners have established a Mobile Competence Centre (MCC) which is hosted by ETSI at its premises in Sophia Antipolis, Southern France. The MCC was created in March 1999 to provide support not only to the 3GPP but also to ETSI's own studies in mobile technologies. A full description of MCC including the financing is given in Chapter 15, Section 3.

9.1.3.2 Who are the Stakeholders in 3GPP?

3GPP has attracted a very strong commitment from organisations and companies around the world, reflecting the truly global nature of the project. There are currently six OPs (in alphabetical order):

- The Association of Radio Industries and Businesses (ARIB), Japan

- The China Wireless Telecommunication Standards Organisation (CWTS)
- The European Telecommunications Standards Institute (ETSI)
- Committee T1, US
- The Telecommunications Technology Association (TTA), Korea
- The Telecommunication Technology Committee (TTC), Japan

ARIB, ETSI, T1, TTA and TTC can be considered as founding OPs of 3GPP who signed the agreement in Copenhagen on 4 December 1998. CWTS signed the 3GPP agreement during the OP meeting #1 in Seoul on 8 June 1999.

In addition to the six OPs mentioned above, there are now seven MRPs (in chronological order of joining 3GPP):

- The UMTS Forum (December 1998 at Antibes/Juan les Pins)
- The Global Mobile Suppliers Association (GSA) (February 1999 at Cannes)
- The GSM Association (OP meeting #1 in Seoul)
- The Universal Wireless Communications Consortium (UWCC) (September 1999 at Geneva)
- The IPv6 Forum (OP meeting #2 at Sophia Antipolis)
- The Multimedia Wireless Internet Forum (MWIF) (during OP meeting #3 in Beijing)
- The 3G.IP Focus Group (during OP meeting #3 in Beijing)

In addition, by January 2000 3GPP had 284 companies participating as individual members, and the numbers continue to grow; their affiliation with the OPs is as follows:

ETSI (Europe) 173 companies   (61%)
T1 (US) 22 companies   (8%)
ARIB (Japan) 37 companies   (13%)
TTC (Japan) 18 companies   (6%)
TTA (Korea) 25 companies   (9%)
CWTS (China) nine companies   (3%)

The brackets at the end indicate the percentage of individual OP members active in 3GPP. From this, one can deduce that the representation from the three continents involved in 3GPP is as follows:

Asia   89 (31%)
Europe   173 (61%)
North America   22 (8%)

9.1.3.2.1 Leadership positions

During the first meeting of the 3GPP PCG and the OPs in Fort Lauderdale, US in March 1999, a lot of effort was undertaken in order to establish a good regional balance regarding the leadership positions within 3GPP. Fortunately, that goal was achieved as can be seen in the following:

PCG for the first 2 years: chairman from Europe, vice-chairmen from Asia (ARIB) and North America.

From 2001 on there will be an annual rotation. For 2001 the PCG chairman comes from Asia (ARIB).

Regarding the four TSGs: CN, RAN, SA&T, with one chairman and two vice-chairmen each, we can register the following situation as of the end of the year 2000:

- Europe has five (of 12) leadership positions;
- North America has four (of 12) leadership positions;
- Asia has three (of 12) leadership positions.

9.1.3.3 3GPP Meetings

Regarding the meetings within 3GPP, we have to distinguish basically between three different levels:

- PCG/OP meetings;
- TSG meetings; and
- WG meetings underneath the TSG level.

PCG/OP meetings are not that frequent. Until April 2001, the following meetings took place:

PCG#1	1–4 March 1999 in Fort Lauderdale, US
PCG#2	6–7 July 1999 at Sophia Antipolis, France
PCG#3	19–20 January 2000 at Sophia Antipolis, France
PCG#4	17 July 2000 in Beijing, China
PCG#5	14 November 2000 in San Francisco, US
PCG#6	10 April 2001 at Sophia Antipolis, France

The composition of PCG, the project co-ordination group, is as follows:

Six OPs with a maximum five delegates each;
Seven MRPs with a maximum of three delegates each;
Five TSGs with one chairman and two vice-chairmen each;
Two observers with one delegate each;
ITU-T and -R with three delegates as special observers,
One secretary, Mr Adrian Scrase.

Thus – ignoring any guests – the PCG may encompass up to about 70 delegates. This high amount of delegates within PCG – although most of the delegations do not send the maximum number of delegates allowed – does not correspond anymore to the "light structure" originally intended.

Decisions within PCG are taken by consensus among the OPs. In "unavoidable cases" a vote may be taken.

Most of the OP meetings were organised in connection with the PCG meetings as follows:

OP#1	27–28 May 1999 in Seoul, Korea
OP#2	18 January 2000 at Sophia Antipolis, France
OP#3	18–19 July 2000 in Beijing, China
OP#4	15 November 2000 in San Francisco, US
OP#5	11 April 2001 at Sophia Antipolis, France

The OP meetings are composed of delegations stemming from the participating OPs. They are normally chaired by a representative of the hosting OP. The delegation number is not fixed, but as most OP meetings are joined with a PCG meeting, the delegations from the OPs do not differ very much from PCG meetings with the difference that the TSG chairmen and vice-chairmen are now part of the OP delegations. Up to now the MRPs have been invited to participate as guests within the OP meetings.

There is a 3GPP Funding and Finance Group, chaired by Mr Phil Davidson, active and reporting to the OP meetings providing appropriate advice.

The TSGs meet quarterly in parallel (CN, RAN, T) and in sequence with SA. The TSGs have met 11 times since the creation of 3GPP.

TSG#1	Sophia Antipolis (December 1998), France
TSG#2	Fort Lauderdale (March 1999), US
TSG#3	Yokohama (April 1999), Japan
TSG#4	Miami (June 1999), US
TSG#5	Kyongju (October 1999), South Korea
TSG#6	Nice (December 1999), France
TSG#7	Madrid (March 00), Spain
TSG#8	Düsseldorf (June 00), Germany
TSG#9	Hawaii (September 00), US
TSG#10	Bangkok (December 00), Thailand
TSG#11	Palm Springs (March 01), US

At those TSG meetings between 400 and 600 delegates can be expected. During the recent ones we have been closer to 600 – quite a challenge from the logistics point of view!

The working groups belonging to the TSGs may even meet more frequently. Therefore, it does not make much sense to refer to them here in more detail. But it should be clearly pointed out that the bulk of the 3GPP work is being done there...

9.1.3.4 What Progress has been Made in 3GPP?

In little over 1 year, the project had already produced the first series of more than 300 specifications and reports in what is called Release 99. The specifications in Release 99 include those that define UTRA, the radio interface, and these have been submitted by the 3GPP OPs to the ITU-R for reference in its IMT.RSPC Recommendation, which forms the compendium of 3G terrestrial radio interfaces.

In addition to UTRA, this initial release of specifications includes the definition of around 50 services including multimedia messaging, plus architectural aspects, other features and enhancements. Release 99 has enabled industry to proceed with the development of the system, which, as noted earlier, is planned to come into service in Japan in 2001, with progressive launches around the world thereafter. These incredibly short timescales have necessitated an extremely efficient, and yet open, specification process. The fact that around 300 vital specifications have been produced so quickly confirms both the commitment of all concerned and the effectiveness of the 3GPP process.

The Project's specification activities will continue as it develops Release No. 4 (scheduled for March 2001) and beyond, providing more enhancements and features including an "all-IP" based network, specification of a UTRA repeater, further refinements of radio access modes, seamless service provisioning, enhancements to security, emergency calls, languages and alphabets, optimisation of power, spectrum and quality of service, and numerous other aspects. The task also entails an updating of the Release 99 specifications as needed (the technology is still evolving), for which a mechanism for handling quarterly updates has been established.

9.1.4 What is the Relationship Between 3GPP and ITU?

The 3GPP has been recognised by the ITU as one of the sources of technical specifications for the IMT-2000 family. There is thus a clear understanding between the two parties, which includes that 3GPP results will be submitted to the ITU where appropriate. However, because of the Project's status it does not contribute directly to the ITU. Formal contributions to ITU Study Groups, based on 3GPP Technical Specifications and Technical Reports, are made by individual members or OPs who are also members of the ITU.

The ITU entrusts the work of developing the standards needed for 3G systems to groups such as 3GPP, 3GPP2, UWCC and ETSI. For its part, the ITU is focusing on the interfaces between IMT-2000 family members to ensure seamless operation for users. A large number of 3GPP specifications, notably those for UTRA, have been accepted by the ITU as an essential component of its IMT.RSPC Recommendation, and 3GPP will continue to contribute to this process as the ITU updates and enhances the Recommendation.

In order to improve the exchange of information with the two sectors of the ITU, ITU-R and ITU-T, 3GPP agreed during its recent PCG/OP meetings in San Francisco to provide a special observer status for the ITU within the PCG of 3GPP.

9.1.5 What is the Relationship Between 3GPP and Other Initiatives?

The technologies that form the terrestrial component of IMT-2000 are being developed in several different communities: 3GPP is producing the specifications for the UTRA FDD (W-CDMA) mode and the UTRA TDD modes (high and low chip rates, the low chip rate mode

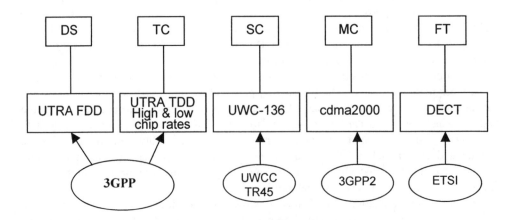

Figure 9.1.1 The five IMT-2000 terrestrial interfaces agreed by ITU-R

being TD-SCDMA, proposed by CWTS. A similar group, 3GPP2, is preparing the cdma2000 specifications, UWC-136 TDMA specifications are being developed principally by UWCC, and DECT specifications are defined by a set of ETSI standards (Figure 9.1.1).

Since the UWCC and ETSI are both partners in 3GPP (as an MRP and OP respectively), there is direct liaison between the Project and these two sources of specifications for other IMT-2000 family members.

When 3GPP and 3GPP2 were created it was felt that the interests of each group were sufficiently different to require them to remain separate but here again many of the OPs and MRPs are common to both groups, so there are natural channels for information exchange. Particular instances of formal co-operation between the groups were the two joint 3GPP-3GPP2 workshops held in 1999 to address the "Hooks and Extensions" issue aimed at ensuring interworking between the respective technologies.

In addition to all this, the 3GPP TSGs and their working groups are encouraged to liase directly with relevant technical bodies within the Project as well as among the Partners.

9.1.6 Conclusions

The attentive reader, and let's assume that you are one, may still be astonished to learn what big efforts have been necessary in order to establish 3GPP.

In the meantime, i.e. in the course of year 2000 nearly the whole work of the former ETSI TC SMG has been incorporated into 3GPP, to a large extent within the new TSG GERAN.

Well, our Japanese friends were not happy about this shift because they feared that it could result in delays in the 3G introduction in their country. They accepted this because a separate TSG for the GSM work was created.

Trying to recall the heated discussions within ETSI about the scope of 3GPP from 1998, one has to admit that such a move would have been violently rejected by a lot of ETSI members, maybe even by the majority. Using a philosophic approach, one can, of course, state that time helps healing (even wounds).

Today, i.e. February 2001, we can conclude from ETSI's point of view that all standardisation work related to further evolution of the second generation mobile system, i.e. GSM, and to UMTS is being done well within 3GPP. Thus, the objective to avoid parallel work within ETSI TBs and 3GPP has been fully achieved.

The ability of organisations and individuals around the world to co-operate and make available a full set of stable, agreed 3GPP specifications in 1 year is a remarkable achievement, one that is unprecedented in the world of standardisation. 3GPP meetings have taken place in many parts of the world, emphasising the strong commitment of the SDOs from China, Europe, Japan, Korea and the US. The MRPs have also lent very strong support to the work. Thanks to this widespread and determined commitment it has been possible to meet the very aggressive targets for 3G.

But Release 99 was only the first step – the workload has continued to intensify this year, as the initial specifications are refined and many new ones added, opening the path to full, seamless, global 3G services, changing forever the way that people communicate.

In the meantime, 3GPP has changed the designation of the releases and dropped to mention the year in order to avoid irritations and wrong expectations. Release 99 was first established in December 1999 and got its maturity in the course of year 2000. In 2000, more than 5700 change requests were implemented. The next big event will be Release #4 as a result of the series of TSG meetings scheduled for March 2001 in Palm Springs, US. This will obviously result in a further stabilisation and extension of the UMTS specifications.

3GPP is running very well. The individual members from the six OPs seem to be very satisfied with this arrangement – and the production line is in full swing.

Without exaggeration, one can give 3GPP the attribute of a success story. One may quote again here Mr Ed Roney who even addressed the 3GPP concept – prior to its realisation – as a "paradigm shift".

As the results of the GSM and UMTS related standardisation work represent a great part of ETSI's deliverables, it might be justified to note here that during the year 2000, ETSI published more than seven new deliverables each working day (Monday through Friday), i.e. one deliverable per hour!

Further information may be found on the 3GPP website at http://www.3gpp.org.

Chapter 9: The Third Generation Partnership Project (3GPP)

Section 2: UMTS in 3GPP (December 1998–May 2001)

Niels Peter Skov Andersen[1]

9.2.1 A Change of Environment

In the period 1982 until end of 1998 the work on the GSM standard, and in the later part of the period on UMTS, had been performed in the same environment, starting under CEPT and later transferred into ETSI. The Technical Committee GSM, during this period renamed to SMG, and its working groups (Sub Technical Committees) had continuously existed and evolved. The same was the case for the working methods and procedures used within the work. Over time with the success of the GSM system more and more interested parties became involved in the work including parties from outside the original CEPT area. However, this was all a relatively slow evolution and no major revolutions in the organisation or the working methods occurred in this period.

After all these years of continuity in the work the discussions around the creation of 3GPP and the decision to establish 3GPP for the initial phase of UMTS[2] naturally created some uncertainty amongst the members of SMG. Especially the resulting split of the GSM standardisation, with the responsibility for the GSM core network transferred to 3GPP, but the responsibility for the GSM radio access Network maintenance remained in ETSI in SMG. This caused some concern amongst many delegates. Also the internal structure for the technical work within 3GPP was different from the well-known structure in SMG. SMG was based on a technical plenary with a number of working groups (SMG1, SMG2, ..., SMG12) performing the detailed technical work. The SMG plenary was the approving authority for the results of the work performed by the working groups. Also the plenary was the group responsible for approval of all new work items and the content of the releases. The structure for the work in 3GPP, as agreed by the partners, was quite different. The project

---

[1] The views expressed in this section are those of the author and do not necessarily reflect the views of his affiliation entity.

[2] The term UMTS is throughout this section used to keep consistency of terminology with the other chapters and sections. The term UMTS do not appear in the in 3GPP agreement, which defines the system as a third generation mobile system based on an evolved GSM core network and UTRAN (including UTRAN (FDD and TDD modes)).

was organised with four equal Technical Specification Groups (TSGs), who had complete autonomy for their area of responsibility, i.e. they were responsible for approval of new work items and final approval of deliverables. The four technical groups originally defined were:

TSG CN	Responsible for the core network development
TSG-RAN	Responsible for the radio access network based on UTRAN (FDD and TDD modes)
TSG-SA	Responsible for services and system aspects
TST-T	Responsible for Terminal and UIM

In addition to the technical groups the 3GPP organisation has a Project Coordination Group (PCG). However, the role of this PCG cannot be compared to the role the SMG plenary played. The SMG plenary was an open technical group with the approving authority in all technical questions including approval of new work items. The 3GPP PCG is a closed group with a defined membership consisting of a limited number representative of each of the partners (SDOs, MRPs) and the leadership (chairman and two vice-chairmen) of each TSG. Thus as a closed group the role of the PCG becomes more like a board overlooking the overall well being of the project.

This structure made many long-term SMG delegates concerned about how the overall coordination of the project could be ensured. This new structure was not introduced to overcome known deficits of the SMG organisation, but in my opinion, by political considerations to ensure than no single individual, individual member, organisational partner could obtain a controlling position in the project.

9.2.2 The First Two TSG Meetings

The inauguration meeting of the 3GPP TSGs was held in December 1998 in Sophia Antipolis, France. In the process of creation of 3GPP this was the first time that the 3GPPs real work force – the technical experts – met. The main objectives for this first meeting was to get the work started. One of the elements of the meeting was a presentation from the different partners on the status of their work on the third generation mobile system, the work, which they now were in the process of handing over to 3GPP.

Listening to the presentations and the discussions during the breaks it was very obvious that the background for standardization amongst the delegates was quite different. As an example, I remember that during the coffee break just after I, as chairman of ETSI SMG2, had presented the status of the UMTS radio work in ETSI, and had ended my presentation by stating that the UMTS radio work would only be on the agenda of one more meeting of ETSI SMG2. This was in order to complete the documentation to be handed over to 3GPP and then the work on UMTS radio in ETSI would cease, a small group of non-ETSI delegates came to me and asked "if all work on UMTS radio in ETSI ceases, how do the Europeans then co-ordinate their views on 3GPP?" Coming from the ETSI SMG background this was a comple-tely unexpected question, as the working procedures for 3GPP were very similar to those of ETSI, it was clear to me that the contributions to 3GPP in general should come from the individual members – the companies, regulators etc. – in their own name and not as regional contributions. I explained this, but I also understood that for delegates with a background in international standardization from, e.g. ITU this was the normal way of thinking. During this first meeting a lot of small explanations similar to this were given over a cup of coffee and already by the second meeting there was a far better common understanding on how the work in the groups was intended to be performed.

Even though the partners, already before the first meeting of the TSGs, had made the principle decision of having four TSGs and had elaborated draft terms of references for the groups, the definition of the area of responsibility for the TSGs and refinement of the terms of references was a key item on the agenda. Each of the TSGs adjusted their terms of references and with some subsequent adjustments at the second meeting, the terms of reference for the TSGs have until now (March 2001) stayed the same except for few minor adjustments.

In order to get the detailed work started and not loose the momentum, which had existed in the SDOs before the creation of 3GPP, it was a very important task at the first meeting of the TSGs to get the detailed work within the TSGs organised so technical work could commence and progress in the period up to the second meetings of the TSGs in March 1999 in Fort Lauderdale. This part of the programme for the first meetings of the TSGs went well, and by the end of the meeting each of the TSGs had established between three and five working groups, outlined their area of responsibility and appointed convenors for the groups. With the establishment of the working groups the detailed technical work was ready to start, and already by the second meeting of the TSGs significant progress was reported.

By the second meeting of the TSGs, which took place in Fort Lauderdale, the complete atmosphere had changed from the general uncertainty and procedural questions to a far more technical focus, even though a few items of a management and organisational nature still needed to be sorted out. In addition, at this second meeting the leadership (chairman and vice-chairmen) for the individual TSGs was elected for the next 2-year period.

As indicated, one of the main differences with the 3GPP organisation compared to the organisation of SMG was the lack of a superior technical group with an open plenary with responsibility for the technical coordination, final decision making, conflict resolution and the project management including adaptation of work items, etc. Already the original description for the role of TSG SA, which was elaborated by partners together with the 3GPP agreement in Copenhagen in early December 1998, contained a paragraph on giving TSG SA the role of – "High level co-ordination of the work performed in other TSGs and monitoring of progress". This role was subsequently reflected in the terms of references for TSG SA agreed at the first meeting of TSG SA (TSG SA#01). At the second meeting of TSG SA the TSG SA convenor Mr Fred Harrison, BT, provide a proposal[3] for how the TSG SA could fulfil its project coordination role. The key principles of the proposal were:

- To establish a project management function to create and maintain a cross TSG project programme including status of technical specification and reports.
- To establish close co-operation with TSG CN; TSG RAN and TSG T. Requiring the chairman or vice-chairman of each TSG to attend the TSG-SA meetings and bring new work items, issues and progress information to the attention of TSG-SA.

At the meeting another proposal[4] was received from a group of companies[5] who suggested that a TSG plenary be created, i.e. a fifth TSG with plenary function similar to that of ETSI

[3] SP-99050: proposals for managing the TSG project co-ordination role.
[4] SP-99068: TSG plenary.
[5] AT&T, BT, FRANCE TELECOM, NTT DOCOMO, TIM, TMOBIL.

SMG. The argument for this proposal was that a TSG plenary would help to ensure overall project coordination and elaboration of a consistent and complete set of UMTS specifications.

After long discussions a compromise not requiring changes to the TSG structure was found and agreed. This comprise[6] was based on the following principles for the TSG SA's project coordination role:

- At least while performing its project co-ordination role, the TSG SA will not meet at the same time as other TSGs.
- At least one representative of TSGs RAN, CN and T and their working groups will attend each TSG SA meeting, to report on the activities of their respective TSG. They shall be responsible for bringing new work items, issues and progress statements on work such as specifications and existing work items from their respective TSGs to the attention of TSG SA.
- The TSG SA plenary will also include reports from its own working groups and facilitate information exchange between those working groups and the other TSGs.
- The TSG SA shall have arbitration responsibility to resolve disputes between TSGs.

As can be seen from the principles, the independence and the rights of the other TSGs was not touched by the compromise. Each TSG maintained its right to approve work items and deliverables, etc. As a result of the way forward on the TSG SA management role, the TSG meetings in Fort Lauderdale were the last meetings where all four TSGs met in parallel. At the subsequent TSG meetings in Shin-Yokohama in Japan at the end of April TSG CN, TSG RAN and TSG T met in parallel followed by TSG SA and the chairmen of TSG CN; TSG RAN and TSG T provided to TSG SA a status report on the work and progress in their respective TSGs. The TSG SA meetings starting from the third meeting in Shin-Yokohama then had a three part structure. A part related to TSG SA internal matters where the different TSG SA working groups report the progress of their work and submit their contributions for approval, this part is similar to the work in the other TSGs. A second part related to the technical coordination with the other TSGs and a third part dealt with general project management issues such as working methods, document handling, etc.

By the end of the second TSG meetings most of the "beginners" difficulties had been resolved, the interaction between the TSGs defined and TSG SA was ready to take on-board is role in the coordination role. Also the second TSG meetings showed that the detailed work in the working groups had got a good start, the work handed over from the partners was well received and progressing well. All in all, the definition and establishment phase of the technical work in 3GPP had been completed successfully and the transfer of work from the partners to 3GPP had been performed without causing any major disruption in the ongoing technical work.

9.2.3 The First Release – Release 99

After the two first two meetings of the TSGs where especially TSG SA had used time to organise the work, the third meetings were into their routine and could fully concentrate on the technical specification work.

The work in 3GPP followed the same basic methodology as was used for the GSM work in ETSI. The specifications generally are based on a three stage approach, with a stage 1

[6] SP-99087: proposals for managing the TSG project co-ordination role.

description containing the functional requirements, stage 2 containing the overall functional description and architecture for a given functionality and stage 3 being the detailed technical specification down to the bit level. Working with this methodology the idea is of course that the stage 1 description is first completed or nearly completed so the requirements are clear. The next step is then to complete the stage 2 description and thereby define the overall architecture and functional split for the technical realisation of the functionality. When stage 2 is complete or close to completion the third step the stage 3 specifications containing the detailed technical specification is complete.

However, it was not possible for 3GPP to do this work serially, because of the very short timescale for completion of the first set of specifications in December 1999 only 1 year from 3GPP's creation in December 1998. Thus the work on stage 1, 2 and 3 specifications had to a large degree to be performed in parallel. Doing so TSG SA WG2, which is responsible for system architecture, quickly became a bottleneck in the process, as it was difficult, especially for TSG CN (core network) to draft the detailed specification before the architectural decisions were made. This problem peaked at the fourth TSG SA meeting in June 1999, when going through the status report from TSG SA WG2, where it became clear to the full membership that an extraordinary effort was needed to ensure that the architectural work was speeded up.

Standardisation by committee is not a traditional project, where the project leader can reallocate resources to the most urgent task. In standardisation the important task is to ensure that all the participants know and understand where additional effort is most urgently needed, so the volunteer work effort is pointed in the right direction. The recognition of the need for an extraordinary effort in TSG SA WG2 helped to speed up the architectural work and minimise the problem of TSG SA WG2 being a bottleneck. The initial delay of course made the work schedule even tougher for the groups responsible for the detailed stage 3 specifications.

As you can imagine it is not possible here to go into the details of the work, which led to the first set of specification from 3GPP in December 1999. In the following I will therefore only provide a few of examples of items, which required resolution by TSG SA.

For UMTS a new ciphering and authentication mechanism providing a higher degree of security has been developed. The SIM card (for UMTS USIM) is involved in the authentication process and calculates the necessary keys for the authentication and ciphering. Thus new SIM cards are required, or to be technically correct, cards with the USIM application are required. In the following I will use the short term USIM to indicate the card supporting the new security algorithms and SIM for the old cards supporting the GSM level of security. At the third meeting of the TSGs there was the question of whether the UMTS networks should only support USIM and thus always provide the highest possible degree of security or whether it should be possible to access a UMTS network with terminals with a SIM only. On one hand a number of delegates believed that it was preferable only to allow the usage of USIMs in the UMTS terminals, this on the other hand was questioned by operators that could foresee a slower roll-out of UMTS, e.g. due to the expected licensing time. For them a requirement for usage of USIM only in the UMTS terminals would leave them with two alternatives; either to issue USIMs even though they did not yet have a UMTS network, or be in a situation where their customers could not roam to, e.g. Japan and Korea with no GSM networks but only UMTS networks. This lead to a long discussion where it could have been tempting to perform a quick vote; however, to keep the good spirit of cooperation and consensus based work I as 3GPP TSG SA chairman considered voting as an emergency solution if everything else failed. As almost always the attempt to find a solution for which consensus could be obtained succeeded. The comprise was found based on the following elements:[7]

- Support access to UMTS access networks while using cards equipped with either the SIM, the USIM functionality or both; and
- Allow a serving UMTS operator the option to block access to the UMTS access network when a card equipped only with a SIM functionality is used.

As usual when compromises of this type were obtained it was the assumption of the meeting that the companies/members who required the capability should do the work to specify the signalling and other mechanisms required.

At the fourth TSG meetings the very rare situation of one of the other TSGs raising an issue to TSG SA for resolution occurred. TSG CN had completed the feasibility study of the Gateway Location Register (GLR). TSG CN had then decided not to start specification work for the GLR. However, as some members of TSG CN had expressed strong interest in the GLR, it had been proposed to let the interested parties elaborate the specifications required for the GLR outside TSG CN and submit the result to TSG CN. This decision had caused some problems and the TSG CN raised the question to TSG SA of how to proceed, e.g. should a vote be taken. I as chairman of TSG SA indicated to the meeting that votes were to be seen as an emergency solution when everything else has failed. First, an attempt should be made to find a solution for which consensus can be obtained. For this explicit case it seemed clear that the resistance to start work on the GLR was coming from operators not seeing the need for a GLR and fearing that the introduction would impact existing networks and other networks without a GLR. On the other hand especially operators with no GSM legacy network showed a strong interest in the GLRs as a way to reduce the amount of international signalling caused by roamers moving around in very densely populated areas. Taking into account the strong interest and the concerns expressed, it was found, that there would be no problem, if a GLR could be done in such a way, that it had no impact on an existing HLR[8] (pre-3G), if a subscriber belonging to a HLR roamed onto a network utilising a GLR. Similarly the support of the GLR in one network should not impact networks not utilising the GLR. Based on this analysis, TSG SA recommended that TSG CN adopt a work item on GLR requiring a GLR to be fully compatible with old and new non-GLR networks. As hopefully can be seen from this example it is and has been a key priority in 3GPP to as far as possible base decisions on consensus as it also was the case for the GSM development in ETSI.

Another type of problem, which every now and then needs resolution at TSG level is the specific national or regional requirement often caused by the local regulation. Requirements that often can cause problems in relation to roaming. One example of this is the emergency call where TSG SA at meeting number 5 received a proposal[9] for national variation on terminals to cater for the differences in emergency call requirements. When GSM was introduced one unique number for initiating emergency calls had been defined (112); this ensured that a roaming user would always be able to perform a emergency call without

---

[7] SP-99208.
[8] HLR = home location register.
[9] SP-99481.

knowing a any specific local situation. When GSM entered into new parts of the world this function had been improved by letting the local operator store a number on the SIM card which should be considered as the emergency call number, and thus, e.g. and American user could use 911 wherever he brought his mobile. However, there are other differences in the handling of emergency calls other than just the number to dial. The GSM solution only allows routing of emergency calls to one central emergency centre and does not differentiate the type of service needed such as ambulance, fire brigade or police. However, some operators had a regulatory requirement to route directly emergency calls to the relevant service and thus needed different numbers per service. Therefore, they had suggested having a national variation of terminals. After some discussion in TSG SA the proposal was rejected. The main reasons for this was that it was seen as essential to avoid local variations of terminals and secondly a solution based on local variation of terminals would not solve the problem of subscribers roaming from other parts of the world with terminals without the specific local variation. Anyhow, the rejection of the proposal did not mean that the problem was ignored; on the contrary the relevant working groups were tasked to find a generic solution, which would satisfy the local regulations without causing problems with roaming or requiring variation in terminals.

That the previous examples from the elaboration of 3GPP Release 99 all come from the TSG SA does not mean that this type of problem does not appear in the other TSGs. As also can be imagined, the specification of a complete new radio access network in TSG RAN in the timeframe of 1 year was one of the most demanding tasks during the elaboration of the first set of specifications from 3GPP (Release 99).

As mentioned earlier, when 3GPP started in December 1998 a target date of December 1999 was set for the first set of specifications. So the sixth meetings of the TSGs in Nice, France in December 1999 were the meetings where the status for the first year of 3GPP was to be made. In order to get a full overview of the status of the work and the degree of completion, the process for documenting the remaining open issues had been agreed amongst the chairs and vice-chairs of all of the TSGs.

The principle for this was relatively simple and building on the assumption and desire that a set of specifications should be completed and frozen at the sixth meetings of the TSGs. The term frozen meant that there should be no functional changes or additions made to the set of specifications, but only strictly necessary corrections of errors or omissions which if uncorrected risk making the system malfunction. The idea behind the principle was that at the next meetings of the TSGs all proposed changes to the specifications, which could not be justified as an essential correction should be rejected, unless an exception for that specific item had been given in December 1999. In order to document these exceptions all working groups and TSGs had prepared and forwarded to TSG SA sheets describing the non-completed functionality for which they wished to have granted an exception from the general rule of no functional changes. In addition to the description of the functionality, the sheet also indicated the consequences if this functionality was completely removed from Release 99.

TSG SA collected the status reports from the different groups and created a relatively large table[10] where on one side was the different functionalities and on the other side the different groups and in the table an indication if a group had requested an exception for completion of the functionality. After having created this table based on the status reports, TSG SA went through the table on a per functionality basis and evaluated the expected completion date and

---

[10] SP-99639.

the necessity of the function in Release 99. In order to maximise the stability of the set of specifications, especially in the case where several groups had items open for the same functionality, specifications were scrutinised in detail and in several cases the functionality was completely removed from 3GPP Release 99. This review led to the removal of function-alities such as Enhanced Cell Broadcast, Tandem Free Operation for AMR, Support of Localised Service Area and a reduction in the location service functionality in Release 99.

At the end of the December 1999 TSG SA meeting approximately 80 exceptions from the rule of no functional changes were granted. At the following meeting of the TSGs in Madrid in March 2000 the status and the list of open items was once again reviewed and the number of open items was reduced from 80 to approximately 30. At the TSG meetings in June 2000 the remaining open items were completed and since then only necessary corrections could be made. However, it is to be understood, that when such a substantial set of specifications for the 3GPP Release 99 have been elaborated in the time frame of approximately 1 year, it is unavoidable that there are some ambiguities and errors in the specifications. It is a very important task to have these errors corrected in the specification as soon as they are discov-ered, as this is the only way to avoid small differences in implementation due to different solutions to errors. Differences which if not avoided could lead to problems of interoper-ability, etc. Also it should be noted that there will continuously be errors discovered in the specifications which need to be corrected, at least until every detail has been implemented and made operational in the field.

9.2.4 Introduction of Project Management

As indicated, one of the main differences with the 3GPP organisation compared to the organisation of SMG was the lack of a superior technical group with an open plenary with responsibility for the technical coordination, final decision making, conflict resolution and the project management including adaptation of work items, etc. Instead the different TSGs approved work items and technical work on their own. Even though they reported the status of their work to TSG SA there was no simple way to for linking a given functionality with the work being performed in the different TSGs. This was clearly a problem during the elabora-tion of Release 99, as it was difficult for the delegates to get an overview of which function-alities were on the critical path for completion. To get an overview actually required that key experts from the different areas sit together and fit the different parts of the puzzle. It there-fore, required quite some effort in and outside the TSG meetings of December 1999 to provide an overview, which allowed the meetings to make conscious decisions.

As this potential problem was clear to me from the start of the project, I had, already at the second meeting of the TSGs in March 1999, had discussions with the chairmen of TSG SA WG1 and TSG SA WG2 on introducing a model for the project co-ordination which would follow the work from the initial requirements to completion. This model was then introduced for initial discussion to the leadership (chairmen and vice-chairmen) of the other TSGs at the third meeting of the TSGs. During the rest of 1999 additional background work was done in order to prepare for the introduction of the model for project co-ordination. At the December 1999 TSG SA the model was presented to TSG SA for approval and became the model for the organization of the work for the following releases and the basis for the overall project plan.

The model was based on the introduction of the Feature, Building Block and Work Task concept, and categorization and linkage of the work items. The model was thought of as a reference model for structuring the work. It was not the intention to rigorously enforce the usage of the model on all ongoing work, but merely to use the model as a common reference model across the TSGs and to structure future work. The model took its origin from the typical flow for creation of a new feature or service and can briefly be described as follows.

TSG SA is through TSG SA WG1 responsible for defining the features and services required in the 3GPP specifications. TSG SA WG1 is responsible for producing the stage 1 descriptions (requirement) for the relevant features and passing them on to TSG SA WG2. TSG SA WG1 can also forward their considerations on possible architecture and implementation to TSG SA WG2, but is not responsible for this part of the work.

TSG SA WG2 should then define the architecture for the features and the system, and then divide the features into building blocks based on the architectural decisions made in TSG SA WG2. TSG SA WG2 will then forward the building blocks to the relevant TSGs for the detailed work. These proposals will be reviewed and discussed in an interactive way together with TSGs/WGs, until a common understanding of the required work is reached. During the detailed work of the TSGs and their working groups, TSG SA WG2 is kept informed about the progress.

The TSGs and their WGs treat the building block as one or several dedicated Work Tasks (WTs). The typical output of a given WT would be new specification(s), updated specification(s), technical report(s) or the conclusion that the necessary support is already provided in the existing specifications.

A part of TSG SA WG2's role is in co-operation with the TSGs and their WGs to identify if synergy can be obtained by using some of the building blocks or extended building blocks for more than one feature. Part of TSG SA WG2's task is to verify, that all required work for a full system specification of the features relevant take place within 3GPP without overlap between groups. In order for TSG SA WG2 to be successful, this has to be done in co-operation with other TSGs/WGs.

About the project scheduling: TSG SA WG1 sets a target, TSG SA WG2 performs a first technical review and comments on the target. TSG SA WG2 indicates some target for time schedule together with allocation of the defined building blocks. The TSGs and their WGs comment back on these targets. TSG SA WG2 tries if necessary to align the new target between the involved parties. TSG SA WG1 and TSG SA are kept informed of the overall schedule.

It was also in the model, it was identified as a task for TSG SA, TSG SA WG1 and TSG SA WG2 to ensure early involvement of TSG SA WG3 (working group responsible for security) to ensure that the potential security requirements, service requirements and the architectural requirements are aligned and communicated to the TSGs and their WGs.

In order for TSG T and its subgroups to plan and perform its horizontal tasks on conformance testing and mobile station capabilities, it was foreseen to invite TSG T to evaluate the potential impact of a new feature. Also work on the horizontal tasks is required to be included in the overall work plan.

With the acceptance of the modeling of the work based on the work breakdown into features, building blocks and work tasks, the next step was to map the work onto the model, create the corresponding work items for the features and building blocks and establish a first version of an overall project plan for 3GPP. In order to kick-start this process a number of Inter Group Coordination groups were establish within TSG SA WG2. The purpose of these groups was to try to establish a first version of a project plan for a given area. To ensure the correctness of the information rapporteurs and representatives from the different working groups were invited to either participate or provide status and planning information, which then was used to establish a "traditional" project plan. Also the groups identified and informed the relevant groups if, e.g. building blocks or WTs were missing.

After the establishment of stable versions of the project plan covering ongoing activities for all of the TSGs and their working groups. The responsibility for maintenance of the project plan, was shifted so each TSG was made responsible for keeping updated the parts of the work plan, which correspond to their work. The practical maintenance of the project plan was then transferred to the MCC, the team of technical experts functioning as technical secretaries for the groups and responsible for implementation of the decisions of the meetings. The MCC corresponds to the Permanent Nucleus later known as PT12 during the elaboration of GSM.

Today the project plan is just another well functioning and convenient tool, which allows delegates and their organizations a quick overview of the status of the ongoing activities. However, this is only possible because the different groups and the MCC make a significant effort in keeping the plan up to date.

In the August 2000 TSG SA held an ad-hoc release planning, which recommended entirely controlling the 3GPP work program via the work plan, and doing this independent of releases. This recommendation, which later was confirmed by TSG SA further proposed that approved work items introduced into the plan are given calendar target dates and not particular release target dates. These "calendar" work item target dates will need to monitored and adjusted as work and knowledge about the work items progress. For this purpose reasonable milestones shall be defined. The work plan calendar should then also indicate planned future release dates with reasonable frequency to allow for stability, e.g. approximately every 12 months, depending on whether there would be enough completed work to justify the issue of a release.

The content of each release could then be easily deduced from the work plan, i.e. those items scheduled for completion by the closing day for the release being included in that particular release, a 3GPP road map. The definition of the content of a release could then be based upon the work plan, with a review of the release content starting approximately 6–9 months before the initial predicted closing date of the release. Work items not completed at the chosen closing time of the release are not included in that particular release. Maintaining the closing date of a release is a priority. Only when it is identified that no substantial new features would be available at the target date, is shifting the date considered to be an option.

In addition, independently of the actual release date, upon completion of a particular work item, the work item is frozen, denying any further functional change on the completed work item, permitting only essential technical corrections. This helps stabilize the specifications and the availability of the draft new release versions of the specifications can assist companies wanting to start developing the new features.

In all, the definition and establishment of an overall project plan was successful and has provided a high degree of visibility of 3GPP's activities. Especially, when the second set of specifications from 3GPP (Release 4) was completed in March 2001. The advantage of having the project plan to identify the completed features showed a major advantage and helped simplify the work compared to when Release 99 was completed. Also the process has changed from a release centric approach to a project plan approach with individual planning for each function or feature. To mark this change the naming of the releases was decoupled from the calendar and changed to refer to the version number on the specification and thus what would in the old philosophy have been called Release 2000 is called Release 4, which then is to be followed by Release 5, etc.

9.2.5 Technical Work in 3GPP Following the First Release

About the first release of specification from 3GPP, one can in short describe the system specified as a core network evolution where the circuit switched domain provides circuit oriented services based on nodal MSCs (an evolution of GSM). Similarly the packet switched domain provides IP-connectivity between the mobiles and IP-networks based on an evolved GSM GPRS core network. In contrast to this the radio access network is a complete revolution with a brand new radio access technology. From this background it was not a major surprise that the most significant changes to come in the next releases are focused on the core network side.

Already when the work after Release 99 was discussed for the first time at the fourth meeting of the TSGs in Miami, this trend was clear. It was at this meeting that 3GPP accepted the idea of specifying an all IP based architecture option, i.e. an architectural option not requiring the traditional nodal MSC. The work on an all IP based architectural option started with a short feasibility study to identify the implications and to plan the time-scales. However, this work progressed so fast and in parallel with the time critical task of completing Release 99 that several organization, especially those amongst the smaller operators had problems following the work. Also the architectural analysis progressed much faster than the work on requirements. Therefore, in order to bring everybody level again, it was, at the TSG meetings in December 1999, decided to hold a workshop on the subject of the "All IP" option. This workshop took place in Nice, France in February 2000.

The "All IP" workshop in February 2000 was organized as a two part event, the first part where members were invited to present their vision for the "All IP" work, being about operational scenarios, technical visions, etc. The second part of the workshop was used to draw up the general trends from the presentations and thereby identify the goals by going "All IP", the requirements for the solutions and the way forward.

From the discussions it was clear that the key motivator for moving toward the "All IP" option was to establish a flexible service creation environment, allowing for quick service/ application creation with well defined APIs allowing for third party applications and thus allowing gain from Internet as well as intranet services. Further, the development should provide for real time applications including multimedia services, this to allow the operators to market new and interesting services allowing the creation of additional revenue streams. Further, the introduction of IP based architecture was seen as providing the option for independence of access type and thus allowing seamless services across different access networks. Also the independence of access type could allow savings through the common development of services for several access types. Clearly one of the key motivators for the operators' interest in an IP based architecture was the expectation of cost reduction due to the possibility of leveraging the IP technology cost factor and the expected gains from the better scalability compared to nodal switched based networks.

From the discussions at the workshop it was also clear that a hybrid circuit switched and packet switched network would exist for a long time. It was also clear that the changes towards the IP based architecture should not be done at any price. Especially, the need for an open multi-vendor environment with at least the same quality and security levels as the "state of the art" mobile networks at the time of introduction. Another requirement identified due to the co-existence of the circuit switched and packet switched domains was the requirement for service transparency across domains. Finally, an important and far from trivial requirement to fulfill, was the need to respect spectrum efficiency. It was noted at the workshop, that the IP header was actually larger than a standard 20 ms speech frame in the cellular system, which on its own clearly made the spectrum efficiency requirement a challenge. During 2000 the need for being economical with spectrum was clearly illustrated by the prices paid at the 3G spectrum auctions, with payments of approximately US$35 billion in the UK and approximately US$50 billion in Germany for the licenses to install and operate 3G networks.

At the workshop in February 2000, there were different opinions about what would be a reasonable and realistic timescale for the specification of the IP based architecture option. Some of the large operators indicated that they felt that a target date of December 2000 was too aggressive and not realistic, whilst other large operators indicated that they believed it could be completed by December 2000 and wanted to keep a target date of December 2000. Even though it was never said, one of the reasons for the aggressive timescale was clearly to ensure that the focus especially from the manufacturers was kept on this development, and not risk unnecessary delays, due to a time schedule, which people might regard as relaxed.

Even though the initial time schedule kept a target date of December 2000, in the further work the size of the task quickly became clear and some more realism appeared in the definition of targets in terms of content and completion dates. With respect to this it should not be forgotten that in difference to when working on the creation of the first release (Release 99), 3GPP now had a major task to perform in parallel to all new developments, that was the maintenance and error correction of Release 99. As mentioned earlier the first years of maintenance of a brand new standard are very time consuming, and thus it was very ambitious to plan for a next release already 1 year after the first. Even though the GSM work in ETSI used an annual release schedule, one should not forget that it took more than 3 years from the stable specification for GSM phase 1 before it was followed by the second set of specification for GSM phase 2.

Anyhow the second release (Release 4) was planned for and completed in March 2001, this without the result of the ongoing IP based work, which is the target for the next release (Release 5) expected approximately 1 year later than Release 4. Thus Release 4 does not contain significant revolutionary news, but instead it contains a number of smaller features and functionalities, which can be seen as an important complement to Release 99.

The work on the IP based architecture for Release 5 is focusing on the introduction of an IP multimedia subsystem, the part of the IP based network providing the capabilities for multimedia services. This choice has been made in order to ensure that the first results of the "All IP" work do not only provide for alternative methods of providing already existing and well known services, but also allow the operators to create new innovative services and new revenue streams which can justify the investment in the IP based architecture. The service drivers for Release 5 have evolved to be compatible with Release 99 and Release 4, with the addition of IP based multimedia services, including efficient support for voice over IP over the radio for the multimedia services. In Release 5 it is foreseen that the circuit switched domain is retained and provides 100% backward compatibility for the circuit switched services. Similarly the existing packet service domain is kept and the IP multi-subsystem is added and provides new IP multimedia services that complement the already existing services.

In the longer term, the IP multimedia subsystem might evolve to the extent to where it can provide all services previously provided by the CS-domain, and thus the specification will need to support all the commercial interesting services from today's circuit switched domain in the packet switched domain in the IP based architecture.

9.2.6 The Transfer of the Remaining GSM Activities into 3GPP

As described earlier, the original terms of reference for 3GPP covered a third generation mobile system based on an evolved GSM core network and UTRAN (including UTRAN (FDD and TDD modes)) and not covering the GSM/EDGE Radio Access Network (GERAN) part. This work together with a few other GSM only items remained in ETSI under the responsibility of SMG. This resulting split of the GSM standardization caused concern when 3GPP was created. However, time showed that it was possible to co-ordinate the work between 3GPP and SMG. For most areas, except for the GERAN specific work, co-locating the meetings of the SMG working groups with their corresponding 3GPP groups enabled the co-ordination. However, it was also clear that there was no longer one single forum with an overall responsibility for GSM as a system. This overall co-ordination was to some degree made during the TSG meetings, in the corridors and in the meetings by delegates, who ensured that the service, architectural and core network decisions would be compatible with the GERAN. However, this way of working reduced transparency of the background for arguments and decisions, both for those interested in the further development of GSM as a system and for those not interested in the GSM legacy.

In September 1999, Committee T1 sent a liaison to its 3GPP Organizational Partners requesting that the terms of reference of 3GPP be expanded to include evolved GSM radio access; that all evolutionary work of GSM should be transferred to 3GPP. The reasoning provided was that for the foreseeable future, the GSM/EDGE radio access would co-exist with the 3G radio access and there would be a clear benefit for all parties in ensuring co-ordination between the further GSM/EDGE development and the work related to the UTRAN access. Also the liaison statement indicated that by including the remaining GSM/EDGE radio work in 3GPP the overall number of meetings, liaison statements, etc. could be reduced and thus the efficiency increased.

At the 3GPP PCG meeting in January 2000, the responses from the other partners was tabled and discussed. ETSI indicated that they could support the proposal from T1 and suggested that the transfer should be effective from June 2000. ARIB indicated that 3GPP activities were based on common interest, meaning that each participating SDO and individual member needs to commit to the 3GPP objective and scope. ARIB continued that unfortunately, ARIB had no requirements to produce standards of GSM radio access including EGPRS in Japan. In conclusion ARIB could not support the request of ARIB individual members to take part in the study related to GSM radio access in 3GPP. Also the response from ARIB indicated concerns regarding potential impact on the timescales for the UMTS work as well as concern regarding financing of the project if not all parties had equal benefit of the work performed. TTA's response was very similar to that of ARIB additionally commenting that the existing process was functioning well.

At the PCG meeting CWTS indicated that they could support a transfer of the GSM radio work into 3GPP. After some short discussions it was agreed to form an "Ad-Hoc Group on Movement of Work into 3GPP" to assess the impacts and appropriate program structure to support the transfer of appropriate ETSI/SMG and T1 programs related to the GSM/EDGE radio access into a 3GPP. It was agreed that the work should be based on the following key assumptions:

- Any proposed new 3GPP work items should have no negative impact on current Release 99/Release 4 schedules, resources and funding.
- Only those parties within 3GPP interested in contributing to 3GPP developments in the area of GSM/EDGE radio access will be required to resource and fund this specific activity.

The ad-hoc group, which was lead by a member of the T1 delegation to 3GPP, meet three times in order to elaborate on a detailed report covering the concerns, potential advantages and disadvantages of the transfer, and the proposal for how the transfer could be performed, in terms of organization, funding, timing, etc. At the final meeting of the ad-hoc group in late March 2000 in Tokyo the report of the ad-hoc group was completed and contained the following proposals:

- A new TSG should be created – TSG GERAN – into which essentially all current SMG2 work would be moved.
- The work of SMG7 would be moved into the proposed TSG GERAN.
- The generic operations and maintenance work of SMG 6 would be transferred to 3GPP TSG SA WG5, while radio-specific GERAN work in SMG 6 would be transferred into the proposed TSG GERAN.
- The work of SMG9 that is specific to GSM and 3GPP systems would be transferred into 3GPP T3.
- The other ETSI SMG groups already have direct SMG-3GPP correlation, and the corresponding groups are already meeting in parallel or at least in close collaboration. Therefore this proposal recommends the formal transfer of this work.

This proposal from the ad-hoc group was accepted by all the partners in 3GPP at the PCG and OP meetings in July 2000 in Beijing. At these meetings also the corresponding modifications to the 3GPP working procedures, project description, and partnership agreement was approved. At this meeting, terms of references for TSG GERAN was approved and I was appointed convenor for TSG GERAN with the task of convening the first meetings of TSG GERAN.

TSG GERAN held its first meeting in Seattle at the end of August 2000 on the days originally planned for the meeting of ETSI SMG2, which held its last meeting in late May 2000. With the transfer of the remaining GSM work from ETSI to 3GPP, the first part of the GSM era in standardization had finished and the forming of 3GPP completed.

The transfer of the GSM/EDGE radio activities to 3GPP went without any major problems and without causing any delays to ongoing GSM/EDGE or UMTS activities. The work in TSG GERAN is now focusing on upgrading the GSM/EDGE radio access network to support the Iu interface as defined for UMTS, as well as supporting the IP multimedia subsystem. This is in order to allow full independence for the core network from the type of radio access network used, being either UTRAN or GERAN. This of course only as long as the required service from the radio access network is within the physical limitations of the radio access network in question.

As a part of the decision of transferring the remaining GSM activities into 3GPP it was agreed to perform an organizational review in a 6 month time frame after the transfer. This review was performed during early 2001 and at the PCG meeting in April 2001 it was as a result of this review concluded, that there was no need for changes to the 3GPP organizations. It was further noted that the current organization of 3GPP had been able to evolve and handle the changes and challenges appearing.

In all, 3GPP is now a mature organization able to continue the good work and the co-operative spirit, which was always the trademark of the GSM/SMG group.

Chapter 10: Services and Services' Capabilities

Section 1: The Early Years up to the Completion of the First Set of Specifications for Tendering of Infrastructure (1982 to March 1988)

Friedhelm Hillebrand[1]

10.1.1 Introduction

This section reports about the early discussions in the 1882–1985 timeframe, the agreement on a service concept in February 1985 and the elaboration of the first set of specifications needed for tendering of infrastructure by the operators in early 1988.

A well structured and future-proof portfolio of tele-, bearer and supplementary services was specified. The principles for charging and accounting of international roamers were agreed. Innovative solutions for mobile station licensing, circulation and type approval recommendations were developed. They needed to be implemented by national or international regulatory authorities.

This intensive services' work provided guidance to the technical work on radio, network and data aspects. During this period the culture of service-lead standardisation was developed in GSM. This was a major achievement in the GSM work, since it ensured market orientation.

10.1.2 Overall Guidance is Provided by the CEPT Mandate and the First GSM Action Plan of December 1982

The mandate of CEPT for the GSM work requests the "Harmonisation of the technical and operational characteristics of a public mobile communication system in the 900 MHz band".[2]

The first action plan prepared by the Nordic and Dutch PTTs was approved at GSM#1 in December 1982.[3] It contained basic requirements. Those related to services were:

---

[1] GSM Working Party 1 "Services Aspects" (WP1) was chaired by Martine Alvernhe (France Telecom) from February 1985 to April 1991 with great engagement. I was an active contributor and participant in the work of WP1 during the period treated in this section. The views expressed in this module are those of the author and do not necessarily reflect the views of his affiliation entity. All quoted GSM Plenary documents can be found on the attached CD ROM. The quoted GSM WP1 documents are not copied on the CD ROM. They can be retrieved from the ETSI archive.

[2] Quoted in GSM 2/82, see also GSM 1/82.

[3] GSM 2/82.

10.1.2.1 Basic requirements for GSM services in the first action plan (December 1982)[4]:

- "Mobile stations can be used in all participating countries, preferably all CEPT countries
- It is expected that in addition to normal telephone traffic, other types of services (non-speech) will be required in the system.
- ..state-of-the-art subscriber facilities at reasonable cost.
- The services and facilities offered in the public switched telephone networks and the public data networks...should be available in the mobile system.
- The system may also offer additional facilities (e.g. special barring functions, rerouting of calls and special message handling facilities).
- It should be possible for mobile stations...to be used on board ships, as an extension of the land mobile service.
- The system shall be capable of providing for portable (handheld) mobile stations, but the consequential impact on the system shall be assessed.
- ..voice security...must be taken into account."

GSM#1 also discussed whether a harmonisation of the emerging analogue 900 MHz interim systems[5] should be studied. It was concluded that this would not be reachable due to the necessary short implementation periods and the commitments made by the PTTs and that the work should be focussed on a new mobile communication system. This was an important decision on a viable work focus.

10.1.3 Discussions from the Beginning of 1983 to the End of 1984 (GSM#2–6)

10.1.3.1 Interactions with CEPT Working Group Services and Facilities

CEPT working group services and facilities produced several versions of a report on mobile services. The report gave rise to many questions by GSM. Towards the end of 1984 GSM came to the conclusion that this dialogue would not lead to results, since the CEPT group services and facilities had too broad a scope and a lack of know-how in mobile communication. It was agreed to make their own efforts in GSM and a special focus on services was agreed for GSM#7 in February/March 1985.

10.1.3.2 The "Coexistence Between Vehicle-borne and Hand-held Stations"

The big theme of the period 1982–1985 was the "Coexistence of vehicle-borne and hand-held stations". This was a hot issue, since most mobile systems had a very limited capacity and used call duration limitations. No existing European system supported hand-helds. This existed only in the American AMPS.[6] Several countries in Europe planned the support of hand-helds in analogue systems.[7] In others, mainly in the centre of Europe, spectrum was much scarcer due to the large demand by the many large armed forces in cold war times.

---

[4] GSM 2/82.

[5] NMT-900 in the Nordic countries, TACS (an AMPS derivative) in the UK and later in Italy and Austria, later the planned Franco-German S900 system.

[6] Advanced mobile phone system.

[7] The Nordic countries planned the introduction of NMT900 mainly for this purpose. The UK planned the introduction of TACS (an AMPS derivative) with an integrated hand-held support. France and Germany planned a 900 MHz interim system IS900 which would also support hand-helds.

Several delegates clearly proposed not to admit hand-helds. Should there be a sufficient demand, a separate system evolved from cordless systems working in another part of the spectrum should be developed.

"... several people believe that the majority of the mobile stations will be hand-held portables."[8] A special working party was set up during GSM#5 and 6. They produced an exhaustive study.[9] It was realised that personal communication would be an important future demand. Issues like in-building coverage were studied. The fears on the negative effect of hand-helds on high-rise buildings and in aircraft on the spectrum re-use pattern and this network capacity were addressed.

In the end "the meeting agreed that the GSM mandate (Doc 2/82) clearly states that hand-held stations should be catered for in the system."[10]

This was to my judgement a decision effected by exhaustion of the participants. But it opened the way to GSM as we know it today, where hardly any dedicated vehicle-borne stations exist any more. But the discussion continued for some time before it was formally closed (see paragraph 10.1.5.4.1)

10.1.3.3 Discussions on Other Issues

Other service related issues which attracted substantial interest were market surveys, the discussion of traffic models and a comparative evaluation of the tariffs in existing systems which showed large differences in the charging criteria and charging levels.[11]

10.1.4 The First Concept for Services Agreed in February/March 1985 (GSM#7)

10.1.4.1 A Concept Proposal by Germany and France

A major step forward in the definition of services was reached at GSM#7 (February/March 1985) in Oslo. Germany and France presented a input document[12] which proposed to use the ISDN differentiation of tele-, bearer and supplementary services and to use the description method using attributes within the GSM framework.

The document clearly stressed that radiotelephony would be the most important service.[13] It proposed realistic quality of service targets (e.g. a delay which should not exceed 80–100 ms) and a low bitrate bearer capability (8–16 kbit/s) to support radio-telephony.

A comprehensive list of supplementary services was given. A classification as E (essential, i.e. mandatory for all networks) and A (additional, i.e. optional) was proposed.

---

[7] The Nordic countries planned the introduction of NMT900 mainly for this purpose. The UK planned the introduction of TACS (an AMPS derivative) with an integrated hand-held support. France and Germany planned a 900 MHz interim system IS900 which would also support hand-helds.

[8] Report of GSM#4 in February/March 1984, pp. 3, 2.j.

[9] GSM 58/84 rev. 1.

[10] GSM#7 Report, Section 13, second last paragraph.

[11] Details can be found in the meeting reports on the attached CD-ROM.

[12] GSM Doc 19/85.

[13] This was an implicit rejection of the ISDN concept of a universal integrated network equally suited for all services.

The document contained an initial list of possible telematic and data services. These were tailored to the market needs and the possibilities in a GSM system.

An overall target of the proposed data service concept was to recognise that GSM is not a mobile ISDN. There were strong proponents of this concept, who wanted to implement an ISDN channel structure (2B + D with 2×64 kbit/s + 16 kbit/s) on the radio interface. This would have severely deteriorated the spectrum efficiency and the system capacity.

The discussions in GSM on speech coding had led to a vision of low bitrate speech codecs (about 16 kbit/s). The system architecture discussion had led to an emerging vision to use ISDN in the core network. The German/French input document was based on these assumptions and proposed a rich data services portfolio, which respected the overall target that GSM must be optimised for telephony and offer attractive data services for the "mobile office".

A range of *circuit switched data services* with rates of up to 9.6 kbit/s was envisaged, since this was the maximum possible on a single traffic channel.

No *packet switched services* were proposed, since packet switching was not possible on the traffic channels of ISDN switches.

Instead the concept of *short message transmission* was proposed. The application envisaged was the following: GSM was seen at that time as a car telephone system (the discussion on the viability of hand portables in the same network was still ongoing). So a typical application scenario was a plumber or other technician doing some repair work in the customers home could receive short messages in his car waiting in front of the home. Another scenario was to enable a user to receive a short notice while he was engaged in a call.

It was envisaged to carry all short messages on signalling links of the system with low priority. This signalling network in an ISDN-type GSM core network and the signalling links on the radio network form essentially an embedded packet switching network in an ISDN. Therefore SMS can be seen as the first packet switched service in GSM. The length of the signalling packets on the GSM radio interface is shorter than in the ISDN in order to allow an efficient transmission over the radio channel with its difficult transmission quality. Therefore the length of short messages was limited. Initially 128 bytes were envisaged. The detailed work allowed it to be extended to 160 characters (seven bit coding).

The document also proposed several *types of mobile stations*: vehicle-mounted stations, hand-held stations, combined vehicle-mounted and hand-held stations, mobile payphones and mobile PBX.

The first draft of this document had been elaborated by me. Very valuable comments and contributions were received from Bernard Ghillebaert (France Telecom). The document was then agreed and presented as an input with source "Federal Republic of Germany and France". This was the first example of a series of co-ordinated input documents, where the lead could lie on either side. They were called "Ghillebrand-Documents" by Philippe Dupuis.

The significance of such a contribution is not the level of innovation, but the provision of a viable market-oriented concept, which can be agreed in the standardisation group and can provide direction to the future work of the group. And these criteria were fulfilled by this contribution.

10.1.4.2 Proposals for Services by the Nordic Countries

A list of *basic functions and capabilities* of the GSNM system was proposed by Denmark,

Finland, Norway and Sweden.[14] Access capabilities made up of a control and a traffic channel with 16 kbit/s were proposed. Services should be telephony and several data services. Network functions were mentioned. A catalogue of types of mobile stations (identical to 10.1.4.1) was added.

A initial list of *supplementary services* to be performed by the network was proposed by Denmark, Finland, Norway and Sweden.[15] It contained, e.g.

- barring services
- absent subscriber services
- mailbox services
- group services: closed user group, group calls, conference calls

In addition services and *functions* implemented *in mobile stations* were proposed, e.g.

- dialling functions: abbreviated dialling, number repetition
- hands-free operation
- barring of outgoing calls
- prevention of unauthorised use

This work was based on the ISDN definitions and experiences of NMT.

10.1.4.3 The first Agreement on a Concept for Services at GSM#7 (February/March 1985)

During GSM#7 a new Working Party WP1 "Services" met and elaborated a document on "Services and facilities of the GSM System" which was endorsed by GSM[16] based on the input documents mentioned in 10.1.4.1 and 10.1.4.2. The output document contains a reference model for services, definitions of tele-, bearer and supplementary services, network connections and types of mobile stations. The reference model introduces Terminal Adaptor (TA) functions at the mobile station and an Interworking Unit (IWU) between the mobile and the fixed network. The diagram is shown in Figure 10.1.1.

The annexes of the document contain lists of *teleservices* including the Short Message Service (SMS). SMS had three services: mobile originated, mobile terminated and point to multipoint. It foresaw a maximum message length of, e.g. 128 octets, and an interworking with a message handling systems. Several other non-voice teleservices were proposed.

A comprehensive range of circuit switched *bearer services* with speeds up to 9600 bit/s was proposed.

The significance of this document was, that it was the first consensus in the Groupe Spécial Mobile on the service concept. The document provided the first "permanent" definition of services which could be enhanced by WP1 "Services". It could be used by other groups, e.g. WP2 "Radio Aspects", WP3 "Network Aspects" or later WP4 "data".

10.1.5 The Work on Services Aspects from March 1985 to March 1988

This work was carried out in WP1 "Services Aspects".

[14] GSM Doc 7/85.
[15] GSM Doc 8/85.
[16] GSM Doc 28/85 rev 2.

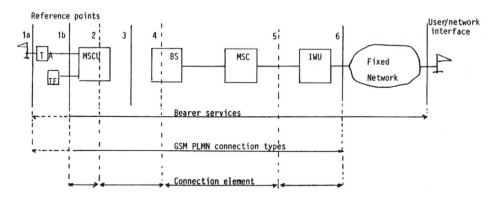

TA: Terminal Adaptor
TE; Terminal Equipment
MSCU:Mobile Station Central Unit
BS; Base Station
MSC:Mobile Switbhing Center
IWU:Interworking Unit

Figure 10.1.1   Reference model for mobile communication (February 1985)

10.1.5.1 Teleservices

10.1.5.1.1 Broad Services Portfolio

The target to have all "services and facilities offered in the public switched networks and the public data networks..." [17] lead to a rich portfolio of teleservices (see Table 10.1.1)).

The great variety in this portfolio was caused to a large extent by the inability in certain fixed network services to agree on one technical solution.

In the category of speech transmission the speech mail access teleservice was deleted later. Also the teleservice "Data Message Handling System Access" was deleted later. In both cases it turned out that a simple transparent dial-up bearer service was sufficient and no need for further standardisation existed.

In Videotex three different standards ("profiles") were used in different European countries since they had not reached an agreement on a single solution. This lead initially to the definition of three teleservices, which would allow only a very limited international roaming. Videotex was in the fixed networks a success only in France and to a lesser extent in Germany. In the end the standardisation of these services was stopped and later on those teleservices were even erased from the list, [18] since a data bearer service was seen as sufficient for access in the relevant countries.

The Teletex service was defined for a circuit switched data network in several countries and for a packet switched network in others. In the end Teletex died in the fixed networks caused by over-regulation and the inability to agree on Pan-European common solutions. The

---

[17] See above paragraph 10.1.2, bullet 4 in list.

[18] This exercise to "prune" services without functionality or with very low likelihood to come into use was initiated by the new privately owned license holder against resistance by some of the old public operators.

[19] GSM WP1 Doc 21/86, Section 4.

Table 10.1.1 Teleservices (status spring 1987)

Category	Individual teleservice
Speech transmission	Telephony
	Emergency calls
	Speech mail access
Short message transmission	Point-to-point
	Mobile terminated
	Mobile originated
	Point-to-multipoint
Data message handling systems	300 bit/s
	1200 bit/s
Videotex access	Profile 1
	Profile 2
	Profile 3
Teletex transmission	Circuit switched
	Packet switched
Facsimile transmission	Group 1
	Group 2
	Group 3
	Group 4

GSM standardisation was stopped. The Teletex development was eventually surpassed by e-mail.

Facsimile foresaw four teleservices, since the ITU had standardised four types of fax machines. It was possible to reduce this in the GSM networks to the support of one type (see paragraph 10.1.5.1.4).

10.1.5.1.2 Emergency Calls

Agreement on Principles

At the time of standardisation and still many years later the emergency numbers were different in the fixed networks of different European countries. A mobile user would need to know the different emergency numbers and from which country he receives coverage. In border areas a user could already be in a foreign country, while his mobile could still receive service from his home network.

Therefore a uniform access in all GSM networks would be desirable for a user friendly service. After an intensive discussion the following principles were agreed in the GSM WP1 meeting in April 1986:[19]

It was considered that the following attributes were desirable for emergency services access:

Instant access without the use of unlock code subscriber identity cards, etc.

A common means of access regardless of the country in which a mobile was working.

[19] GSM WP1 Doc 21/86, Section 4.

Use of the same means of access as used in the fixed network of the country in
which the mobile was working.

Following the discussion it was agreed that a common means of access should
be implemented by the provision of an "SOS" button in conjunction with a
"SEND" button. Therefore when the mobile was switched on a user need only
press "SOS" followed by "SEND" to initiate an emergency call. It would then
be the responsibility of the GSM network operator to make suitable provision
for delivering this call to the relevant emergency services.

It should also be possible to initiate emergency calls by means of the standard
PSTN access method over the GSM system.

The question of automatic transmission of the user identity was also discussed
but was considered to conflict with the need for instant access.

Later on after consultation with the UK emergency services the UK delegation requested a
better protection by the need to overcome a mechanical barrier (e.g. breaking a glass cover) or
dialling a preceding function key.[20] The Netherlands was opposed to it.[21] Later on no additional protection means were specified.

It was possible to find an agreement on the principle "...that in addition to established
national dialling procedures a single mandatory uniform method of access to emergency
services will be defined" and "The single method will require the customer to input a unique
key-sequence..."[22]. This was used as guidance for a dedicated group. The discussion was
fertilised in 1988, when the fixed network operators agreed on 112 as a uniform emergency
number. Then this was standardised for GSM and implemented from the start of service,
much earlier than in most fixed networks (see Chapter 10, Section 2).

Emergency Calls With or Without SIM

The debate was whether a mobile station without a SIM should be able to place an
emergency call. Charging of such calls was possible only with the SIM inserted.

Italy requested that all calls should be chargeable. This meant as a consequence that a
mobile station without a SIM could not place an emergency call.

The majority in GSM thought it would be unacceptable that, e.g. in a traffic accident a
vehicle-mounted mobile station would not be able to call help without the SIM inserted in the
mobile station.

It was not possible to reach a unanimous decision. Therefore both options were specified
for the operator to choose and approved in early 1988. Later it was agreed that all mobile
stations must be able to place emergency calls whether a SIM is inserted or not. As a
compromise those countries who initially insisted on retrieving the identity of the subscriber
who dials an emergency call later accepted the possibility of identifying the mobile station by
retrieving its International Mobile Equipment Identity (IMEI).

The free-of-charge emergency calls were misused as free-of-charge test calls in Germany
in 2000. The emergency services operators requested that emergency calls should be charged.
But the standard was not changed.

---

[20] GSM WP1 Doc 44/87.
[21] GSM WP1 Doc 88/87.
[22] GSM Doc 190/87

10.1.5.1.3 Short Message Service

Based on the concept (see paragraph 10.1.4.3) a more detailed description was developed in the early draft of recommendation GSM 02.03 Teleservices of September 1985.[23]

The next small progress reached in April 1986 was to define a bit more precisely the service "short message, mobile terminated, point to multipoint" by the term "cell broadcast". This described the ability to send a given message to all mobiles in a particular cell or several cells.[24]

The next big step in the definition of the short message service came from a comprehensive common Franco-German contribution to the November 1986 meeting of GSM WP1.[25] It dealt with the teleservice "short message, mobile terminated, point-to-point". It proposed a more detailed definition of the service and proposed implementation aspects such as the functional architecture and protocols. The contribution was also submitted to WP3 "Network Aspects".

WP1 accepted the service related proposals with small modifications:[26]

- Basic service definition including:

  - message length, alphabet
  - levels of acknowledgement (delivery, user)
  - definition of a short message service centre
  - delivery mechanisms, if a mobile is not reachable

- Principles of international operation

The main boost for the progress of the short message service came from the technical work in WP4 in 1987–1999, which implemented the service requirements agreed in WP1 (see Chapter 16, Sections 1 and 2).

10.1.5.1.4 Facsimile

Fax was becoming popular in the fixed networks in the mid-1980s, since the ITU had completed successfully several standards on fax (group 1, 2, 3 and 4). The service concept (see paragraph 10.1.5.1.1) had foreseen to define four teleservices for all four ITU standards.

In order to progress this matter I studied it. A joint contribution of Germany and France was submitted to the November 1986 meeting of WP1. It proposed not to consider fax group 1 and 2 using analogue transmission. Half of the installed machines belonged to these groups, but nearly all new machines belonged to group 3. In addition group 3 machines used digital transmission and were better standardised and therefore world-wide compatible. It could be expected that in the early 1990s the vast majority of fax machines would belong to group 3. Group 4 machines were foreseen for ISDN. It was therefore proposed to focus the GSM teleservice fax on group 3 machines and foresee group 4 as an evolution possibility to be studied later. Several implementation scenarios were discussed.[27]

---

[23] GSM WP1 Doc 12/85.
[24] GSM WP1 Doc 21/86, para 3 and attached draft recommendation with data sheet.
[25] GSM WP1 Doc 66/86.
[26] GSM WP1 Doc 86/86, Section 4 and Annex 5.
[27] GSM WP1 Doc 47/86.

GSM WP1 endorsed this service concept and asked the other working parties to study the implementation aspects.[28]

Later on intensive studies were made in WP4 "Data Services" in the 88–90 time frame on a technical support offering a sufficient quality of service (see Chapter 16, Section 2). But despite all efforts in standardisation, the GSM teleservice fax did not become a big success in the market.

10.1.5.2 Bearer Services

10.1.5.2.1 Broad Services Portfolio

The target to have all "services and facilities offered in the public switched networks and the public data networks..."[29] lead to a rich portfolio of bearer services with data rates of 300, 1200, 1200/75, 2400, 4800 and 9600 bit/s duplex asynchronous and synchronous and inter-working scenarios with many network types: other GSM networks, Public Switched Telephone Networks, ISDNs, Public Circuit and Packet Switched Networks. This allowed the support of a wide range of existing terminals.

When GSM data bearer services became available the normal terminal connected to a mobile station was a laptop computer. Users wanted the maximum possible speed of 9600 bit/s.

Several attempts have been made to simplify the broad portfolio. I tried to propose and promote a single asynchronous duplex 9.6 kbit/s high speed service with high quality (inte-grated ARQ protocol, later called non-transparent). But such concepts were seen as too narrow. On hindsight I would say that such a solution could have been available earlier and would have allowed a higher volume.

Finally in 2000 the issue was re-discussed. Everybody recognised that the lower bitrates were not used and hence not needed. But there was agreement not to change the standard and the products, since nobody was ready to bear these cost.

10.1.5.2.2 Support of 9.6 kbit/s

Several delegations wanted this bitrate as a mandatory service, so that it could be used by international roamers. But there were doubts in Italy whether the necessary quality of service could be achieved in the whole coverage area. This was a discussion from 1986 to early 1988. Since unanimity was needed, the majority gave in and accepted a classification as A (addi-tional), when the specifications for tendering were approved in early 1988.

10.1.5.3 Supplementary Services

Supplementary services "supplement" the basic telecommunication services (tele- and bearer services).

10.1.5.3.1 Harmonisation with ISDN Standardisation

In order to maintain a harmonised services portfolio with fixed ISDN networks it was tried to

---

[28] GSM WP1 59/86, Section 3.2 and Annex 5.
[29] See above paragraph 10.1.2, bullet 4 in list.

take over as many ISDN supplementary services as possible and not to change them unnecessarily.

This was difficult, since the ISDN side saw the GSM service as a small service compared to ISDN and did not co-operate intensively. The ISDN colleagues had their own rhythm of specifications and deadlines. This was influenced very much by ITU time tables. The market needs of GSM had no weight in the definition of such timetables.

In addition the need to agree on single solutions was lower in ISDN than in GSM since the subscribers were connected to one access line in one network and did not roam to other networks. All ISDN supplementary services were executed by the subscriber's network. Foreign networks were not involved. As a consequence the ITU ISDN standards contained too many options for a direct use in GSM. Therefore GSM had to choose between these options in order to enable international roaming.

This situation lead to incompatibilities between ISDN and GSM supplementary services.

10.1.5.3.2 Mobile Specific Supplementary Services

In addition there was the need for mobile specific supplementary services. Examples are:

- Conditional call forwarding on mobile not reachable
- Call barring of incoming or outgoing calls, when roaming

10.1.5.3.3 Conclusion

GSM WP1 elaborated a fairly comprehensive list of supplementary services in recommendation GSM 02.04 in the set of specifications for tendering of infrastructure in March 1988. But the detailed technical specification work was cumbersome and time consuming. Therefore GSM phase 1, the specification used for the opening service in 1991/1992 contained only some barring and call forwarding services. The rest was specified and implemented in GSM phase 2 only.

10.1.5.4 General Services' Aspects

10.1.5.4.1 Hand-held Station Support Mandatory

After the initial discussion and conclusion (see paragraph 10.1.3.2) the impact of hand-helds on the system was studied in depth.[30] The final conclusion came when the revised strategy targets was agreed at GSM#8 in June 1985.[31] This document made the support of hand-helds mandatory.

But for a long time it remained unclear, whether the basic technology choices would really allow the building of small and cheap hand-helds with a low power consumption. A dedicated hand-held ad hoc group provided a report on the viability of hand-helds in the GSM system in 1987[32]. A complexity review was performed in 1988 (see Chapter 19, Section 1, paragraph 19.1.4). Such studies created some certainty that viable GSM hand-helds could be expected.

---

[30] GSM 27/85, rev. 3.
[31] GSM 73/85 rev. 1, superseding GSM 2/82, Section 3.
[32] GSM 102/87

10.1.5.4.2 Network Selection

The first generation networks had normally a fixed coupling of a mobile to a network (e.g. by the frequency band used in AMPS). Networks offering international roaming like the Network B in Germany, Austria and Benelux or NMT in Scandinavia had country selector switches, which allowed the user to choose manually the network he/she wanted to use.

In GSM the situation became more complicated, since the system should support full international roaming to all countries in the GSM system area. In addition it should be possible to have several competing networks in a country. The situation is especially tricky in border areas. There are some places in Europe where three countries come together at one point, e.g. Germany, Belgium and Netherlands near Aachen. It was also evident early that some type of an automatic user friendly procedure was needed. On the other hand it should not be excluded that the customer might make a personal decision.

This lead to the need to have a network and country indicator which showed in which country and network the mobile station was registered. A network selector function would allow the user to choose manually, which network he wanted to use. Both features were mandatory (see mobile station features, see paragraph 10.1.5.6.3.3).

Some stable agreements were possible in regard to the automatic network selection procedure. An agreement was reached, that the mobile should register on its home network (where the subscription existed), if this network was available. Another agreement was that in his/her home country the user should have access only to the home network and not to competing networks, in order not to destroy the competition for coverage and quality. Only much later some national roaming agreements were concluded (see paragraph 10.1.5.4.4).

Then there were debates, what to do in a foreign country. Should there be a list of preferred networks "prescribed" by the home operator? This was seen as too limiting for the customer. Should the user be offered all available networks for a personal choice? This was seen as too user unfriendly. So it was agreed that the mobile should register on the network offering the best quality as a fall back solution. But the user was given the possibility to override this automatic decision by a personal selection or he could program his/her preference on the SIM. In cases where the mobile had registered on a (foreign) network there was a discussion on how to bring it back to the home network. Should there be a regular check for the availability of the home network or was it sufficient to check this when the mobile was switched on? (For more details see Chapter 10, Section 2, paragraph 10.2.8).

10.1.5.4.3 Dialling While Roaming

Calls to a Mobile

A major step forward compared to the pre-cellular system was the fact that a user of the fixed network could reach the mobile station by dialling its telephone number without knowing the actual location of the mobile user. This was performed by the mobility management function. This function had to be expanded, so that calls could be routed also to users who were roaming in foreign mobile networks. This was a GSM service requirement.

Calls Originated by the Mobile

The coverage areas of base stations and the network areas in the fixed networks are not aligned. Therefore a mobile cannot use local numbers to dial a subscriber in the fixed network, since it is not known, in which network area the mobile is. At least the area code and the local number of a user in the fixed network have to be dialled. This was done already in first generation systems.

In GSM, a second generation system with international roaming, new problems arose. When a subscriber is, e.g. on a German network he/she dials a line in the German fixed network by dialling the long distance access code, the area code and the local number. If this user is registered on a Dutch mobile network, he/she needs to dial the Dutch access code to the international network, the German country code, the area code and the local number. There were different access codes to the international network in different countries. France had two codes, one in the Paris area and another in rest of the country. In addition in border areas the customer needed to use the conventions of the network that provided coverage to him, whether he was in the home country of the network or in a neighbouring country in a border area. This was seen as extremely complicated and user unfriendly.

Therefore the idea was born, to create a uniform dialling format, that would free the user to bother about all the a.m. peculiarities. It was proposed to dial the format, which is used on international business cards " + country code/area code/local number". The mobile network, which the user uses would convert the number into the format the fixed network of the country needed. This format could also be stored on the SIM or in the mobile and be used in all networks including the home network.

A similar issue existed for emergency calls. There were different numbers in all European countries. It was proposed to use an SOS button at the mobile station. Later this proposal was converted to dialling 112, when the fixed networks had agreed to use this number in the long-term everywhere in Europe.

This work was done in 1986/1987. The first document, however, I found was a draft specification "GSM 02.07 Mobile station features"[33] of August 1987, which lists the service requirements for the use of the + key ("International Access Function") and the SOS key. I recall some early work of my team, but all documents are lost.

On hindsight it can only be regretted that this concept was not extended to some other cases, e.g. the access to a voice mail system. But voice mail systems were not popular in these years since memory was still expensive.

10.1.5.4.4 No Automatic Hand-over Between Base Stations Belonging to Different GSM Networks

The question of handover between different GSM networks in different countries was discussed and decided at GSM#10 in February 1986 and GSM#13 in February 1987. It was rejected, since the numbering plans and the charging systems were different in different countries. Such internetwork handovers were implemented much later in national roaming situations between competing operators:

- In Denmark in the late 1990s at request of the regulator based on consumer protection arguments.
- In Germany in a commercial agreement between T-Mobil and Viag Interkom in 2000.

  However these implementations were not covered by the GSM specifications.

---

[33] Meeting report of WP1, August 1987, WP1 TDoc 168/87, Annex 8.

10.1.5.4.5 Use of Identities

There was a long discussion about the "great number of emerging identities in the system". In the end it was agreed to use both the mobile station identity (later called International Mobile Equipment Identity (IMEI)) and the subscriber identity (Mobile Station ISDN Number (MSISDN)), since they serve different purposes. This was agreed at GSM#11 in June 1986.

10.1.5.4.6 Lower Bitrate Codecs

The support of lower bitrate codecs was proposed and agreed already in June 1985.[34] This opened the way to the later half-rate and multi-rate codecs.

10.1.5.4.7 Subscriptions

There was also a long lasting discussion whether besides subscriptions which entitled the user to international roaming, subscriptions which were restricted to the home country should be specified. Due to the unanimity required it was not possible to exclude this fully. It was agreed in March 1988 (GSM#17bis) that all defined subscription types must entitle the user to international roaming. But the technical specification of the home location register foresaw the possibility to provide restricted subscriptions should a regulator request this.

In the 1992/1993 timeframe some operators used this restriction possibility as a temporary measure to combat fraud. A large number of customers had acquired SIM cards and exported them immediately to a foreign country, sold them there and paid no bill. Operators did not discover this fraud immediately, since the charging data for traffic were transmitted by tapes and arrived often late. Some mechanisms to bar users did not work in foreign networks in the beginning. Therefore the operators used the barring of international roaming if there were doubts about the credit worthiness of customers. But this emergency use disappeared soon. The regulators executed substantial pressure, since they saw international roaming as an essential ingredient of the GSM service.

10.1.5.4.8 Quality of Service

There were intensive discussions and studies about quality of service. In June 1986 a differentiation was made between quality of service as seen by the user and network performance. Parameters and values for quality of service were elaborated and a complete recommendation GSM 02.08 was produced by the end of 1987.

However the recommendation was not approved in early 1988. GSM felt uneasy about certain definitions and values. No consensus was possible. Therefore the recommendation was converted into a report in 1988.

The advent of competition led to the understanding that quality of service was a key competitive area which required no recommendation or specification. Therefore the report on quality of service was deleted eventually.

10.1.5.4.9 Inspections of Mobile Stations

This issue sheds light on the situation in existing mobile networks in the mid 1980s. Several

---

[34] GSM 45/85 by the late Frieder Pernice (the document is not available).

existing mobile networks requested that the subscriber presented mobile stations for regular inspection to ensure that they still fulfilled the type approval criteria. This was discussed controversially in GSM. But GSM concluded that this would not be practical in the expected mass market. It was therefore rejected in June 1987 (GSM#14).

10.1.5.4.10 Security Aspects

Security in mobile networks has two main aspects:

* protection against misuse
* protection of the user data

The service requirements were specified in the beginning by WP1. But due to the highly specialised matter a Security Experts Group was created early. The related issues are treated in Chapter 15.

10.1.5.5 Charging and Accounting for Roaming Subscribers

The issue of charging of users and international accounting was a hot potato, since there were very large differences in the charging criteria and charging levels of existing systems (see paragraph 10.1.3.3). There was very little experience with international accounting for subscribers roaming abroad. The two existing cases (Network B between Germany, Austria and Benelux) and NMT (Denmark, Finland, Norway and Sweden) existed between friendly administrations and used simplified procedures.

GSM WP1 "Services Aspects" started to address the difficult issue in the meeting in May 1986 finally. The participating administrations described their existing regulations and tariffs and found great complexities in some cases and large differences from one country to another. WP1 decided to seek solutions in the next meeting.

After this meeting in May 1986 I did not know how such a complex problem area could be solved or even be approached. Therefore I paid a visit to the late Alfred Schwall in the German Ministry of Post and Telecommunication, who had great experience with charging and international accounting for telecommunication services in the fixed networks. He strongly recommended to use the existing CCITT[35] work as a basis. He pointed to may existing definitions, which might be applicable to GSM. Then he mentioned a telegraph service, where somebody could buy a card and present it in all countries of the world and get telegraph services. The users account generated in the foreign country would be transferred to his home administration and he would be charged at home. Alfred Schwall recommended to study whether this world-wide recognised model could not be used for the GSM international roaming service.

It turned out that the CCTTT D series of recommendations dealing with international accounting in the fixed networks provided a set of very valuable definitions and that the principles of the Transferred Account Telegraph Service were usable to a large extent.

Based on this input from CCITT and considering the GSM features I drafted two concept papers.[36] The papers proposed to realise that traffic originating from or terminating in a GSM network is always routed via fixed networks in the relevant countries concerned and that this

---

[35] CCITT was the Comité Consultatif International de Télégraph et Téléphone of the ITU, later replaced by ITU T.
[36] GSM/WP1 53/86 and 54/86.

traffic is included in the existing international accounting for these fixed networks. Therefore it was proposed not to modify the existing international accounting between the fixed networks of the world.

It was proposed that a new international accounting procedure the "Transferred Account Procedure" should be defined between GSM operators. Visited networks would calculate collection charges for the visiting roamers and transfer such accounts back to their home networks. These home networks would charge the subscribers on their normal invoice. For this purpose it was necessary to standardise the charging criteria for the collection charges.

Therefore it was proposed to develop three recommendations:

- Collection charges in GSM networks
- Transferred account European communication service
- International accounting for telecommunication services of GSM networks

In the area of collection charges it was proposed not to include the network access charges (initial fee and subscription rentals) in the transferred account procedure. This was in line with ITU principles, where the monthly rental of a telephone access is not included in the international accounting. This gave rise to highly controversial discussions in GSM since several operators wanted to receive a share of the rental of incoming roamers in order to have a monthly fixed income. But the principle not to share this revenue remained stable. It is user friendly and encourages roaming.

Therefore out of all the collection charges only the network utilisation charges, also known as connection charges were considered in the transferred account procedure. These charges for roamers would be calculated by the visited network and the account would be transferred to the home operator. The home operator would remunerate the visited operator and collect the amount from its subscriber.

In order to make this viable two fundamental principles were proposed:

- A visited GSM network treats the external visitor like its own subscriber (i.e. same set of services, no discrimination, e.g. in case of scarce radio resources)
- The home GSM network operator guaranties vis-à-vis the visited operator that the charges for its subscribers with a valid subscription will be paid.

The equal treatment principle gave rise to substantial debates, since some operators preferred their own subscribers, e.g. in overload situations. But all such initiatives were turned down since the participants wanted a Pan-European service providing a Pan-European infrastructure.

The financial guarantee of the home operator was made viable and bearable by appropriate authentication key exchanges (see Chapter 15), when a roamer registered in a foreign visited network. In addition principles were agreed later on a fast transfer of the charging data back to the home network by electronic data exchange, so that the home operator could detect and combat fraud early.

This method of transferring charges was based on the model of "The transferred Account Telegraph Service" according to CCITT recommendation F41 (=D 98). This concept allowed an elegant solution, since the existing international accounting between all fixed networks did not need to be changed and the additional accounting took place between two mobile operators who were involved in providing service to the user.

WP1 accepted the principles in their meeting on 8–10 September 1986 in Rome "in principle" and sent the information to GSM#12 in Madrid in September 1986.[37] GSM noted this concept and forwarded it to the commercial groups in CEPT for comments. These groups had few comments.

The concept was refined by GSM WP1 and the three relevant recommendations were approved in early 1988:

- GSM 02.20 collection charges
- GSM 02.21 transferred account procedure
- GSM 02.22 international accounting

Later around 1990/1991 the responsibility for these documents was transferred to the GSM MoU Group since these were agreements on commercial principles which could be better handled in the environment of the GSM MoU Group. The only major change to the principles was made in the late 1990s. Then the transferred collection charges were replaced by accounting rates negotiated between the operators. This was called an inter operator tariff.

10.1.5.6 Licensing, Circulation and Type Approval for Mobile Stations

10.1.5.6.1 The Situation in Europe in the Mid-1980s

The national regulations for licensing, circulation and type approval of mobile stations were very different between most countries in Europe in the mid-1980s. This area needed a harmonisation in order to enable a Pan-European roaming service and also the high volume production of mobile stations.

Most countries saw the subscriber equipment not primarily as a mobile telephone but as a mobile station emitting radio waves. This equipment needed a radio license for legal operation which had to be carried by the user like a driver's license. Such a license was valid only in the country which issued it. If the mobile station was to be brought into another country, customs would require either a license of the visited country or the station needed to bring it into a status where it could not be used. Users of hand-held equipment needed to take out batteries. Car-mounted stations needed to be removed from the car. Then the equipment had to be sealed.

Pragmatic exceptions from these rigid rules existed between the countries participating in the Network B roaming service (Germany, Austria and Benelux) and the NMT roaming service (Denmark, Finland, Norway and Sweden).

10.1.5.6.2 Licensing and Circulation of Mobile Stations

A Concept Proposal

France and Germany had discussed this area also during the S900 analogue system co-operation[38] and agreed on a pragmatic solution. Based on this model an input document was submitted to WP1 in December 1985. This was restructured and re-submitted to WP1 in February 1986.[39] It proposed that the use of GSM mobile stations should be covered by a general license, which would be published in an official bulletin. The user would not be required to carry license document(s). If his/her mobile station had a valid type approval and

---

[37] GSM 89/86 contains the concept and raw drafts of the three recommendations.
[38] Chapter 2, Section 3, paragraph 2.3.3.
[39] GSM WP1 6/86.

a valid subscription of a GSM network in his/her home country, he/she would be entitled to use it in his/her home country without individual radio licensing documents.

It was furthermore proposed that countries operating GSM networks should grant a general license to mobile stations which were temporarily in that country. As a pre-condition such a mobile station had to have a general license by another CEPT country, a valid type approval and a valid subscription of its home network

GSM mobile stations covered by a general license as described above having a valid type approval and a valid subscription should in principle be entitled to operate in all GSM networks. Users should be able to carry their mobile stations through customs without additional procedures (e.g. sealing). CEPT administrations were invited to initiate appropriate measures in their countries.

For CEPT countries without a GSM network it was proposed that users should be able to carry their equipment without administrative procedures.

The Cumbersome Process to Solutions

When the proposal was presented to WP1 in February 1986, many administrations pointed to the very different national regulations. The next meeting in April 1986 brought no progress. The May 1986 meeting[40] confirmed the large differences from one country to another. The September 1986 meeting discussed the issue in substance and came to some general conclusions:[41]

- Individual licenses were not required for each mobile
- A "National Only" license may be required
- Procedures were required to allow free circulation of mobiles
- Reciprocal type approval arrangements would be needed
- A procedure for the acceptance of licenses between countries would be required

A drafting group was set up to elaborate draft recommendations for the next meeting. The group presented a very lean focussed draft GSM 02.12 "Licensing".[42] It recommended that no form of individual licenses "shall" be required in CEPT countries for

- GSM mobile stations, subscribers or users of national and foreign GSM networks
- the transportation or use of GSM mobile stations

This found some objections in the November 1986 meeting, since it was not in line with present legislation in some countries. A footnote was added pointing to this fact. Then the draft recommendation was agreed at the WP1 meeting in January 1987.[43] The drafting group presented a very focussed draft GSM 02.15 "Circulation of Mobile Stations".[44] It recommended that:

- all CEPT administrations make appropriate provisions for the free circulation of GSM mobile stations across their borders;
- in CEPT countries with one or more GSM networks, it shall be possible for users to use GSM mobile stations;

---

[40] Updated concept paper in GSM WP1 34/86.
[41] GSM WP1 report in WP1 Doc 59/86, paragraph 5.
[42] GSM WP1 Doc 85/86.
[43] GSM WP1 Doc 37/87, 7.3 and Annex 11.
[44] GSM WP1 Doc 86/86, Annex 9, the final version is in WP1 Doc 123/87, Annex 8.

- in CEPT countries without a GSM network it should be allowed to use GSM mobile
  stations in border areas within the coverage of a GSM network of a neighbouring country.

This draft recommendations were agreed in the WP1 November 1986 meeting.

The recommendations GSM 02.12 on licensing and GSM 02.15 on circulation were
approved after considerable consultation at the GSM plenary in December 1987.

10.1.5.6.3 Mobile Stations' Specifications, Testing and Type Approval

Introduction

Mobile stations were critical for the success of GSM. In order to achieve low cost, an
effective competition of manufacturers and high production volumes were needed. For international roaming every mobile station needed to be able to function in every GSM network.

These targets made it necessary to harmonise the mobile station types and features and to
specify a single radio interface as an open interface with a very high specification quality. The
operators wanted a comprehensive testing of the mobile station in the type approval process
in order to guarantee the functioning and the quality of service. This was seen as very critical,
since GSM was planned as a service for very high user numbers. This philosophy reduced
also the risk of the manufacturers.

Types of Mobile Stations

Already in the first service concept agreed in early 1985[45] a differentiation of several
mobile station types had been made: vehicle-mounted stations, hand-held stations, combined
vehicle-mounted and hand-held stations, mobile payphones, mobile PBX. This was a differentiation according to usage types, later other attributes were also studied and defined.[46]

Usage of Hand-held Stations   The UK pointed to the need to improve the definition for
hand-held stations in April 1987.[47] The contribution stresses the fact that hand-helds are
becoming increasingly popular on existing networks, that they are used when walking or
in trains or cars and that most existing units are not small enough to carry them in a pocket.
The contribution stresses that GSM as a digital system offers a new opportunity to overcome
the disadvantages due to very large scale integrated circuits. It believed that due to the greater
convenience hand-helds will become the most popular type of GSM mobile stations.
Predictions were quoted which forecast that hand-helds will comprise well over half the
market in some countries.

It was therefore requested to maximise the performance of hand-helds in the networks.
This should include ensuring that they would work at high speeds (and not at walking speeds
only as the existing draft recommendation requested). Furthermore battery saving techniques
such as a sleep mode should be supported.

It should be recalled, that many people still believed at that time that GSM would be
primarily used for vehicle-borne mobile stations. However such interventions as the UK
contribution were crucial to open up the way to an efficient support of hand-helds.

---

[45] See paragraph 10.1.4.3, GSM Doc 28/85, rev.2, Annex 3.

[46] A more developed version can be found in GSM WP1 Doc 4/86, a first approved version in GSM WP1 Doc 30/
88.

[47] GSM WP1 Doc 45/87.

Targets for Hand-held Stations   The targets for hand-helds in the draft recommendation GSM 02.06 were stable since early 1987:[48]

> Hand-held stations: they are self-contained units incorporating their own power source and antenna. They are intended to be easily carried by a person and therefore have specific characteristics:

- The total weight is less than 0.8 kg;
- The volume is less than 900 cm^3;
- The power source is expected to provide at least 1 hour of call duration or 10 hours in the state of being able to set up or receive calls.

These targets were seen as ambitious at that time. Technology surprised us once more.

Mobile Stations Features

A mobile station feature is a function or a piece of equipment which directly relates to the man machine operation of the mobile station. Three categories were distinguished:

- Basic (directly related to tele- and bearer services): display of called number, network and country indicator and selector, call progress signals
- Supplementary (directly related to supplementary services): display of a calling line number
- Additional: hands free operation, switch-off timer, reception quality indicator

All mobile station features were qualified as mandatory or optional for the manufacturer. This initiated a process of harmonisation for mobile station features, mainly for the mandatory features. This was essential, since the features of mobile stations varied greatly in the existing analogue networks.[49]

Principles of Type Approval for Mobile Stations

Type approval of mobile stations shall ensure that the produced equipment fulfils the technical specifications. The main targets are the functioning, no interference to other users, saving of spectrum efficiency, etc. The procedures and specifications differed greatly between the European countries.

In order to reach the targets of international roaming, it became inevitable to use a single mobile station specification and a single mobile station test specification.

In such a situation the technical tests needed to be done only once, in order to save cost and shorten time to market. This lead to the concept of recognised test laboratories and a Europe-wide recognition of the test results.

Then the type approval could be an administrative act performed by one administration. This could be recognised by all other participating administrations.

These were ideas in several heads of the GSM delegates in 1986. They were well ahead of the European Commission's type approval regime. GSM WP1 asked a drafting group to produce a draft for a recommendation GSM 11.01 "Principles of the mobile station type

---

[48] See e.g. GSM WP1 Doc 31/87, the text is quoted from the first approved version in GSM WP1 Doc 30/88.

[49] An early version of the specification can be found in GSM WP1 Doc 5/86, a developed version in the Annex to WP1 Doc 168/87, the first approved version in GSM WP1 Doc 17/88.

[50] A first draft was in GSM WP1 Doc 61/86.

approval procedure".[50] This lead to a lengthy discussion about basic questions in the WP1 meeting in November 1986[51] The draft was amended[52] and the drafting group was asked to continue the improvement of the draft. But the draft contained key principles mentioned above.

The drafting group elaborated the document further. This was agreed in GSM WP1 in January 1987.[53] Besides the general principles the most fundamental sections were:

3 Mobile Station Type Approval Specification

The type approval of the GSM mobile station is carried out according to the requirements of the "GSM mobile station type approval specification" contained in recommendation GSM 11.10. It sets out the full and precise requirements that must be satisfied by a mobile station to be granted type approval. There are no additional (e.g. national) requirements. It is a common specification for all CEPT countries having a GSM network. It also includes administrative requirements and the associated verification methods.

4 GSM mobile station conformity testing

Such a test is conducted by an "accredited testing laboratory" in accordance with the regulations laid down in the CCH recommendations...If the result of the test is positive, a certificate of MS conformity is issued which in accordance with CCH recommendation...is recognised by all CEPT-countries.

5 Granting and mutual recognition of GSM mobile station type approvals

Actual type approval is granted by the national approval authority. It is based on the certificate of mobile station conformity according to Section 4.

The approval authorities should mutually recognise type approvals carried out in accordance with the procedures in Sections 3 and 4.

The principle agreed in Section 3 was highly controversial, but remained stable in the end. It ensured the compatibility of GSM mobile stations internationally.

The concepts of conformity testing by recognised laboratories and mutual recognition of the test results in Section 4 were taken from an emerging regime of mutually recognised conformity tests developed by CEPT under pressure from the European Commission who wanted to open the national telecommunication markets.

Section 5 recognised the competence of national authorities to issue type approvals and recommended that type approvals should be mutually recognised by all national authorities.

This draft recommendation raised several comments in several GSM WP1 meetings and GSM plenaries. But it stayed stable and provided the framework for GSM type approval. It had even the flexibility to support GSM expansion into many countries outside Europe.

---

[50] A first draft was in GSM WP1 Doc 61/86.
[51] GSM WP1 Doc 86/86, p. 5.
[52] GSM WP1 Doc 83/86.
[53] Report GSM WP1 Doc 37/87, Section 7.3, text of the recommendation in Annex 12.

10.1.5.6.4 Conclusions

The area of licensing, circulation and type approval was a very difficult matter for the Technical Committee GSM since it involved beyond strategic matters a lot of legal and regulatory matters. Indeed it required revolutionary solutions. GSM did not delegate the matter, but treated it and achieved a pragmatic opening.

In several areas the output papers were really CEPT "recommendations" asking national and European authorities to do certain things in a defined way. Many elements needed an implementation in national and international regulations by the European states and the European Commission. With the growing success of GSM this was achieved.

10.1.5.7 Production of the Specifications for Tendering of Infrastructure (Mid-1985 to early 1988)

Based on the agreed service concept (see paragraph 10.1.4.3) a structure for the specifications (CEPT called them "Recommendations") to be produced in the services area was agreed in summer/autumn 1985. The complete list is given in Table 10.1.2. The list initially agreed in summer 1985 had half the number of documents. During the work it turned out to be more appropriate to cover certain areas with additional dedicated documents.

Table 10.1.2   Specifications on services aspects

GSM number	Title of recommendation (later called specification)
01.06	Service implementation and evolution
02.01	Principles of telecommunication services
02.02	Bearer services
02.03	Teleservices
02.04	Supplementary services
02.05	Simultaneous and alternate use of services
02.06	Types of mobile stations
02.07	Mobile station features
02.08	Quality of service
02.09	Security aspects
02.10	Provision of telecommunication services
02.11	Service accessibility
02.12	Licensing
02.13	Subscription
02.14	Service directory
02.15	Circulation of mobile stations
02.16	Equipment identities
02.20	Collection charges
02.21	Transferred account procedure
02.22	International accounting
02.30	Man/machine interface of the mobile station
11.01	Principles of type approval procedure

Table 10.1.3   Meetings of GSM WP1 "Services Aspects"

Meeting number	Date	Venue	Report in GSM Doc	Report in WP1 Doc	Comments
1	25 February–1 March 1985	Oslo, GSM#7			
2	10–14 June 1985	Paris, GSM#8			
3	4–6 September 1985	Paris	90/85	8/85	
4	30 September–4 October 1985	Berlin, GSM#9	123/85		
5	9–12 December 1985	Paris		20/85	
6	3–5 February 1986	Bonn	19/86	13/86	
7	2–4 April 1986	The Hague		21/86	
8	6–9 May 1986	London	36/86	40/86	
9	8–10 September 1986	Rome		59/86 ?	
10	4–7 November 1986	Berlin		86/86	
11	19–23 January 1987	Paris		37/87	
12	6–10 April 1987	Paris		80/87	
13	22 -25 June 1987	Helsinki		123/87	
14	31 August–4 September 1987	Copenhagen		168/87	
15	2–6 November 1987	Athens		212/87	
16	30 November–4 December 1987	Amsterdam		2/88	
17	25–28 January 1988	Florence		47/88	
18	7–10 March 1988	Bonn		89/88	

The elaboration of all these documents during the period from mid-1985 to early 1988 was a formidable task. It required a pretty good management, a lot of contributions and many interactions with other groups, mainly WP2 Radio, WP3 Network, WP4 Data and SIMEG. WP1 needed 18 meetings of 2–3 days duration to complete this task (see Table 10.1.3). Some documents (e.g. 02.21) were approved only in mid-1988[54]. The details can be retrieved rom the meeting reports and status reports mentioned in Table 10.1.2 from the attached CD-ROM. The number of WP1 delegates varied between 10 and 18 during the period from 1985 to early 1988.

10.1.6 Conclusion

This work phase provided all documents necessary for tendering and built the base for all services definitions as well as regulatory and commercial aspects related to services. It established also the services working party as the lead body to elaborate requirements. This resulted in a service lead development of GSM.

There were numerous contributors to the work in WP1 in this early period. Especially active were M. Alvernhe (France Telecom) chairperson, D. Barnes (Cellnet), C. Benard-Dendé (France Telecom), A. Cox (Vodafone), U. Grabolle (DETECON), B. Ghillebaert (France Telecom), F. Hillebrand (DETECON), D. Hoefsloot (PTT NL), H. Sandberg (Telenor), H. Thiger (PN), B. Waijer (PTT NL).

---

[54] GSM 31/88 rev. 1 shows the result of the first round of approvals in GSM#17 in Florence in February 1988

Chapter 10: Services and Services' Capabilities

Section 2: The Years from Mid-1988 to Early 2001

Alan Cox[1]

10.2.1 Scope

"Top-down or bottom-up"? That is the question.

The approach to developing new standards can vary from engineers inventing exciting new technology, with no certainty of a market for these, to a market led approach which may or may not take advantage of new opportunities.

There are plenty of examples of the former, where there has been considerable investment in technologies which then don't perform in the market place. The slow acceptance of ISDN in many markets may actually have been such an example, which was a cause of concern, since the initial service requirement for GSM was often thought of as a "mobile ISDN for Europe".

On the other hand, a market requirement that is isolated from new possibilities is likely to be quickly overtaken by more innovative approaches. Thus we must choose a combination of market need with technical awareness, with the market always dominant. This is the approach that we have aimed to apply for GSM and UMTS and may be considered to be one of the keys to the enormous success of GSM.

10.2.2 Personal History

Perhaps a little personal history will be permitted at this stage, since it gives a flavour of how different things were in the early days of GSM from the 21st Century.

I worked for many years in Racal, initially as a pioneer of frequency synthesisers for military use, so with a fairly broad experience of design and development in manufacturing electronics. In May 1986, I transferred to Vodafone, then a division of Racal, as their first full-time employee on the GSM and I sat on the GSM committee. The difference between telecoms and my previous experience of electronics came as quite a shock – in those days,

---

[1] The views expressed in this section are those of the author and do not necessarily reflect the views of his affiliation entity.

we were modelling GSM on ISDN, but there were no textbooks on ISDN then – the knowledge was mainly in the PTTs.

In April 1987, I attended my first meeting of Working Party 1 "Services and Services Capabilities" in Paris, chaired by Martine Alvernhe of France Telecom. The group was quite small – around 10–18 persons. The task was to provide the service requirements for GSM, in practice based on adding mobility to ISDN.

10.2.3 Early Working Methods

All members of the CEPT/GSM committee were from operators, in fact all were state-owned PTTs except for those from the UK, where DTI, Cellnet and Vodafone all attended. However, we were officially national delegations, so had to hold preparatory meetings at home to determine policy before each meeting of GSM or its working parties. This not only added to the time needed to work on the task, it also removed much spontaneity from the meetings, slowing progress.

The task was to prepare "stage 1 descriptions" for all the services to be specified. These were to include the imported ISDN services – tele-, bearer and supplementary services – and also new ones related to mobility and roaming – describing the mobile stations, power classes, network selection, etc. There were also high-level requirements for security and the SIM card. These were all to be contained in a series of "*Recommendations*" in the 02-xx series. (The topics for each series had been determined in 1986.)

Some working methods from that time have survived to the present in 3GPP, while others have been consigned to history! Individual delegates take on tasks as editors (later known as rapporteurs) of each recommendation (specification), preparing draft texts for the consideration of the committee. Once each is considered almost complete, it is submitted to the plenary meeting (GSM, later SMG or 3GPP) for approval, where it is placed under change control. All this has changed little in over 15 years, except for change control which did not arrive until the first recommendations were ready for approval. At each plenary, working parties report their progress and seek assistance on difficult questions, especially those affecting other working areas.

Early documents were often prepared hand-written, although we preferred them typed for clarity! Certainly, all was paper based, so huge photocopiers were always required at meetings to copy all the documents – a real contrast with today's all electronic meetings. The input documents were numbered with the meeting name and year, similar to today. Meeting invitations and some documents were circulated by post, but this tended to be very slow, especially when there were postal strikes in some countries! Thus we soon used facsimile for this, although it took a while for all delegates to become equipped with this modern technology. Indeed, in the early days, we even used Telex on occasion!

There was an interesting mix of cultures amongst the early delegates – some were from old-fashioned PTTs who did not permit most staff to use mobile phones since they were for the bosses only! Thus some delegates had never used a mobile before, yet were being asked to specify how the next generation would work. Luckily, there were others who were very familiar with their use – important when considering how to specify the man–machine interface. Again there was a mixture of those who were from a user background, while still being technical, and others who understood the technology in more detail – a necessary balance as mentioned above.

Although GSM had been conceived as being the mobile version of ISDN, there was little liaison between the fixed and mobile groups, at least at the service requirements level. However, manufacturers did not want two different ways of specifying similar services (i.e. fixed and mobile) except where necessary due to the requirements of mobility. Hence the editors themselves would use their knowledge of progress in the fixed committees to avoid clashes. As closer working was attempted, there was a problem of both fixed and mobile groups aiming to complete their work in similar timescales, so at each meeting we would be chasing a moving target.

10.2.4 Phase 1

The initial aim was to have a complete working GSM system specified in time for start of commercial service in 1991, a target specified as the first act of the GSM MoU (now GSM Association) in 1987. This meant that service definitions would need to be complete by early 1988 to allow the remaining working parties to base their work on the work of WP1 and also to have these ready for manufacture. In December 1987 manufacturers joined the GSM group and in 1989, GSM transferred across to the new ETSI. Delegates were now able to represent their own companies, rather than national delegations – a great improvement.

We decided that in order to keep to the timetable, we would phase the project and phase 1 would include what we considered to be the most important features for start of service – *if* GSM succeeded, we could then add extra features and corrections later! Thus a number of features were marked "For Further Study", but we concentrated on telephony, emergency calls, basic data services including facsimile and the short message service. Initially, we had just two classes of supplementary service, call forwarding and call barring. See Table 10.2.1.

Telephony is clearly a basic service which everyone expects and the original codec was close to state of the art for manufacture at the planned start of commercial service in 1991. (The codec is the component that converts the analogue speech signals into digital streams

Table 10.2.1 Final list of phase 1 recommendations

GSM 02.01	Principles of telecommunication services supported by a GSM PLMN
GSM 02.02	Bearer services supported by a GSM PLMN
GSM 02.03	Teleservices supported by a GSM PLMN
GSM 02.04	General on supplementary services
GSM 02.06	Types of mobile stations
GSM 02.07	Mobile station features
GSM 02.09	Security aspects
GSM 02.11	Service accessibility
GSM 02.16	International mobile station equipment identities
GSM 02.17	Subscriber identity modules, functional characteristics
GSM 02.20	Collection charges
GSM 02.30	Man-machine interface of the mobile station
GSM 02.40	Procedures for call progress indications
GSM 02.82	Call offering supplementary services
GSM 02.88	Call restriction supplementary services

that are then sent across the radio interface, as needed in all digital systems.) From the start, it was recognised that technology would improve, so a half rate codec was called up in phase 1, to ensure that the signalling would support it, although the actual codec was not specified until phase 2.

GSM also supports data services and these included a range of standards at different rates to support different applications. They are described in more detail in Chapter 16, but in general allow terminals to be connected to modem and packet services. In addition, there was support for facsimile, in those days perhaps the most important data service, although now its use is far less. All these services were familiar to users, being largely mobile access to existing fixed applications.

However, GSM did create a pair of totally new services, the *short message service* and *cell broadcast*. SMS allows text messages of up to 160 characters to be sent, generally between mobile terminals. This was rather slow to take off commercially, perhaps due to the rather unfriendly way of entering alpha characters from a numeric keyboard. However, recently, perhaps due to the availability of pre-pay phones for children, this service has proved immensely popular, with billions of messages sent each month. Cell broadcast is another unique service, somewhat similar to teletext on television, where a series of information messages can be sent to users, based on their location.

One difference between fixed and mobile terminals was our use of the *Subscriber Identity Module* (SIM, GSM 02.17) which allowed the subscription to be taken and plugged into different terminals. The primary purpose of this was to provide security, but it also had some useful customer features. In particular, most early terminals were rather large, often built into vehicles, so one could travel extremely light and take just the SIM card. The problem arose that it might then be fitted into a strange terminal, quite likely set to a foreign language. We therefore agreed that although we wanted to give as much flexibility as possible to manufacturers to design attractive-to-use terminals, there were certain basic operations which we specified must be supported by the *Man-Machine Interface* (MMI, GSM 02.30). The most important feature was our insistence that all terminals support, at least logically, a "+" key for the international prefix. At the time, there were about 15 different prefixes in use across the networks of the world, which caused great confusion. This made it impossible to store numbers in the SIM card, the terminal or the network (for call forwarding), since they would generally be specific to a given country – no use for a roaming system. By replacing this with the + function, the problem is solved. We also aligned ourselves, as far as possible, with the fixed network and PABX standards for supplementary service control. This has proved less necessary since then, as the MMI of terminals and its sophistication have improved. On the other hand, if the user is faced with a terminal set in a foreign language, if he remembers his required codes, he may still find this feature useful. Similarly, these codes are needed if the user wants to store frequently used supplementary service requests in the SIM, e.g. to set or unset a call forwarding repeatedly.

10.2.5 Progress

Examples of questions which arose in the early reports of WP1 to GSM plenary include: the need for transparent DTMF tone signalling; whether all mobiles should support full roaming; the difficulties of specifying network selection procedures (still a problem 15 years later!); whether packet data should be "essential" at 9.6kb/s or would 4.8kb/s suffice? In those early days, 9.6 was considered much faster than possible in the fixed network and therefore was it needed?! Another difficult question was the need to support emergency calls in the absence of a valid SIMcard – still being debated in 2001!

The first block of six (less controversial) recommendations was approved by the plenary meeting in December 1987, with a further six at the following meeting in February 1988. These included a number of recommendations of a more commercial or regulatory than technical nature that were later transferred to the GSM MoU – covering topics such as policy on roaming, charging aspects, subscriptions, free circulation of mobiles, essential or optional services, etc. The final recommendations identified as needed for procurement of networks were agreed by March 1988, although subsequently there were a number of changes and updates!

In December 1987, it was recognised that work on the SIM was of increasing importance and that it required specialist skills. A new group was set up called SIMEG as a subsidiary group of WP1. This was only promoted to being a group in its own right in March 1994.

Later in 1988 and again in 1989, a few more recommendations were approved, including 11.11 for the SIM, and change requests were debated. The GSM committee would only approve changes considered essential, to minimise the impact on product being developed for the first release – all other improvements and additions would be considered only for a new "phase 2". However, one large change in 1989 was of a more editorial nature – the splitting of 02.04 on supplementary services. This was becoming unmanageably large and inflexible, as we planned to add further services. Thus 02.04 became to cover just the principles and relationship of supplementary services, while the individual definitions became the 02.8x (and later also the 02.9x) series.

An on-going difficulty was to reach agreement on the actual number to be used for dialling emergency calls. The EC had proposed 112, but this took some time to be ratified by all member states who were concerned at the cost of changing their fixed number plans. However, it was essential for GSM to have a common, Europe-wide number, since it had to be burnt into the mobile station and manufacturers were developing these. Eventually in January 1990, it was agreed to adopt 112.

10.2.6 Phase 2

In March 1991, we learned that our chairman, Martine Alvernhe, was leaving (to have a baby). Her replacement was Gunnar Sandegren, at that time the technical director of the Swedish operator Comvik. He later moved back to Ericsson, where he had previously worked. Thus it turned out that Martine was chairman for 7 years, primarily for phase 1 and Gunnar for phase 2.

The freeze on any changes other than the most essential corrections, imposed as far back as 1988, was a practical approach to achieving commercial service in 1991–1992. However, throughout this time, there were a number of proposals for improvements to the existing features and for new services. These included several topics which had been planned for phase 1, but which were not ready in time. Sometimes features were specified as requirements in the 02 series, but the corresponding stages 2 and 3 were not ready. Clearly a service needs to be completely defined, or not there at all! (see Table 10.2.2).

The most visible aspect of phase 2 from the service perspective was the addition of new supplementary services. These were: 02.81, the *calling and called line identities and restric-*

Table 10.2.2   Final list of phase 2 recommendations

GSM 02.01	Principles of telecommunication services supported by a GSM PLMN
GSM 02.02	Bearer services supported by a GSM PLMN
GSM 02.03	Teleservices supported by a GSM PLMN
GSM 02.04	General on supplementary services
GSM 02.06	Types of mobile stations
GSM 02.07	Mobile station features
GSM 02.09	Security aspects
GSM 02.11	Service accessibility
GSM 02.16	International mobile station equipment identities
GSM 02.17	Subscriber identity modules, functional characteristics
GSM 02.24	Description of charge advice information (CAI)
GSM 02.30	Man–machine interface of the mobile station
GSM 02.40	Procedures for call progress indications
GSM 02.41	Operator determined barring
GSM 02.81	Line identification supplementary services – stage 1
GSM 02.82	Call forwarding supplementary services – stage 1
GSM 02.83	Call waiting and call holding supplementary services – stage 1
GSM 02.84	Multiparty supplementary services – stage 1
GSM 02.85	Closed user group supplementary services – stage 1
GSM 02.86	Advice of charge supplementary services – stage 1
GSM 02.88	Call barring supplementary services – stage 1
GSM 02.90	Unstructured supplementary service data – stage 1

tions; 02.83, *call hold and waiting*; 02.84, multi-party; 02.85, *closed user group*; 02.86, *advice of charge*. All of these had been planned for the first phase and are similar to fixed line (ISDN) supplementary services. Indeed, many of the supplementary services may be applicable across fixed and mobile networks. For example: a call forwarding may be to a fixed or mobile line; an incoming call, whose number is to be displayed, may be from a fixed or mobile network; a closed user group may contain both fixed and mobile numbers. Thus the fixed and mobile version must be compatible, although not identical.

Interestingly, the most attractive service in the market place turned out to be 02.81. It may be because CLI is generally offered free on mobile services, although often charged for in fixed networks. In reality, customers like to know who is calling before they answer the call and this has proved most popular. The service was originally modelled on the fixed network service, but we tried to simplify it with the restriction part of the service. Unfortunately, just afterwards, the European Commission published a Directive on Privacy which required us to reinstate the parts we had left out, and this delayed the start of that service by over 1 year. We all agreed that full privacy should be provided, although in reality it is rarely used!

Some services were developed with some differences from the ISDN specification, either due to technical reasons of mobility, or for commercial reasons. The *multi-party service* was originally conceived to be identical to the ISDN *conference call*. However, this is a service controlled by an operator where the subscriber pre-books his conference, with a known number of participants, up to 12. The multi-party service works by the subscriber adding

extra participants one-by-one, and is limited to a total number of six. This is a more convenient approach in the mobile environment and it is felt that more than six parties is going to be
difficult to sustain with a likelihood of dropped calls!

Closed user group provides another interesting case history. In principle, the requirement
is very similar to that for the ISDN CUG service. However, this is quite complicated, so
SMG1 decided to keep it simpler for the user to use, in particular from the MMI point of view.
It is intended to be automatic for mobile applications, so that if call conditions are applicable,
then the CUG applies, whereas with ISDN in many cases it was necessary to state that a CUG
was intended, along with its conditions. Interestingly, it turned out that CUG can be provided
using Intelligent Network (IN) techniques, much of which do not need to be standardised.
Thus, all that must be agreed is inter-network signalling. Thus the arguments about the MMI
became irrelevant since each network can choose how they wish to implement the service for
their customers.

Advice of charge is specified rather differently in GSM from ISDN, since it does not have
to be compatible. In ISDN, there are three variants of the AoC service, covering before,
during and after the call. In GSM, on call set-up, if AoC is active, the network sends
information on the rate at which the call is to be charged (in considerable detail, with
seven parameters) and the calculation is made in the mobile station. This contrasts with
ISDN, where the calculation is made at the network and then sent on request to the terminal.

Unstructured SS data (USSD-GSM 02.90) was also a feature poorly described in phase 1
and much better established in phase 2. It is not a customer service but provides a mechanism
for use of the keypad to be sent to the network for controlling special services and also for
receiving replies for display to the user. This is a form of stimulus signalling, especially
allowing older terminals to handle new services. By contrast, standardised supplementary
services in GSM are all signalled in a way similar to ISDN, known as functional signalling.
Functional signalling is more flexible, especially with friendly MMI, except that it cannot
handle "new" services unknown to the terminal. USSD is the first example in GSM of a tool
for service creation.

Most of the remaining changes were to improve the detailed specification for the data
services, including facsimile, short message and cell broadcast. However, these were all
present in phase 1. A new feature for phase 2 was 02.41, *operator determined barring*.
This allows an operator to control the amount of service permitted to a customer when
roaming, aimed at gradually reducing the liability for further expense, especially when he/
she has overspent on his/her bill, or is perhaps a new customer without a proven financial
record. Clearly this service was not needed in ISDN since there is no roaming.

The half-rate codec was specified for phase 2, although a special group was created to
determine the specification for the codec itself.

GSM phase 2 also integrated the DCS 1800 specifications which had in phase 1 been
considered separate and parallel. The impact of this on SMG1 was minimal – mainly to
describe both frequency versions of mobile stations in GSM 02.06.

It was clear that all the features of phase 1 were also carried across into phase 2, although
there were a number of changes to improve the performance or description. Only GSM 02.08,
Report on *quality of service* was not carried across to phase 2. This caused debate for many
meetings since it aimed to set the requirements for the performance of the specifications and
also the construction of the networks. The first of these was logically valid but it was
considered very difficult to determine the performance of the specifications. Thus it should

be considered as giving guidelines between the committees of GSM to determine how they might write their specifications, rather than being of long-term value as to what a customer might expect from a service. The quality of construction of commercial networks, such as the coverage or dropped call rate, was outside the scope of GSM. Thus, although GSM 02.08 gave some useful guidance, it was considered not to be that accurate – the worst case was very much worse than typical – so was never developed for phase 2.

It was clear that the service requirement (stage 1) specifications needed to become stable well before the completion of the stages 2 and 3. Thus as with phase 1, we decided to freeze these for phase 2 in the second half of 1992 and generally only essential corrections were to be allowed thereafter. After this time, new ideas for improvement or new services were considered only for what became known as phase 2+, to emphasise the continuing compatibility of terminals and networks as new features were added to phase 2. This is in contrast with the original phase 1, where most features were assumed to be present and there was no dynamic version control.

10.2.7 Impact of SMG5

When the GSM committee was given the responsibility not only for GSM but also "UMTS", the third generation cellular standard, it changed its name to SMG to reflect that it was not dedicated solely to GSM. The committee SMG5, a continuation of a previous group, which handled all this work, was transferred to SMG in January 1992. The services aspects of UMTS at that time remained within SMG5 and were only considered by SMG1 much later.

Throughout this time, SMG1 was also responsible for the work on the SIM card, through the SIMEG group which reported to them, until it was promoted to being SMG9 in April 1994. Also, any work on security might impact the stage 1 for this (GSM 02.09) which at this time belonged to SMG1 but was later transferred to SMG10. The progress of both these topics is described in the relevant chapters of this book.

10.2.8 Phase 2+

In April 1994, Gunnar Sandegren handed over the chairmanship of SMG1 to me. Thus Gunnar's contribution can be considered as primarily for phase 2 and preparation for phase 2+, while my contribution was for phase 2+ and later the initial releases of UMTS.

Phase 2+ became a series of annual releases so separate versions of specifications needed to be maintained for each release – indeed they are still maintained. Changes may affect all releases or (now more usually) just later ones (see Table 10.2.3).

The first phase 2+ release was version 5, known as Release 96. The first items of phase 2+ approved were: GSM 02.67, 68 and 69 – a priority service, voice broadcast and group call. These were typical Private Mobile Radio (PMR) services and could be used in a variety of applications but were specifically required by the European Railway Project (UIC). Collectively, these became known as Advanced Speech Call Items (ASCI) – although this term is not officially defined.

Phase 2+ also included:

- GSM 02.33 on lawful interception which was later transferred to SMG10;
- GSM 02.34 on High Speed Circuit Switched Data (HSCSD), allowing up to eight traffic channels to be used and thus correspondingly faster data;

Table 10.2.3   Final list of phase 2+ recommendations, with release of introduction[a]

GSM 02.01	R96	Principles of telecommunication services supported by a GSM PLMN
GSM 02.02	R96	Bearer services supported by a GSM PLMN
GSM 02.03	R96	Teleservices supported by a GSM PLMN
GSM 02.04	R96	General on supplementary services
GSM 02.06	R96	Types of mobile stations
GSM 02.07	R96	Mobile station features
GSM 02.11	R96	Service accessibility
GSM 02.16	R96	International mobile station equipment identities
GSM 02.24	R96	Description of charge advice information (CAI)
GSM 02.30	R96	Man–machine interface of the mobile station
GSM 02.34	R96	High speed circuit switched data; stage 1
GSM 02.40	R96	Procedures for call progress indications
GSM 02.41	R96	Operator determined barring
GSM 02.42	R96	Network identity and time zone – stage 1
GSM 02.43	R98	Support of localised service area – stage 1
GSM 02.56	R98	GSM cordless telephony system, service description – stage 1
GSM 02.57	R98	Mobile station application execution environment – stage 1
GSM 02.60	R97	General packet radio service – stage 1
GSM 02.63	R96	Packet data on signalling channels service – stage 1
GSM 02.66	R98	Support of mobile number portability – stage 1
GSM 02.67	R96	Enhanced multi-level precedence and pre-emption service – stage 1
GSM 02.68	R96	Voice group call service – stage 1
GSM 02.69	R96	Voice broadcast service – stage 1
GSM 02.71	R98	Location services – stage 1
GSM 02.72	R98	Call deflection service, stage 1
GSM 02.78	R96	Customised applications for mobile network enhanced logic – stage 1
GSM 02.79	R96	Support of optimal routing, stage 1
GSM 02.81	R96	Line identification supplementary services - stage 1
GSM 02.82	R96	Call forwarding supplementary services – stage 1
GSM 02.83	R96	Call waiting and call holding supplementary services - stage 1
GSM 02.84	R96	Multiparty supplementary services – stage 1
GSM 02.85	R96	Closed user group supplementary services – stage 1
GSM 02.86	R96	Advice of charge supplementary services – stage 1
GSM 02.88	R96	Call barring supplementary services stage 1
GSM 02.90	R96	Unstructured supplementary service data – stage 1
GSM 02.91	R96	Explicit call transfer
GSM 02.93	R97	Completion of calls to busy subscriber – stage 1
GSM 02.95	R96	Support of private numbering plan – stage 1
GSM 02.96	R97	Name identification supplementary services – stage 1
GSM 02.97	R98	Multiple subscriber profile – stage 1

[a] Note that a few specifications in the 02 series were later transferred to other groups, such as SMG9 and 10. These are still service requirements, but under different "ownership".

- GSM 02.42, Network Identity and Time Zone (NITZ) which allows serving networks to update the name displayed in terminals and also to indicate the time zone – useful when changing zones, e.g. in Europe or within Australia;
- GSM 02.43, support of local service area, which allows different services, especially charging rates, based on location;
- GSM 02.56, cordless telephony system was developed but not fully exploited;
- GSM 02.60, general packet radio service was a very important late development of GSM and a pre-cursor of the approach being adopted for UMTS. This took a long time to stabilise, since it is technically difficult, because GSM was originally designed for continuous circuits, rather than sending packets as required;
- GSM 02.63, Packet Data on Signalling Channels (PDS) which never became popular and was later dropped;
- GSM 02.66, number portability was a requirement from some regulators so that a customer was not trapped on a network but could change to another, perhaps cheaper or offering better service, taking with him his telephone number. This seemed important but the UK and the Netherlands regulators refused to wait for common standards to be developed. Thus SMG1 did agree service requirements which were eventually supported by the remainder of SMG, but not until well after non-standard commercial service! In fact, this feature has not proved very popular with subscribers, perhaps because the advent of pre-pay has meant there were some very attractive financial deals where the customer accepted having a new phone with a new telephone number;
- GSM 02.72, call deflection, a PABX-type service similar to forwarding;
- GSM 02.78, CAMEL, introducing Intelligent Network (IN) technology;
- GSM 02.79, optimal routing, important for avoiding tromboned calls when roaming, although networks have tended to favour a CAMEL approach since;
- GSM 02.87, user to user signalling is an ISDN service which has been copied across into GSM, but not yet used very much;
- GSM 02.91, explicit call transfer, another PABX-type service;
- GSM 02.94, follow-me, is a PABX based service. However, it has proved difficult to establish a simple way to use the service in a secure way, to avoid misuse. Its main application has therefore been for railway use, where security is different since it is a private network;
- GSM 02.95, support of private numbering plan, similar to the ISDN service but later taken as one of the requirements for CAMEL, so no stages 2 or 3 were ever completed;
- GSM 02.96, Calling Name Provision (CNAP) is an important feature for the North American market, yet probably unknown outside that region. This was developed to support a feature well used in the fixed networks there, so also needed for mobile. The rest of the world tends to use the address book stored in the terminal (SIM in GSM) so is easier to maintain and update;
- GSM 02.97, Call Completion to Busy Subscriber (CCBS) took up a lot of resources, since it was very popular on fixed networks and especially PABX. However, it is very complicated to support in mobile networks, where the caller may have roamed to another network before the call has been completed. Thus it has not (yet) made much impact in the market place.

We also worked for some time on a joint *DECT/GSM* standard. However, this suffered from project DECT having worked on old phase 1 versions of our specifications without consulting SMG while we were well advanced on phase 2. Although a specification set was published, it formed part of DECT, rather than GSM.

SMG1 was also asked to prepare a report on *radio local loop* – could GSM be commercially suitable for this application? There was also a companion report from RES3 who had identified some weaknesses in the GSM standard for this application. SMG1 considered that GSM could be used, although since it was designed for cellular use, it was somewhat over-specified and not optimised for radio local loop. However, the cost saving in removing unnecessary features would be very little and the advantages of large production runs of a single design GSM would outweigh this. This has probably proved to be true since by May 2001 there were 500 million subscribers to GSM, far more than had been anticipated earlier.

As work progressed on the ASCI services, there was a desire to add the last important PMR feature, direct mode, where two terminals can communicate with each other, if in radio range, without use of a network. However, this proved too difficult for the GSM standard since it is a full duplex system and both terminals would be transmitting at the same time on the same frequency. It was concluded that if this feature were really needed, there would have to be a solution outside the GSM standard.

Another topic studied for many years in the fixed network community was known as *UPT*, providing them with some degree of terminal mobility. The idea was to go to a (fixed) terminal and indicate that you wished to register there. The user could then receive and make calls there until he re-registered on another terminal. The problem was that all attempts at a simple MMI, where the user entered a simple PIN, proved to have poor security. SMG firmly believed that the only solution was to use a SIM, compatible with GSM. This has the problem of the cost of having to convert all fixed terminals to accept a SIM! In reality, the market proved that the best solution was to use GSM itself and indeed in many countries, mobile penetration is now higher than the fixed networks. Thus, SMG never endorsed a change to the standard to facilitate UPT.

Network selection was specified in GSM 02.11 in phase 1. However, some time after this was frozen, severe problems were identified. These were corrected to a minimum extent in phase 1 to allow it to work, but were not optimised until phase 2. Indeed, some phase 1 mobiles would look for networks listed on the SIM, and if these were out of coverage, would close down, rather than look for another suitable network! Perhaps the biggest problem was how to handle poor coverage in a home network where there was coverage from an adjacent country. This border problem was perhaps best exemplified by Copenhagen, which can receive a good signal from neighbouring Sweden. The problem was that home (Danish) subscribers would generally not wish to switch to a foreign network with higher charges and different numbering plan. If they considered that coverage was essential – e.g. to make a call – how do they then return to their home network while still in Swedish coverage? The terminal can always look for a signal from home, but to do so in idle mode consumes more battery power and while searching is unable to receive incoming calls. Since in most situations when roaming, the terminal is well away from home, the best case is not to search for home, even though there may be situations where it should. This has been debated repeatedly by several generations of standards engineers, generally with the same conclusions, although with the development of location services, it may be possible to tailor applications to optimise network selection without altering the standard!

A specification GSM 02.22 (*Personalisation of the Mobile Equipment*) was developed to ensure that terminals could only be used together with a SIM registered with, e.g. a specified home network. This specified not only the service requirements but also how it was to be achieved, so eventually responsibility was transferred to SMG10. However, much of the original work was done in SMG1. This feature is sometimes (incorrectly) called "SIM Lock", but in fact it locks the mobile equipment to the SIM, not the other way round. The reason for this feature is that in some markets the terminal is highly subsidised, so long as it is sold with a subscription for a SIM for typically 1 year. Indeed, the price may be as low as a few euros, or even a negative amount. There was a danger in the market place of customers then not paying their subscriptions, but selling on the terminals at a profit, which may be considered a form of fraud. With GSM 02.22, the terminal can be locked so that it will only work with a SIM from a specified network operator, or some similar categories. At first, the European Commission expressed concern at the possibility of this being "unfair" to customers, but after some explanations and promises by operators of fair treatment, it did enter commercial service, being especially popular with North American operators.

By early 1996, there was a long list of new services for phase 2+ and a few new network features such as CAMEL whose service requirements were either finalised, or at least well advanced. However, there is a limit to the resources of the other groups in SMG who have to design the actual services, especially the signalling specifications. A stage 1 description cannot generally be implemented on its own – a complete set of specifications for each service is required. In particular, with CAMEL, there were many more features requested by SMG1 than could be provided in its first release, so the most important aspects were treated as CAMEL phase 1, with later phases planned to add extra functionality. Indeed, even in 1996 it was recognised that not all the features desired could be accommodated in CAMEL 2 and now in 2001, we are currently working on CAMEL phase 4!

It should also be noted that some new features were specified *without* creating new stage 1 specifications. Examples include the higher data rate at 14.4kb/s and the speech codec at half-rate and Extended Full Rate (EFR). In fact, the original date rate of 9.6 kb/s became 14.4 kb/s and then later a range of maximum rates significantly higher became feasible by using more powerful channel coding, although it works over a smaller cell radius. This technique became known as EDGE.

10.2.9 The Way Ahead

Early GSM standards provided an excellent start to commercial service, offering what were seen as the most useful commercial features. However, the range of features was fairly limited and most operators offered an almost identical range of services. This did not matter so much then since, except for the UK, all countries' networks were run by a state-controlled monopoly. However, times change!

Under the influence of the EU, all countries of Europe developed competing operators, sometimes as many as four, all with national coverage. Clearly they wanted to compete on features as well as price, service and coverage. The timescales for developing new services became quite long, not just in the committee but also for the manufacturers. Also, many operators would buy their infrastructure from more than one supplier, but could not offer some new feature until all their suppliers had it available. By then, the operator's competitor also had the same feature since he is likely to use some of the same suppliers.

Two solutions for this problem appeared on the market:

- Some felt that ETSI was too slow and that they could move faster with private groups so various fora were set up. Perhaps the best example was the WAP forum, which developed its own "private standard" for use by members of its forum. However, problems can arise and it is becoming recognised that public bodies such as ETSI can work quickly if needed – the work is always by voluntary contribution so is dependent on the effort provided by members.
- An alternative approach is to allow operators, service providers and even customers to design their own services, either for all customers on a network, or even individually. This "service creation environment" has proved popular and is the way most new services are being developed – as applications using tools developed within the standard.

The first major toolset was Intelligent Networks (IN), developed by the fixed networks for services such as freephone. This formed the basis for CAMEL, as mentioned above. Here the most important feature is number translation, so for example an active user, even when roaming, can dial a home based number plan even when abroad. This can be extended to cover security checking, flexible barring of expensive calls and even as a cheaper alternative for features such as closed user group and optimal routing which already have standards, although these tend not to be used. CAMEL is proving popular and a number of the later features allow market led requirements such as pre-paid roaming.

While CAMEL is useful for network based features, there are times when a terminal based solution is preferred. Mobile Station Execution Environment (MExE – GSM 02.57) supports Java and similar features in a mobile terminal, so that terminals can perform services irrespective of their manufacture.

SIM application toolkit (GSM 02.38) specifies a dialogue between the network and the SIM, so that, e.g. new codes for features can be downloaded into the SIM, or via that to the terminal itself. Unlike earlier phases of mobile, this can allow the SIM to act as master "on its own initiative", rather than only as a slave to the mobile equipment. So it can have the intelligence to act at a pre-arranged time, or when some situation arises, such as moving to a particular region. Again, services maybe applications written by the network operator, a service provider or possibly a well educated customer!

As early as 1996, it was becoming clear that at the highest level, the user wanted to be able to create his/her own "home feeling" through a mixture of settings of his/her own terminal, together with the environment of his/her home network and any applications he/she chose. He/she needs this to accompany him/her when he/she roams, with as little disruption as possible. We coined this requirement the "Virtual Home Environment" or VHE. Sometimes there is confusion as to what this is exactly, but the intention was for it to be not much more than a philosophy, guiding how we were to provide the necessary tools to achieve this for the customer.

The final main tool was OSA. This originally stood for open service architecture, but it is not the role of the services group to specify an architecture, so it was renamed by us as open service access. This is still being developed but provides an open interface for applications to run on a terminal so that third party applications should always work on any suitable terminal. Even the largest network operators are finding that they do not have the resources to write all the application software that the market desires, and this should open the way for others to assist!

10.2.10 The Third Generation

The work of SMG5 and much of the third generation project is described elsewhere, but SMG1 did play an important role, which will now be covered. The initial work on service requirements in SMG5 seemed to adopt the ITU approach of defining an ever increasing range of services. This appeared logical – customers want more services, so the standard should specify them. However, in the section above, this is shown to have been replaced with the concept of service creation, which is far more flexible, quicker and responsive to market forces.

There were also concerns that SMG5 had been assuming that *new* operators would run UMTS networks, not the existing GSM operators. However, although this was not finally established until the spectrum auctions of 2000, it did eventually become clear that UMTS would in reality be an addition to existing networks which themselves would be upgraded to offer similar services, albeit at lower data rates.

Thus GSM and UMTS services would be largely indistinguishable, except for the different radio interface. This work started in SMG1 and was later continued in 3GPP SA1, gradually merging the GSM and UMTS specifications, leaving out some of the less relevant specifications, especially those specific to a given radio access method.

It became apparent that many of the current supplementary services could alternatively be provided by service creation tools, based on VHE. However, this was not always the most economic choice for services that had already been designed and were in popular use.

The approach we chose for specifying the service requirements in this new environment was to create an overall top-level specification named 22.101 – UMTS service principles. This gave a brief description of all our requirements, many of which would then be elaborated in more detail in the remaining specifications of what became the 22 series. We also kept much of the old GSM specification base.

Life tends to follow history and inevitably, we found that we needed a deadline for start of service and hence for the specifications for the first release to be frozen. This was Release 99 and for the first time, this contained new specifications for UMTS (22.1xx series) and also renumbered GSM specifications which were transferred, often with minimal change, becoming dual purpose GSM and UMTS specifications (22.0xy series were equivalent to the GSM 02.xy). There were some GSM specifications that were no longer relevant to UMTS. Some were specific to the radio interface, such as 02.34 (HSCSD), 02.68 and 22.69 where multicall would be handled quite differently between GSM and UMTS. Others like 02.06 and 02.07 were no longer needed since they were in effect subsumed into 22.101.

Release 99 service requirements were stabilised early in 1999, but then updated as necessary with changes which were essential or more often those which were needed for alignment with the remainder of the UMTS project. This was the same pattern as established for phase 1 and later phase 2 of GSM. As before, it was also found that some of the features we had planned for the first release were too ambitious (or even no more thought to be required). Thus we had to constrain the first release to what could be achieved as a consistent set of specifications for Release 99. This we did by creating 22.100 as an overview of Release 99.

At the end of 1998, UMTS had become so international that it was decided to create the Third Generation Partnership Project. This carried on the work of UMTS, taking it out of ETSI SMG. The services group of 3GPP was called SA1 and was also chaired by me and had mainly the same delegates as SMG1, which retained the GSM work for a while. In reality, they were joint SMG1/SA1 meetings, but different parts of the agenda would cover either GSM or UMTS, and technically the rules for procedure were slightly different.

This process lasted for 18 months until it was agreed to simplify things by moving the work on GSM also into 3GPP. This was really inevitable, since the radio interface affected only minor details of our specifications and the rest really was a common project. Thus the procedural problems began to get in the way of progress. Once SMG had reached the milestone of 32 plenary meetings, a binary number and exactly the number of GSM meetings beforehand, it was agreed to close SMG and hence SMG1 also!

The work continues in 3GPP at a great pace, with an increasing emphasis on service creation environments and Internet protocol based operation. Thus applications now are getting much closer to those on the Internet and much software is likely to be written by application providers. Whereas the main application in GSM was for voice, UMTS has more emphasis on multi-media, of which voice is just one component (although still probably the most important). Transport for traffic and signalling will be more likely to use IP than traditional circuits. How things have changed from the early days of a mobile ISDN! SA1 continues with roughly annual releases – Release 99 was followed by 4 and then 5 (predicted end 2001). Release 6 is already being considered, but perhaps we need time to slow down and consolidate – we are now, before the start of commercial service for UMTS, writing specifications that will not be implemented until around 2006.

On 9 February 2001, I handed over the chair of SA1 to Kevin Holley of BT, after 14 years in GSM1, SMG1 and SA1, 7 as chairman.

The early vision of GSM was very clear, although the extent to which it would take over the world and the more than 500 million customers was never even a pipe dream in the early days. Even in 1987 there were arguments between operators as to whether hand-helds would be viable or should we concentrate on vehicle phones.

The picture has changed unbelievably within the my career of 15 years. It seems impossible to predict where we will be in 15 years' time! Probably information terminals, but will people *speak* into them? Who knows!

Chapter 11: System Architecture

Michel Mouly[1]

11.1 Scope

The decisions that were taken in GSM3/SMG3 were centred on the specification of signalling protocols. Signalling protocols can be understood as languages for exchanging control information between distant nodes, such as network nodes or mobile stations. The exchanged information is quite variegated. It includes data such as identities, called numbers, nature of call, description of allocated resources, result of measures, and so on.

11.2 Architecture

Before the specification of a protocol can be started, it is necessary first to determine the two entities that are exchanging the information. This in turn requires a description of the overall system in terms of entities, such as mobile stations, terminals attached to mobile stations, radio base stations, switches, … The analysis and description of a system in terms of entities, of interfaces between such entities and of dialogs flowing over these interfaces, in short of the system architecture, was originally a derived task of GSM3; this task was eventually given to a sub-group of SMG3, and to a spawn of SMG3, SMG12.

The choices driving the specification of the architecture of a telecommunications system such as GSM are determined by a variety of factors. Some splits are derived naturally from fundamental requirements, such as the split between the user equipment and the infrastructure, with radio as the means for communication between the two sides. Roaming, interfacing with non-GSM telecommunications networks, interfacing with non-GSM user terminals, are other requirements leading to natural splits.

Some other splits have been introduced to answer particular requirements from the operators, for example to centralise some functions and thus to reduce costs. For instance, the separation between switches and radio site equipment allows the procurement of high capacity switches (i.e. able to handle the traffic from a vast area, covered by many radio sites). Another kind of operator requirement that has been a leading factor in some architectural decisions is the wish to increase the possibility for competition between equipment manufacturers. An example in this category is the split between the radio resource controlling node (the BSC) and the switches.

---

[1] The views expressed in this chapter are those of the author and do not necessarily reflect the views of his affiliation entity.

Finally, many architecture decisions arose from the progressive enhancement of GSM past the first phase. Adding new functions was in some cases done by introducing new nodes, to ease manufacturing and/or deployment.

On the other hand, splitting a system, which has a unique goal (allowing end users to communicate), in distinct and often distant entities leads to needs for co-ordination of the actions of these entities so that the common goal can be achieved. The additional requirement to allow different manufacturers to compete on the equipment market (both for infrastructure and for user equipment), as well as giving some level of independence between operators, leads to the need for common, publicly available, specifications of the key protocols ruling the communications between distant entities determined by the architectural choices.

This sums up the main activities of SMG3/SMG12: to choose and to describe the general system architecture, to decide for which of the interfaces so determined a set of standard protocols should be produced, and to specify among those protocols the signalling ones.

11.3 General Trends

The *a posteriori* analysis of the decisions taken by GSM3 distinguishes between two main approaches: on one side the adoption and, if needed, adaptation of techniques and protocols existing in the non-mobile telecommunications world, and on the other side the creation of original solutions, usually to answer problems specific to mobile telecommunications. If the latter approach illustrates the pioneering work and creativity of SMG3, the former has shaped more fundamentally the design of GSM.

GSM was not designed independently from the rest of telecommunications. On the contrary, the services provided by GSM are primarily an extension to mobile users of the services available to users accessing networks through wirelines. For this reason, the technical choices made in wireline telecommunications have had a deep influence on most key technical choices in GSM.

Two periods can be distinguished when looked at from the wireline point of view. The initial, foundation building, technical decisions for GSM were taken around 1985–1988, at a time when the telecommunications "paradigm" was ISDN, characterised by integration of voice and data, digital transmission of voice, and packet transmission for signalling. The second period started around 1992–1993 and was full-blown around 1995, and corresponded to the shift to all-packet transmission, with a focus on the Internet.

The areas where SMG3 has been the most creative were, not astonishingly, those where no on-the-shelf solution was available from wireline techniques. From a signalling and architectural point of view, the essential difference between cellular access such as in GSM and wireline access is that a given user is not tied to a single access point, but on the contrary, is allowed to access from many points, spread geographically at the scale of the world, and from an administrative point of view, between many operators. This "roaming" capability has a very important impact on how to route calls towards users, since their location is known only imperfectly. Mobility during calls leads to dynamic routing modifications, the handovers.

For all these points, and many others, SMG3 built on the experience gained in NMT and other first generation cellular networks, using original ideas from its members to set down principles for digital cellular networks that remain valid through the next generation.

11.4 Historical Perspective

11.4.1 Before 1985

The examination of the reports and participant lists of the GSM meetings in those first years shows that the topics that were to be those of GSM3 were rather marginal. The main advocate of those topics, and of their importance, was Jan Audestad, who was to become the first chairman of GSM3. Important guidelines were established, if not officially, during this period, in particular the idea that GSM was to be designed mainly to interwork with ISDN.

11.4.2 1985–1988

GSM3 was created officially in 1985, in the first batch of creation of technical sub-committees of the GSM standardisation group. Activities related to network architecture and protocols had started some months ago, as part of the GSM meetings, and the newly acquired independence allowed the working pace to increase.

At that time, the radio transmission was not defined. The main tasks of GSM3 were then the general architecture, with as a main guideline the adaptation of ISDN switching and routing, as well as the protocols involved in call establishment and management. The main difference with ISDN was mobility, and in particular roaming. This led to the development of the Mobile Application Part (MAP), a protocol suite to be fitted in the larger protocol suite that is Signalling System No. 7 (SS7, the packet network used between ISDN switches for routing and call management). The other area of work was the adaptation of the ISDN protocols between the switch and the user equipment. From the start there was a distinction in GSM3 between "network" related activities, meaning aspects related to switches and routing, and then roaming, and "radio" related activities, meaning what happened more locally, between the switch and the user equipment. A third activity was identified, namely the adaptation of supplementary services. These activities were allocated to sub-sub-groups, named WPC, WPA and WPB (WP for working party), in the order of presentation. This working organisation was essentially unchanged up to 1995.

WPC, dealing with switching network architecture, set down very early the main concepts, in particular the very important notion of Home Location Register (HLR) and the technical principles for roaming, with such notions as location registration (allowing users to register with the local GSM switch) and two-branch routing of calls towards GSM users (so that a single number can be used to call a GSM user, wherever he/she is). This was a formalisation and clarification of ideas already existing in the analogue system NMT. The result has been one of the key factors for worldwide adoption of GSM, since with this technical basis (plus the financial aspects, treated by the MoU) roaming could be offered with not too much effort by any new operator.

In WPA, dealing in particular with the radio interface, work started on the protocol architecture, but was somewhat limited due to several factors, in particular the fact that the main aspects of radio transmission were set down only during 1987. Several key decisions were taken during this period, decisions which are still acting for instance in the third generation standards. They deal with the architectural separation between what relates to the transmission means over the radio interface, and what transits over that interface, but is essentially independent of it. The two key decisions concern first the system architecture, and second the radio interface protocol architecture.

It was clear from the start that the cellular concepts implied that there would be much more emission/reception geographical points (base stations) than switches, if scale economies were to be obtained by using high capacity switches. This led naturally to the notion of switches and base stations as distinct nodes. The more in-depth studies of the architecture showed quickly that there were tasks that fit well in neither node, in particular those related to handing over connections from one base station to another when users moved too far away during calls. Some centralisation over several base stations seemed needed, and it appeared best not to overload switches with such computing hungry radio-related tasks as those required to determine when a handover is needed and to which base station. Thus came the three node architecture, the base station, the switch and the base station controller in between.

This decision led to another, more difficult one, about whether protocol suites for the newly created interfaces (between base station and controller, and between controller and switch) were to be standardised, or left to node manufacturers. Such decisions are not technical, but the result of strategic equilibrium between on one hand operators, which favour open interfaces, i.e. interfaces for which one standardised protocol suite exists, so to allow them independent choices for sources of the different nodes, and on the other hand manufacturers, which prefer to avoid the competition thus created.

In the case of the GSM radio sub-system, the operators, no surprise, favoured the standardisation of protocol suites for both interfaces, whereas manufacturers tried to limit the standardisation to a single one, with some more discussions on which one. To make the story short, the position that prevailed was that of the operators (no surprise once again, the GSM group being dominated by operators at that time), with a qualification that the effort was to be put mainly on the protocol suite for the switch to controller interface (the A interface).

The other key decision taken by SMG3/WPA was the split of the so-called "layer 3" protocol over the radio interface in three parts, one protocol for call management, one for mobility management and the other for radio management. Though this was not that obvious at the time, this decision is consistent with the system architecture, since it distinguishes the protocol that terminates in the base station controller (for radio management) and those terminating in the switch. In the short-term, the main advantage was to separate the part that could be done by a simple adaptation of ISDN protocols (the call management protocol) from those that were specific to cellular access (radio management and mobility management) and thus required a specific focus.

With hindsight, those two decisions were excellent, because they set as a foundation the division between the radio access part (base stations and their controllers) and the network part (switches, location registers). Though the protocol suite developed by GSM3 was not fully universal, there are clear indications now that the A interface is the right point for which a universal protocol suite could be developed. Other cellular networks followed this route, as well for instance systems offering mobility with transmission through satellites. The Iu protocol suite developed for third generation is fully in the same spirit, and one step further in the direction of a universal protocol suite.

WPA had many other tasks. Within this period the main lines for such aspects as handover support, access to the system from idle mode, network and cell selection, and paging over the radio interface, were established, among others.

11.4.3 1988–1992

In 1988, a complete first draft of the main protocol suites (MAP, A interface and radio interface) was available. Thanks to the decision to split the work between phase 1 and phase 2, the following years were devoted mainly to transforming these drafts into stable documents.

Curiously, there is little of interest to describe in this period, though it was one of very intensive work. The work load was very high, and amount of documents dealt with and generated during this period was enormous. However, they concerned myriads of decisions concerning details; none of the founding decisions taken beforehand were challenged, and no new requirements led to decisions important to note here.

11.4.4 1992–1995

The years 1992–1995 correspond to the development of the phase 2 program. The main work was simply to modify the protocols to fit a number of new functionalities, all fitting rather naturally into the framework. Many of them were in the area of supplementary services, and the corresponding group, WPB had a lot to do.

The introduction of some of those new features revealed some weaknesses in the capacity of the protocols to incorporate them with "upward compatibility". By this term is meant the possibility to interwork a node which has been upgraded with the new features and a node which has not. This is not really a problem when both nodes are under the control of the same operators, and then not of major importance in telecommunications so far. On the other hand, a consequence of roaming and of procurement from multiple, competitive, manufacturers, is that a network has to accept users with any equipment, old or new, and has to interwork with many other networks worldwide. Since it is difficult (!) to coordinate upgrading of all user equipment and of all networks in the roaming web, upward compatibility is a strong require- ment for the radio interface and for the interswitch protocol suite, the MAP.

As a consequence, it was accepted that the transition between phase 1 equipment and phase 2 equipment would be done the hard way, but that everything should be done so that further enhancements can be introduced smoothly. This led to a number of modifications, very technical in detail, of both the radio interface protocols and the MAP.

11.4.5 1995–1999

Up to 1995, the work of GSM3 (SMG3 at that time) had been well delimited and consistent. With the advent of the phase 2+ program, the work spread in many directions and the group expanded correspondingly, with unfortunately a loss of focus. A number of new topics arose that were too big to be dealt with together with the maintenance of the existing standardised features. Moreover, many were by nature weakly linked with these existing features. Another factor was the devolution of full work items to T1P1. The result was that the work in SMG3 was split into many independent tasks, minor and major. Some examples of interest are dealt with below.

11.5 Supplementary services

Many phase 2+ work items were supplementary services in the ISDN meaning of the term, and thus fell to SMG3. The adaptation to GSM of services offered with wireline ISDN was more and more difficult, because the easy one were done first. Some services such as Call Completion on Busy Subscribers (CCBS) proved to be tricky to combine with mobility.

11.6 GPRS

The GSM architecture designed in the 1980s was heavily influenced by ISDN. In the ISDN perspective, packet switching is to be used only for signalling. Transmission of user data is limited to circuit. As a consequence of this limited view, ISDN switches, and hence GSM switches, since they are usually derived from them, have a high capacity for circuit switching, and a small one for packet switching. The latter was sufficient, however, so that the very low rate packet service called Short Message Service (SMS) was acceptable, but widely insufficient for the emerging packet services, in particular for wireless access to the Internet. The consequence of this was that those services could be offered only using a circuit to cross the switch, and, unfortunately, the A interface.

The topic of one major phase 2+ work item was to provide some improvement of this state of affairs, and that was GPRS. It fell to SMG3 to study the system architecture for GPRS. However, because the members of WPC, the group in which switching and routing are dealt with, were mainly ISDN experts, and because the experts sent for dealing with GPRS architecture were mainly Internet experts, WPC was considered by both parties as unsuited to study the item, and a special technical sub-group was created.

Other factors led in the same direction to estrangement between the original ISDN core of GSM and the GPRS work item. One was the weight of in-the-field equipment. Both operators and manufacturers felt that it was important to minimize the impact of the introduction of any new feature on the existing infrastructure, and on this basis they were reluctant to accept, or even to study in depth, an integrated architecture. Another factor was the preference of Internet people for something totally separate from the ISDN and PSTN. That was what happened for wireline telecommunications, and there was seemingly no reason why this should not be suitable to cellular. And a third factor was a unexamined trust in packet multiplexing as a source of major improvement.

All this resulted, after discussions that lasted over years, in an almost totally parallel architecture, with the existing series of nodes essentially unmodified, and a new series of nodes (in particular a switch) and of protocol suites entirely devoted to GPRS. Moreover, the new part was designed on the simple, and seemingly natural idea that the Internet transmission techniques have "just" to be extended down to the cellular user equipment. In practice, the feature was technically very ambitious, and a long time was necessary to cover all the issues.

The difficulties came from two directions: the contradiction between the lack of integration between the GPRS infrastructure and the original infrastructure and the need for such an integration in the user equipment; and the impact of mobility. The double architecture led in particular to two mobility managers, one in the circuit switch and the other in the GPRS switch (plus the idea of a third mobility manager, with mobile IP). Also, the protocol suite developed for the switch to radio controller interface (in correspondence with the A interface), influenced by Internet approaches to transmission, took a narrow view to the handover issue. It is interesting to note that in the third generation, the development of the Iu protocol suite did not follow the GPRS approach but a more integrated view.

11.7 UMTS

One consequence of the explosion of SMG3 in many parallel working activities was that the lack of a group dealing with the system architecture per se was more and more visible. In the early years, SMG3 plenary meetings were the main place where architectural consistency was checked. With the increase in the number of topics in phase 2+, time was insufficient for that, and it was difficult to attract the required experts in meetings dealing mainly with preparing a synthetic report of what all the sub-groups did.

More or less at the same moment, the need was expressed to integrate third generation activities into the main SMG groups. A sub-group of SMG3 was then created to answer both problems, and was nominally in charge of decisions related to architecture, both for GSM and the ongoing phase 2+ features, and for the UMTS. In 1998 this group became a technical sub-group of SMG, under the name of SMG12.

In some way, this group happened too late for GSM, and too soon for UMTS. On the GSM side, it was too late to come back on past architectural decisions, in particular regarding GPRS. On the UMTS side, the group discussed many interesting concepts such as the enhancements of roaming known as "Virtual Home Environment" or a universal switch to radio access system protocol suite. However, recent events showed that such big steps forward were not compatible with the short-term objectives of the 3GPP group.

11.8 Onward

With the advent of 3GPP, the organisation of the work dealing with the subject matter of SMG3 and SMG12 was drastically changed. The architectural aspects are dealt with 3GPP/SA, the next avatar of SMG12, and have been set nearer to the group dealing with services. All technical aspects involving the switches have been grouped together and separated from the radio-subsystem aspects. The protocol suite for the radio interface, which was henceforth dealt with in a single group, has been split so that the protocols terminating in the switch are dealt with together with other switch-related topics, and the protocols terminating in the radio sub-system are treated in the group dealing with radio transmission.

Despite these reorganisations, there is a true continuity from GSM to UMTS in the areas of architecture and protocols. On the technical side most of the key developments and ideas of GSM architecture and protocols are still living in UMTS. In the switching and routing area, the standards are common to GSM and UMTS. The continuity is also clearly visible in the working methods, and, not least, the membership.

11.9 Key people

The output of the SMG3 and SMG12 amounts to tens of thousands of pages of specifications. Over the years, several hundred people participated in the different meetings, representing the work of possibly thousands of contributors spread over dozens of companies. Some contributors will be named, though it should be kept in mind that their contributions represent only a very small part of the overall work.

As in any project, the influence of individuals is more marked at the beginning. This is particularly true for the switching aspects, which brings to mind the names of Jan Audestadt and Christian Vernhes who participated in the founding technical work.

On the radio sub-system side, a small group of people has animated the work for quite a long time, mainly Roland Bodin (whose influence is deeply missed since his premature death), Michel Mouly, Chris Pudney, and also Niels Andersen and François Courau.

However, SMG3 and SMG12 key people are the first of hundreds of remote contributors, who provided the substance on which the chairpersons, the meeting secretaries and the participants worked.

Chapter 12: Radio Aspects

Section 1: The Early Years from 1982 to 1995

Didier Verhulst[1]

12.1.1 From Analogue Car Telephone to Digital Pocket Phone

CEPT took a very much forward looking decision when it decided to create, as early as 1982, the GSM Group with the mandate to define a second generation harmonised cellular system in Europe. At that time, the true market potential for mobile systems was not known. Also many technologies, which became key to the GSM radio design, were just emerging. This is true particularly of cellular networking, digital signal processing and real-time computing.

In fact, GSM work started when the telecom industry was experiencing a fundamental shift between the "circuit switched analogue" world and the "packet switched digital" world. Microprocessors had just been introduced a few years before and it was the time when the first PC was created. The PTT administrations were introducing digital switches in their telephone network to replace mechanical switches, and they were developing their first packet switched data networks. We know today that this "digital" revolution ultimately lead to the Internet as we know it today, but this was not at all clear at the time. As we shall see, even the decision to select a digital rather than analogue modulation was not obvious and it took almost 5 years to be settled!

This technological "turning point" was also a wonderful opportunity for a young generation of engineers who had just learned in school the beauty of digital transmission and packet switching, and had therefore the opportunity to contribute actively to the creation of a new standard.

12.1.1.1 Marketing Requirements

In the early 1980s, the market for a second generation cellular system was perceived as primarily radiotelephone in vehicle. In a study called *"Future mobile Communication Services in Europe"*, prepared for the Eurodata foundation in September 1981, PACTEL introduced the concept of "Personal Service" as opposed to "Mobile Service" and explained that the total market for vehicular mobile could be as high as 20 million in Europe while the demand for low-cost hand-held service could ultimately reach 50% of the European popula-

---

[1] The views expressed in this section are those of the author and do not necessarily reflect the views of his affiliation entity.

tion. But it was also concluded in the same report that a single system could not realistically serve both vehicle-mounted and portable terminals, because vehicle terminals would require high power to insure continuous coverage and complex control function to ensure seamless handover at all speeds, while low-power hand-held terminals would use a network of non-contiguous small cells and should not be considered as truly mobile.

The work of GSM started initially with the objective of providing service primarily to vehicles, but it was recognised in the process that there should also be a proportion of portable devices as these started to appear even in first generation systems. Ultimately, the radio interface selected in 1987 by GSM turned out to be efficient enough to allow a true personal service, with continuous service anywhere, and a number of users largely exceeding the most optimistic early market projections. The first commercial GSM terminals in 1992 were vehicle-mounted or bulky transportable terminals, but the terminals in use a decade later are almost exclusively very compact hand-held devices!

12.1.1.2 Technical Background

The cellular concept was first described in the 1970s by the Bell Labs, and the first pre-operational cellular network was launched in 1979 in Chicago. In Europe, the NMT system started operation in the Nordic countries in 1981. The key radio features of a cellular network, i.e. its seamless handover between base stations and the reuse of frequencies between distant cells were being implemented for the very first time in commercial networks when GSM work started. We were therefore to design a second generation when the true performances of the first generation were not yet known!

Around 1980, we were just seeing the first practical implementations of digital processing in commercial domains such as microwave transmission and digital switching but, while the principles of digital encoding and modulation were already well known, there was still some doubts about the amount of processing which could be implemented in cellular base stations and, particularly, in mobile terminals.

When the first descriptions of the GSM radio interface were published around 1987, showing the mobile stations monitoring in parallel a large number of logical traffic and signalling channels multiplexed in time, we heard sometimes the comments that this was far too complicated and could never be implemented in low cost terminals. Retrospectively, it is amazing to see the amount of software found today in low cost devices such as toys and also to realize that current GSM hand-held terminals already have more processing power than most micro-computers produced only a few years ago!

12.1.1.3 Building a New System

As research engineers, we faced the ideal situation whereby we had to define the most advanced system possible with a very limited number of constraints. It was really the perfect "blank page" exercise. We were given access to a completely new 900 MHz spectrum, up to 25 MHz in each direction, without any requirement to ensure upward compatibility with the first generation. In fact, because there was already several incompatible analogue systems in preparation in various countries of Europe, it became clear instead that only a truly innovative and more efficient system could be adopted by all administrations.

In comparison, the US mobile industry tried in the late 1980s to define their own digital second generation system, with a constraint of upward compatibility, in terms of channel spacing and signalling protocols, with their analogue first generation "AMPS". This was seen as a key advantage to allow the production of dual-mode analogue/digital mobiles, thus allowing the operators to digitalise their network progressively according to traffic demand while maintaining continuous coverage.

As it turned out, this compatibility constraint delayed the introduction, in the "digital AMPS" standard, of advanced features such as detailed measurements by mobile terminals, advanced handover control, and flexibility for innovative frequency allocations schemes while these features were in GSM from day one. As a consequence it took several releases – and many years – for the American second generation cellular standard to seriously compete with GSM in terms of performance. In fact, the digital AMPS standard never became a real threat to GSM in the world market, and it was even challenged in its own market by the IS-95 CDMA proposal in the early 1990s.

12.1.2 GSM Initial Work on Radio Specifications

At the beginning of GSM, some essential decisions had to be made concerning radio para-meters, including the choice between analogue and digital modulation. In the case of a digital system, there was also a number of key specifications to be produced concerning the trans-mission bit rate, the type of modulation, the multiple access principles, the source and channel coding schemes, as well as the frame structure and the detailed mechanisms to handover from one cell to another.

12.1.2.1 Establishing "Working Party 2" (WP2) on Radio Aspects

At the GSM 03 meeting in Rome in early 1984, it was decided to set up three specific working parties to progress on key technical subjects: services, radio and networking. The second working party, WP2, was mandated to investigate radio transmission aspects. I was asked to chair that group and we worked on a general model of the radio transmission channel applicable to a digital mobile system. The activity of WP2 continued in 1984-1985 during specific sessions in parallel to the GSM plenaries, and focused on early comparisons between various digital multiple access options. Early 1985, I was leaving the French Administration and I handed over WP2 responsibility to Alain Maloberti.

During 1985, it was decided by GSM that due to the increasing amount of work required, the Working Parties would hold dedicated meeting every three months during the interval between GSM plenaries. The years 1985-1987 were very important for WP2 as they allowed the selection of key parameters for the radio subsystem. This process was lead by radio experts from various PTT administrations involved in WP2, and it was supported by experi-mental programs involving manufacturers from various European countries. In February 1987, when the main options had been decided and the work was focusing on the finalization of the first release of GSM recommendations, ETSI allowed manufacturers and research institutes to contribute also directly to the work of GSM plenary as well as working parties. The production of the GSM standard was therefore a truly European effort involving all concerned parties of the industry.

12.1.2.2 Analogue Versus Digital

On the comparison between analogue and digital options, it was by no means obvious at that time that the quality of encoded speech could be equivalent – not to mention better – than plain analogue FM when considering (i) the limitation in terms of bit rate to accommodate an average spectrum utilization of about 25 kHz per carrier as in analogue systems and (ii) the fact that gross bit error rate over the fading mobile radio carrier could be as high as 10^{-2}. Also, while we were evaluating digital options, analogue systems such as NMT were even able to improve their capacity with channel spacing reduced from 25 to 12.5 kHz while maintaining a good quality of speech.

From the early days, we did work however with the "working assumption" that the GSM system would be digital, but this assumption was only formally confirmed in 1987 when we could prove, including with field trials, that a digital system would really outperform all analogue systems.

12.1.3 The Choice of the Multiple Access Scheme

One interesting advantage of digital transmission is that there exists a variety of methods to multiplex several users over the same radio carrier, namely "Frequency Division Multiple Access" (FDMA), "Time Division Multiple Access" (TDMA) and "Code Division Multiple Access" (CDMA). In comparison analogue systems are restricted to the "one carrier per active user" FDMA scheme. I remember some meetings during the early years where the basics of TDMA had to be explained to experienced radio engineers who had always thought of radio resources in terms of "frequency carriers" and never in terms of "time slots". The fact that several mobiles could coexist without interference on the same frequency connected to the same base station was truly intriguing for a number of delegates!

During 1984, WP2 had already identified the three multiple access options FDMA, TDMA (narrowband or wideband) and CDMA (frequency hopping or direct sequence) which would be the object of much debate until 1987 when the "narrowband TDMA/frequency hopping" solution was selected. Interestingly enough, several years later, quite similar discussions took place to compare wideband TDMA with CDMA in the context of UMTS.

12.1.3.1 The Selection Process

GSM decided to launch a series of experimental digital systems to facilitate the selection of the radio transmission and multiple access scheme. This trial activity was initially supported by the French and the German administrations who decided to collaborate in the definition of an harmonized system, and agreed in 1985 to focus their efforts towards the selection of a digital second generation system. Accordingly, manufacturers were selected from both countries to develop prototype systems implementing different digital radio subsystem concepts. Soon after, the Nordic administrations also decided to join and additional proposals were submitted to the experimentation program which took place in Paris from October 1986 to January 1987, at the France Telecom research centre, CNET, under the auspices of the GSM permanent nucleus.

Eight different system proposals, corresponding to nine different radio subsystem solutions (one system proposal having two different multiple access solutions for mobile-terminated and mobile-originated links) were proposed for the Paris trial. Sorted out by multiple access type, they were the following:

FDMA

- MATS-D (mobile to base), by Philips/TeKaDe, Germany

"Narrowband TDMA"

- S900-D, by ANT/BOSCH, Germany
- MAX II, by Televerket, Sweden
- SFH900, by LCT (now Nortel Matra Cellular), France
- MOBIRA, by Mobira, Finland
- DMS90, by Ericsson, Sweden
- ADPM, by ELAB, Norway

  "Wideband" TDMA (combined with CDMA)

- MATS-D (base to mobile), by Philips/TeKaDe, Germany
- CD900, by Alcatel SEL + ATR and AEG and SAT, Germany and France

---

GSM needed to compare on the same basis all these different radio subsystems. Quantitative measurements were therefore performed with identical environmental conditions created with propagation simulators, designed according to the specifications agreed by the COST 207 Working Group on propagation. The work of this group have been very important since all the previous mobile channels propagation models could be simplified assuming narrowband transmission, while a more general model and simulators, applying as well for wideband transmission had to be elaborated. In order to crosscheck the behaviour of the radio subsystems with the propagation simulator and in a real environment, qualitative field measurements were also organised in Paris around the CNET.

Before the experimental program took place, GSM had decided that any new system would have to satisfy five minimum requirements and that the comparison between multiple access options would be based on eight additional comparison criteria.

Minimum requirements

1. *Quality:* the average speech quality must be equal to that of first generation compounded FM analogue systems
2. *Peak traffic density*: the system must accommodate a uniform traffic density of 25 Erl/ km^2, with a base station separation equal or greater than 3.5 km
3. *Hand-held stations*: the system shall be able to accommodate hand-held stations
4. *Maximum bandwidth*: the maximum contiguous bandwidth occupied by one implementable part shall be less than or equal to 5 MHz
5. *Cost*: the cost of the system, when established, shall not be greater than that of any well established public analogue system

---

System comparison criteria[2]

1. Speech quality
2. Spectrum efficiency
3. Infrastructure cost
4. Subscriber equipment cost
5. Hand portable viability
6. Flexibility to support new services
7. Spectrum management and coexistence
8. The risk associated with their timely implementation

The results of the Paris trial lead to two fundamental conclusions (for more details refer to Doc GSM 21/87 and GSM 22/87):

1. A digital system could satisfy all the minimum requirements set by GSM, and in fact a digital system would do better than any analogue system for all five criteria;
2. With respect to the comparison criteria, the radio experts of WP2 agreed on the following comparison table (Table 12.1.1) regarding the three main "broad avenues" of system options.

Table 12.1.1 Comparison results for the 3 main system options

Preferred option (= for comparable)	Analogue/ digital	FDMA/ TDMA	Narrowband/ wideband TDMA
1. Speech quality	=	=	=
2. Spectrum efficiency	=	=	Narrowband
3. Infrastructure cost	Digital	TDMA	Narrowband
4. Mobile cost	Digital	TDMA	Narrowband
5. Hand portable viability	Digital	TDMA	Narrowband
6. New services flexibility	Digital	TDMA	=
7. Risk	Analogue	FDMA	Narrowband
8. Spectrum management	=	FDMA	Narrowband

The conclusions of the WP2 work were:

1. A digital system can exceed the minimum requirements compared with an analogue system;
2. TDMA has advantages over FDMA;
3. Narrowband TDMA is preferred to wideband TDMA although both can meet the minimum requirements.

There was a majority of countries supporting the choice for narrowband TDMA. But it was also apparent that wideband TDMA was a viable option and, as recalled by Thomas Haug in

---

[2] GSM required that, for any selected digital scheme, performance with respect to criteria 1–6 would have to be at least equal to that of analogue systems and would be significantly better in at least one criteria. Candidate systems were also compared with respect to criteria 7 and 8.

Chapter 3, a lot of additional technical and political discussions took place before GSM could finalize in May 1987 its choice for narrowband TDMA with frequency hopping.

12.1.4 Tuning the Details

Undoubtedly, the ability of GSM to agree in 1987 on the "narrowband TDMA" broad avenue was a very important achievement. But it was by no means the end of our efforts: rather it was the beginning of very intense activity which lead to the finalisation of the details to be specified in the GSM radio interface. The exact definition of the physical layer of the radio interface by WP 2 was also a prerequisite before the functional specifications and the detail protocol design of the logical layers could be progressed by WP 3.

It was decided that the key radio aspects would be documented in the 05.xx series of "Recommendations" (later called "Technical Specifications") describing the different channel structure, the channel coding scheme, the modulation, the transmitter and receiver characteristics, the measurement and handover principles, the synchronisation requirements, etc.... In comparison to many other recommendations produced by GSM, the 05.xx services may appear pretty thin: in total it was less than 200 pages! But each parameter specified was often the result of very thorough analysis and had to be supported by detailed simulations and experimental measurements.

12.1.4.1 Channels Structure

It was defined that the "physical layer" would support a variety of traffic and signalling channels. Hence a large number of new acronyms were created: TCH/FR, TCH/HR, BCCH, CCCH, SDCCH, SACCH, FACCH, AGCH, PCH, RACH, etc.,... To make things more confusing, it was decided also that control channel BCCH/CCCH multi-frames would be made of 51 frames (of 4.615 ms) while the traffic channel multi-frames would be made of 26 frames (these two numbers being chosen as "prime" to allow a mobile in traffic having an idle time slot every 26 frames to "slide" across the complete BCCH multi-frame every $51 \times 26 = 1326$ frames). The exact structure of each frame, composed of eight time slots of 148 bits each, was also decided together with the allocation of each bit, including those for training sequences and those for actual data, with a specific channel coding and interleaving scheme for each type of traffic and signalling channel.

12.1.4.2 Modulation and Channel Coding

The experimental program performed in Paris, and the additional analysis performed by WP2, had shown that particular care had to be given to the selection of the modulation and channel coding schemes. It was clear also that the performance of the radio link, under mobile propagation conditions characterized by severe multi-path conditions, would also depend strongly on the equalisation algorithms implemented in mobile station and base station receivers. In fact, as reported by Thomas Haug in Chapter 2, Section 1, the final choice of the modulation scheme was difficult to reach and the initial preference of WP2 for ADPM was finally changed to GMSK. The reason for this choice was that the latter modulation method did not include any redundancy; therefore all the redundancy could be used for channel coding, which was much more efficient by taking advantage of interleaving and frequency hopping schemes, particularly for slowly moving terminals. It was also decided by WP2 that the equalisation method should not be specified in the standard, thus leaving freedom of implementation in the base station and mobile receivers. As it turned out, for all manufacturers, we saw significant improvements in terms of receiver performance between early versions and more stabilised versions of their products (which ultimately performed better in terms of sensitivity than the minimum requirements set in the recommendations). GSM was right not to over-specify and to simply define minimum performances rather then decide on exact implementation.

In addition to the choice of channel coding for full-rate speech 13 kbit/s, a substantial effort was dedicated very early to defining several types of data traffic channels, including full-rate at speeds ranging from 2.4 up to 9.6 kbit/s, and also half-rate from 2.4 up to 4.8 kbit/s with different levels of error protection. Retrospectively, we probably defined too many types of low bit rate data channels as the demand for high speed became quickly dominant and there is today very few mobile data applications with speed less than 9.6 kbit/s. A few years later, as part of the GSM phase 2+ program, WP2 efforts would in fact be redirected towards enhanced data coding at speeds of 14.4 kbit/s instead of 9.6 kbit/s, associated with the use of multiple time slots (HSCSD and GPRS). More recently, the "Edge" modulation was also proposed to increase the bit rate over a single carrier above 300 kbit/s! It is quite remarkable that a channel structure initially designed to squeeze many low bit rate data circuits on a single carrier could be adapted later, without too many difficulties, to allow bursty traffic to be transmitted at a much higher bit rate.

Concerning speech, GSM decided in 1993 to introduce half-rate coding as an option to increase capacity. In more recent versions of the standard, "Enhanced Full-Rate" (EFR) and "Adaptive Multi-Rate" (AMR) speech coding options were also defined to provide other trade-off's in terms of quality versus spectrum efficiency. These various options are all compatible with the initial definition of the GSM radio channels, and provide a good demonstration of the flexibility we obtained with our digital foundation.

12.1.4.3 Handover Mechanisms

In first generation analogue cellular systems, the decision to handover from one base station to another is a central process made by the network based on "uplink" signal strength received at base stations. The second generation GSM system, being digital and time division multiplexed, had the flexibility to introduce innovative schemes for handover. In particular GSM Mobile Stations (MS), which are not transmitting or receiving all the time, have the capability to (i) perform measurements of the "downlink" signals received from the serving as well as the neighbouring Base Transceiver Stations (BTS) and (ii) report these measurements regularly to the network. The handover decision algorithm, implemented in the Base Station Controller (BSC), utilizes both "uplink" measurements performed by the BTS and "downlink" measurements reported by the MS. This technique is referred to as "mobile assisted handover" which proved to be a very efficient and future proof feature of GSM.

GSM being digital, the measurements of radio transmission performance could be based not only on signal strength but also on estimates of the Bit Error Rate (BER) and Frame Erasure Rate (FER) for speech. WP2 dedicated big efforts to the definition of the precise radio measurements to be performed by MS and BTS, and to the detailed mechanism for the decision to handover. There was at that time a considerable debate on whether we needed to specify the complete handover algorithm, but the GSM group took the wise decision to define only the radio measurements, and not to specify the handover algorithm itself which was to become a proprietary implementation of each base station system manufacturer. In doing so, GSM left a lot of freedom for competitive innovation and, indeed, we saw a lot of new ideas introduced in the 1990s when the GSM networks had to cope with ever increasing traffic demands. It turned out that the initial radio subsystem specifications, and therefore all the mobiles produced during the early years, could support advanced radio mechanisms introduced later such as "concentric" cells and multiple layer "micro cell\umbrella cell" handover.

12.1.4.4 Spectrum Efficiency Features

While selecting the parameters of the GSM radio subsystem, priority was given to overall network spectrum efficiency as opposed to the efficiency of a single base station. GSM had indeed derived precise modelling of the maximum number of users per cell as a function of various parameters such as the carrier spacing, the voice activity radio, the availability of various diversity schemes, etc.,... It was concluded during the early definition stages that a good system, be it FDMA or TDMA, should always include a good level of inter-cell interference rejection even if it would be with added coding redundancy and therefore less carriers per cell.

Also GSM took the decision to introduce from day 1 Slow Frequency Hopping (SFH) as a mandatory feature for terminals: this feature added some initial complexity but it turned out to be very useful many years later when GSM operators were able to implement high-capacity cellular reuse strategies taking into account the interference diversity effects provided by SFH. Examples of such innovative strategies include the utilisation of "fractional" reuse clusters whereby GSM cellular planning with SFH is based on a minimum reuse distance which is non uniform for all frequencies, or the option to use all the hopping frequencies in every cell, controlling the interference level by the load of the cells.

With features such as voice activity detection, interleaving, channel coding and frequency hopping, GSM had introduced very early in its TDMA design some advanced functionalities which differentiate GSM from other more traditional FDMA or TDMA systems. These advanced features also allowed GSM to compete well with the IS-95 CDMA standard when it was proposed in the early 1990s by American industry.

12.1.4.5 New Frequency Bands

In the early 1990s the "PCN" initiative was promoted by the UK Administration to extend the spectrum utilisation of GSM from the 900 to the 1800 MHz band. Accordingly, some adaptations of the radio subsystem were made to utilise 75 MHz of additional spectrum in the so-called "DCS 1800" band. New types of mobile stations were defined, with reduced power to allow easier implementation of hand-held devices in such "Personal" Communication Networks. As it turned out, the majority of GSM terminals produced today are in fact dual-band as many cellular networks have increased their capacity by combining the utilisation of both 900 and 1800 MHz spectrum.

The flexibility of the GSM standard to adapt to new spectrum was very attractive, and the exercise was reproduced again later to accommodate, in the US, the so-called PCS spectrum at 1900 MHz (accordingly some terminals became tri-band 900-1800-1900). Other extension bands have also been studied, including specific frequencies allocated in the 900 MHz band for railways applications, as well as more recently extensions of GSM also in the 450 MHz and the 800 MHz bands.

12.1.5 Group work towards a single standard

It would be difficult to name only a few people as the main contributors to the definition of the GSM radio interface.

During these early days of GSM, there was a truly open collaboration between many European organisations, originally limited to PTT's, then rapidly extended to their industrial partners as discussed above. The discussions were very open and in a true collaborative spirit.

At that time, in comparison with more recent practices in standardisation forums, we were also less concerned by the need to protect the Intellectual Property of our technical contributions and, as a result, we were able to exchange rapidly and openly a lot of new ideas between many contributors. In that respect I am not sure that, today, the creation of an innovative standard like GSM could be organised again so efficiently.

In the radio interface definition, we could probably isolate contributions from many specific GSM participants. But the more remarkable result is that, even with a large number of inputs, the group was able to converge quite rapidly towards a consistent, well optimised and future proof foundation for its radio interface.

Chapter 12: Radio Aspects

Section 2: The Development from 1995 to 2000

Michael Färber[1]

12.2.1 The Work in SMG2 was Influenced by the Growing Success of the GSM Standard

Reflecting on the time period I was asked to describe, the growing success of GSM and the work of UMTS mostly influenced the standardisation work. In 1995 Alain Maloberti withdraw from the SMG2 chairman's position after having done this job since 1984 in an excellent way. Niels Peter Andersen, who was with Tele Denmark mobile at this time, was elected as the new SMG2 chairman. Niels had experience in mobile development, and had served several years as SMG3 secretary. Furthermore, he was experienced in the way the ETSI works. The long-term SMG2 vice chairman Henrik Ohlson was also with Tele Denmark and the ETSI rules made it necessary to go for a new vice-chairman, due to the rule that the officials should not be connected to the same company. In the Biarritz meeting in September 1995 I was nominated and elected as the vice-chairman of the group, and I kept the position until SMG2 closed its activities in 2000. However, Niels and I continued to serve in similar positions in the 3GPP GERAN structure.

When I was involved in the phase 2 activities of SMG2, a manager suggested to me, that with the finalisation of phase 2 the standardisation work could be stopped. The estimation was that there would be no future enhancements, and it would be sufficient to produce a product. This prediction was not entirely correct. The growing success of GSM inspired people to get new functions and features for the concept. Instead of a dry out of the work, a multitude of ideas where launched. Looking at my notes during this time, I found it hard to describe the work in a comprehensive way along a time line, because of the multitude of activities. Therefore, I selected work items of the period, to structure the chapter. I also considered issues, which failed to find their way into a product, and also the work items, which amounted to big leaps of the concept. In 1995 the basic definitions of the GSM concept were 10 years old, however the system concept showed a remarkable flexibility of integrating new functions and features never considered 10 years ago. The key element of backward compatibility requires sometimes painful work around solutions, but finally enables mobiles from phase 1 to still work in networks which may have a release 99 functional content.

---

[1] The views expressed in this section are those of the author and do not necessarily reflect the views of his affiliation entity.

A not insignificant part of the work was influenced by the UTRA selection process and the work on UTRA throughout 1998 before the work was continued in the 3GPP structure.

But even after the work transition of UMTS to 3GPP, SMG2 did important work related to the third generation. For the conclusion of the 3GPP Release 99, UMTS contained a feature handover from UMTS to GSM. To have a useful functionality, this process has to work in both directions. SMG2 started therefore in 1999 to specify the conditions for a GSM to UMTS handover. This feature was available on time when also the UMTS standard achieved a level of completeness. As an addendum in the 3GPP Release 4 the GSM to narrowband TDD handover was incorporated.

Other ETSI groups steadily worked on improvements for the speech coders. The largest leap in the late 1990s was the Adaptive Multirate Codec (AMR). This codec has the property to align its coding rate or resource usage according to the link quality. The coder itself is a task of another group, however SMG2 was in charge of defining the needed channel codings and the performance thresholds. The AMR is part of Release 98, although a part of the work penetrated into 1999.

Not all activities were entirely covered by the SMG2. Sometimes other standardisation bodies took the lead for special needs. In that sense the US T1P1 took care of the frequency adaptation to 1900 MHz. Later this work was integrated to the ETSI core specification. The same applies for the location services LCS. The initial requirement was set up by the requirements of the US FCC authority. T1P1 took the lead work to develop the concepts for GSM positioning. It was not a simple task, because from its properties, the GSM system is not in any sense prepared for such a feature. Several positioning methods were allowed in the concept, and after drafting the required CRs the discussion in SMG2 started together with the people from T1P1. This review process was fruitful in that it catered for aspects of compatibility and removed ambiguities. LCS based on the A-interface is part of the Release 98 functional content, although parts of the work spread largely in 1999.

For Release 5, enhancements of the LCS are planned, to allow operation via the Gb interface and the Iu interface. This work is still ongoing.

12.2.2 Working Methods

The working methods changed significantly in the second half of the 1990s, and I think it is an aspect worth mentioning. It also reflected the technical progress at this time, and the increasing demand for data service capabilities. This may illustrate the peak of the iceberg of service expectations for the wider consumer market of the future.

When I started my participation in GSM standardisation, just the secretary had a laptop computer. A kind of 286 PC in the shape of a portable sewing machine, and likely of the same weight. All documents where copied in large quantities, to have sufficient copies for all delegates. More than one meeting followed the rule of copy availability, rather than the agenda. Hosts were terrified by copy machine break downs, or late submission of input documents. A part of the meeting room or parts of the corridors were filled with tables carrying piles of documents. The delegates rushed off at every break to collect new available documents. Experts had little notepad sheets with a matrix for the document numbers ticking off the numbers of the documents successfully collected.

In the beginning revisions of documents resulted in a re-submission for the next meeting. Or the document was so handsome, that a handmade version could be generated at the meeting. An example of such a handmade document can be found on the CD-ROM.[2] Bengt Persson, Franco Pattini and I drafted this document, and the final copy was done by me.

Over the years more and more delegates used notebook computers, mainly to draft their individual meeting reports online, to make revisions of input documents, and to draft liaison statements. The capability to generate new and revised documents during the meeting accelerated the working speed of the committee. However, this is one reason that the number of documents in a meeting starts to increase. Further, more participants showed up, therefore also more potential contributors. Another aspect, the split in working parties, especially in the plenary sessions, resulted in all the approved change requests being published a second time. The consequence was that the delegate carried huge amounts of paper as additional travel luggage. Meetings in 1995 had about 100 documents for a week, today it is usual to deal with about 500 documents in 1 week.

The introduction of e-mail exploders allowed a convenient submission of the documents; there was no need to collect a hardcopy. Over a certain period, this was at least an improvement for the delegates, who just collected the papers only available as hardcopy. For the host of the meeting, there was no choice. Each document of the meeting had to be copied, irrespective of the availability by the reflector. It turned out that the copy cost was taking up a significant part of a meeting budget. In a meeting hosted by Siemens in 1997, with a large plenary with 150 delegates dealing with GSM and UMTS, 80 000 pages were printed in 1 week. In 1999 the last paper meeting took place. Now all meetings are electronic meetings, and CD-ROM or LAN does the document distribution. For March 2002 it is scheduled to have all meetings as wireless LAN meetings.

Another characteristic of the work is the travel the delegates do. In a faked ETSI abbreviation document (see CD-ROM[3]) which was available at the SMG2 meeting in Madrid 1994, the term ETSI was defined as European Touring and Sightseeing Institute. People outside the standardisation work see the delegates mainly travelling. Usually people not involved overestimate the "pleasure" travelling. In principle the work could also be done by having the meeting always in the same place. But the ancient working procedures were introduced to allow the involved parties take care of the hosting of the meetings. I can personally identify one advantage of this principle. In my memory I can connect discussions and documents to dedicated meeting places and dates. Therefore I find it quite helpful in retrieving data from my mind.

Personal relations are a key element of the work, and should not be underestimated in its worth to building consensus decisions. Although all delegates are in charge to represent their companies' interests, they are also committed to a higher goal, to build a good standard. This work of intensive discussions, in coffee breaks, side meetings and evening sessions conducted over years helps to build up good relations with the other long-term delegates. These relations may not apply for the quality of friendship, but it is far more than just working together.

12.2.3 System Scenarios

The GSM system scenarios were a result of DCS 1800 activities in 1990. But in a lot of the activities in the second half of the 1990s it paid off to have the system scenarios available. At the GSM2 meeting in May 1990, the creation of specifications for DCS 1800 was decided.

---

[2] TDoc GSM 2 189/91.
[3] Draft prETS 800 069-2 "Christmas Edition".

One evening after the plenary session, a subgroup stayed together to define system scenarios to be used for the definition of the DCS 1800 radio parameters. Some experienced people, having suffered long-winded discussions in phase 1 suggested this. What we mainly did, was to define coupling losses and the procedures on how to calculate the effects between MS/BTS; MS/MS; BTS/BTS. As a rookie in this business, I was on the one hand fascinated by the ability of these folks to define the scenarios; on the other hand I was puzzled how this would help in defining a 05.05 radio specification for DCS 1800. But it was becoming quite clearer for me during the discussions in the first DCS 1800 ad-hoc meeting which took place in Borehamwood (UK). From the input documents I got a picture of how to use the scenarios, and in the following ad-hoc meeting in Bristol, I was able to give my first contribution based on system scenarios. The importance of the parameter definition is simply to enable uncoordinated system scenarios, i.e. two operators can co-exist in the same geographical area, without any co-ordination of spectrum, TX power, etc. Of course, a little guard band is needed between the operator band slots, however, GSM is also quite efficient in this view.

With the conclusion of the work the documents were filed the usual way in PT 12. In the upcoming discussion about micro BTS a new need for scenario calculations was given. Some of the initial documents were based on the original structure, where others started with their own considerations. In the discussions it was found most useful to base all new considerations on the known structure. Further, the idea was born, to collect all relevant background papers in an ETR. This was the start of the 05.50, which was compiled from a number of useful background papers. Over the years, with other frequency extensions, Pico BTS discussions, compact EDGE and mixed mode operations, the 05.50 was stepwise enhanced. The 05.50 is a very valuable paper for people trying to understand how the parameters in the 05.05 were created. However, the paper is somewhat difficult, due to the fact that the basic papers were not updated. This means that changes of parameters that for some reasons happened later, are not reflected in the original calculations. The best solution is to use the 05.50 for the calculation principle, the calculated parameters will differ between 05.05 and 05.50. In merging the DCS 1800 and GSM phase 1 specification in the GSM phase 2, it was realised that some figures did not fit together. A re-calculation of the GSM figures according to the system scenarios took place. Some figures were aligned, others not, because the operators were afraid that this would cause harmful relaxations to the system. By experience it is best to use the DCS 1800 05.05 parameters and the 05.50 to understand how the principles work.

The value of a system scenario document is common sense in the standardisation work today. The GSM scenarios may be questioned in some assumptions, but today it has been proved that they enabled a stable system operation.

When the activities on UMTS radio, i.e. UTRA started it was immediately decided to set up a scenario paper. Of course for wideband systems such as UMTS, the straightforward worst case calculations are not suitable. If you calculated UTRA system scenarios in such a manner, you would simply find that the system would not work. Therefore the system scenarios for UTRA are based on simulations, describing a defined loss of capacity which is acceptable.

12.2.4 Frequency Extensions

GSM railway was the frequency extension we had to cater in 1995. In the US T1P1 had started activities to adopt GSM for the US market and this was going along with a GSM definition for the PCS frequency band in the 1900 MHz range. The initially independent work from T1P1 concluded in a reuse of channel numbers for 1900 MHz, which had already a definition in the 1800 MHz band. This is troublesome for multi-band mobile stations comprising 1800 and 1900 MHz. The ARFCN is not an unique definition in this case.

Later the US version was integrated into the GSM core specification. In 1998 a 450 MHz version was defined, inspired by the idea of having a replacement technology for the NMT 450 system. In 1999 the US 850 MHz band was defined, mainly for EDGE purposes. The availability of additional spectrum in the 700 MHz domain required a further extension. With the last frequency enhancement, it was becoming obvious that the channel numbering structure of 1024 channel numbers was short. Therefore activities were started, to generate a new backward compatible channel numbering system, to overcome the limits of the actual scheme. The principle currently under discussion, is based on the UMTS principle. Instead of an absolute number, an absolute start value is defined, and then the channel numbers as relative information are used. This concept which is not yet concluded, will remove all current borders in the channel numbering system.

12.2.5 Micro/Pico BTS

The growing success of the GSM concept called for solutions to handle the increasing traffic, especially in the cities. There is of course quite a variety of spectrum efficiency improving methods. But methods where changes of the mobile station are necessary have little impact in a market with a high penetration of existing mobile stations. After a short attempt to create micro cell operation with special low power mobile stations, the common sense approach was that the existing mobile station should be suitable for all kinds of cells, macro, micro and pico cells. This of course called for the introduction of the micro cell base station definition. By use of the system scenario concept the micro BTS TX power and the receiver dynamic range was adapted to the operational scenario needs. Three proximity scenarios were used in the calculations, and resulted in three micro BTS classes. In 1997 the idea to adapt the BTS to a special scenario was brought up again as the pico BTS concept. In contrast to the micro the pico is intended for indoor use only. This calls for the consideration of a much tighter MS to BTS proximity, which was subsequently used for the calculations. For the indoor use also the frequency stability requirements for the BTS were to be reconsidered, due to the fact that indoor significant Doppler effects cannot be expected.

In UTRA we find the same strategy now. Specifying a multi-purpose mobile station and then defining different BTS classes.

12.2.6 Mobile Station Power Classes

The GSM standard consisted of four power classes, later enhanced to five classes. Class I was then removed from the standard, due to the consideration that a 20 W class is not state-of-the-art anymore. The discussion about the impact of radio waves on the human body made it necessary to remove this class, which was quite simple, because no MS of this class was ever built. The class V (0.8 W), which was added later, is also a relic of the idea, that for special scenarios low power MS may be needed. Class II and III (8 W and 5 W) mobiles where around, especially as vehicle mounted devices. Also car modules or adapters used this power classes. However, class II and III stayed a minority in the market. The mainstream GSM mobile is the 2 W class IV mobile station. The first suitable hand-held class ruled the market, and dictated also, that the radio network planning enables contiguous coverage for the class IV devices.

For DCS 1800 three classes are specified, but as for GSM, only one class succeeded as the major class. For DCS 1800 and for PCS 1900, this is the 1 W mobile station. The lower power class, and also the high power class are just relics of the standard. The lesson learnt is that the definition of many classes does not necessarily mean that the market will carry all the classes. The optimal hand-held suitability turned out to be an important issue. Today, of course, multimode considerations strengthen the strategy to align power classes for new bands to existing power classes.

The strategy to shape mobile stations for special applications, e.g. micro cells failed, due to the need to have multipurpose mobile stations. It was found more suitable to define the infrastructure equipment for special needs, rather than having special terminals.

12.2.7 GSM Railway

GSM railway could be easily grouped to frequency extensions, if not, a bundle of services would come in conjunction with railway. In the beginning it was simply called UIC. UIC stands for Union International des Chemises. This is a railway organisation, which first looked into the subject of a new generation of railway communication systems. GSM was one of the candidate technologies and succeeded finally in the selection process. For GSM railway, a new 2 × 4 MHz bandslot was defined, attached to the lower band of E-GSM. This guardband free allocation of course called for a careful look into the system scenarios. The railway application also called for new services such as group call, etc. Especially customised for the communication between railway engines and service support teams along the tracks. After treating the frequency extension and the new services as one work item, it was realised that the services can also be used in classic GSM. Therefore the work item was split into the frequency adaptation and the Advanced Speech Call Items (ASCI) features. These feature sets were on schedule for Release 96 and were sufficiently well concluded. As mentioned before the system scenarios had to be considered carefully to avoid harmful effects between GSM and the railway bands. As an off meeting activity Jesper Evald (this time with Danish Rail) and I re-calculated the GSM-R system scenarios to enable the preparation of the needed change requests, and to have the input for the 05.50. Once again it was found very valuable to have the 05.50 as a material collection available.

12.2.8 Evolution of Circuit Switched Data Services

Circuit switched data services were a part of GSM phase 1. A large variety of services was defined, not all of them showed success in their later live. For example, a service class was defined with data rates the same as the French Minitel. FAX and the 9.6 kbps service succeeded. Especially for mobile data applications the 9.6 kbps service was the choice, providing the highest possible data rate. Knowing what computer applications need today, it is easy to understand, that a 9.6 kbps data rate was not satisfactory for long periods. As a first idea for improvement, the increase of data rate per timeslot was considered. The idea is simply to trade off between the payload bits and the redundancy bits. This resulted in the 14.4 kbps data service. The 14.4 kbps data rate service is also an important building block for HSCSD. Introduced in some markets, and also available in mobile stations, this feature did not receive too much public marketing to the end users.

Concatenation of time slots is the basis for the HSCSD concept. We found a similar idea in GPRS. The difference is mainly that the HSCSD still uses a circuit connection, i.e. after the call establishment the up- and downlink connection is exclusively allocated to a single mobile. The concept does not consider how far the data traffic is actually downstream or upstream oriented. In applications like a file download, the upstream data needs may be far less than the allocated resource enables. On the other hand the exclusive allocation of the link gives no dedicated delay constraints as in packet oriented systems. This is a clear advantage of HSCSD in support of real time services. Another advantage compared to GPRS was that the feature required less fundamental work to get part of the standard. This enabled an availability in advance of GPRS.

The concatenation of timeslots is not without an impact on the terminal implementations. There are some magic thresholds, when for example more than one synthesiser is needed, or a duplex filter has to enable simultaneous transmission and reception. Additionally leaving the concept of single slot, opens the variety of terminal classes according to their multi-slot capabilities. HSCSD and GPRS ended in 22 multi-slot classes, in a retrospective view this was likely going to far. It can be expected, that as for the power classes, only some kind of mobile classes will rule the market.

ECSD was a logical consequence of the EDGE activities, to use the new modulation scheme also for circuit switched data. The idea was launched in the 1998 activities in EDGE. From the network side an upper limit of 64 kbps per CSD connection was given. With ECSD a two-slot configuration can go for this, which of course is the lowest complexity in multi slot configuration you can get. ECSD was consequently developed in parallel with the EGPRS services, and was approved as part of EDGE feature content in Release 99.

12.2.9 General Packet Radio Service (GPRS)

In the time period of SMG2 which I describe in this book, the work on GPRS had already started. In 1995 there were mainly debates about the functional principles, and how such a concept should be integrated in existing GSM networks. The detailed work happened at ad-hoc meetings from SMG2 and SMG3. Individual companies were working on concepts on how to allocate the new functions in the existing network elements. Furthermore, a protocol stack discussion took place, because a protocol stack as now planned did not exist in classic GSM. In a busy phase end of 1995 interested parties worked on concepts, to have them available for the ad-hoc meeting in Tampere in January 1996. In Tampere four proposals were put on the table by Ericsson, Nokia, Nortel and Siemens. The proposals differed mainly on the function split and the degree of changes in the network elements. After the presentation of the different concepts, the discussion about the pros and cons started, without a conclusion, but with a number of questions for clarification for each of the concepts. Further simulations had to be provided for the next ad-hoc meeting, to give an order of magnitude about data throughput and data latency.

The following ad-hoc meeting took place in Burnham just 1 month later, and the discussion of the concepts continued in this meeting without a conclusion in sight. The concept promoters did not see a reason to give up, neither was there a potential to create a consensus solution out of all the proposals. Also the simulation results did not help, due to the fact that the concepts with highest impact on existing hardware showed the best result. The discussion sparked off a debate on how such drawbacks will balance with the performance benefits.

To avoid a deadlock of the activities, the chairman of the ad-hoc group, Han van Bussel from T-Mobil, managed to get all operator representatives to an evening session at the hotel bar. During this session the operators compiled a document outlining their view of the solution, what is allowed to be modified and what performance should be achieved. The next day the discussion continued, but now with an operator's requirement paper at hand. This allowed a reduction of the number of proposals to be considered. Finally it was feasible to draft a working assumption at the end of the meeting. The working assumption was that the GPRS activities should be further developed on the basis of the concepts from Ericsson and Nortel. This was a big step forward, but still 2 years of intense work was ahead to get the GPRS release together.

GPRS was likely underestimated when the idea was kicked off. It was simply treated as another new function. However, this function enabled the connection of the BSS to another type of core network. The connection to the packet world required the establishment of a full OSI protocol stack, which was in the circuit switched domain not present from the past. The circuit switched part was optimised with the voice services in mind. For such a function the RR function was more or less directly in touch with the physical layer. This is feasible and efficient for the voice services, but does not fit to well for data services.

Looking to the UMTS protocol stacks of today, they are more generic, for the real and non-real time services. On the other hand, a simple voice call carries a remarkable overhead.

But the large architectural leap which was made with the packet services in GSM, is of course one reason why it took a while to get the standard frozen. At the Geneva SMG2 plenary meeting the chairman asked the plenary delegates insistently if they were confident that all the proposed changes had a level of maturity, that the CRs can be given to SMG for final approval. Nobody objected, although nobody was really confident. In the aftermath it showed that the level of maturity was not sufficient. In 1998 and 1999 SMG2 was fighting with a flood of change requests, and it needed a while to sort out the essentials and the non-essentials. It needed almost 2 years to stabilise, and still people find the need to write CR on Release 97. However, in the meantime the threshold for such CR is very high. The change must be of essential character to avoid a malfunction. Nice to have, or removing an ambiguity does not qualify for an approval.

The lack of stability in the first GPRS release was often a complaint afterwards, and it can be questioned why it happened. I think it was the combination of various effects, the ones I think that had an impact are listed here.

The work needed for GPRS was underestimated. The work started in a time when no system architecture group was present, to define the fundamental parts. The upcoming work on UTRA involved a loss of experts in the working groups. Too many features and options were incorporated in the release. Only a few months before freezing, new options found the way into the concept, screwing up parts which were already drafted. None of the involved parties were finally courageous enough to postulate their concerns, when the chairman asked if the changes were really complete.

But I think the community learnt a lot from GPRS, and it will help to avoid such problems in the future.

GPRS bears a lot of capabilities to provide new types services to the end user, and I think it is the incubator for the service needs which shall be covered by UMTS in the future. This close relationship to UMTS is becoming visible in the intense activities of DTM as part of Release 99. Release 99 is the first UMTS release, and can provide simultaneous real and non-real time services. Operators having GSM and UMTS developed an interest in having the same combination also feasible for Release 99 GSM. GPRS provided the class A mobile station, which was defined as having the functionality to provide two services in parallel. However, the definition was so generic, that it practically required the design of a mobile station, which consists of two independent transceivers and signal processing units. It was obvious that this class will only rarely available. To get a mobile, which was viable to support these services in a more economic manner, the dual transfer mode was defined. A reduction in the multislot capabilities, and the requirement of the use of adjacent timeslots allowed the definition of a more practical class A type implementation. Although the work was first started in late 1999, a high joint interest and by means of frequent ad-hoc meetings SMG2 was able to finish this work for Release 99 with only a slight delay. But the evolution for the GPRS concept is not standing. Just recently features were introduced in Release 4, e.g. to support the IP transport on the Gb interface, or the network assisted cell change procedure to reduce transmission stalling during cell change.

12.2.10 EDGE

The start of the EDGE activities was at a time, when the new feature was hardly noticed. The idea and the draft work item description were presented at SMG#20 in Sophia Antipolis. This was the plenary, which happened in conjunction with the first UMTS radio access workshop, and the major attention of the delegates was on the UMTS subject. Nevertheless the work item received sufficient support, and the work on the issue kicked off in the form of a feasibility study which had been conducted throughout 1997.

The work item received in the beginning quite a large portion of sceptics. The change to a higher order modulation scheme obviously required higher C/I values in operation. The change of the modulation system was initially coupled to the provision of non-real time services EGPRS. Later enhanced CSD (ECSD) was introduced. Today we discuss the combination of EDGE modulation with the AMR codec, seeing benefits in an advantageous combination of the source code and the ability to add redundancy bit based on a clever coding structure. The expectation is that this combination allows a more robust overall system, although the modulation threshold of 8PSK in principle requires a higher interference threshold.

Within the feasibility study the understanding about the basic idea of the concept improved over time. The usual concepts applied for a cell planning, which allow a minimum acceptable quality at the cell fringe, i.e. the link budget has to provide under the consideration of a number of statistical effects, a defined level of availability and quality at the cell border. To illustrate this by numbers, we can target as an example a 98% probability of service at the cell fringe. This means that along the cell border for example the C/N drops only in 2% of the locations and time beneath a defined threshold. This borderline requirement applies for the area of the cell on a higher degree, e.g. 99,9%. If you take a look at the higher C/N or C/I thresholds within the cell, they will show that for achieving the borderline requirements, within the area, large areas provide higher quality. This circumstance allows for a non-real time data transmission to speed up by means of trading off the redundancy bit versus the payload bits. Further the modulation can be adopted to the properties of the radio link.

During the feasibility study different modulation schemes were discussed, but no conclusions were drawn. The purpose of the study was mainly to show that it is feasible to integrate in a existing GSM system a new modulation scheme, without mixing up existing network planning, and avoid any extra interference.

After approval of the feasibility study in December 1997, the work had to be started to get specific proposals on how to modify the standard. For the detailed work on these matters, ad-hoc meetings were proposed. Frank Müller from Ericsson had conducted the feasibility study. Being himself involved in proposals, the chairman suggested to Frank, that he should ask me, if I would like to chair this working sessions. Frank asked me, and I agreed. Frank took the job as the secretary, and we started off in Versailles in spring 1998, with a working session on the modulation question. The working session was scheduled for 2 days, and the only agenda item was to find consensus on the modulation matter. The working session was really a small 15 person group, the number of documents to be discussed was in the same range. However, the discussion achieved quite a great level of detail, and the different parties showed high engagement for their proposals. The debate was about constant envelope versus non-constant envelope modulation. In brief, the constant envelope concept allowed the power amplifier to use peak powers comparable with GMSK, where the 8 PSK concept needed PA back off to keep the modulation mask requirements. But on the receiver side, the 8PSK needed less interference threshold as the constant envelope concept, compensating the gains of the higher transmit power capabilities. 8PSK has in principle the quality to operate also in QPSK, which can look like a GMSK when the parameters are carefully chosen. Superimposed was the discussion with an increasing interest of US operators which were in favour of combining EDGE with IS 136 systems. In these systems 8PSK was already introduced. In terms of multimode terminals or equipment reuse, 8PSK received here an advantage. The problem was that for some companies the implementation of 8PSK transmitters was critical on the actual terminal chip set platform. Supporting the 8 PSK on the receiver side was seen as less critical by all parties. After some discussion, we started to draw a matrix of possible combinations on a flipchart to structure the discussion. The requirement to support non-real time services led to a discussion about the asymmetry of such services, allowing higher downstream capabilities as upstream capabilities. This turned out to be the key for a consensus solution. 8PSK was defined mandatory in downlink, i.e. the mobile receiver has to support it, where in uplink the transmitter could only use GMSK as a mandatory feature. The support of 8PSK in uplink was defined as optional. This was a compromise for all vendors to have a fair chance of designing an EDGE mobile. This was taken as the proposed working assumption, because not all vendors could just agree from scratch. But in the following SMG2 plenary the working assumption was approved. We started off with other working sessions, with an increasing number of participants and input papers. The next step was mainly related to the channel coding and the link adaptation. As a competitive proposal to the link adaptation an incremental redundancy concept was proposed. For a while we struggled to come to a conclusion, but in autumn 1998 it was clear that both concepts had their advantages. However, incremental redundancy required the re-transmission of radio blocks, which was not acceptable for all PCU locations within the BSS. To maintain compatibility with the existing GPRS specification and its freedom in terms of functional split within the BSS, the conventional link adaptation was getting the mandatory approach. The incremental redundancy concept can be optionally used in conjunction with the mandatory algorithm. ECSD was also coming up as a new idea in the discussions, and quite a number of contributions dealt with this matter. To avoid segregation of GPRS and EGPRS on different time slots, a concept was incorporated which allows a blind detection for the receiver to identify on the physical layer domain, if a burst is GMSK or 8PSK modulated. Also the segmentation of the data via the protocol stack, and the needed changes in the GPRS protocol stack were analysed. At the end of 1998, the first draft CR were available, indicating the needed changes in the core specification. Some changes in my responsibilities made it impossible to continue as ad-hoc chairman for EDGE. Frank Müller changed from secretary to chairman, and continued in this role until the freeze of EDGE with Release 99.

In 1999 a lot of drafting took place, to get all the needed changes for the different specifications settled. As a new requirement the need for compact EDGE was introduced, to enable EDGE a suitable system for US operators having less than 2×5 MHz spectrum available. This was tight together with the need for frequency adaptation to the AMPS 850 band, and the considerations of a mixed mode AMPS and EDGE operation in the same band. Although these requirements were first fully established in early 1999, the SMG2 managed to integrate all these additional features into the EDGE 99 release.

I found the naming conventions interesting when the compact concept was being introduced. To distinguish between compact and the other EDGE it received the prefix "Classic". This was a noble step for a concept not yet frozen in the field. On the other hand it illustrates how high the pace was for new ideas.

After the large modifications for GPRS in the protocol domain concluded as part of Release 97, a large modification to the physical layer was made for Release 99. The suggested EDGE phase 2 concluded in the GERAN activities.

12.2.11 Cordless Telephony System (CTS)

CTSwas based on GSM. The CTS idea was launched in early 1997, and the work continued until 1998. The charm of the idea is that a single terminal operates as a mobile, and if it enters the area of its dedicated home base station it acts like a cordless phone. However, there is no specific spectrum for a GSM cordless application. Therefore the system was intended to be underlaid to existing spectrum allocations. This turned out to be the biggest weakness of the concept. The operators using the spectrum where frantic at finding an uncoordinated system underlaid, which added additional interference to their system, i.e. loss of capacity. This is especially difficult for the mobile operators when the cordless connection is handled by another operator which earns the money.

The work item progressed in the beginning, by the support of mainly two manufacturers, and few operators showed interest in the idea. Later, one manufacturer suspended the contributions to the work item, and also the operators' interest vanished. The work item was finished, but in the late phase the operators pushed the RF parameters in such a way that the impact on their networks would be minimal. The consequence of this influence was lowering of the applicability of the CTS concept. CTS is a completely implemented part of the standard. However, no products followed, and it must be considered that this work ended in a dead end street.

12.2.12 Repeater, Feeder Loss Compensator and Other Auxiliaries

These entities were not originally considered as being part of the mobile network, and therefore not considered to be specified. However, when such equipment was being presented in the systems, problems occurred showing the need for a radio parameter definition. For repeaters, the discussion on how to specify the repeaters benefited from the condition that the repeaters be controlled by the network operators. After first interference problems, we settled the problem so that a framework for repeaters was specified in the normative annex in 05.05.

For the so-called antenna feeder loss compensator the whole process was more difficult. Such devices are designed to compensate in vehicle mounted installations the sometimes remarkable antenna feeder losses. The pain is that this cable length is almost unknown, therefore the front parameters of the device are hard to define. We could of course make the consideration that the cable is integrated to the compensator and can be seen as part of the compensator. Even if the equipment was made in such way, nobody could ensure that, at installation in the car, the cable would be shortened, or simply cut away and replaced by another cable. We also created an annex for this type of equipment in 05.05, but knowing that the equipment is under control of the users, and there is no type approval, the effect of the specification is limited.

A meaner experience was the twinkling antenna. This gadget is a replacement antenna, which gets screwed onto the terminal instead of the original antenna. In the aerial LEDs are connected, i.e. the TX power is used to lighten the LEDs. Younger users, not knowing that these antennas have terrific performance find this quite attractive. It is not the fact that the efficiency of these antennae is so lousy. The real thrill is that the non-linearity of the diodes produces harmonics hitting, e.g. in flight navigation radio bands. The amplitude of such harmonics are sometimes remarkably high. The standardisation body is practically helpless to sort out such a problem. There is no protection in terms of the standard if users modify type-approved equipment afterwards. The only measure which was left to us was to write a letter to all kinds of bodies in standardisation and regulation, to warn them about the effects of such antennae (see CD-ROM).[4]

12.2.13 UTRAN Work in SMG 2

The UMTS work in ETSI was monitored by SMG2 by receiving liaison statements and documents for information. For the actual work in SMG2 it had for this phase practically no impact. It changed in 1996, when as described in other chapters of the book, a re-location of the work towards the working groups started. For the work and the delegates in SMG2 this was not too visible in the beginning. Some discussions about transfer of documents took place in the Bristol meeting in March 1996, but this was more for a small group of interested parties. In May 1996 we had parallel meetings of SMG2 and SMG5. On a joint social event, an unofficial TDoc. called the UTRA selection procedure was distributed, which contained a sheet of paper with the two-dimensional representation of a cube. Each plane of the cube was labelled with a radio access technology. When cut out, folded and clued a dice was created, to gamble on the appropriate UMTS radio access technology (see CD-ROM).[5] This was the

---

[4] TDoc ETSI SMG 2 2000/99.
[5] Draft DTR/SMG-50402.

sense of humour at least SMG2 was able to develop at this moment in time about the UMTS matters. Most of the delegates would have wondered that just 1 year later the issue of radio access technology for UMTS was one of the major activities of SMG2.

The kick-off for the UMTS work in SMG2 happened in a workshop scheduled 2 days before the regular SMG2 plenary meeting in December 1996 in Sophia Antipolis. The form of a workshop was chosen, to enable all interested organisations to present their concept proposals, and was also open for non-ETSI members. The meeting gave us the flavour of the upcoming meeting style. SMG5 had closed in the meantime, and the interested SMG5 delegates joined the SMG2 group. This enlarged the number of participants easily by about 100%. During the Sophia Antipolis workshop not only concepts were discussed. Also the discussion started about the time schedule and the working procedure for the UTRA selection process. This caused large debates, which were not really conclusive during this event. But it was quite clear, that the regular SMG2 meeting was not suitable to carry out the detailed work needed, and therefore a SMG2 ad-hoc meeting was created.

The first SMG2 ad-hoc meeting on UMTS took place in Le Mans in January 1997, mainly dealing with the timetable and working procedure matters. The need for evaluation systems was discussed, and finally it was proposed to base the work on simulations, and not on any physical available testbed. This is of course a major difference to the GSM process, with different evaluation systems tested in Paris in 1985. The progress in link and system simulation capabilities was estimated to be as good as the availability of test systems. This is true in principle, but is built on the need of a very accurate description of the simulation guidelines. At the end of the process it was found that the simulation results were just an indicator how far a system fulfilled the evaluation requirements, but they did not allow for a direct system A versus system B comparison just by using the figures. The point was simply that the evaluation criteria did not describe to the latest detail how the simulations should be made. Especially on the system simulation side, individual implementations obstructed the straightforward comparison of the results.

The author does not quote for evaluation hardware, but would like to point out why it was so difficult to shake out concepts by simulations. But coming back to the meeting in Le Mans, the working procedure proposed by the chairman did not consider a beauty contest type of selection.

The concept was that after the initial phase, when the different individual concepts were presented sufficiently, concept groups would form by individual organisations. A concept group would bundle the concepts, which contain fundamental similarities, for example a TDMA access scheme or CDMA access scheme, etc. The concept groups were intended to create a single concept out of the concept ideas of the individual contributors. Performance results would be delivered based on the concept groups' proposal. The results of the concept groups would then enable discussion of the concepts to identify the more or less performing properties of the concept. Each radio access method can be sub-divided into building blocks, and the idea was to create a single concept using the best performing building blocks from the various concept groups. This method bears in general the advantage of avoiding the shake out of whole concepts and the related supporting companies. Further a concept considering building blocks from various concepts also provides a certain averaging of intellectual property among the whole group of contributing individual members. The ETSI IPR policy is well known, and SMG2 as a technical body does not deal with IPR matters, but it is an important aspect hidden behind a lot of technical discussions. This was in coarse words the working procedure we wanted to follow. The ad-hoc meeting in Le Mans finally concluded on proposals for the time schedule and the working procedures. Another major proportion of the discussions was dedicated to the evaluation criteria document. This document was originally based on the ITU REVAL document. But for ETSI, it was refined and published as an ETSI document named 30.03. The other important work was the definition of a high level requirements document, which should be a brief paper enabling the quick comparison between concepts, and should help the groups to structure the deliverables to cover all matters of interest.

These draft results were presented to the SMG2 plenary in Copenhagen in March 1997. During this meeting we saw a number of discussions about the contributions from Le Mans, but no major concern to follow this schedule. In the coffee breaks and in the evenings, extensive discussions between individual organisations took place, to find appropriate partners for possible concept groups. But no concept group was defined during that meeting.

This was up to the next UMTS ad-hoc meeting in Lulea in April 1997. During this meeting five different concept groups were defined. To avoid lengthy and useless debates about the group names, the chairman allocated simply Greek letters to the groups, by following the order the in which groups officially showed up. The group structure in April was as follows:

- Alpha: wideband Code Division Multiple Access (CDMA)
- Beta: Orthogonal Frequency Division Multiple Access (OFDMA)
- Gamma: wideband TDMA
- Delta: wideband TDMA/CDMA
- Epsilon: Opportunity Driven Multiple Access (ODMA)

In the SMG2 meeting #22 in Bad Aibling, the concept groups were accepted at the SMG2 level, and a workshop was scheduled for the last week of June. The purpose was to create the concept groups finally, to elect officials for the concept groups, and to get the individual concept group meeting schedules settled. In Bad Aibling the discussion of the evaluation criteria document and high level requirements continued. Still in the paper meeting mode, bulky documents such as the 30.03 had to be printed in several revised versions for 150 delegates. The number of delegates at the SMG2 plenary meeting had now more than doubled, the meeting itself was split into a GSM part and UMTS part, where during the GSM part for example parallel drafting sessions for the evaluation criteria were held.

In front of the official inauguration meeting scheduled for the end of June in London, the members of the concept groups had their unofficial preparation meetings. The meeting in London was formed as a workshop, dedicating 1 day per concept group. Therefore it was quite practical to have not more than five groups, because this helped us to deal with it in 1 week. Each concept group started with the election of the officials. For the start of each working group I conducted the elections in my role as SMG2 vice-chairman. After that, each group had to organise the 1-day meeting by themselves.

Since the start of the UMTS activities in SMG2, there was an obvious increasing involvement of Japanese companies in the work. In Japan the national standardisation body ARIB had conducted similar activities earlier, and the group was also heading towards defining a third generation standard. UMTS contains the word universal, and to keep this claim, it was found useful to avoid diverting work in different regions of the world. SMG had therefore started to build up connections with ARIB, and to discuss how a way of co-operation could be found.

For the technical part of the discussion a joint ARIB/SMG2 workshop was allocated for 30.6.97 until 1 July 1997. The first planned venue was Sophia Antipolis, but the chairman pulled the meeting to his hometown Copenhagen for good reasons. His wife was expecting a baby, and the calculated delivery coincided with the workshop. This was really more than a good reason to have the meeting in Copenhagen. To be prepared for all cases, Niels asked me to be available for the meeting, to have sufficient time for a joint preparation, enabling me to step in if necessary. This was of course no question for me to follow this request. I had a full understanding of the worth and need of the back-up measures. Anyhow I estimated the likelihood that the baby would come on the calculated date as low, by my own personal experience. The first day worked as planned. Niels and I had a preparation session in the morning; the workshop started in the afternoon and went until the evening. After the meeting I went to Copenhagen to my Hotel, what was by the way quite difficult due to a heavy thunderstorm hitting the city.

The next morning I found an SMS on my phone, telling me that Niels and his wife went to hospital that night. Not being sure if Niels was trying to scare me, I went on with my usual morning business. But than I received a call from Niels confirming that they were in hospital and that the whole thing would need more time, and that I would have to run the meeting. This was not the first time I had to chair a meeting, but it was the first one of that size and multi-cultural structure. Further the mandate of SMG2 was really limited to the technical discussion, and all discussions about co-operations of the bodies, etc. should be kept out of this group, because that was the duty of ETSI SMG.

But the meeting went reasonably well, thanks to our joint preparations. From an organisation point of view I am still grateful to the Tele Denmark staff, which supported me as an alien in their premises and took care that everything from the logistics end worked. Finally Niels found the opportunity to contact me over the day, to listen and to discuss with me the progress of the meeting. During the meeting we received a message that the baby had arrived, and the whole auditorium gave a big salute. Talking about the next generation, this generation at least was coming up to date.

Besides this personal story, the workshop itself was a good opportunity to develop a better view of the activities in the different standardisation organisations and the status of their individual work. ARIB had already passed a selection phase, and was at that moment in time working only on one concept. ARIB used the ITU REVAL documentation unchanged. In contrast to ETSI they had not identified any need to add on top of REVAL specific requirements. We did so, for example for the analysis of re-farming in other spectrum, or for dual mode terminal viability, i.e. the inter-working with second generation GSM system. It was becoming clear in the discussion, that the way ETSI tried to get to a UMTS radio access scheme also differed from the procedures used in ARIB. Although the meeting had no decisive character, it helped to have a better understanding of the increasing Japanese community in the SMG 2 meetings, and their perception of the work.

The first UMTS ad-hoc meeting with full-established concept groups took place in Rennes in August 1997. In this meeting the first simulation results were presented and the competition between the concept groups increased. It is important to mention the fact that in this meeting a US delegation presented the WCDMA2000 concept, which was based on IS-95, and treated in the TIA standardisation body. After a lively discussion, there was quite an expectation that there would be an attempt to join the alpha concept group. However, this did not take place, at least in public. Until the ETSI decision in January 1998, this CDMA2000 involvement remained an unique event. Later in 1998 we entered into a harmonisation discussion between these concepts, which resulted in long-winded side discussions in the SMG plenary in Dresden in November 1998.

In the SMG2 plenary in Bad Salzdetfurth in October 1997 the concept groups presented their latest achievements. Where on the one hand the concept groups succeeded in melting together their concepts, the relationship between the concept groups developed a more confrontation oriented character.

In the meantime the involved organisations spent thousands of CPU hours on the simulation efforts, to catch up with needed deliveries according to the evaluation criteria. The operators also expressed during that meeting an increasing need to get answers to aspects, they found not perfectly covered in the evaluation criteria, nor in the high level requirements. This set of questions was published during this meeting. A question and answer workshop was scheduled for late October 1997, which would then help to clarify these questions. Further a first version of the concept papers should be available for discussion.

The Copenhagen question and answer workshop showed again an increase in confrontation between the groups. It was also becoming visible that a certain focussing on the alpha and delta concept took place. The delta concept group had received a number of new supporters, which used this workshop to make themselves visible.

It was becoming obvious that the idea to select the best performing building blocks out of the different concept groups, to merge them to a joint system was likely to fail. A lot of political and commercial interests were going to superimpose the technical discussion. The interest of the individual organisations in making their concept ideas flourish was stronger than the will to follow the merging concept. The usual work style in SMG2 was to build consensus on contradicting matters. But the creation of a new air-interface and the vital but heavily differing interests of the involved organisations was not a good basis for consensus building.

There were high expectations of the concept group team members by their organisations, and the people tried to fulfil these expectations. This caused an increasing brusque tone between the groups. The chairman steered against this effect in balancing between the groups, and if a speaker started to hit below the belt, he turned around the arguments in some ironic manner, indicating indirectly that this strategy would not work. I was myself closely involved in the promotion of the delta concept, and I can still recall the increasing stress and tension during this phase of the work. Between the question and answer workshop and the last UMTS ad-hoc meeting in Helsinki, I participated in a WBP meeting in Edinburgh. This session "just" dealing with GSM matters was like a realm to me, down to earth on technical discussion without the tensions of the UMTS sessions.

At the UMTS ad-hoc meeting in Helsinki the last big exchange of arguments took place at the ad-hoc meeting level, which gave room for rather detailed discussions. The concept groups had built up their concept papers, made their simulations, and took their positions for the showdown. It was becoming obvious that any consensus solution out of the different concepts was impossible. Each group prepared now for a voting on the SMG level.

SMG2 #24 in Cork was the largest SMG2 meeting ever held. This plenary meeting dealt with all GSM relevant parts, which should be part of Release 97. This included major steps such as GPRS. However, the UTRA debates superimposed heavily on all the discussions. The groups presented their results and concepts in the large auditorium. Operators were invited to evening events, to provide them with more detailed inside views on the concepts. Officially the report to SMG had to be prepared, and the way SMG2 would present the concept groups' results to SMG had to be settled. The chairman proposed some kind of comprehensive management summary structure. After the concept groups had drafted the first versions, the group members were sitting together in corridors and lobbies, reviewing and criticising the other groups' drafts. Several revisions were created and reviewed again, finally fighting about phrases and even single words for hours. Finally each group had a version, which was acceptable for the other concept groups, and was presented for approval in the plenary. In this meeting it was also decided to put a kind of disclaimer on the coversheet of the concept groups detailed reports. This disclaimer explained that the simulation results in the reports were not suitable for direct comparison between the concepts, due to the differences in simulation environments.

When SMG2 #24 was finished, the UTRA selection phase was also finished, the next steps were all the responsibility of the SMG plenary. Refer to the other chapters of this book to read how the decision developed in that phase.

Looking back to this work which was executed in only 1 year, I have to admit that it was something very exceptional that I took part in. On the other hand I do not dare to have this kind of work style as the normal mode. The standardisation meetings, promotion meetings, and concept group work forced the involved people to extreme frequent travelling. In the latest phase of the activities, I think all people were happy to see the end of the process, because in an open ended process no one would have been able to continue with that kind of work style.

As mentioned before, Niels Andersen as the chairman did a great job in managing this process, which was influenced strongly by the interests of the different groups. He stayed neutral and fair. He also took care that in the debates no personal attacks occurred. This was an important fundamental for the continuation of the work after the "big" decision because this work could then be based on consensus building again.

The next SMG2 plenary took place in Geneva after the Paris SMG24bis meeting, which created the consensus solution between the alpha and the delta concept, and avoided a continued voting to reach a decision. Under this perspective, finally a consensus was found. However, it was based on allocating the two remaining concepts on the bands, which differ in their Duplex properties.

From the UMTS point of view this meeting was not too technical. It helped mainly to digest the Paris decision, to develop an understanding on how to interpret the decision from the SMG2 work tasks. When I recall the atmosphere, it was like a hangover after a big party. After the climax of the concept group activities, the delegates had to find their way back to routine work.

The structure of the UMTS ad-hoc group was retained, but the concept groups had closed their activities in Cork. Instead of the individual concept groups, three working parties were established. WG 1 dealing with physical layer matters, WG 2 for RLC/MAC and RRC matters, and WG 3 for architecture and interface matters. The SMG2 work comprised now the GSM working parties, the UTRA working parties, the UMTS ad-hoc group as a communication level atop the UMTS working parties, EDGE ad-hoc meetings and finally the SMG2 plenary which had to keep all these things together.

From the UTRA perspective in Geneva the general design rules were received. First it was clarified that we have one UTRA consisting of two modes of operation. The modes shall try to achieve a maximum on harmonisation to enable best dual mode terminal viability. Further the inter-working with GSM was still a major goal.

Another large open workshop took place in Sophia Antipolis, similar to the one we had in December 1996. This workshop, open to all interested parties, was used to present the technical status ETSI was starting off. It was also a platform for the presentation of the activities from other standardisation organisation activities.

The work in 1998 was marked by the ITU delivery deadline in 1999. This created again quite a tight timetable for the UMTS working groups. In WG 2 and 3 a lot of baseline work had to be carried out. In the UTRA evaluation phase in 1997 the focus was clearly oriented on the radio access technology. However, even in WG 1, the amount of work was huge. Although the FDD concept was largely derived from the alpha concept, in quite a number of cases solutions had to be designed. To build the first set of stage II specifications, concept papers were created to allow a better tracking of the function evolution. The TDD start was a little more difficult. The delta concept in its original form was intended for FDD mainly. So the concept had to be re-worked to be optimised for the TDD operation. Secondly the harmonisation requirement had to be fulfilled. This altogether required a lot of concept work again, and a special ad-hoc meeting on TDD took place before the WG 1 meetings. To accomplish the work, the working group meeting was held about every 4 weeks.

In 1998 we got back into discussion with the promoters of the CDMA2000 concept on the result of US TIA standardisation activities. Quite controversial discussions took place in Marseilles during the SMG plenary meeting. As a result of the meeting, the US delegation claimed they were badly treated in the discussions. But it is likely the different work style in the individual bodies had amplified this impression. To avoid further irritations in the future, for the following SMG2 plenary in Dresden the agenda was styled in such a way to keep sufficient room for discussions of the subject.

In Dresden we had now a big discussion about chiprate, synchronisation of node B and other issues in the plenary. The discussion was continued in a side session, to work on an output paper describing the findings and the potential for further work. As described later, the UTRA standardisation process moved to 3GPP, and the discussion followed to 3GPP. To prevent a further diversion the operators founded an Operators Harmonisation Group (OHG), to look from their perspective at how far these standards could be brought closer, to at least allow a viable implementation of both concepts for the handsets. The outcome of the OHG activities was a good match of the findings we made at the Dresden meeting. From a radio interface point of view, the major change caused by the discussion was the change of the chiprate. The chiprate was being lowered from 4.096 to 3.84 Mcps. This is not an alignment with the CDMA 2000 chiprate but they are no closer together. On the core network interface side the discussion continued into the second half of 1999. Several so-called hook and extensions workshops took place. By means of these workshops it was possible to operate the different RAN with different core network, whether based on GSM MAP or ANSI 41.

The achievements in 1998 were quite impressive, and the effort made by the delegates and their companies were quite high. As described in other chapters of the book, activities took place to come to a global initiative of standardisation work. This was quite useful in keeping the universal character of UMTS in mind, and even in having better efficiency of the work.

In December 1998 the 3GPP kick-off meeting took place in Sophia Antipolis, forming the structure of 3GPP and defining the convenors of the various TSG and working groups. This moved the responsibility of the work from SMG2 to 3GPP. Within TSG RAN four rather than three working groups were defined. The new working group was a split of the radio-related matters from the physical layer group. In general for the ETSI working group members, the change was not drastic. The working procedures, the support and the structure were very similar to what the delegates were used to. Finally it must be said, that the actual transition of the work caused only low losses, and the overall timetable was not badly effected as suspected by some groups in the beginning. US standardisation is also a partner in the 3GPP, but this did not prevent the interested groups around the CDMA 2000 group founding the 3GPP2 group. Both groups were now working independently on their radio interface and their preferred core network base.

Looking from December 1996 to December 1998 huge progress was made in the UMTS domain, before the work was transferred to 3GPP.

SMG2 remained active to continue the work on the GSM air-interface. Having all these UMTS groups disappearing, SMG2 used the chance to re-structure its own working procedures. SMG plenary and working parties were co-located within a 1-week period. The meeting week started and finished with a plenary day, where on the days within the week, the different working groups meet in parallel. This schedule proved to be a very efficient work style, reducing the amount of travel, and practically removing any need for inter-group liaison statements.

The SMG3 integrated themselves sooner or later entirely into the 3GPP structure. This meant finally that, for example, for core network or architecture matters, SMG2 had now always to deal with 3GPP groups. This applied for all problems which could not be solved inherently in SMG2. From a formal point of view, this can cause some problems. Further the increasing interest of the US in GSM and EDGE in general, was also building a desire from the US organisations to contribute to the standard in an organisational environment such as the 3GPP organisation. This lead finally to the request to integrate SMG2 into the 3GPP organisation. It needed some time to sort this out, but in May 2000 we had the last SMG2 meeting in Biarritz. By accident the same city Niels Andersen and I started off in our responsibilities as chairman and vice chairman, respectively of SMG2.

12.2.14 The GSM EDGE Radio Access Network (GERAN)

GERAN is the name for the network evolution of GSM towards the third generation networks, based on the 200 kHz TDMA system, providing full backward compatibility to GSM.

GERAN is further the name of the TSG in 3GPP which continues the SMG2 work.

To get a better understanding it is worth comparing the UMTS and the GSM architecture. The function split in GSM is different than for UMTS. As an architectural requirement, UMTS is designed to keep mobile specific functions mostly out of the core network. Table 12.2.1 shows the differences in functional split among the Gb and Iu interfaces.

The introduction of the UMTS networks leads to an evolution of the core network elements which inspires the network operators to have an unique core network in the future, which is able to be connected to an UTRAN and a GERAN which is based on the existing GSM infrastructure. As a consequence this requires an Iu interface for the GERAN, and an adoption

Table 12.2.1   GERAN high level block diagram

Function	Functional split with I_u interface	Functional split with G_b interface
Ciphering	RAN	CN
Compression (IP header and payload)	RAN	CN
Termination of LLC and SNDCP	RAN	CN
Buffer management	RAN	CN
Flow control over the interface	No	Yes
Radio resource handling and cell level mobility	RAN	CN + RAN

of the UMTS function split. To maintain the backward compatibility to pre-GERAN mobiles, the A and Gb interface has to be retained. This leads to an architecture as it is shown in Figure 12.2.1.

The Iur-g interface is new providing a control plane connection between BSCs or even to an RNC. However, this interface is discussed as an optional feature. The GERAN architecture shall support among the Iu interfaces the similar kind of bearer classes and grade of service definitions as for UTRAN. This will allow for easy transition of services between the two radio access technologies. The GERAN concept of course will be limited by the maximum bitrates achievable.

Another requirement for GERAN is the support of IP multimedia services, and optimised voice as a voice over IP solution. These features require header compression and header removal techniques. Further compression for the SIP signalling is under discussion.

The first GERAN release is intended for the 3GPP Release 5. The work load is again intense, and the level of modifications needed comparable to the GPRS modifications. Other aspects of the GERAN work such as the introduction of the wideband AMR codec and

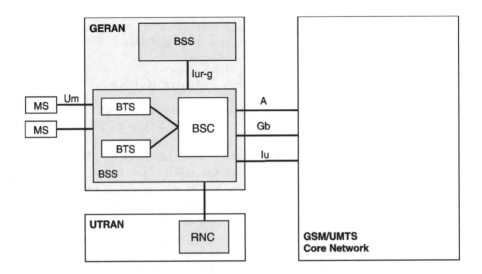

Figure 12.2.1   GERAN high level block diagram.

improvements to the physical layer are discussed. Further a handover function for packet services shall be introduced, which can work in all previous releases with the cell reselection mechanism.

The work for beyond Release 5 is not yet defined. A 3GPP workshop scheduled for October 2001 will help to clarify the possible overall work items for GERAN and UTRAN for Releases 6 and higher.

Chapter 13: The Subscriber Identity Module: Past, Present and Future

Klaus Vedder[1]

13.1 Introduction

Since its conception in 1988 the Subscriber Identity Module (SIM) has undergone continuous development extending its technical and functional capabilities. Initially, it was defined as a security module to authenticate the user to the network providing, at the same time, some very limited amount of memory for network and private user data. In those days, smart cards were still in their infancy. The technological and market requirements of GSM, its need for a global solution and its growing market power shaped the face of the SIM and changed the world of the smart card. Today's SIMs offer more than just security. They are a secure platform for operator defined services allowing operator differentiation by exploiting the power of the microcomputer in the SIM. All this could not have been achieved without the close co-operation of all parties involved in its standardisation.

13.2 The Committee

From a dozen participants at the first meetings in 1988 to about 70 delegates at the last meeting in January 2000 reflecting the growing importance of the SIM:

SIMEG, SMG9 "SIM Aspects" and SMG9 "IC Card Aspects".

Three names and only two chairmen for one committee which in 56 ordinary plenary and hundreds of working party and ad-hoc meetings:

- defined the Plug-in SIM, which has become the best-selling smart card;
- required the use of high temperature plastic material to support the new environment;
- specified new and advanced electrical and physical parameters such as low power consumption and the 3 V smart card interface; and
- advanced the SIM from a security and storage device to a secure platform for value added services.

---

[1] The views expressed in this chapter are those of the author and do not necessarily reflect the views of his affiliation entity.

The first meeting of the Subscriber Identity Module Expert Group (SIMEG) was hosted by the Dutch PTT in January 1988. SIMEG had been established as an expert group by GSM-WP1[2,3] to deal with all issues concerning the SIM and its interface to the mobile, thus covering a specific issue within the scope of its parent body, Working Party 1 (WP1), which was responsible for developing the GSM services. The first change in name and status was in April 1994 when, reflecting the importance of the SIM for the GSM system, the plenary meeting of TC SMG[4] held in Regensdorf, Switzerland "decided to raise the group of experts dealing with the specifications of the Subscriber Identity Module to the status of an STC". As a Sub Technical Committee (STC), SMG9 reported now directly to TC SMG, the technical committee responsible for the specification of GSM and UMTS.[5] SIMEG had become STC SMG9 "SIM Aspects". Four years later in June 1998, SMG9 changed its name from "SIM Aspects" to "IC Card Aspects". The scope of SMG9 had been extended to include work on smart cards of a generic nature which had been part of the responsibility of the disbanded TC ICC (Integrated Circuit Cards). With the foundation of the Third Generation Partnership Project (3GPP) in late 1998, the UMTS work of SMG9 was transferred to the USIM group within 3GPP and the future of SMG9 seemed to be quite predictable. It was assumed that all the work of SMG9 would, eventually, be transferred. The emerging need for a common smart card platform for the various mobile communication systems created a completely new scenario. SMG9 was "closed" by the ETSI Board on 29 March 2000 to be succeeded by ETSI Project Smart Card Platform (EP SCP). This committee inherited the generic work of SMG9 as part of its task to specify such a common smart card platform. The GSM specific work of SMG9 was handled by the new committee on an interim basis until a few months later, with the closure of TC SMG, all GSM work was transferred to 3GPP.

13.3 The Early Years

The first SIMEG plenary meeting in January 1988 was attended by nine delegates from France, Germany, The Netherlands and the UK. These countries formed the core of the plenary meetings of SIMEG for the first few years with Finland and Sweden contributing extensively through their work in ad-hoc meetings. The chairman of the first meeting was Gérard Mazziotto from France Telecom – CNET. He held this position for 5 years until his resignation at SIMEG#31 in March 1993. By then the number of delegates attending the usually 1 or 2 day plenary meetings had stabilised to around 20 from seven countries.

The plenary reports of those days often refer to the opinion of a national delegation. Industry experts attended courtesy of the (national) operators and only two industry delegates per country were allowed at a meeting. Industry contributions often stated the respective national administration as the source. This was, for instance, the case with the first document proposing what today is known as the Plug-in SIM. The description of a semi-permanent SIM

---

[2] Folder B4 of the attached CD ROM contains a content list covering all documents quoted in this chapter sorted according to footnote numbers. Folder B4 contains most documents quoted. For the others a folder number is given in the content list.

[3] GSM-WP1 217/87 rev1 (GSM 159/87): Draft terms of reference of the SIM expert group.

[4] With the inclusion of UMTS into its work program, ETSI TC GSM (the former Group Spécial Mobile) became the Technical Committee TC SMG (Special Mobile Group). GSM1 (the former GSM-WP1) became SMG1 etc. The first SMG plenary (SMG#1) was held in January 1992.

[5] Scope of SMG9 "SIM Aspects" in SMG 477/94: Terms of references of TC SMG and its STCs.

based on the IC card SIM was presented at SIMEG#5 in Munich in September 1988 by the author; it was sourced "Deutsche Bundespost".[6] This situation gradually changed and, with the transfer of the GSM specification work from the European Post and Telecommunication Authorities (CEPT) to ETSI in the summer of 1989, industry could attend the meetings in their own right.

The first years were obviously dominated by the need to specify the basic functionality and physical features of the SIM and its interface to the mobile. The original role of the SIM was that of a security device as defined in the report of SIMEG#1[7]:

> A SIM is the physically secured module which contains the IMSI, an authentication algorithm, the authentication key and other (security related) information and functions. The basic function of the SIM is to authenticate the subscriber identity in order to prevent misuse of the MS (Mobile Station) and the network.

This understanding of the function of the SIM underwent quite an evolutionary process which was, however, not always reflected in the requirement specification, GSM 02.17 *Subscriber Identity Modules (SIM), functional characteristics*. There was often the question of what was first: the technical realisation of a feature or the definition of the functional requirement. Though the document was revised intensively several times, it is no easy reading to gain an overview of the full functionality of the SIM.

13.4 Functionality, Form and Interface

To fulfil its role as an authentication device bringing a new dimension of security to mobile communication networks,[8] the SIM had to be able to execute internally the operator specific authentication algorithm (A3) and to store securely the subscriber specific key (Ki) and other parameters needed for this task. Additional functionality, the physical forms of the SIM and its interface to the mobile were the three main topics for quite a few years.

13.4.1 Functionality

At the third meeting in May 1988, SIMEG discussed for the first time that the SIM should also store data related to services and not only those data necessary for the security functions. Advice on this issue as well as on the creation of a new specification on SIM communication – which was later to become GSM 11.11 – was sought from the parent committee. GSM-WP1 supported the request from SIMEG and decided that the SIM should offer the capability to store information elements for the following features: Short Message Service (SMS), Advice of Charge (AoC), Abbreviated Dialling Numbers (ADN), Fixed Dialling Numbers (FDN), barring of outgoing calls, pre-programmed PLMN selector and language of announcement.[9] New data-fields on the SIM as well as new procedures for the interface between the SIM and the Mobile Equipment (ME) needed to be specified, while the constraints imposed by the memory provided by the special chips used in a SIM were a constant source for discussion.

---

[6] SIMEG 72/88: Semi-permanent SIM.
[7] SIMEG 28/88: Report of the 1st SIMEG meeting, The Hague, 19-20 January 1988.
[8] See Chapter 15: Security.
[9] WP1 162/88, SIMEG 47/88: Letter from WP1 at SIMEG#4.

13.4.1.1 The Influence of the Memory

Memory was a scarce resource. The chips of those days offered less than 10 kbytes of memory in total. Though the functionality and internal organisation of the chip in a SIM with its own CPU and on-board memory is comparable to a PC, it is not surprising that the performance of such a microprocessor chip is limited. State of the art chips had about 6 kbytes of mask-programmed Read Only Memory (ROM). This type of memory is used for data which are the same for a large number of cards as it cannot be changed. It typically contains the operating system and the GSM application including authentication algorithm(s) and application proto-col. The 2–3 kbytes of programmable memory, so-called EEPROM, would hold all the subscription and user (thus SIM) specific data such as the IMSI[10] and the secret, subscription specific key (Ki) for the authentication of the subscriber to the network as well as all the subscriber information now to be stored in the SIM. A typical SIM of those days supported the storage of about 20 abbreviated dialling numbers with 6-10 characters for each name and five short messages in addition to those data. The Random Access Memory (RAM) which is the "working memory" was not even big enough to store all of a short message when this was transferred from the memory in the mobile equipment to the EEPROM in the SIM. The skills of the engineers when writing SIM software were constantly challenged and more than once it was a question of just a few bytes as to whether an extension of the functionality would fit on the chip or further optimisation of all the software would be required.

Neither the huge success of GSM, nor the role of the SIM as the driving force for the smart card industry were foreseen in those days, rather the opposite opinion prevailed. The attitude of "order a few million microprocessor chips and we may think about developing a product" may have been an extreme position around the turn of the 1990s but certainly an indication of the general feeling. So it is not surprising that chip manufacturers were reluctant to introduce new technologies and that it took until the mid-1990s for chips with just 8 kbytes of EEPROM to become available. Those chips offered storage for about 100 abbreviated dialling numbers and 20 short messages. The tremendous increase in the number of GSM subscribers at that time and GSM becoming by far the largest smart card consumer had a significant impact on the development of new smart card chips. Today's chips offer over 100 kbytes of ROM, 64 kbytes of programmable memory, a few kbytes of RAM and often extra hardware for the execution of public key algorithms. This makes, in particular, the implementation of public key solutions for mobile commerce now a reality. Furthermore, the introduction of new types of memory such as flash memory will break up the "historic" separation into ROM and EEPROM and lead to new ways in the handling of SIMs and its software management.

13.4.1.2 PIN Management and Other New Security Features

March 1988 (SIMEG#2) saw the cipher key generation algorithm (A8) become part of the security functionality of the SIM. A8 generates the cipher key, Kc, which is then used by the cipher algorithm (A5) residing in the ME for the ciphering of the radio interface. It was now up to the operator, as the party responsible for the SIM, to choose the algorithm and thus the quality of the cipher key, Kc. Typically, A8 and the authentication algorithm A3 are combined into one algorithm denoted by A3/8. As 10 of the 64 bits of Kc were set to zero the "effective key length" was 54 bits (though from a cryptographic point of view a key with

---

[10] IMSI: International Mobile Subscriber Identity. This number uniquely identifies a network and the subscriber.

several bits set to zero is not necessarily more "effective" than a "full length" 64 bit key). As the mobile and the base station system were not supposed to manipulate Kc but use it as received from the SIM and the Authentication Centre of the subscriber's home network respectively, the "effective key length" could be controlled by the operator. Some equipment manufacturers had, however, interpreted the specifications differently. The issue was finally clarified by SMG#30 in November 1999 where the original meaning was confirmed.

User access to the SIM as a medium to provide GSM service is controlled by a Personal Identification Number (PIN). This number can be changed and freely chosen by the user within the range of 4-8 digits. The introduction of new features created a new security scenario as the user of the SIM might not be the subscriber. Typical examples at the time were lorry drivers using SIMs with fixed dialling numbers controlled by the fleet manager and SIMs supporting advice of charge with a spending limit set in the card by a parent. As the driver should not necessarily be able to edit the numbers or the child to reset the spending limit, a "super-PIN" needed to be specified to protect the contents of the new data-fields against unauthorised changes. As neither of these two features was completed for phase 1, the super-PIN became a phase 2 item. Before its introduction in September 1991 at SIMEG#23, it was renamed PIN2 to emphasise the fact that it was not superior to the normal PIN, but of a similar nature, accessing data-fields compared with accessing an application.

PIN Unblocking Keys (PUKs) had been introduced by SIMEG as another hitherto unknown feature. The PUK provides the user with a means to reactivate the corresponding PIN which had previously been blocked by wrong PIN entries. The actual process of keying in a PIN with the subsequent verification by the SIM is, however, not mandated by the specifications. These allow disabling of the check of the PIN altogether (though not of PIN2) subject to the discretion of the operator who has to find the right balance between security and ease of use for its specific clientele or group of subscribers. As this feature is programmed during the personalisation of the SIM, it can be set on a per SIM basis as specified by the operator.

An interesting interaction between PIN check and security is the order in which the PIN check and the authentication of the SIM by the network are performed. Doing the PIN check first, as introduced by SIMEG#32 in May 1993, has two advantages. The prompt for the user to key in the PIN comes immediately after the mobile has been switched on, and not after the log-on to the network which may take some time in particular when roaming. It also mitigates the possibility of a cryptographic attack against the (SIM specific) secret subscriber authentication key of a stolen SIM as the correct PIN of the "interrogated" SIM would have to be presented to the SIM prior to the delivery of the authentication challenges (unless the PIN check is disabled). Such attacks briefly surfaced in spring 1998.[11]

13.4.2 Physical Form or Realisation of the SIM

The functional splitting of the MS into ME and SIM was described in GSM-WP1 document 173/87.[12] Three different types of SIMs had been identified for specification: fixed, removable and contained in an IC card.

---

[11] SMG 475/98: Statement by SMG9 and SMG10 chairmen; see also Chapter 15: Security.

[12] GSM-WP1 173/87: Functional Split of MS into ME and SIM.

13.4.2.1 The ID-1 Card

The first mobile network to employ a smart card for the authentication of the subscriber to the network was the analogue network Netz-C of the Deutsche Bundespost (later Deutsche Telekom). This subscriber card had evolved from a "magstripe device" just like a credit card, via a card having a memory chip containing the subscription details, to an IC card containing a microprocessor chip for authentication and other purposes. As such smart cards were already deployed in the field, it was not surprising that this solution was also adopted for GSM.

Looking at it from today's perspective, it is interesting to note how much time was devoted to certain aspects of the IC card or ID-1 SIM (ID-1 is the standardised name for cards having this format). It was generally assumed that most SIMs would have this format and that they might also be used as, say a payment card outside the mobile.

For this reason ID-1 SIMs were allowed to be embossed like a credit card and even today the slot of an ME supporting an ID-1 SIM has to be designed to cater for an embossed SIM. No such SIM saw a subscriber. With the change of the business model, the SIM manufacturers stopped the extensive and mostly unsuccessful testing. The new card material needed to satisfy the high temperature requirements of GSM imposed on the SIM, was not really suited for embossing. The tests resulted, in most cases, in pretty warped cards.

A much discussed and thought to be typical scenario for the SIM was its use in a car phone. How much contact pressure needed to be exerted by the card reader in the phone to the SIM contact area so that communication between the card and the car phone would work under "extreme driving conditions"? It was also assumed that people might own a SIM and no phone, or just travel with a SIM. A subscriber enters a taxi, which has a GSM phone in the boot with an in-built plug-in SIM, and uses his or her own ID-1 SIM in the telephone receiver in the back of the taxi to make a call to be charged to this SIM. These thoughts and scenarios are behind the requirement that an ID-1 SIM takes precedence over the plug-in SIM as stated in GSM 02.17 until June 1998 when, in connection with the specification of a second card reader driven by the SIM application toolkit, the choice for the precedence was left to the user. Interestingly enough, similar ideas are now discussed by 3GPP for car pooling and access to multi-media devices from each seat in a car.[13]

The reality was different. More and more ID-1 SIMs were delivered "pre-punched" so that the user could break out the plug-in SIM. Though overall market figures are not available, it can be deduced from the numbers of a major operator that as early as 1995 the number of pre-punched SIMs exceeded the number of ID-1 SIMs, with this number approaching nearly 100% of the total SIM market in 1998.

These were also the days of the manufacturers of punching machines and Plug-in adapters whose products transformed ID-1 SIMs into Plug-in SIMs and vice versa. These adapters were clearly outside the relevant GSM specifications with respect to thickness, bending and torsion. Nevertheless, they were widely used. SMG9 was even asked by the Terminal Working Group (TWG) of the GSM MoU to specify an adapter. This was rejected by the SMG9 plenary meeting in March 1996 on grounds of violation of its core documents and potential liability issues. Also outside the GSM specifications was the use of a "pre-punched" SIM as an ID-1 SIM in a mobile requiring the latter. Special punchings were, however, developed around 1996 to minimise the risk of damaging the card reader in the mobile when a pre-

---

[13] TP-010066: UE functionality split over physical devices, TSG-T#11, Palm Springs, March 2001.

punched SIM was inserted or removed. The potential damage consisted of the contacts of the card reader falling into the gap between the plug-in part and the remaining part of the ID-1 SIM thus getting torn, resulting in an unusable telephone. Operators had quite an interest in such solutions as this drastically reduced their logistic problems and cost – one instead of two types of SIM.

13.4.2.2 The Fixed Solution

The "fixed" SIM was a major topic at SIMEG#2 in March 1988.[14] This solution meant that all functions of the SIM including the (secret) operator specific authentication algorithm and the secret subscription specific key used for the authentication of the subscriber, would be an integral, thus fixed, part of the mobile. Such mobiles would be operator and even subscriber specific.

SIMEG agreed that this solution would have severe disadvantages with respect to flexibility and security. Apart from the question of whether secret keys could be stored securely in a mobile, considering all the issues around the storage of the IMEI, a fixed solution would require a loading mechanism for the authentication algorithm and the secret subscriber key as well as a mechanism to replace such an algorithm or the original key. These concerns were also expressed in a letter to SIMEG by MoU-BARG, the billing and accounting rapporteur group within the GSM MoU.[15] The letter further points out "the commercial impact of the fixed SIM solution with respect to the possibility of free trade with mobile equipment. ... Thus (commercial) barriers would be raised in relation to the trade of mobile equipment." Other concerns were related to potential security issues when a mobile was repaired and to the handling of personal user data when the subscriber replaced the mobile. GSM-WP1 followed the conclusions of SIMEG and the concept of the fixed SIM was dropped altogether in early 1988.

13.4.2.3 The Plug-in SIM

The form of the removable plug-in SIM was discussed quite controversially for nearly 9 months while agreement on the lower layers had already been reached at SIMEG#3: "The electrical and logical interfaces for IC card SIMs and plug-in SIMs will be identical in principal, and according to ISO 7816". The ISO/IEC 7816[16] series of standards forms the core reference for all smart card applications.

The first proposals for the physical form of the plug-in SIM were discussed at the two following meetings where the UK and Germany presented their solutions. The first proposal by the UK, later modified to a more compact 28 pin J-lead package, was the use of existing electronic components in the form of a 24 pin DIL socket with only eight pins connected as ISO 7816 specified just eight contacts. In the German proposal already mentioned above, the plug-

---

[14] SIMEG 43/88: Report of the 2nd SIMEG meeting, Paris, 16-17 March 1988.

[15] SIMEG 12/88: Letter from MoU-BARG meeting to SIMEG (prior to the foundation of the GSM association, the GSM operators were organised within the GSM MoU – the name derived from their memorandum of understanding).

[16] ISO: International Organisation for Standardization; IEC: International Electrotechnical Commission. ISO/IEC 7816, Information technology – Identification cards – Integrated circuit(s) cards with contacts. Prior to the formation of the Joint Technical Committee 1 (JTC1), Information technology by ISO and IEC in 1988 these standards were published by ISO and still today people refer to them as ISO standards.

in SIM was a "cut-down IC card" obtained by simply cutting away the "excessive" plastic of an ID-1 SIM and thus reducing the size to 25 × 15 mm. This realisation would allow the use of existing technology for production and personalisation and the interface to the ME would be identical to the that of the ID-1 card.

The discussions about the advantages and disadvantages of the two proposals centred sometimes around interesting aspects of removable, though potentially rarely removed components. Concerns were raised about the handling of the cut-out version which was, however, equally applicable to a DIL package. Would a little tool coming with the DIL package and similar to the one used by a dentist for testing a filling, satisfy the requirement in Recommendation GSM 02.17 that the SIM is a removable module which can (easily) be inserted and removed by the subscriber? Would the consistent pressure and connection cause gold wandering between the contacts of the cut-out version and the card reader? It was clarified that voltage and current would clearly not be high enough to cause any such problem.

As GSM-WP1 wanted the final say in this issue, SIMEG was requested to elaborate a decision document outlining the advantages and disadvantages of the two proposals. Respective documents were elaborated by both delegations but in the end not required. The matter was resolved by SIMEG itself at its eighth plenary meeting which took place in Issy-les-Moulineaux (Paris) in January 1989. As no delegation no longer supported the 24 pin DIL package or the modified proposal, unanimous agreement was reached in favour of the cut-out version. GSM-WP1 endorsed the proposal at its meeting in Madrid a month later in February. The statement about the SIM being a removable module was also clarified at that meeting for the Plug-in SIM by inserting the following text in GSM 02.17: "It is intended to be semi-permanently installed in the ME". The precise meaning of "semi-permanently" was left to the manufacturers who have come up with a lot of good and compact solutions since then (Figure 13.1).

Figure 1.3.1   Early (hand-made) samples of Plug-in SIMs

The final form of the Plug-in SIM realises the UK proposal that the Plug-in SIM shall be positioned in the mobile by means of a cut-off corner and not by a hole, as originally proposed. This simplified the manufacturing process and SIMEG#10 agreed on the final form as contained in document SIMEG 60/89 for incorporation into GSM 11.11.[17]

13.4.2.4 Mini-SIM and Mini-DAM

In September 1990, the Association of European PCN Operators[18] proposed a third size for the SIM card, a third of the size of the ID-1 SIM: "Large enough to be frequently insertable and removable by handset end users. The ability to do this easily and reliably will maximise smart card use. ... Small enough to not impact on handset design and shape/style."[19] This mini-smart card appeared in the report of SIMEG#20 in January 1991 as a phase 2 work item only to be removed from the same by the GSM1 meeting in Bonn as reported at SIMEG#22 in May 1991.

The idea resurfaced in September of that year with an explicit reference to the mini-card of the DCS 1800 operators as one of the requirements for the realisation of the DECT Authentication Module (DAM).[20] The concept of this module was similar to that of the SIM with the actual specification work commencing in October 1991 in an expert group chaired by me. The dimensions of the mini-DAM were different to the original proposal of the DCS 1800 operators. The size of the Paris metro ticket competed with a card of dimensions 66×33 mm being the top left part of an ID-1 DAM (or SIM). The latter was eventually chosen as it allowed to construct card readers which could accept both an ID-1 DAM and a mini-DAM. The fate of a third card size was finally sealed in early 1994. For reasons of compatibility with GSM, the mini-DAM was dropped from the specification in response to requests by the national standards bodies of France and the UK made in the public enquiry preceding the publication of the DAM as a European telecommunication standard.[21] The DAM group rejected the additional French request to delete the plug-in DAM. Among the reasons given for the deletion were "the difficulty in handling the plug-in", "printing restrictions" and that "the state-of-the-art in GSM handsets shows that it is no more a problem to integrate an ID-1 card in the handset as some mobile manufacturers provides now a full ID-1 card interface (as) part of the original design". To follow the request would have been a de-alignment with GSM, also affecting the planned DECT-GSM interworking.

13.4.3 GSM 11.11 – The SIM-ME Interface Specification

Would the functionality of the SIM laid down in GSM 02.17 not be sufficient for operators to write their own interface specifications based on the relevant international standards of ISO/IEC for smart cards? It certainly would, but, what about interoperability? International standards often contain numerous options due to the wide range of applications they have to

---

[17] SIMEG 45/89: Proposal for outline of semi-permanent SIM; SIMEG 60/89: Plug-in SIM (drawing).

[18] Personal Communications Network, later DCS 1800 (Digital Cellular System 1800) and then GSM 1800.

[19] GSM1 171/90: Mini smart card.

[20] RES 3S 37/91: Requirements for DECT authentication module specification. DECT: Digital Enhanced (then European) Cordless Telecommunications.

[21] ETSI Public Enquiry (PE 47) closing 31 December 1993. The DAM specification was published as the European Telecommunication Standard "ETS 300 331, Radio Equipment and Systems (RES); Digital European Cordless Telecommunications (DECT); DECT Authentication Module (DAM)" only in November 1995.

cater for, and to the conflicting interests of the parties involved in their creation. There is usually no specific application driving the standardisation process. It is thus not too difficult to specify smart card systems which are fully compliant with the same international standards but not compatible with each other.

Rephrasing the question highlights the issue and one of the factors behind the success of GSM:

- Shall every SIM work in every mobile independently of the issuing operator, the mobile manufacturer and the SIM manufacturer and thus enable a global market for mobiles; or
- Shall there be operator specific mobiles, at least from a software point of view, and thus a fragmentation of the mobile market?

The third SIMEG plenary in May 1988 agreed to seek advice from its parent committee on the creation of a new specification on SIM communication. At the following SIMEG meeting it was reported that GSM had created "Recommendation GSM 11.11: SIM specifications" to "define the internal logical organisation of SIMs and it specifies its interface with the outside world. As a consequence, this recommendation also specifies the part of the ME which communicates with the SIM." For years to come the work of SIMEG was dominated by the completion and the enhancements of this document. The first milestone was the finalisation of the phase1 version. Phase 1 documents were going to be frozen in early 1990 as the first networks were supposed to go on air in mid-1991. To achieve this milestone and to advance the document to a stable level, numerous specialised meetings were called for.

13.4.3.1 The Electrical Interface and the Environment

One major issue was the communication protocol itself. Not surprisingly, the battles known from ISO/IEC were also fought out at SIMEG. The French delegation promoted "their" byte or character-oriented transmission protocol T = 0, the German delegation tried to introduce the block-oriented T = 1 protocol. It was a lost cause. T = 0 had been specified in the first edition of ISO/IEC 7816-3 "Electronic Signals and Transmission Protocols" in 1989, the core document for all smart card work, while T = 1 was published only in 1993 when the GSM system was already up and running. As a compromise SIMEG had agreed that "The transmission protocols to be used between SIM and ME shall at least include the choice of the character per character protocol specified and denoted by T = 0 in IS 7816-3".[22] This left it open to manufacturers to include, in addition to T = 0, the transmission protocol T = 1. With such a wording, it is inevitable that no SIM or ME ever had the choice to communicate with their counterpart by means of T = 1. Eleven years later, the support of both protocols became mandatory for all terminals being compliant with the new smart card platform specification. The choice is now left to the application on the card (such as a USIM) which may communicate with the terminal using either protocol.

SIMEG#9 also saw the first deviation of an electrical parameter from the core standard. In recognition of the special environment of mobile communication with a limited power supply, SIMEG restricted the maximum power consumption of a SIM to 10 mA, compared with 200 mA then allowed by ISO/IEC 7816-3. The requirement was a challenge to chip manufacturers, in particular when incorporating special, power consuming hardware to support public key cryptography. The challenges were solved. Also the other new parameters

---

[22] SIMEG 83/89: Report of the 9th SIMEG meeting, The Hague, 29-30 March 1989.

such as extended tolerances for the voltage supply and the duty cycle have become industry standard by now and were incorporated into the second edition of the core standard, ISO/IEC 7816-3, which was published in 1997.

The environment – a mobile lying on the dashboard of a car in the heat of the day in the Sicilian summer – caused long discussions about the temperature resistance of SIMs. While this situation did not look like too much of a problem for the chip itself fears were expressed that the large cards may warp and get stuck in the mobile while the Plug-in SIMs might just melt away inside the mobile. The standard material for credit cards in those days and today is PVC, not known for a high temperature resistance. Agreement was finally reached in the SIMEG plenary forcing manufacturers to invest in new material satisfying the higher requirements of GSM set at 70°C with "occasional peaks of up to 85°C".

13.4.3.2 The First Version of GSM 11.11

"SIMEG agrees to present the draft of Recommendation GSM 11.11... for approval as it was asked by GSM. However, it is a common view in SIMEG that this recommendation is far to be complete, especially further editorial improvements will be necessary." This disclaimer contained in the meeting report of SIMEG#10 held in May 1989 did, however, not deter the approval of the document by GSM#23 in Rønneby. Industry needed a stable basis on which to develop its implementations. All further changes to GSM 11.11 had now to be approved by the GSM plenary. The meeting in Rønneby was, incidentally, the first meeting of GSM as a Technical Committee (TC) of the European Telecommunications Standards Institute (ETSI).

13.4.4 Future Work of SIMEG

With the approval of GSM 11.11, the question was raised what the future tasks of SIMEG were to be. Clearly, the specification had to be completed and bugs had to be fixed. What else? The specification of type approval procedures for the SIM was not going to be one of these tasks, as SIMEG received its directives from GSM-WP1 and GSM, and neither of the two were in favour of type approval, which they considered to be a matter for the GSM MoU. The SIM-ME interface tests specified for the type approval of the ME were done by a different group. So the house had been built by July 1989 and only a few issues remained:[23]

> It is clear that the main task of SIMEG in the future will remain the management of Recommendations GSM 02.17 and GSM 11.11. However it seems that GSM and MoU network operators might need some expertise from SIMEG as it concerns the definition of acceptance tests for the SIM and the elaboration of SIM administrative management procedures. It is clear that such expertise could be provided by reports to GSM, for guidance only, and not as mandatory implementation. ... Concerning the administrative management of the SIM, it is recognised that the parts of Recommendation 11.11 which deal with the administrative procedures cannot lead to any type of approval tests for the ME. Since there is no SIM type approval, they will never be checked. ... SIMEG agreed that a consistent report on all administrative management of SIM is necessary before taking any decision about changing GSM 11.11.

One of the very first changes to GSM 11.11 was the removal of all those sections purely related to the administrative management phase of the SIM. Some operators considered the personalisation of SIMs their very own matter and outside the scope of the GSM committee.

[23] SIMEG 143/89: Report of the 11th SIMEG meeting, Lund, 18-19 July 1989.

They pursued their own, sometimes quite elaborate procedures. In June 1994, problems in the field forced at least the inclusion of an informative annex in GSM 11.11 containing recommended default values for the coding of data-fields at personalisation. Some pre-settings of, in particular, the ciphering key, Kc, in brand-new SIMs had just not worked in some mobiles. It took a further 5 years until, in 1999, SMG9 started, as part of its mandate for generic smart card specifications, work on a document containing administrative commands and functions for IC cards. The "standardless" time resulted in manufacturers implementing proprietary commands and features for the personalisation of SIMs and their administration over the air. Software developed for the personalisation of SIMs of one manufacturer could consequentially not be used for personalising those of another supplier.

Removal of the administrative sections, technical enhancements and correction of errors, observable in nearly 50 Change Requests (CRs), were the topics until GSM 11.11 was "frozen" for phase 1 (i.e. no new technical features, only error fixing) by GSM#26 in Sophia Antipolis in March 1990.

13.5 Phase 2

By May 1990, SIMEG had compiled a first list of work to be done for GSM phase 2. Apart from the maintenance of the "frozen" GSM 02.17 and GSM 11.11 specifications, SIMEG intended to introduce 3 V technology and "discuss and specify SIM requirements which have already been discussed during the phase 1 specification work, but are not fully specified in the recommendations up to now". These included fixed dialling numbers, the advice of charge feature and the "super-PIN" already discussed above.

SIMEG believed that, given more time, a SIM based solution could be found to satisfy the need of the operators for fixed dialling numbers. This time was indeed needed. A phase 1 mobile would not understand this feature and would allow the user to make any call, not just to those numbers listed in the SIM. A mechanism had to be invented which would prevent the use of a fixed dialling SIM in such a mobile. Though the task sounded fairly easy, a data-field specifying the phase of a SIM was a mandatory phase 2 feature, the technical realisation grew more and more complex with the years. It took until December 1992 when a CR to introduce the required functionality in GSM 11.11 was finally agreed by SIMEG#29. This solution does not use any information about the phase of the mobile or the SIM. The SIM is invalidated at the end of a GSM session and will refuse to work by denying access to its IMSI which is needed for the authentication procedure of the SIM to the network. Only a mobile supporting the mechanism can "rehabilitate" such a SIM.

In January of that year SIMEG celebrated its fourth anniversary at its 25th plenary meeting. The meeting was hosted by France Telecom in Paris who "were happy to offer to the SIMEG participants during the meeting free operational national and international calls through the GSM experimental network infrastructure". By that meeting SIMEG had about completed the work on the advice of charge feature including a field for the coding of the currency. The usefulness of this feature for purposes other than advice was still a matter of concern. The statement in the report of SIMEG#4 that the accuracy of the AoC counter may not be sufficient as the SIM might be removed during a call (or the open interface between the SIM and the ME might be manipulated maliciously) was still valid. One such attack was the insertion of a thin piece of foil between the contacts of an ID-1 SIM and those of the card reader in the mobile. This attack foiled quite a few of the mechanical or other devices all mobiles were required to have to detect the removal of a SIM. To counteract such attacks an additional electrical and logical check was specified in August 1993.

13.5.1 The Incorporation of DCS 1800

In June 1990, SIMEG welcomed for the first time a delegate from a DCS 1800 operator. In December of that year, a proposal was presented to incorporate DCS 1800 on the GSM-SIM though as a completely separate application. This would allow plastic roaming between GSM 900 and DCS 1800 operators. One and the same SIM could be used in both systems. The proposal did not meet with general enthusiasm. A statement made at the meeting emphasised the fact that the decision of whether to issue cards supporting both systems would remain a decision of each network operator. No requirement was foreseen for a common GSM-DCS SIM for phase 1, these would still be two totally separate SIMs. The data-fields of the DCS 1800 directory in the phase 2 SIM would mirror those of GSM with the exception of the BCCH coding. Differences between GSM 11.11 and the DCS requirements were going to be contained in a phase 1 delta specification.

With the acceptance of DCS as part of the GSM community, roaming between such networks became an issue. SIMEG was to look into this for phase 2+ from a SIM point of view. The easiest solution seemed to be to abolish DCS 1800 as a separate application and to merge it into the GSM directory on the SIM. The operating system of the SIM could take care of backwards compatibility issues by pretending to support the DCS application on the SIM. The task sounds far more complex than it actually was. The only action required was the translation between the DCS and the GSM identifiers in the communication with a DCS mobile. The solution would even work with a phase 1 DCS mobile. This suggestion made by the author at the SMG1 meeting in Helsinki in August 1993 was eventually accepted by SMG at its meeting in Regensdorf in April 1994 for phase 2. It had been clarified in the meantime that there would be no problem if the BCCH parameters for both systems would be stored in one and the same data-field. At the meeting itself objections were, however, raised to allow the manipulation of the identifiers by the SIM operating system as such methods were not "in line" with the international standards of ISO/IEC. As the proposal met with broad support from, in particular, other operators and such a behaviour of the operating system was not "outlawed", a compromise was reached and a new specification was written during the plenary.[24] As before, the operating system could ignore a DCS mobile, manipulate the identifiers, or, as a new third solution, store them in the rudimentary directory specified in the new document. The author does not know of any implementation of the third solution, as it was consuming several hundred bytes of scarce memory and the switching between the directories could cause security problems in SIMs supporting fixed number dialling.

13.5.2 Aligning the SIM with the DAM

When looking at a phase 1 or an early phase 2 version of GSM 11.11 it is not immediately obvious that they are just a previous edition of today's document. GSM 11.11 had grown from a collection of input papers in a natural way to a core specification. The necessity to improve the document editorially was already seen at SIMEG#21 and several attempts were made and failed. A major obstacle proved the language itself. Improving just the language required

---

[24] GSM 09.91: Interworking aspects of the SIM/ME interface between phase 1 and phase 2.

detailed technical background. The editorial update was only one issue, the other issue was the harmonisation and alignment with the specifications developed by TE9 (see below). So in July 1992, SIMEG#28 set up an alignment group for this task which consisted of a handful of experts. The group was also to take into account the work done for the DAM specification as this had been modelled after the respective TE9 document. The group chose to completely rewrite GSM 11.11 based on the DAM specification.[25] The editorial cleaning had become a review process as well. While the inconsistencies were tabled as CRs to the existing version, the group had incorporated them into the new version assuming they would be approved by the SMG plenary. This way, the "new style" GSM 11.11 could be approved by SMG as a purely editorial change to the existing version. This happened at SMG#6 in late March 1993.[26]

13.5.3 Phase 3 and Other Issues

The new style GSM 11.11 had been agreed by SIMEG#31 in March 1993 for presentation to and approval by the SMG plenary. This meeting was also the end of an era. Gérard Mazziotto who had chaired the SIM Expert Group for a good 5 years guiding it through its years of infancy, resigned as he had announced at the previous meeting. The author who at the time worked for GAO, a subsidiary of his present company Giesecke & Devrient, was elected the new SIMEG chairman by the SMG1 plenary in Düsseldorf in April 1993. He held this position until the closure of SMG9 in March 2000.

Time had shown that, in particular, the technical issues of the SIM-ME interface were a specialists topic. At SIMEG#32 the new chairman could report that an agreement with the SMG1 chairman had been reached "that CRs which do not affect the functionality of the SIM and are of a technical nature may be presented directly to TC SMG (once they have been passed to the SMG1 chairman and PT12)[27] without them being presented to SMG1", a first step towards SMG9 which was established a year later.

After phase 2 there was phase 3. By July 1992 most phase 2 topics were well advanced and SIMEG#27 compiled a list of phase 3 issues it intended to work on, for comment by SMG1. The main outstanding topic of phase 2 was the specification of the 3 V SIM-ME interface. This reappeared in the list for phase 3 together with a third format (again), relationship between GSM service and UPT services and payphone applications based on the SIM. Phase 3 was never realised nor was the third size or the payphone applications. The success of GSM made such applications superfluous. Phase 3 and the intended further phases became over the summer of 1992, phase 2+ to emphasise the fact that the specification of GSM had been completed with the completion of phase 2 and that everything else was just on top of this and optional. The 3 V interface was, however, considered an exception and eventually approved as a phase 2 specification.

13.6 The World's First Low Voltage Smart Card Specifications

Who should take the initiative to introduce 3 V technology: SMG, TE9 or industry? This

---

[25] SIMEG 146/92: Report of the 1st meeting of the Alignment Group, London, 1-3 December 1992.

[26] SMG#6 saw three different versions of GSM 11.11. The CRs resulted in version 4.6.0, the new style GSM 11.11 was then 4.6.1 to which a CR was agreed resulting in version 4.7.0.

[27] The ETSI project team PT12 provided the technical support to TC SMG and its sub-technical committees.

query to the question asked by the parting chairman at SIMEG#31 of whether it was time for a subgroup of SIMEG to specify a low voltage interface, was answered at the following meeting. The formal work item for phase 2+ was agreed at that meeting and approved in June 1993 by SMG#7. Though the work on the new specification was progressed well in several ad-hoc meetings it took until July 1995 for GSM 11.12 *Specification of the 3 V SIM-ME interface* to be approved by SMG#15 in Heraklion as the world's first low voltage smart card interface specification. The reason for this delay was backwards interoperability.

The new 3 V SIMs would work at both 3 V and at 5 V. At which voltage level should a SIM be activated by a mobile supporting both 3 V and 5 V at the interface to the SIM? This, at a first glance, not too difficult question, became a very major issue. From a long-term view it made sense to activate all SIMs at 3 V. SIMs with chips operating at 5 V only would just die out. Only if failing to activate a SIM at 3 V, the mobile would try again at 5 V (if the mobile supported this frequency). What about even lower voltages in the future? Would 5 V damage a 1 V chip? So why to prolong the use of a 5 V interface? The problem at the time was however that some of the "old 5 V only" SIMs would accept an activation at 3 V and the operating system of the SIM would write information in the non-volatile memory of the chip during this process. The question was whether this information would be correct as the low voltage (3 V) may not have been sufficient to produce the necessary charge to program the memory cells. The decision process involved several questionnaires sent to chip manufacturers and endless discussions at SMG9 and SMG. The long-term solution prevailed only in the end. The original decision taken in mid-1995 to reset a SIM at 5 V first, was turned over in May 1996 at SMG#21 against objections by several operators and SIM manufacturers. The reversal had been caused by a compatibility issue. ISO/IEC had finally come up with a draft for the low voltage interface and this mandated, differently to earlier versions, to activate a card first at the lower of the voltage levels supported. Other (new) arguments were that the number of potentially endangered SIMs had decreased and that the problem existed anyway for 3 V only mobiles (though this was not a new argument). The number of potentially affected SIMs seemed to be small in comparison with the long-term advantages. Indeed, no major problems in the field were reported to SMG9.

Having a 3 V SIM specification did not mean having 3 V SIMs. The technology was not quite keeping up with the standardisation. The delivery of quantities started in mid-1996, well after the early and optimistic announcements of some chip manufacturers. By that time the number of GSM subscribers had reached 30 million and far more 5 V only SIMs had been delivered to the market. The idea behind the introduction of 3 V technology at the SIM-ME interface had been the transition of the mobile industry to an all 3 V mobile. To have a charge pump just for the interface to a 5 V SIM was not an appealing thought to mobile manufacturers. Could on the other hand all these GSM subscribers be "forced" to replace their old 5 V SIMs when buying a new 3 V (only) mobile? When should 3 V only mobiles be allowed in the market? This question was discussed at all political and technical levels. In the end, it boiled down to commercial issues between operators and mobile manufacturers. The mass introduction of 3 V only mobiles was well after the date set by the European commission for the protection of the consumers.

Less than 1 year after the introduction of 3 V SIMs to the market, SMG9 started with the work on the next generation of low voltage interfaces. GSM 11.18 *Specification of the 1.8 V SIM-ME interface* was approved as a phase 2+ specification by SMG#28 in February 1999. This time the question was not at which voltage level the mobile should reset the SIM; SIM software engineers had learned their lesson. The question was whether the new 1.8 V smart card chips will have to work over the whole range from 1.8 to 5 V, or just at 1.8 and 3 V. The implications for GSM would go beyond the immediate commercial (and logistical) effect that SIMs with a 1.8/3 V chip would not work in (old) mobiles supporting only 5 V at the interface to the SIM. There is a growing smart card market outside the world of mobile communications. Those smart cards may have to support 5 V as the (fixed) terminals may not support a lower voltage for years to come. This would imply a segregation of the supply market with the obvious impact on price and availability.

13.7 Phase 2+ and the New Specifications

Phase 2+ saw the introduction of new features in GSM 11.11 and the development of several new specifications which were going to change the role of the SIM. As these features are optional their introduction in the "field" was not always a success.

An example of such a feature is service dialling numbers. They are similar to abbreviated dialling numbers with one major difference, the operator determines who can change these numbers and how. Some envisaged applications saw the operator "selling" (part of) the memory space of the SIM to other companies for just a certain period of time and then "re-sell" the space and update the numbers over the air. The idea to specify a separate data-field for service dialling numbers had already been presented to SIMEG in December 1992 but was not taken up by SMG1. Some operators introduced this feature a few years later by storing their service numbers in the data-field for the abbreviated dialling numbers and, in breach of GSM 11.11, did not allow the user to update them. Neither solution was a success. The non-specified implementations confused the user or the mobile, depending on the reaction of the SIM to an attempt by the user to update the numbers. The standardised feature had been introduced too late (January 1996) and the support by mobiles was limited.

A successful and heavily used new feature were and are the group identifiers, better known as "SIM lock", which were introduced formally in January 1995. "Two reasons are given for this feature: one is to prevent MEs which have been stolen, from being used, and the other is to stop people from getting a subsidised ME by signing up for a subscription with no intention of honouring the agreement." [28] Such an ME would read the contents of the group identifier field in the SIM and, if this did not contain the expected coding, would just refuse to work. The specially coded SIM would however work in any "normal" ME. It would thus have been more logical to talk of a "mobile lock" rather than a "SIM lock". The subsidy aspect of the feature called the European Union to the scene. After long discussions and several meetings, the wording in the requirement specification was clarified which had, however, no effect on the implementation of the "lock" or its use.

An important technical improvement was the so-called interface speed enhancement which was introduced in the specifications in October 1995. A major drawback of the SIM-ME interface had been its fixed (and low) data rate depending just on the frequency of the clock provided by the mobile driving the chip in the SIM. Multiples of the ratio clock/baud rate were now allowed. This new technical feature proved particularly useful for the value added services to be introduced a few years later.

There are numerous other "singular" phase 2+ features which have been taken up by the market. We will, however, limit the remainder of this chapter to the new specifications. The

[28] SMG9 42-1 rev1/94: Meeting report plenary no. 1/94 held in Dublin and 1/94 bis held in Düsseldorf.

impact of one of the new specifications, was far beyond any presumption. This is, for instance, manifested in the firm believe that a phase 2+ SIM, though there is no such SIM by definition, is a SIM supporting the SIM application toolkit specifications.

13.7.1 The SIM Application Toolkit

The idea of using the SIM as a personal computer which acts on information received over the air was first discussed within SIMEG as early as 1991. It was suggested that a SIM supporting the advice of charge service would act "pro-actively" and refuse to work if the value stored in the SIM had been used up. The SIM would at each update check the accumulated charges stored by the mobile in the SIM against the maximum value available to the user. As soon as the accumulated charges reached or exceeded the pre-set value in the SIM, the SIM would refuse to process the next authentication request. As a consequence, the mobile would drop the call and no further calls were possible.[29]

The other building block of the SIM application toolkit is "data download". The Telepoint subscriber Identity Module Expert Group (TIMEG, see paragraph 7.1 below) discussed at its eighth meeting in May 1991 a document[30] which describes a mechanism for downloading and storing sensitive subscriber and network data into the TELEPOINT Subscriber Identity Module (TIM) or handset in a secure way using public key cryptography. Two years earlier, SIMEG#9 saw proposals to use protocol transparency and end-to-end connections between a SIM (or the mobile) and a remote service centre to offer additional services to the user or to remotely control the keyboard or display of the mobile. No conclusion was reached at the meeting. GSM 02.17 did not contain a service requirement and ISO/IEC, as the body responsible for the international standardisation of a smart card, was considered to be a more appropriate place for such a general feature. The topic was revisited a few times with no prevail.

The time was just not ripe for standardising such an approach, the SIM was still seen as a secure storage device and not as a "pocket computer". The ideas were, however, further developed on a proprietary basis by a number of operators. Downloading data such as the subscriber's MSISDN number into the SIM over the air, or creating a new data-field in the SIM to support services which were not available when the SIM was originally programmed, soon became a reality. Though these services were by comparison with today's applications rather simple, they required operator specific mobiles and SIMs, limiting the potential market. The first approach to overcome this situation was the development of the Common PCN Handset Specification (CPHS) in the early 1990s by some DCS 1800 operators. Unfortunately, the use of this document was restricted to a group of operators and the document did not find its way into SIMEG or SMG9.[31]

13.7.1.1 The Standardised Approach

To exploit the full potential of these ideas, tools and mechanisms needed to be standardised on a broad basis. In late 1994, SMG9 finally started work on the these topics. It soon became

---

[29] SIMEG 120/91: Report of the Ad Hoc Group "Advice of Charge".

[30] TIMEG 26/91: Public key cryptography and "Over The Air Registration".

[31] Some CPHS features were incorporated into Release 4 of the USIM specification and subsequently into Release 4 of GSM 11.11 in early 2001.

apparent that it would be beneficial to merge the two work items "pro-active SIM" and "SIM data download" to a (single) SIM execution environment. The SIM application toolkit was born.[32] A year and a half later, the first version of GSM 11.14 *Specification of the SIM-ME interface for the SIM application toolkit* was approved at the 50th plenary meeting of GSM/ SMG in Bonn in April 1996. GSM 11.14 defines a set of commands and procedures which are complementary to those specified in the basic SIM-ME interface standard (GSM 11.11) for "normal" operational use. Applications designed in accordance with this specification do not need any special mobiles, they just have to support the functionality of the (SIM) toolkit. (Most people refer to the specification as the SIM toolkit or even toolkit.) Examples of these tools include the provision of local information such as cell number and timing advance, the setting of timers in the mobile equipment, the interaction with a second card reader to, for example, load money over the air into an electronic purse card, the support of (colour) icons, call control and sending of short messages to communicate with network applications.

13.7.1.2 Applications

The ideas for issuer specific toolkit applications allowing issuer differentiation in a highly competitive market are certainly too many to list them all. Three typical areas are location dependant services, mobile banking and Internet access.

Location services can be used to provide the user with, for instance, information about the films shown in the vicinity. The SIM application requests the ME to provide the area information and then sends a short message to the network requesting all the films shown in that area or just those whose title was given to the SIM by the user. The communication between the user and the SIM via the display and keyboard of the mobile equipment is part of the interactive SIM toolkit application. This service can be extended from a purely informative one to one where the user not only orders the tickets but actually "pays" for them on the spot using the same short message service. The actual payment method used, be it a virtual account with the operator in a pre-paid scenario, through the normal telephone bill, or in combination with a payment institution, as well as the technical realisation lead to interesting aspects of mobile commerce and its feasibility.

A key requirement for successful *mobile banking* services is a comprehensive security solution and this is what the SIM can provide. Such an application would typically allow the user to check the balance of the account and transfer money from their account to any other account by means of a short message set up by the SIM toolkit application in an interactive (mobile equipment independent) session with the user. As in the example above, the SIM controls the display and derives all information necessary from input received from the user. Clearly, money transfer requires a high level of security. So the access to the SIM toolkit application needs to be protected. This could be achieved by a separate banking PIN which the user is requested to key in. Without such a valid PIN, the SIM will refuse access to the application. This is, of course, not sufficient for protecting the actual transaction. The short messages containing the information need to be protected. A typical solution would be the enciphering of the contents of the short message using triple DES, a method commonly used in the banking world. The keys would be subscription specific, only known to the SIM and the

---

[32] SMG 488/95: Status report of SMG9 at SMG #15, Heraklion, 3-7 July, 1995. This document contains as an annex a first description of the SIM application toolkit. Of interest is, in particular, also another annex on the interoperability issues with respect to the introduction of 3 V technology as discussed above.

bank. When the short message centre of the network operator receives such a message, it cannot decipher the actual contents, as the operator is not in possession of the keys. Instead, the message will be forwarded to the bank which deciphers its contents. If, in addition, a message authentication code and a counter against replay attacks are used, the bank also checks and validates them prior to executing the money transfer (or refusing to do so). The bank could then send a ciphered message back to the SIM informing the subscriber of the successful (or not) execution of the payment. This way end-to-end security between the SIM and a server in a bank can be realised without anybody in the information chain being able to eavesdrop or interfere with the information without this being noticed.

While in the previous examples the applications are residing on the SIM, a *SIM Internet browser* provides a generic means for active Internet communication and the provision of the required application information on a case by case basis. This has the advantage that the application itself is not static but depends on the actual contents of the web pages and the databases residing on the Internet. These and thus the application presented to the subscriber may be changed without any modification being required to the basic application on the SIM itself. Applications up and running today include on-line brokerage which again requires a high level of security.

One could argue "why not use WAP for such purposes?" as this or similar services will, in the future, provide greater computing power especially in conjunction with, say GPRS. One clear advantage of the SIM is its security. The (operator issued and controlled) SIM can store and handle (subscription specific) security data and run the relevant security processes required, in particular, in financial transactions or applications involving data privacy, as the SIM is a secure device, "personal" to the subscriber. The question is thus not "either WAP or SIM application toolkit" but "WAP and toolkit and how to find the right form of interaction".

13.7.2 Securing Toolkit Applications, the First API and CTS

To achieve a common, high level of security for all these applications, a joint working party of SMG9 and SMG10, the GSM security committee, started work on a new specification in late 1996. The aim of GSM 03.48 *Security mechanisms for the SIM application toolkit*, which was approved by SMG#24 in December 1997, is more a design and an implementation guideline than a rigorous down to the bit coding instruction. It does, for instance, not mandate the use of specific algorithms as different applications may require different solutions. An important part of application security is secure administration, in particular when it is done over the air. A downloading mechanism based on the open platform originally developed by VISA was therefore added in November 1999 as an amendment.

Loading of applications into a SIM requires a programming interface. This is no problem if the SIM vendor has written the code for the application as the code will, naturally, work with the SIM vendor specific operating system. To achieve interoperability between SIM vendors with the aim that one and the same code runs on all the cards of all vendors, SMG9 developed an Application Programming Interface (API) for the SIM application toolkit. The functional requirements defined in GSM 02.19 *SIM Application Programming Interface (API) – stage 1* are now "language neutral" and applicable to all APIs to be developed for the SIM. The stage 2 documents, i.e. the implementation details, had to be language specific and would have to be based on existing work. SMG9 could not develop its own API. With the support of the

Java[TM] [33] Card Forum, an industry organisation to promote the use of Java[TM] for smart cards, SMG9 developed GSM 03.19 *SIM Application Programming Interface (API) – stage 2 – (Java[TM])*. The first version was approved by SMG in mid-1999, the publication was however delayed for an outstanding IPR issue. Java[TM] is, of course, not the only language on which a SIM application toolkit API could be based today. Other candidates are MULTOS and Microsoft. Standardising a second or even a third API was, at the time, considered to be counterproductive to the idea of having a standard one. This opinion, which was very much supported by MoU SCAG, the smart card application group of the GSM MoU, prevailed until mid-2000 when the 3GPP USIM group agreed on a work item for a MULTOS API. This has now been combined with a proposed Microsoft API for a C-language binding. A specification does, however, not automatically guarantee interoperability between the solutions provided by different vendors. The discussion about writing an SMG9 test specification for GSM 03.19 fell into the time when the work was about to be transferred to 3GPP. The USIM group took up the issue and developed a test specification for certain parts of the API as outlined below.

CTS is not CTS. One uses a SIM while the SIM is subjected to the other. The use of a SIM in the cordless telephony system is specified in GSM 11.19 which was completed and approved in February 1999.[34] So far no such SIMs have been delivered to the market as the system itself has not been deployed. The same can be said about the DAM cards. In 1996, there were millions of SIMs in the field but no test specification for the SIM itself. On the other hand there was a conformance test specification for the DAM though there were (and are) no such cards. In October of that year agreement was finally reached on developing a test specification for the SIM. Re-inventing the wheel was never part of the "scope" of SIMEG and SMG9. So the test specification GSM 11.17 *SIM conformance test specification*, the first version of which was also approved by SMG in February 1999, was based on the DAM document.

13.8 The Changing Role of SMG9

The role of SMG9 as the group dedicated to just the SIM changed with the success of GSM itself and its future was largely determined by three events:

- the closure of ETSI TC Integrated Circuit Cards (TC ICC) in June 1998;
- the creation of the 3GPP in December 1998;
- the need for a common smart card platform for all (mobile) communication systems.

13.8.1 Broadening the Scope – a Variety of Specifications and TC ICC

The early 1990s saw the creation of a number of smart card committees which developed specifications for particular telecommunication systems as well as for general telecom purposes. Within ETSI, smart card specifications were developed for DECT, TETRA and UPT.[35] Though all these specifications were closely related at some stage to GSM 11.11 or even derived from it, incompatibilities could not be ruled out. The documents had evolved over the years under the responsibility of different committees.

---

[33] Java is a registered trademark of Sun Microsystems.

[34] GSM 11.19 CTS SIM for FP and MS (FP is the fixed part or home base station).

[35] Digital Enhanced Cordless Telecommunications, TErrestrial Trunked Radio, Universal Personal Telecommunications.

There were also national activities such as the specification of the TIM for the cordless telephony system CT2 (Birdie) of German Telekom. The TIM was based on the SIM and by July 1991 a fairly comprehensive specification had been developed in co-operation with industry. TELEPOINT did however not advance beyond the (card-less) field trials and no TIMs were ever delivered.

The European Norm EN 726 *Telecommunications integrated circuit(s) cards and terminals* series of standards was developed by ETSI in collaboration with the European Standards Committee (CEN). This standard is usually referred to by the name of the ETSI committee having developed the series as the "TE9 standard" (Terminal Equipment, Sub Technical Committee 9). GSM 11.11 and EN 726 were developed in parallel and the relations between the two groups feature in nearly every SIMEG and SMG9 meeting report for several years from late 1989. The spirit was, however, more of a rivalry than a co-operation. This was to some extent caused by the intended use of EN 726 as the core document for all telecommunication smart card applications, and all of these applications were supposed to be based on EN 726. It is interesting to note that in the mid-1990s quite a few GSM operators asked for SIMs to be compliant with EN 726, a request which could not be satisfied for technical reasons. Removed from the direct exposure of the market, EN 726 had evolved more towards a European standard competing with the International Standards of ISO/IEC in this field but specifying, at the same time, detailed application specific features.

The situation of a number of (not necessarily compatible) specifications lead to the formation of the Card Experts Group (CEG) by ETSI in late 1995. Its task was to analyse the various smart card specifications developed by ETSI committees, to "Develop a specification for a common hardware and software platform..." and to "Influence ETSI card groups to introduce phased chances in order to align card specifications towards an enlarged common platform". This group reported to TC SMG and was, after four meetings, elevated to a Technical Committee, TC Integrated Circuit Cards (TC ICC). The very nature of the work of TC ICC was part of the reason for its failure. Companies were not providing enough resources as there were too many parallel activities in the field of smart card standardisation and the usefulness of the analysis also dwindled away with time – the SIM had become the dominant product. The lack of industry support meant the closure of TC ICC in June 1998. Most of the work programme was transferred to SMG9 which, in recognition of this, changed its name from "SMG9 SIM Aspects" to "SMG9 IC Card Aspects".

Among the documents inherited by SMG9 was the technical report containing application identifiers. Such identifiers are used in a smart card to inform the terminal of the applications residing on the card and how to find them. This document was updated to contain the relevant GSM information and became the ETSI Technical Specification TS 102 220 *Application identifiers for telecommunications applications*.

As part of its extended mandate, SMG9 developed the previously mentioned new specification for the administrative phase of IC cards. This contains the commands and functionality typically used for the personalisation of a (SIM) card prior to its issue as well as for remote administration of, say a SIM, over the air. The document was approved by SMG#31 in Brussels in February 2000 as TS 102 222 *Administrative commands for telecommunication applications*.

13.8.2 Reducing the Scope – The Transition to 3GPP

The 1998 agreement of Standards Development Organisations of Europe, Japan, Korea and the USA[36] to jointly develop a mobile communication system for the next, the third generation (3G), created new committees and new working structures. It also created a mix of feelings among GSM experts – from "enthusiastic" to "we are better off on our own" and "it won't work". The work of the Third Generation Partnership Project (3GPP) was going to be based on GSM and ETSI's work on the Universal Mobile Telecommunications System (UMTS). It was thus obvious that at least the UMTS part of SMG's work program would be transferred to 3GPP. The outright willingness to share the results of 10 or more years of "their" work with a much larger community and for a queried advantage was not that widespread at the working group level. The question of whether 3GPP would be short-lived was certainly there and discussed.

The 3GPP had created four Technical Specification Groups (TSGs) to specify Service and System Architecture (TSG-SA), the Core Network (TSG-CN), the Radio Access Network (TSG-RAN) and Terminals (TSG-T) of the new system. The need for a working party to look after the UMTS SIM of SMG9 was outlined in a paper presented at the first meeting of the TSGs in Sophia Antipolis in December 1998. This document[37] also contained the position paper of TC SMG on the USIM, revised by SMG#28 in February 1999 to emphasise the point that the UMTS SIM is a removable module, and a proposal for the terms of reference of a new working group within TSG-T to deal with the UMTS SIM. This new group became TSG-T3 (USIM).[38] The detailed functional requirement specification for the USIM elaborated by SMG9 was the first specification to be transferred to T3:

The USIM (UMTS SIM) of SMG9 became the USIM (Universal SIM) of T3.

Prior to the closure of TC SMG in July 2000, when all (remaining) GSM specific documents were transferred to 3GPP, only four other SMG9 specifications changed ownership.[39] The reasons for this, compared with other committees, relative slow process were manifold. T3 was developing its own core specifications for the USIM. They were based on GSM 11.11 but technically and from a service point of view different in a fundamental way. Though some of the key players in SMG9 also attended the T3 meeting, it has to be acknowledged that initially the overall know-how in T3 was not at the same level as that of SMG9. Not all companies sent their most experienced people, some saw this more as a training ground for their younger experts. This attitude did change during the first year and T3 established itself as a competent group in its own right.

13.8.3 The New Challenge

"Considering that SMG9 has years of experience as the custodian of a widely used mobile

---

[36] The Third Generation Partnership Project Agreement was signed by ETSI (Europe), ARIB and TTC (Japan), TTA (Korea) and ANSI T1 (USA) on 8 December 1998.

[37] TSG#1 063/98: proposal for USIM working party in TSG terminals (source: SMG9 chairman, SMG9 UMTS WP chairman).

[38] TSG-T TP-99019: Draft term of reference for TSG-T WG3 (produced by T3#1).

[39] SMG9 248/99: Transfer of specifications from SMG9 to 3GPP (SMG9#19, Munich, September 1999); the proposal is based on the agreement of SMG9 officials reached in Munich during the total eclipse of the sun.

telecommunications smart card specification, it is proposed that SMG9 extends its role to act as the central focus point for all next generation telecommunications smart cards."

This statement agreed by the 16 delegates to the 2G/3G SIM, USIM & R-UIM officials meeting held on 1 November 1999 in Austin, Texas, was going to change the role of SMG9 altogether. It was also an expression of the trust SIMEG and SMG9 had built up over the years in the world as a group dedicated to the advancement of smart cards and their use in mobile communications. The sentence is contained in the Committee Correspondence[40] which was sent as an output document of that meeting to all the major players in the telecommunications sector involved in specifying the use of a smart card for their systems.

The document continues: "SMG9 should be invited to extend its scope and take the role of managing common aspects of mobile telecommunication smart card in the following areas:

- the physical interface specifications;
- the common logical interface;
- file ID allocation (DFs) at the common level, respecting existing directory structures;
- shared data that is technology independent (e.g. the phone book).

The specification of the structure and content of the technology specific files, procedures and protocols will remain in the domain of the formulating standards committee."

The meeting also identified several requirements which had to be satisfied before SMG9 could take on the new role. These were discussed in detail at the follow-up meeting in Rome on 17 January 2000. By then, nearly all committees had already responded in a positive way to the Committee Correspondence endorsing the approach, with the remaining groups following suit as soon as they had had a meeting at which the issue could be discussed.[41] The report of the Rome meeting further identifies several specifications of common interest and contains a detailed list of the necessary requirements mentioned above.

The proposal was, of course, a topic at the SMG9 plenary meeting held in Rome during the remainder of the same week. SMG9#20 welcomed the initiative and stated that it would be happy to take on the new role.

The matter was then brought to the ETSI Board which in February 2000 agreed in principle to the change in responsibility of SMG9. One month later on 29 March 2000, the ETSI Board created "newSMG9" as the successor to SMG9. As a first step, "newSMG9" was going to be an ETSI Project focusing on the elaboration of technical specifications for a common platform.

To facilitate this work, the chairman of the "newSMG9" was encouraged to exercise the chairman's rights to invite participation from companies involved in the standardisation work for mobile communication systems in 3GPP, 3GPP2, ARIB, GAIT, T1P1, TR45 and others to be identified. As a second step, it was intended to transform EP "newSMG9" into a kind of partnership project similar to the 3GPP. The ETSI Board further agreed that the location of venues of "newSMG9" should reflect the distribution of participants and that all "newSMG9" documents were to be freely available on an open area on the ETSI server. As the common specifications developed by EP "newSMG9" would be published as ETSI Technical Specifications they would no longer have a GSM specification number, another requirement expressed at the Rome meeting. So, SMG9 had ceased to exist.

---

[40] Meeting document 006/99 (SMG9 0012/00): Committee correspondence on "common mobile telecommunications smart card standard".

[41] SMG9 0116/00: Report of the (U)SIM R-UIM officials meeting #2, Rome, Italy, 17 January, 2000.

13.9 The Universal Subscriber Identity Module

The first meeting of T3 (USIM) was hosted by ETSI in January 1999. The meeting was convened by me and I was elected chairman at the second meeting a month later. A major topic at the first meetings was whether the USIM was a removable module. GSM delegates considered this to be a matter of fact. This view was, however, not shared by all delegates. Not everybody considered an open terminal market to be an advantage for the operator. The issue was finally resolved at the plenary meetings of TSG-T, TSG-SA3 and TSG-SA in Fort Lauderdale, in March 1999, where it was agreed that the "USIM is a removable hardware module like the SIM is for GSM".

The new setting also provided the opportunity to look at the SIM with the experience of more than 10 years and to create a new tool not totally bound by backwards compatibility issues and the limitations of the past. Though one could not start from scratch, it was the chance to move from a mono-application SIM, supporting value added services as part of the GSM application, to a true multi-application USIM supporting independent applications which can even run in parallel.

To facilitate this approach, T3 was going to concentrate on the completion of the requirement specification and the timely development of the new core documents. TS 21.111 *USIM and IC card requirements* was approved by TSG-T in April 1999 as its first deliverable. At the same meeting, TSG-T agreed to the proposal of T3 to split the core specification into an application-independent part and an application-specific part. The goal had been set and the workload was enormous. T3 was going to have 11 plenary meetings in 1999 to develop the Release 99 editions of the two core specifications TS 31.101 *UICC– Terminal interface; physical and logical characteristics* and TS 31.102 *Characteristics of the USIM application*. As priority was given to these two documents, work on the 3G version of the SIM application toolkit (GSM 11.14) did not start before the second half of that year. Despite this late start, the USIM application toolkit (TS 31.111) was published as part of Release 99.

13.9.1 The Split of GSM 11.11 and Multi-application

The proposal to split GSM 11.11 into a lower layer part and an application specific part had been discussed within SMG9 for some time, and a work item was approved by SMG#29 in Miami in June 1999 as part of SMG9's mandate on generic IC card standards for ETSI. The original idea was to combine, in a new specification, the sections of GSM 11.11 describing the electrical and mechanical characteristics, with the two low voltage SIMME interface specifications GSM 11.12 (3 V) and GSM 11.18 (1.8 V). This way, all physical interface parameters were contained in one document which could then also be used by applications other than GSM, a step towards reducing the incompatibilities between potentially competing and diverging standards.

The work was not advanced in SMG9 but was brought into T3. There the work item was expanded to include the general logical characteristics and commands as well as the underlying security scheme of a smart card as required for a true multi-application platform adhering to the OSI layers. In such an environment it would, for instance, be possible to have on the card a number of 3G applications (even from different operators) and, say a banking application. Different to applications employing the (U)SIM toolkit, the banking application would not require the 3G interface of the terminal for communication. It could use the terminal's infrared, Bluetooth or other interface for communicating with, for example, a cash register in a supermarket (while the terminal may not even be logged onto a 3G network).

When talking of the SIM, people usually do not distinguish between the hardware, the "SIM card", and the application itself. The SIM is just everything, the card and the application. The SIM, being in essence a mono-application (card), could well be identified with its "carrier", the IC card. With the new concept, there was now a clear separation between the underlying hardware and logic on one side, the UICC, and the application specific details as, for instance, defined for the USIM.

GSM: SIM = physical card + GSM "application" (GSM 11.11)
3G: UICC = physical card and basic logical functionality (TS 31.101)
       and USIM = 3G application on a UICC (TS 31.102)

UICC was originally an acronym for "UMTS IC card". The incorporation of the ETSI UMTS activities into the more global perspective of 3GPP required a change of this name. The analogous change as for the USIM was, however, considered to be a bit too pretentious by T3. So the term UICC is now just a term for the carrier as the platform for multi-application.

In a multi-application environment the terminal has to know which applications are contained in an IC card and, therefore, a way of selecting them. This is solved for the UICC by the introduction of application identifiers which are stored in a specific (standardised) file together with a path to the application itself. In general, it can be said that the new 3G specifications were aligned with the international standards of ISO/IEC whenever possible. As the selection by application identifier is not known to existing GSM mobiles, the operating system of the card has to take care of this scenario for the case that the 3G card "contains a SIM" as a separate application. This is done in a similar manner to the technique used for ensuring backwards compatibility with DCS 1800 mobiles when DCS 1800 was abolished as a separate application on the SIM.

A major issue for multi-application cards is secure access to independent applications and their interworking. The example of two USIM applications from (potentially) different operators residing on a UICC and sharing the phonebook may be an extreme case, the design of the security architecture has to be laid out for this and has to specify the behaviour if a PIN securing access to an application is blocked by wrong PIN entries. There are several possibilities as to how the security system should react and to which applications access will be denied. The other extreme would be a universal PIN which works at card level and allows access to all applications. As the choice of the particular scheme is up to the card issuer a very elaborate PIN management system had to be developed.

13.9.2 New Features and Specifications

The most prominent new feature is probably the phonebook. This goes well beyond the functionality of Abbreviated Dialling Numbers (ADN) and associated names known from GSM. The "second name" entry for a number provides the user, for instance, with the possibility to store the second name in a different language even using a different alphabet. Additional numbers can be linked to an entry for storing associated fax numbers and e-mail addresses. Numbers can be grouped using group names to reflect say, the association of the respective person to an organisational unit in a company. It is also possible to have "hidden" entries which means that the name corresponding to a number is not shown on the display of the terminal. It is of course also possible to check and synchronise the entries against an external database. Clearly, memory may be an issue if one wants to support all of these features. Five hundred entries with all their respective associate information would use all the (programmable) memory of today's smart cards.

Records for incoming and outgoing calls are another new feature of the USIM which was also specified for Release 99, the first release of all 3G specifications. This release contains, of course, all the basic functionality of the USIM application such as the security provisions and the transmission protocols. As already mentioned, a terminal compliant with the 3G specifications has to support both transmission protocols $T = 0$ and $T = 1$, while a USIM has the choice of communicating using either protocol.

3G security is quite different from GSM security. As part of the enhancements, mutual authentication and sequence numbers were introduced. In addition to the authentication of the USIM to the network, the network now has to authenticate itself to the USIM counteracting false base station attacks. To achieve the operation of a USIM in a 2G network, conversion functions were introduced. The complex issue of the case when the USIM application provides GSM specific parameters in order to access GSM radio networks as well as the co-existence of a GSM and a USIM application residing simultaneously on a UICC are described in detail in a technical report on SIM/USIM interworking.[42]

As part of its work for Release 4, which constitutes the second release of the 3G specifications which was finalised in early 2001, T3 incorporated in the USIM specification several features of the Common PCN Handset Specification (CPHS) developed in the early 1990s by the PCN operators for DCS 1800. The new features are:

- storage of the full and short name of registered PLMN to supplement those in the terminal;
- the name of the home network operator for display when roaming on "partner" networks;
- inclusion of files for voicemail, message waiting and call forward indication feature.

The major new feature for the USIM Application Toolkit (USAT) was the introduction of Bluetooth as a bearer independent protocol for local links. This allows (U)SAT applications to communicate with local devices using the local connectivity capabilities of the terminal. As no input had been received on other candidates such as RS232, IrDA and USB, their specification is left for future releases.

A work item only partially completed for Release 4 is the toolkit interpreter. Proprietary solutions for a SIM resident microbrowser had been deployed by several operators. To improve the market acceptance of secure Internet access by means of a microbrowser and to allow interoperability between (SIM) vendors, T3 started work on this multi-part specification in mid-2000. While the functional specification was completed for Release 4 and approved by TSG-T at its meeting in March 2001,[43] the documents on the overall architecture for such a system and the specification of byte codes for efficient access are still under development.

---

[42] TR 31.900 SIM/USIM interworking (Release 99 version approved by TSG-T#12 in Stockholm, June 2001).
[43] TS 22.122 USAT interpreter; stage 1.

13.9.3 Test Specifications

The need for accompanying test specifications and the lack of resources in T3 for their development was pointed out early during the specification of the core documents. In October 1999, T3 proposed to set up a funded task to adapt and extend the SIM-ME interface tests contained in GSM 11.10 and the SIM Conformance Test Specification (GSM 11.17) to fit the requirements of UICC and USIM. Once the funding had been approved by the 3GPP co-ordination body, it was towards the end of the financial year, and after another few months to find the experts to do the job, the three specifications covering Release 99 of the UICC and the USIM were completed in a very timely manner.[44] In all developments, the definition of tests and their elaboration as well as the testing itself seem to be a major obstacle.

This becomes even more of a problem if some documents cannot be referred to as there is a question about them being "publicly available", resulting in either testing only part of the spectrum or in re-inventing the wheel with an enormous effort. TS 11.13 *Test specification for SIM API for Java*[TM][45] *card* was developed to check the compliance of applications with the Application Programming Interface (API) specified in GSM 03.19. The test specification, which was approved for Release 98 in March 2001, contains test cases for interoperability at the application level, it does, however, not include test cases for interoperability at the SIM API framework level.

13.9.4 The Transfer of Documents to the Smart Card Platform

TS 31.101 *UICC – Terminal interface; physical and logical characteristics* was a platform specification and as such not specific to 3GPP nor to GSM. It was thus natural that in June 2000, following the agreement within 3GPP on a common approach for smart cards, TSG-T endorsed its transfer to EP SCP, the new committee responsible for specifying a smart card platform. Clearly, the accompanying test specification (TS 31.120) had to follow suit. It was transferred after its first release in March 2001 together with the 3GPP specification for application identifiers (TS 31.110). The latter has in the meantime been merged with the equivalent ETSI document so that there is now only one such document in the telecommunications area. The transfer of other documents to EP SCP such as the specification of a general API (GSM 02.19) is under discussion.

13.10 The Smart Card Platform

The first meeting of the committee created as the central focus point for all next generation telecommunications smart cards was still dominated by the SIM. The change in name and status did not relieve the group of its obligation to maintain and update those SIM documents not yet transferred to T3. The promotion to technical committee level with the right to approve its own specifications created an interesting situation. EP "newSMG9" was at the same level as TC SMG to which SMG9 had previously reported and which had the final say in the approval of changes to the GSM specifications. As an interim solution, it was agreed

---

[44] TS 31.120 UICC-terminal interface; physical and logical tests; TS 31.121 UICC-terminal interface; USIM application tests; TS 31.122 USIM conformance test specification (all approved by TSG-T in December 2000).
[45] Java is a registered trademark of SUN Microsystems.

between the two chairmen in early April 2000 that newSMG9 would still report to SMG with respect to changes to the SIM specifications.[46] At that time it was already commonly expected that TC SMG would be closed by the middle of the year and that the ownership of all GSM specifications, including those for the SIM, was going to be transferred to 3GPP.

"newSMG9" continued in the spirit of SMG9. Its first meeting took place in Visby, on the island of Gotland in Sweden, from 22 to 24 May 2000, as had originally been planned for the 21st SMG9 plenary meeting. Apart from its work on the GSM specifications, newSMG9 drafted its terms of reference, discussed proposals for an appropriate name to replace the interim solution of newSMG9 and looked into a way to handle the transfer of the UICC-terminal interface specification (TS 31.101) from T3 to itself. The search for a new name and the typical quest for the acronym to be a meaningful name resulted in quite a few, in general funny proposals. They ranged from Smart Card Open Application Platform (SOAP) via Cards Dedicated to Mobile Applications (CDMA) to Smart Card Platform (SCP). The last one was eventually selected by the second plenary meeting as it best reflected the scope of newSMG9[47]. This meeting, now referred to by EP SCP #2, also elected the author, who had convened the first meeting as interim chairman, to be its chairman.

Corrections and enhancements to the platform specification, now referred to as TS 102 221 *Smart cards; UICC-terminal interface; physical and logical characteristics* reflecting in its title its origin as a T3 document, constituted a major part of the work during the first meetings. These changes were often complex in nature to safeguard the multi-application approach. True multi-application clearly requires applications to run in parallel. A user would not accept being cut off in the middle of a call or loosing the network connection with subsequent log-on and authentication procedures, just to pay at the check-out in a supermarket. To allow several applications residing on a UICC to be open and executable at the same time, logical channels were introduced in early 2001 as the last outstanding Release 4 feature. Four applications can be open at the same time using securely and independently the interface of the UICC to the terminal. This makes paying in a supermarket while making a phone call a reality, subject, of course, to the provision of the infrastructure and application providers such as network operators and banking institutions agreeing on issuing a "joint" smart card.

The core part of the smart card platform has been established. Also completed is, as already mentioned, the document on application identifiers derived by merging the respective documents of ETSI and 3GPP. Of great importance not only to the telecommunications world, is the recently completed specification of a *generic* Card Application Toolkit (CAT). Though based on the (U)SIM Application Toolkit documents it is independent of the specific applications for which those were written. Work on generic API issues and secure messaging is in full progress. Work to be started includes the specification of a PKI platform and general security aspects, further enhancements to the administrative specification, improvements to the UICC-terminal interface as well as application management. To advance all these topics EP SCP has created three Working Parties dealing with Architecture, Security and Toolkit/ API.

The interested reader is invited to monitor the work of EP SCP and of TSG-T3 (USIM) by logging onto the ETSI server.

---

[46] newSMG9 170/00, SMG 239/00: Future Organisation of Smart Card Work (4 April 2000).
[47] EP SCP#3 9-00-0402: Terms of Reference of EP SCP (ETSI Project Smart Card Platform).

13.11 Outlook

In January 1998, SIMEG/SMG9 commemorated its 10th anniversary at its 50th plenary meeting. The motto of the event summarised the efforts and the result of the first 10 years:

Billions of Calls
Millions of Subscribers
Thousands of Different Types of Telephones
Hundreds of Countries
Dozens of Manufacturers...

and only one Card: The SIM

Since then, the USIM, based on the new smart card platform, has been introduced, taking on the role of the SIM. The idea of the removable module, which developed from a security device to a secure platform for value added services and now into a platform for true multi-application smart cards, facilitating a global terminal market and ensuring the security of the system, proved to be the right one in the light of the last decade. The increase in computing power, going to 16 bit and 32 bit processors, memory becoming "abundant" (compared with the early days), the ongoing developments in the two committees on topics such as the use of public keys, extended administration and remote management of the card, standardised interpreters for Internet access and use of additional bearers for communication show that we are in a process of further advancing the smart card towards a personal mini-computer.

Acknowledgements

The author would like to thank Nigel Barnes and Michael Sanders for reviewing drafts of this document.

Chapter 14: Voice Codecs

Kari Järvinen[1]

14.1 Overview

Five voice codecs have been standardised for GSM. These are:

- Full-Rate (FR) codec
- Half-Rate (HR) codec
- Enhanced Full-Rate (EFR) codec
- Adaptive Multi-Rate (AMR) codec
- Adaptive Multi-Rate Wideband (AMR-WB) codec

All voice codecs include speech coding (source coding), channel coding (error protection and bad frame detection), concealment of erroneous or lost frames (bad frame handling), Voice Activity Detection (VAD), and a low bit rate source controlled mode for coding background noise. The codecs operate either in the GSM full-rate traffic channel at the gross bit rate of 22.8 kbit/s (FR, EFR, AMR-WB), or in the half-rate channel at the gross bit rate of 11.4 kbit/s (HR), or in both (AMR). AMR and AMR-WB have also been specified for use in 3G WCDMA.

The FR codec [1] was the first voice codec defined for GSM. The codec was standardised in 1989. It uses 13.0 kbit/s for speech coding and 9.8 kbit/s for channel coding. FR is the default codec to provide speech service in GSM.

The HR codec [2] was developed to bring channel capacity savings through operation in the half-rate channel. The codec was standardised in 1995. It operates at 5.6 kbit/s speech coding bit rate with 5.8 kbit/s used for channel coding. The codec provides the same level of speech quality as the FR codec, except in background noise and in tandem (two encodings in MS-to-MS calls) where the performance is somewhat lower.

The EFR codec [3] was the first codec to provide digital cellular systems with voice quality equivalent to that of a wireline telephony reference (ITU G.726-32 ADPCM standard at 32 kbit/s). The EFR codec brings substantial quality improvement over the previous GSM codecs. EFR was standardised first for the GSM based PCS 1900 system in the US during 1995 and was adopted to GSM in 1996. The EFR codec uses 12.2 kbit/s for speech coding and 10.6 kbit/s for channel coding.

A further development in GSM voice quality was the standardisation of the AMR codec [4] in 1999. AMR offers substantial improvement over EFR in error robustness in the full-rate

---

[1] The views expressed in this chapter are those of the author and do not necessarily reflect the views of his affiliation entity.

channel by adapting speech and channel coding depending on channel conditions. Channel capacity is gained by switching to operate in the half-rate channel during good channel conditions. The AMR codec includes several modes for use both in the full- and half-rate channel. The speech coding bit rates are between 4.75 and 12.2 kbit/s in the full-rate channel (eight modes) and between 4.75 and 7.95 kbit/s in the half-rate channel (six modes). The AMR codec was adopted in 1999 by 3GPP as the default speech codec to the 3G WCDMA system.

The AMR-WB codec [5] is the most recent voice codec. It was standardised in 2001 for both GSM and 3G WCDMA systems. Later in 2001, rapporteur's meeting of ITU-T Q.7/16 choose the AMR-WB speech codec for the new ITU-T wideband coding algorithm of speech at around 16 kbit/s. AMR-WB is an adaptive multi-rate codec like the AMR (narrowband) codec. AMR-WB brings quality improvement through the use of extended audio bandwidth. While all previous codecs in digital cellular systems operate on narrow audio bandwidth limited below 3.4 kHz, AMR-WB extends the bandwidth to 7 kHz. Wideband coding brings improved voice quality especially in terms of increased voice naturalness. AMR-WB consists of nine modes operating at speech coding bit rates between 6.6 and 23.85 kbit/s.

A voice codec related development in GSM and 3G WCDMA was the definition of in-band Tandem Free Operation (TFO). This feature was completed including TFO for AMR in March 2001 [23]. TFO brings improvement in speech quality for MS-to-MS calls by avoiding double transcoding in the network. TFO can be employed when the same speech codec is used at both ends of the call.

The speech coding part in all the voice codecs is based on the use of Linear Predictive Coding (LPC). All except the FR codec belong to the class of speech coding algorithms generally known as Code Excited Linear Prediction (CELP). All codecs operate at the sampling rate of 8 kHz except AMR-WB which uses 16 kHz sampling rate. Channel coding in all codecs is based on convolution coding for error correction combined with Cyclic Redundancy Check (CRC) for error detection. Three protection classes are typically used: bits protected by the convolutional code and CRC, bits protected by the convolutional code alone, and bits without any error protection.

The voice codec specifications define the speech codec bit-exactly to guarantee high basic voice quality. For bad frame handling, only an example solution is given to allow the possibility for implementation-specific performance improvements in error concealment.

Tables 14.1–14.3 give a summary of the GSM voice codecs: standards, implementation complexity, and algorithmic delay.

14.2 Codec Selection Process

The development of GSM voice codecs has been carried out in ETSI SMG11 and in its predecessors. Finalisation of channel coding has taken place under SMG2. The AMR-WB codec was developed jointly by SMG11 and 3GPP TSG-SA WG4.

All the voice codecs have been chosen through a competitive selection process among several candidate codec algorithms.

Before the codec selection process starts, speech quality performance requirements and codec design constraints (e.g. implementation complexity and transmission delay) have to be defined. For the most recent codecs (AMR and AMR-WB), the launch of standardisation has been preceded by a feasibility study phase to validate the new codec concept.

Table 14.1 Voice codec standards

Codec	Year of standard	Speech coding bit-rate (in kbit/s)	System/traffic channel	Speech coding algorithm
FR codec	1989	13.0	GSM FR	Regular Pulse Excitation – Long Term Prediction (RPE-LTP)
HR codec	1995	5.6	GSM HR	Vector-Sum Excited Linear Prediction (VSELP)
EFR codec	1996	12.2	GSM FR	Algebraic Code Excited Linear Prediction (ACELP)
AMR codec	1999	12.2, 10.2, 7.95, 7.4, 6.7, 5.9, 5.15, 4.75	GSM FR (all eight modes), GSM HR (six lowest modes), 3G WCDMA (all modes)	Algebraic Code Excited Linear Prediction (ACELP)
AMR-WB codec	2001	23.85, 23.05, 19.85, 18.25, 15.85, 14.25, 12.65, 8.85, 6.60	GSM FR (seven lowest modes), EDGE (all modes), 3G WCDMA (all modes)	Algebraic Code Excited Linear Prediction (ACELP)

Table 14.2 Implementation complexity of voice codecs

Codec	Speech codec complexity				GSM channel codec complexity			
	WMOPS	Data RAM (16-bit kwords)	Data ROM (16-bit kwords)	Program ROM (1000 assembly instructions)	WMOPS	Data RAM (16-bit kwords)	Data ROM (16-bit kwords)	Program ROM (1000 assembly instructions)
FR codec	3.0	1.2	0.1	0.9	1.7	1.7	0.8	0.3
HR codec	18.5	4.4	7.9	4.1	2.7	3.2	0.9	1.3
EFR codec	15.2	4.7	5.3	1.8	See complexity of the FR channel codec			
AMR codec[a]	16.8	5.3	14.6	4.9	5.2 (FR), 2.9 (HR)	2.6 (FR), 2.4 (HR)	5.2	1.3
AMR-WB codec[a]	35.4	6.4	9.9	3.8	3.5	2.9	3.2	0.6

[a] Complexity of channel quality measurement and mode control is counted as part of channel coding.

Table 14.3   Algorithmic transmission delay components of the speech codecs

Codec	Frame length (ms)	Lookahead in LPC analysis (ms)
FR codec	20	0
HR codec	20	4.4
EFR codec	20	0
AMR codec	20	5
AMR-WB codec	20	5

The selection process typically consists of two phases: a qualification (pre-selection) phase and a selection phase. During the qualification phase, the most promising candidate codecs are chosen to enter the selection phase. The qualification is usually based on in-house listening tests. In the selection phase, the codec proposals are tested more comprehensively in several independent test laboratories and using multiple languages. The codec proposals are implemented in C-code with fixed-point arithmetics. For both phases, the codec proponents need to deliver documentation of their proposal including a justification of meeting all design constraints. The codec selection is based both on the speech quality of the candidate codecs and on fulfilling other design requirements.

After codec selection, a verification phase and a characterisation phase will take place. An optimisation phase may be launched to improve some key performances of the codec if there is sufficient promise of improvement. During the verification phase, the codec is subjected to further analysis to verify its suitability for the intended systems and applications. A detailed analysis of implementation complexity and transmission delay is also carried out during this phase. The final phase of codec standardisation is the characterisation phase. This is launched after the approval of the codec standard to characterise the codec in a large variety of operational conditions. The output is a technical report on performance characterisation which provides information on codec performance.

14.3 FR Codec

In the FR voice codec, the speech coding part is based on the Regular Pulse Excitation – Long Term Prediction (RPE-LTP) algorithm [6]. The frame length is 20 ms, i.e. a set of codec parameters are produced every 20 ms. The speech codec operates at 13.0 kbit/s while 9.8 kbit/s is used for channel coding. FR is the default codec to provide speech service in GSM.

The FR speech codec carries out short-term LPC analysis once every frame (without any lookahead over future samples). The rest of the coding is performed in 5 ms sub-frames. The short-term residual signal, after LPC analysis, is further compressed by using Long-Term Prediction (LTP) analysis. LTP removes any long-term correlation remaining in the short-term residual signal. The long-term residual is then decimated into a sparse signal in which only every third sample has a non-zero value. The non-zero samples are located on a regular grid. The grid starting position is determined separately for each sub-frame based on the energy of the sub-frame. This Regular Pulse Excitation (RPE) approach results in rather efficient coding. Only the non-zero samples in the long-term residual need to be quantised

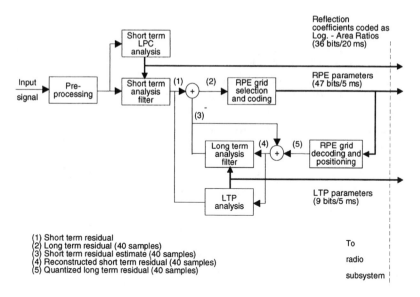

Figure 14.1   Block diagram of the GSM FR speech encoder

and sent to the decoder. The parameters for each 20 ms frame consist of a set of LPC-coefficients (reflection coefficients) and a set of parameters describing the short-term residual for each sub-frame (LTP parameters, RPE parameters). A block diagram of the encoder is shown in Figure 14.1.

The FR channel codec uses convolution coding for protecting the 182 most important bits out of the 260 bits in each frame [11]. A 3 bit CRC is employed for bad frame detection. The CRC covers the most important 50 bits.

The FR codec, like all GSM and 3G WCDMA codecs, includes a low bit rate source controlled mode for coding background noise only (voice activity detection with discontinuous transmission). This saves power in the mobile station and also reduces the overall interference level over the air-interface.

The complexity of the FR speech codec is about 3.0 WMOPS (weighted million operations per second). The complexity has been estimated from a C-code implemented with a fixed point function library in which each operation has been assigned a weight representative for performing the operation on a typical DSP. The channel coding requires about 1.7 WMOPS [13].

14.4 HR Codec

The HR codec employs the Vector-Sum Excited Linear Prediction (VSELP) speech coding algorithm [7]. VSELP belongs to the class of CELP codecs. The codec uses 20 ms frame length. The speech codec operates at the bit rate of 5.6 kbit/s while 5.8 kbit/s is used for channel coding.

Like most CELP codecs the HR VSELP employs two codebooks: a fixed codebook and an adaptive codebook. The adaptive codebook is derived from the long-term filter state (and therefore the content of the codebook changes frame-by-frame). The adaptive codebook is used to generate a periodic component in the excitation, while the fixed codebook generates a random-like component. The excitation sequence is coded by choosing the best match from each of the two codebooks. The excitation that produces the least decoding error is chosen (analysis-by-synthesis). The codebook indices and gains are computed once for every 5 ms subframe. A lookahead of 35 samples is employed in the LPC analysis.

A specific feature in VSELP is the structure of the fixed codebook. The fixed codebook is constructed as a linear combination (vector sum) of only a small amount of basis vectors. There are four modes based on how voiced each 20 ms speech frame is. For the least voiced mode, the adaptive codebook is not used at all, and a second fixed VSELP codebook is used instead.

The HR channel codec employs convolution coding protecting 95 out of the 112 bits in each frame [11]. A 3 bit CRC covers the most important 22 bits.

The HR codec provides the same level of speech quality as the FR codec, except in background noise and during tandem (two encodings in MS-to-MS calls) where the performance lacks somewhat behind the FR codec.

The computational complexity of the speech codec is about 18.5 WMOPS. The complexity of the channel codec is about 2.7 WMOPS [13,14].

14.5 EFR Codec

The EFR codec gives substantial quality improvement compared to the previous GSM codecs. EFR is the first codec to provide digital cellular systems with quality equivalent to that of a wireline telephony reference (ITU G.726-32 ADPCM standard at 32 kbit/s).

The EFR codec standardisation started in ETSI in 1995. Wireline quality was set as a development target because GSM had become increasingly used in communication environments where it started to compete directly with fixed line or cordless systems. To be competitive also with respect to speech quality, GSM needed to provide wireline speech quality which is robust to typical usage conditions such as background noise and transmission errors.

A similar development of enhanced quality full-rate codec was carried out for the GSM based PCS 1900 system in the US during 1995. An EFR codec was standardised for the PCS 1900 system already in 1995. The PCS 1900 EFR codec was one candidate considered for the GSM EFR standard and it was adopted to GSM through a competitive selection process. In addition to voice quality performance, the advantage of using the same voice codec in PCS 1900 and in GSM was one factor in favour of this particular solution. The GSM EFR codec standard was approved in January 1996 (at SMG#17).

The EFR speech codec is based on the Algebraic Code Excited Linear Prediction (ACELP) algorithm [8,18]. The speech coding bit rate is 12.2 kbit/s whereas 10.6 kbit/s is used for channel coding. The codec operates on 20 ms frames which are divided into four 5 ms subframes. Two sets of LPC parameters are calculated for each 20 ms frame with no lookahead over samples in the next frame. EFR employs an adaptive and a fixed codebook. The codebook parameters are computed once for each 5 ms sub-frame. The name ACELP refers to the type of fixed codebook where algebraic code is used to populate the excitation vectors. The ACELP codebook contains a small number of non-zero pulses with predefined interlaced sets of positions. In EFR, the 40 positions in each 5 ms sub-frame are divided into five tracks where each track contains two pulses. Each excitation vector contains ten non-zero pulses with amplitudes of -1 or $+1$. Figure 14.2 gives a block diagram of the EFR speech encoder.

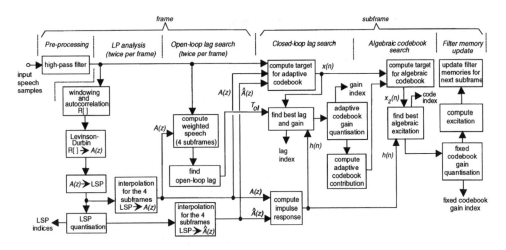

Figure 14.2   Block diagram of the GSM EFR speech encoder

The EFR channel codec is almost the same as the FR channel codec because a key design aim was to keep it as similar as possible. During the GSM EFR codec standardisation, the use of the existing FR channel codec (or existing GSM generator polynomials) was encouraged since this minimises hardware changes in the GSM base stations and thus potentially speeds up the introduction of the EFR codec. In the PCS 1900 EFR codec standardisation, the use of the existing FR channel codec was a mandatory requirement. Therefore, the FR channel codec was included in the EFR channel codec as a module together with additional error protection [11]. The additional 0.8 kbit/s error protection consists of an 8 bit CRC to provide improved detection of frame errors and a repetition code for improved error correction.

The implementation complexity of EFR is lower than that of the HR codec. Computational complexity of the speech codec is about 15.2 WMOPS. The complexity of the channel codec is about the same as for the FR codec [13,16].

Figure 14.3a,b shows the performance of the EFR codec compared to the FR codec and to a wireline quality reference G.726-32 (32 kbit/s ADPCM) [15,16]. Figure 14.3a shows the performance for clean speech under transmission errors. The dotted line shows the performance of (error-free) G.726-32. The EFR codec gives substantial improvement over the FR codec in the error-free channel and in error conditions down to a carrier-to-interference ratio (C/I) of 7 dB. EFR provides wireline quality still at approximately 10 dB C/I. Figure 14.3b shows the performance in background noise for the error-free channel with four types of background noise (home 20 dB SNR, car 15 and 25 dB SNR, street 10 dB SNR, and office 20 dB SNR). The results demonstrate substantial improvement over FR. In many test cases EFR exceeds the performance of the wireline quality reference of 32 kbit/s ADPCM.

14.6 AMR Codec

The EFR codec was the first codec to provide digital cellular systems with quality equivalent to that of a wireline telephony reference. However, it still left some room for improvements. In particular, the performance in severe channel error conditions could be improved by

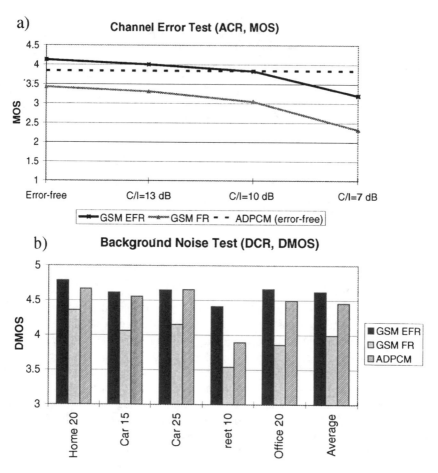

Figure 14.3 GSM EFR performance: (a) in transmission errors for clean speech, and (b) under background noise

employing a different bit allocation between speech and channel coding. This led into development of a new type of codec that is able to adapt to the channel quality conditions. The concept of adaptive multi-rate coding was born.

The standardisation of AMR was launched in October 1997 (at SMG#23). Already before that a one-year feasibility study had been carried out to validate the novel AMR concept. The selection process consisted of a qualification and a selection phase. The qualification phase was carried out during spring 1998. Altogether 11 candidate codecs were submitted. In June 1998 (at SMG#26), the five most promising candidates were chosen to enter the selection tests. The selection phase took place from July 1998 until the selection of the codec in October 1998 (at SMG#27). After the selection, a short optimisation phase took place. The optimisation was focused on making improvements for the channel coding part and bringing corrections to the codec C-code. During the optimisation, the complexity of channel coding was reduced while at the same time obtaining some performance improvements. The AMR codec standard was formally approved in February 1999 at SMG#28 (speech coding part) and in June 1999 at SMG#29 (channel coding part). A detailed description of the AMR codec standardisation process can be found in [19].

The main principle behind AMR is to adapt to radio channel and traffic load conditions and select the optimum channel mode (full-rate or half-rate) and codec mode (bit rate trade-off between speech and channel coding) to deliver the best combination of speech quality and system capacity. AMR provides good overall performance and high granularity of bit rates making it suitable also for systems and applications other than GSM. In 1999, the AMR codec was adopted by 3GPP as the default speech codec to provide speech service in the 3G WCDMA system.

The AMR codec contains a set of fixed rate speech and channel codecs, link adaptation, and in-band signalling. Each AMR codec mode provides a different level of error protection through a different distribution of the available gross bit rate between speech and channel coding. The link adaptation process bears responsibility for measuring the channel quality and selecting the optimal speech and channel codecs. In-band signalling transmits the measured channel quality and codec mode information over the air-interface. The in-band signalling is transmitted along with the speech data.

The AMR speech codec utilises the ACELP algorithm employed also in the EFR codec. The frame length is 20 ms which is divided into four 5 ms sub-frames. A 5 ms lookahead is used. The codec contains eight codec modes with speech coding bit rates of 12.2, 10.2, 7.95, 7.4, 6.7, 5.9, 5.15 and 4.75 kbit/s [9,20]. As seen in Figure 14.4a, all the speech codecs are employed in the full-rate channel, while the six lowest ones are used in the half-rate channel. All the modes provide seamless switching between each other. The GSM EFR, D-AMPS EFR, and PDC EFR speech codecs are included in the AMR as the 12.2, 7.4 and 6.7 kbit/s modes, respectively. Some minor harmonisation to other modes (e.g. in the post-processing) has been carried out for these codecs when used within AMR. The AMR 12.2 kbit/s speech codec was later defined as an alternative implementation of the EFR speech codec.

Error protection in GSM is based on Recursive Systematic Convolutional (RSC) coding with puncturing to obtain the required bit rates [11]. Each codec mode also employs a 6 bit CRC for detecting bad frames. All channel codecs use convolution polynomials previously specified for GSM (either for speech or data traffic channels) to maximise commonality with the existing GSM system. For 3G channels, the general channel coding toolbox of the 3G WCDMA system is used.

Figure 14.5 shows a basic block diagram of the AMR codec in GSM. The Mobile Station (MS) and the Base Transceiver Station (BTS) both perform channel quality estimation for the receive signal path. Based on the channel quality measurements, a codec mode command (over downlink to the MS) or a codec mode request (over uplink to network) is sent in-band over the air-interface. The receiving end uses this information to choose the best codec mode for the prevailing channel conditions. A codec mode indicator is also sent over the air-interface to indicate the current mode of operation. The codec mode in the uplink may be different from the one used in the downlink on the same air-interface, but the channel mode (FR or HR) must be the same.

The network controls the uplink and downlink codec modes and channel modes. The mobile station must obey the codec mode command from the network, while the network may use any complementing information, in addition to codec mode request, to determine the downlink codec mode. The mobile station must implement all the codec modes. However, the network can support any combination of them, based on the choice of the operator. In GSM,

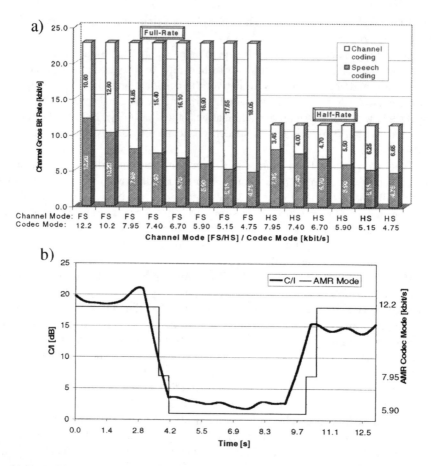

Figure 14.4 (a) Bit rate trade-off between speech and channel coding in AMR. (b) AMR codec mode adaptation in GSM full-rate channel under dynamic error conditions

the in-band signalling supports adaptation between up to four active codec modes. The set of active codec modes is selected at call set-up (and in handover). Codec mode command/ request and codec mode indication are transmitted in every other speech frame in GSM (alternating within consecutive frames). Therefore, the codec mode can be changed every 40 ms. In 3G WCDMA, AMR can adapt between all the eight modes and can switch modes every 20 ms. To obtain interoperability with GSM AMR under TFO, the 3G AMR adaptation rate can be limited to 40 ms in uplink.

Link adaptation is an essential part of the AMR codec. It consists of channel quality measurement and codec/channel mode adaptation algorithms [12,21]. Link adaptation in AMR is two-fold: it adapts the bit-partitioning between speech and channel coding within a transmission channel (codec mode), and the operation in the GSM full- and half-rate channels (channel mode). Depending on the channel quality and possible network constraints (e.g. network load), link adaptation selects the optimal codec and channel mode. Figure 14.4b shows an example of how the codec mode adaptation operates in the GSM full-rate channel under dynamic error conditions. Channel quality varies between about 22 and 2 dB C/I. Based

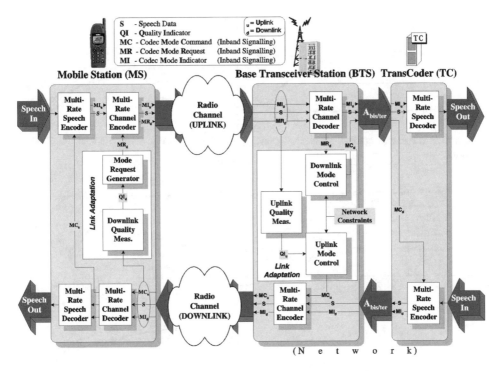

Figure 14.5 Block diagram of the AMR codec in GSM

on estimated channel quality, one out of three codec modes (12.2, 7.95 or 5.9 kbit/s) is chosen.

The computational complexity of the AMR speech codec is about 16.8 WMOPS which is only about 10% higher than for EFR. The complexity of the AMR channel codec is 5.2 WMOPS in the GSM full-rate channel and 2.9 WMOPS in the GSM half-rate channel [13,17].

In the GSM full-rate channel, the AMR codec extends the wireline quality operating region from about C/I ≥ 10 dB in EFR to about C/I ≥ 4–7 dB. In poor channel conditions, AMR gives substantially improved robustness over EFR. At 4 dB C/I, AMR achieves about 2 units of Mean Opinion Score (MOS) improvement over EFR. In typical dynamic error conditions in the full-rate channel, AMR provides up to over 1 MOS improvement compared to EFR. In the half-rate channel at low error conditions (C/I ≥ 16 dB), AMR provides close to EFR quality (equivalent to G.728 for clean speech, and to G.729 and error-free GSM FR for speech under background noise). Figure 14.6 shows examples of the performance in GSM full-rate channel taken from GSM AMR characterisation tests [17]. Figure 14.6a,b shows the performance of the best codec mode for each C/I condition for clean speech in the full-rate and half-rate channels, respectively. Figure 14.6c shows the performance curves for each AMR FR codec mode (corresponding to Figure 14.6a). Figure 14.6d,e shows the performance of the best codec mode for each C/I condition under background noise. (Note that the MOS values in each test depend on the test setting and conditions and are not directly comparable between tests.)

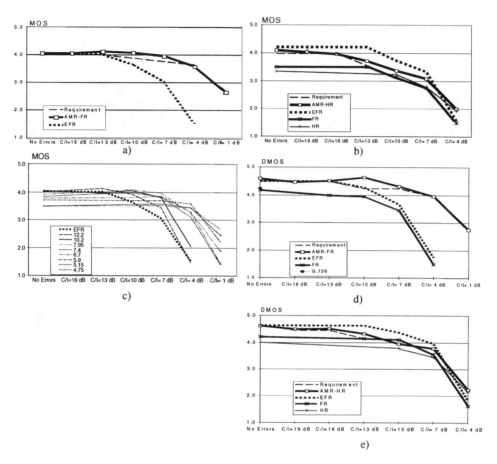

Figure 14.6 (a) AMR FR performance for clean speech, (b) AMR HR performance for clean speech, (c) AMR FR performance for clean speech, (performance curves of each codec mode), (d) AMR FR performance for speech under 15 dB SNR car noise, and (e)AMR HR performance for speech under 20 dB SNR office noise

14.7 AMR-WB Codec

The possibility to develop a wideband speech codec for GSM (with audio bandwidth up to 7 kHz instead of 3.4 kHz) was addressed already during the feasibility study of the AMR codec. When the AMR codec standardisation was launched in October 1997 (at SMG#23), the work was focused on developing narrowband AMR codec. Wideband coding was set as a possible longer-term target. ETSI SMG11 carried out a feasibility study on wideband coding by June 1999. The results showed that the target is feasible. The study considered development of AMR-WB not only for GSM full-rate channel, but also for GSM EDGE channels, and for 3G WCDMA. Based on the results, ETSI SMG launched in June 1999 (at SMG#29) a standardisation of AMR wideband codec. 3GPP TSG-SA had approved a work item on wideband coding already a couple of months earlier in March 1999, but the effective start of the work was pending on the results of the feasibility phase in SMG11. The AMR-WB codec development work was then carried out jointly by SMG11 and TSG-SA WG4 under a joint work item.

The AMR-WB codec selection process consisted of a qualification and a selection phase. The qualification phase was carried out during spring 2000 and the selection phase from July to October 2000. From altogether nine codec proposals, seven codecs participated in the qualification phase. Five of the codecs were accepted to enter the selection phase. The codec selection was carried out in December 2000 (at TSG-SA#10) and the speech codec specifications were approved in March 2001 (at TSG-SA#11).

In July 2001, at the rapporteur's meeting of ITU-T Q.7/16, the AMR-WB speech codec was chosen through a competitive selection process as the new ITU-T wideband coding algorithm of speech at around 16 kbit/s. This results in one harmonised low bit rate wideband codec to be used for many systems and applications.

The AMR-WB codec brings major quality improvement over narrowband telephony through the use of extended audio bandwidth. The introduction of a wideband speech service (audio bandwidth extended to 7 kHz) offers significantly improved voice quality especially in terms of increased voice naturalness. With AMR-WB, the GSM and 3G WCDMA systems will provide speech quality exceeding that of (narrowband) wireline quality. The AMR-WB codec is intended for operation in multiple applications: in the GSM FR traffic channel, in the GSM EDGE 8-phase shift keying (8-PSK) circuit switched channels, in the 3G WCDMA channel, and also in packet based VoIP (voice over IP) applications.

The AMR-WB codec, like the AMR (narrowband) codec, is an adaptive multi-rate codec consisting of several codec modes. Link adaptation chooses the optimal mode based on the operating conditions of the radio channel. This provides similar high robustness against transmission errors as in AMR. Speech coding is based on the ACELP algorithm like in the EFR and AMR codecs. The codec uses a 20 ms frame length, divided into 5 ms subframes for coding of codebook parameters. A 5 ms lookahead is used. Coding is carried out separately for two frequency bands (50–6400 Hz and 6400–7000 Hz). The codec operates at the speech coding bit rates of 23.85, 23.05, 19.85, 18.25, 15.85, 14.25, 12.65, 8.85 and 6.6 kbit/s [10].

Error protection in GSM FR and EDGE channels is based on recursive systematic convolutional coding. Channel coding uses polynomials specified previously for GSM to maximise commonality with the existing GSM system. The codec modes use a 3, 6, or 8 bit CRC for bad frame detection. For 3G channels, the general 3G WCDMA channel coding toolbox is used.

In-band signalling in AMR-WB is exactly the same as in AMR. Link adaptation is similar to AMR (with modifications only in the adaptation thresholds). Codec mode can be changed every 20 ms in 3G WCDMA. However, for uplink the adaptation speed is limited to 40 ms to obtain TFO interoperability with GSM. In GSM, the codec mode can be changed every 40 ms.

The computational complexity of the AMR-WB speech codec is about 35.4 WMOPS [22]. The complexity of the AMR-WB channel codec is 3.5 WMOPS in the GSM full-rate channel.

In the applications that enable the highest bit rate speech codecs of AMR-WB to be used, i.e. 3G WCDMA and EDGE, the AMR-WB codec provides error-free quality at least equal to ITU-T 64 kbit/s wideband codec G.722-64k. In the GSM full-rate channel, AMR-WB gives error-free quality at least equal to the 56 kbit/s wideband codec G.722-56k. When restricted to codec modes with bit rates capable of 16 kbit/s submultiplexing over the A-bis/A-ter interface (14.25 kbit/s and below), quality in the GSM full-rate channel still exceeds that of the

48 kbit/s wideband codec G.722-48k. The multi-rate operation provides AMR-WB with high error robustness in all the applications.

The introduction of AMR-WB brings a fundamental improvement of speech quality, raising it to a level never experienced in mobile communication systems before [22].

Acknowledgements

The GSM and 3G WCDMA voice codecs have been developed within ETSI SMG11 (including its predecessors) and 3GPP TSG-SA WG4. The codecs are a result of the efforts of many companies and persons contributing to their development and standardisation.

References

[1] TS 06.01, Full-rate speech processing functions; general description.

[2] TS 06.02, Half-rate speech processing functions; general description.

[3] TS 06.51, Enhanced full-rate speech processing functions; general description.

[4] TS 26.071, adaptive multi-rate speech codec; general description.

[5] TS 26.171, AMR wideband speech codec; general description.

[6] TS 06.10, Full-rate speech transcoding.

[7] TS 06.20, Half-rate speech transcoding.

[8] TS 06.60, Enhanced full-rate speech transcoding.

[9] TS 26.090, AMR speech codec; transcoding functions.

[10] TS 26.190, AMR wideband speech codec; transcoding functions.

[11] TS 05.03, Channel coding.

[12] TS 05.09, Adaptive multi-rate link adaptation.

[13] Reproduction of AMR narrowband document AMR-9: Complexity and delay assessment, v1.3, Tdoc S4/SMG11 71/00 (S4-000071).

[14] TR 06.08, Performance characterization of the Half-Rate (HR) speech codec.

[15] COMSAT test report on the PCS1900 EFR Codec, ETSI SMG2-SEG TDoc 47/95, June 1995.

[16] TR 06.55, Performance characterization of the Enhanced Full-Rate (EFR) speech codec.

[17] TR 26.975, Performance characterization of the Adaptive Multi-Rate (AMR) speech codec.

[18] Järvinen K. et. al. GSM enhanced full rate codec. Proc. IEEE International Conference on Acoustics, Speech and Signal Processing, Munich, Germany, 20–24 April 1997.

[19] Järvinen K. Standardisation of the adaptive multi-rate codec. Proc. X European Signal Processing Conference (EUSIPCO 2000), Tampere, Finland, September 4–8, 2000.

[20] Ekudden E. et. al. The AMR speech codec. Proc. IEEE Workshop on Speech Coding, Porvoo, Finland, June 1999.

[21] Bruhn S. et. al. Concepts and solutions for link adaptation and in-band signaling for the GSM AMR speech coding standard. IEEE Vehicular Technology Conference, 1999.

[22] Results of AMR Wideband (AMR-WB) codec selection phase, 3GPP TSG-SA TDoc SP-000555, Bangkok, Thailand, December 2000.

[23] TS 28.062, In-band Tandem Free Operation (TFO) of speech codecs.

Chapter 15: Security

Michael Walker and Tim Wright[1]

15.1 Introduction

Security was perceived by some in the first days of GSM as an unnecessary expense. Certainly, initially, all involved considered protection of user data from eavesdropping as more important than authentication of the user, though some questioned whether the perceived complexity of introducing encryption over the radio interface was justified. However, as fraud losses from cloning of analogue phones rocketed in the US and the UK, as the Dutch PTT withdrew all its NMT phones so a form of authentication could be added, and as the Germans introduced simple authentication on its C-Netz system, it became apparent that authentication of user identity was also very important.

In recent times, the security of GSM has been attacked as too weak. These criticisms are often made without knowledge of the design goals of GSM security nor the regulatory context in which the designers had to work. This section aims to show that GSM security met its design goals in a simple and elegant way, and has provided more than adequate security for most of its users. Indeed, GSM offers more "access network" security than fixed phones in most countries (taking the phone to the local exchange link as the "access network for fixed line systems). GSM has never been subject to the commercial cloning that was visited upon analogue NMT, AMPS and TACS systems. Moreover, GSM represented the first time ever that encryption functionality had been provided in a consumer device, and played its part in the liberalisation of policy on encryption that today's security designers enjoy.

15.2 Origins of GSM Security

The security of GSM was developed by the Security Experts Group (SEG) which was formed by CEPT in 1984. There was a lot of concern in CEPT regarding protection of communications systems in general at that time. The origin of the SEG could be said to be a joint meeting of the three CEPT groups CD (data), CS (signalling) and SF (services and facilities) in Berne in January 1984, in which land mobile systems were discussed for the first time. In November, 1984, a proposal from CD to set up a joint CD-GSM group on security (SEG) was accepted by GSM and the first meeting of SEG was held in Malmoe, Sweden, in May 1985. This was a memorable meeting for the delegates as the Swedish air traffic controllers were on strike at that time, forcing the delegates to fly to Copenhagen and travel by boat to

---

[1] The views expressed in this chapter are those of the authors and do not necessarily reflect the views of their affiliation entity.

Malmoe. The SEG was initially a joint CD/GSM activity, but gradually, the CD part vanished so it was in fact a subgroup of GSM.

The SEG was chaired by Thomas Haug of Swedish Telecommunications Administration, now called Telia. The membership, like that of CEPT and GSM was drawn from national PTTs and from those organisations that had won a mobile network licence in their country.

15.3 Design Goals

The security functionality within any system is a balance between the likelihood and impact of threats, user demand for certain security features and the cost and complexity of security measures. A security mechanism that is impervious to attack by any organisation over any timescale would generally not be appropriate for a system transporting public, largely non-sensitive data. System designers must therefore set appropriate goals for the security of their system prior to beginning detailed design. SEG undertook this task and came up with the following simple goal for GSM security:

> It would provide a degree of protection on the radio path which was approximately the same as that provided in the fixed network.

SEG were concerned with security on the radio interface only – there was no attempt to provide security on the fixed network part of GSM.

Before describing how this simple goal was translated into more formal security requirements, a few definitions are given:

Confidentiality is the property data has when it cannot be read by parties not authorised to read it. Confidentiality is provided by encryption in GSM.

Authentication of user identity is the process of establishing that the claimed identity of an entity really is their identity.

Integrity protection is the property of data whereby modification to the data can be detected. This is not explicitly provided by GSM but is provided implicitly by the use of ciphering along with the use of non-linear checksums (as stream ciphers are used in GSM, stream ciphering alone does not provide integrity protection).

The following requirements for GSM security were developed over the course of the design exercise. These are listed in [1]:

- Subscriber identity authentication. This protects the network from unauthorised use.
- Subscriber identity confidentiality. This provides protection against the tracing of a user's location by listening to exchanges on the radio interface.
- User data confidentiality across the radio interface. This protects the user's connection orientated data from eavesdropping on the radio interface.
- Connectionless user data confidentiality across the radio interface. This protects user information sent in connectionless packet mode in a signalling channel from eavesdropping on the radio interface.
- Signalling information element confidentiality across the radio interface. This protects selected fields in signalling messages from eavesdropping on the radio interface.

There was not general agreement on the issue of identity confidentiality within the group. Some members felt it was very important, particularly the German delegates. Others felt it was not a real requirement, and since the subscriber must in some circumstances reveal their identity anyway, that the requirement could not be robustly met in any case.

There was also some debate during the design of GSM security as to whether user data should be given "privacy" or " confidentiality". "Privacy" was taken to mean protection from a determined "amateur" attacker but not necessarily a large organisation – "confidentiality" was taken to mean protection from attack by the latter. The final conclusion was to try and provide confidentiality.

It should be noted that right from the start, there was concern within the group of providing too much security and thereby bringing unnecessary export problems upon GSM. The security was therefore designed with this constraint in mind, and also two further constraints:

- GSM did not have to be resistant to "active attacks" where the attacker interferes with the operation of the system, perhaps masquerading as a system entity. Active attacks are in contrast to passive attacks where the attacker merely monitors inter-system communications and does not interfere).
- The trust that must exist between operators for the operation of the security should be minimised.

15.4 Choosing the Security Architecture for GSM

Many people see security for communications systems as a matter of algorithms and attacks on the security of communications systems as a matter of attacks on algorithms. However, in designing security for a system, the choice of algorithms is often one of the last choices. The first task is to decide the goals of the security, which for GSM we have already talked about. The second choice is to determine the security protocols that will be used to achieve these goals. Usually, only after this point can the algorithms be decided. However, if the choice of algorithm is going to involve the choice between the use of secret or public key cryptography, then this basic choice must be made sooner, as it may dictate the whole security architecture. The choice of public or secret key cryptography should still not be taken until the security goals have been decided though, in all the interesting debates about algorithm security, the designers may lose sight of the goals and why they are actually engaged in a design process at all. Having said all this, there were contributions at SEG meetings which proposed particular architectures without any rationale or reference to claimed goals.

A security protocol is an interaction between two or more (but usually only two) entities following pre-determined steps that achieve some security goal. These protocols may involve proving that certain parties have certain items of secret information (this occurs during authentication or proof of claimed identity) and also may involve distribution or generation of secret keys for protection of communication. For instance, the widely known protocol, SSL achieves authentication of the server (generally a web server), generation of a shared secret key to protect the communication of data between the client (usually a browser on a PC) and the server, and the subsequent protection of that data for the duration of the SSL session. The emphasis on server authentication (though client authentication is possible) in SSL was provided so that users would have confidence in who they were sending data, e.g. credit card numbers, to. Client authentication is not mandatory in SSL as the use of the channel is generally "free" or the user has already been authenticated for charging purposes prior to the start of the SSL session (and as authentication is public key based in SSL, there is the complexity of provisioning clients with key pairs and certificates). This emphasis on server authentication can be contrasted with the emphasis on client or user authentication in GSM, where the use of the channel is not free, and the user must therefore be authenticated so that they can be charged.

There are many ways of satisfying the goals for GSM security and the process of designing the GSM security architecture reflected the many possibilities open to the designers – many candidate architectures were proposed by the participating parties. BT, for instance, proposed the use of public key cryptography along with their own secret, symmetric encryption algorithm, BeCrypt. The reader might be interested to know why Public Key Cryptography (PKC) was not used. There were three main reasons:

- Implementations at the time were immature, the impact on the terminal for the provision of PKC functionality was therefore not accurately known;
- Messages would be longer, as PKC requires longer keys than symmetric cryptography, for provision of the same cryptographic strength;
- There was no real gain from the use of PKC. The authentication protocol runs between a subscriber and the network operator the subscriber has chosen to use. There is therefore a well established relationship and the one to many authentication possibility of PKC is not therefore required.

When SMG10 were designing security for third generation phones in 1999, they likewise decided that the use of PKC could not be justified and adopted a symmetric key approach.

The large number of proposals caused problems for the group's progress, in that there were just too many protocols to examine properly. A security protocol should not be accepted until it has been examined thoroughly by a good number of experts. Flaws in communications systems security often occur because of a weakness in the protocols involved, and not in the strength of algorithms. However, these flaws are often subtle and take time and careful analysis to uncover. A small group therefore decided to trim the number of proposals down to a manageable level. This small group was, as is often the case in such situations, not elected or formally tasked with trimming the number, it was just a small number of people taking an initiative.

15.5 The Architecture Chosen

The architecture finally chosen was a simple and elegant one. It is based on secret key and not public key cryptography as stated above.

The architecture is centred on a long-term secret key, K_i, which is possessed by both the subscriber's mobile phone and subscriber's operator only. Authentication of the mobile phone by the network consists of proof by the mobile phone that it possesses the K_i. As part of this process, cipher keys used for encryption during a call are also derived from K_i.

Before describing these operations in detail, a couple of design principles/constraints must be given. The first is that it was decided that the K_i must remain with the subscriber's home operator and must not be passed to another operator if the subscriber roams to that network. This is because the K_i is such a sensitive piece of information. With the K_i of a particular subscriber an impostor can pretend to be ("masquerade as") that subscriber and they can eavesdrop on all that subscriber's calls. K_i should therefore not be revealed to more entities than is strictly necessary and SEG found a way for it only to be known by the minimum number of entities, the mobile phone (actually the "SIM", see below) and the home network. The second constraint is that long distance signalling should be minimised. It was therefore not acceptable that the authentication process should involve the home operator for every call made by a roaming subscriber of that operator.

K_i should not even be known by the user themselves either, as this would allow the self-cloning of phones, and subsequent denial of the calls made by the cloned phones that had already occurred in analogue networks. A secure module within the phone, that could be programmed by or under the control of the operator, and in which K_i was stored and all operations involving K_i carried out, was therefore required. A smart card was the obvious choice for such a security module, and the GSM Subscriber Identity Module (SIM) was born (see Chapter 13 for details). The K_i is stored and used in the SIM and not in the terminal (or to use GSM terminology, the Mobile Equipment (ME)).

The security architecture is now described, in two stages.

K_i in the home operator is held in the operator Authentication Centre (AuC). (GSM Phase 1 did allow the K_i to be sent from the AuC to a VLR for use there, but this was only allowed for VLRs in the same PLMN as the AuC, the specification advised against the option, and the option was dropped in GSM Phase 2.) The AuC generates a random number, RAND for each subscriber. Random challenges are commonly used in security protocols to guarantee that a particular run of the protocol is "fresh" and entirely new and that an impostor who has captured some parameters from a previous run of the protocol cannot masquerade as the genuine subscriber or operator or interfere (either actively or passively) with the current run of the protocol. As shown in Figure 15.1, for a particular subscriber, each RAND is passed as a parameter, along with the K_i for that subscriber, through an algorithm named A3. A3 produces as an output, an expected response, XRES. The use of a challenge-response mechanism was not a proposal of a particular delegate in SEG. Once it was decided that a secret key mechanism would be used, a challenge-response mechanism was the obvious choice.

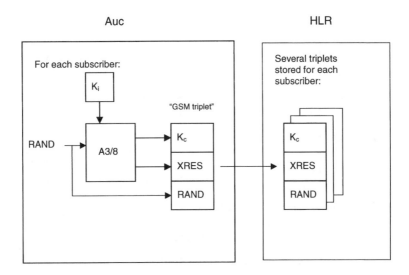

Figure 15.1   Diagram of triplet generation

RAND and K_i are also passed to another algorithm A8 which produces a cipher key, K_c. Typically, algorithms A3 and A8 are combined into one, called A3/8, and we shall consider them as such from now on. A RAND and the resulting XRES and K_c produced by A3/8 are called a "triplet". An AuC will normally produce a batch of triplets for a particular subscriber all at once and pass these for distribution to the HLR. This separation of triplet generation in the AuC from triplet distribution and subscriber management in the HLR, means that the AuC need only communicate with the HLR. In theory, therefore, greater access control can be placed on the AuC since it only ever communicates with one, known, entity, the associated HLR of the same operator.

When a subscriber attempts to make a call or a location update in either its home operator's network, or in a network it has roamed to, the SIM passes its identity to the VLR serving that subscriber. The VLR makes a request to the subscriber's HLR for a batch of triplets for the identity claimed by the subscriber (i.e. the SIM) and the HLR responds with a batch of triplets for that claimed identity. The VLR authenticates the SIM by sending a RAND from the batch to the mobile phone, as shown in Figure 15.2. The ME passes RAND to the SIM where K_i and A3/8 are held. The SIM passes RAND and its K_i through algorithm(s) A3/8 residing within the SIM as was done in the AuC. The "signed response" produced by the SIM, SRES, is passed back to the VLR. The VLR compares SRES with the expected response, XRES, for that RAND, and if they match, the SIM/mobile phone's claimed identity is deemed to be authenticated. A3/8 in the SIM also produces K_c and if the SIM is authenticated, the VLR passes the K_c from the triplet to the Base Transceiver Station (BTS, the "base station") serving the mobile. The SIM passes K_c to the ME and the BTS and mobile can then begin ciphering communications using K_c. The algorithm used for ciphering is termed A5.

With the use of the triplets, authentication can be performed in the serving network without the serving network operator having knowledge of K_i. When the serving network has run out of triplets, it should request more from the home operator (though the serving network is allowed to re-use triplets if it cannot obtain more).

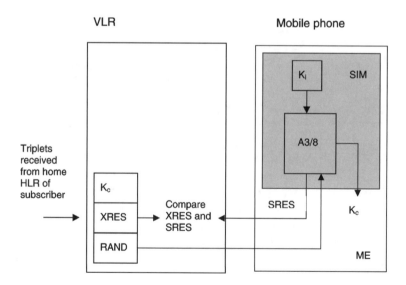

Figure 15.2   Diagram for authentication in the serving VLR

The inquisitive reader may be wondering when the use of algorithms A1-4, A6 and A7 is to be described. Their use will not be described though, as they, and also algorithms A9 to A12 came up in initial versions of the architecture, but were not required in the final version. A3, A5 and A8 were not renamed A1 to A3. A4 and K4 are used by some operators to denote the encryption algorithm and key protecting the personalisation data of a SIM (including the IMSI and K_i) between the personalisation centre and the AuC, but they are not specified in any GSM specification.

The security architecture described above was specified in GSM specification 03.20 [2].

15.6 Authentication Algorithm Design

The effectiveness of authentication relies on a number of algorithm requirements not yet given. The first is that it is statistically near impossible for an impostor to guess what the correct SRES should be and therefore masquerade as another subscriber. As parameters SRES/XRES are 32 bits long, and the mobile has only one chance to return SRES for a particular RAND, provided that the algorithm has been so designed that SRES is indistinguishable from any other 32 bit number that might be returned instead of SRES, such an impostor has only a 1 in 2^{32}, or 1 in approximately 10 billion chance of guessing SRES correctly. This was felt sufficiently improbable as to not represent a realistic attack.

The second assumption is that, as RAND and SRES are passed un-encrypted between the mobile and the base station, an impostor cannot derive K_i from collecting a number of RAND-SRES pairs. This means that A3/8 must be designed to resist a known plaintext attack where the attacker knows what is ciphered as well as the ciphered result. Further, as an attacker could steal a SIM for some time, and send whatever challenges he liked to the SIM, and collect the SRESs given, A3/8 must be resistant to a *chosen*plaintext attack. This latter requirement was shown not to be satisfied by the algorithm COMP128, used as A3/8 by many operators.

A third requirement is that, again as RAND and SRES are passed un-encrypted between the mobile and the base station, an impostor cannot derive a particular K_c from the RAND and SRES in the same triplet as that K_c or by collecting a number of RAND-SRES pairs. This means that SRES and K_c must be completely unrelated though derived from the same RAND and K_i.

It has been mentioned that an important design consideration was that K_i was not to be shared with the serving network. A by-product of this decision is that algorithm A3/8 does not need to be known by the serving VLR, as A3/8 is only used where K_i is present, that is, in the AuC and the SIM. It should be noted that the VLR does not need any cryptographic algorithms, as A5 is not used in the VLR either but in the BTS, the security functionality in the VLR is therefore a simple comparison and distribution of parameters. This differs from systems based on ANSI-41, as used in many US networks, where the VLR must possess cryptographic capability. As A3/8 is only present in the SIM and AuC and the use of A3/8 is a protocol between a subscriber's SIM and the AuC of that subscriber's operator (albeit with the HLR and VLR as intermediaries), A3/8 does not have to be standardised. However, as the parameters of the triplet are passed via the HLR, VLR and ME as well as the SIM and AuC, the lengths of the parameters in the triplets must be standardised in the absence of a flexible encoding method. Each operator can therefore have a different A3/8 and operators were encouraged to take advantage of this possibility.

SEG felt it was an advantage that A3/8 did not need to be standardised. One claimed advantage of this is that less standardisation work must be done. However, in response to this it could be said that now each operator must develop their own A3/8, so though the amount of standardisation has gone down, the amount of development will go up. A second purported advantage is that each operator can also keep their A3/8 secret. However, this is also a moot point, because, as has been mentioned previously with regard to protocols, flaws in algorithms can be very subtle, and keeping an algorithm secret necessarily means there will be less potential examination of the algorithm. A clear advantage is that operators can gracefully bring in a new A3/8 on a SIM by SIM basis - the AuC knows which subscriber a request for triplets is for and can therefore use an updated A3/8. A clear disadvantage of there being different A3/8 is that AuC manufacturers must cope with different requirements from different operators.

However, in spite of the arguments for and against standardisation of A3/8, it was recognised that an example algorithm would be required, for implementation tests, and for those operators that did not possess or wish to possess the capability to obtain such an algorithm. This algorithm was COMP128, designed by a research wing of Deutches Telecom. The use of COMP128 amply illustrates the disadvantages mentioned above. This minor but salutary controversy is described later in this chapter.

15.7 Identity Confidentiality

Subscriber identity confidentiality on the radio interface was one of the security requirements of GSM as given in [1].

A robust identity confidentiality mechanism is in fact quite a difficult thing to achieve. Confidentiality usually involves encryption and encryption requires, for symmetric ciphering, a shared secret key. However, generation of a shared secret key should generally be done in combination with, or after authentication, or how does one entity know who they are sharing a secret key with and who therefore they are revealing their identity too? Authentication, however, is authentication of a particular identity, but identities have not yet been exchanged! Mechanisms involving public key cryptography can be used but only if one of the entities (e.g. a "server" or a network operator) does not mind revealing its identity to eavesdroppers. A simple mechanism involving public keys might be that one entity (the "server") transmits a certificate for its public key and the other entity encrypts its identity using the received public key. The transmitted identity can then be authenticated by a variety of means that do not reveal the identity to passive eavesdroppers.

Public key cryptography was not available to the GSM designers, so a simple mechanism using temporary identities and the basic facilities of GSM security was designed.

When a subscribers attempts access with an operator with which it is not presently registered (so, first access in a roamed to network, or the first access for some time in its home network) it must reveal its identity, and request access using its permanent identity, the International Mobile Subscriber Identity, or IMSI. The IMSI is then authenticated, a process which results in the sharing of K_C. The subscriber is then assigned a Temporary Mobile Subscriber Identity (TMSI, pronounced "timsy") which is sent to the subscriber encrypted with K_c. The next time the user attempts access in that network, it uses the TMSI to identify itself and the network looks up its table of TMSI to IMSI mapping to find the subscriber's permanent identity and the triplets with which it can authenticate the subscriber and begin encryption. So that a subscriber cannot be followed around, it is frequently given a new TMSI (if the same TMSI were used for a while, a subscriber previously identified by some out of band means could be recognised by the TMSI). Theoretically, the IMSI should only have to be used on a subscriber's first ever registration with any network, and it should be possible for the TMSI to be used even across different networks. In practice, however, the IMSI must be revealed on first registration in a new network at least, and in some networks, more frequently than this.

The GSM identity confidentiality is simple and efficient, but is not robust. The IMSI must be revealed on first registration with a network, and the mechanism as a whole can be compromised using a "false base station" as described later in this chapter.

15.8 Ciphering in More Detail

"Architectural" aspects of ciphering were considered by the SEG. A separate group, the Algorithm Experts Group (AEG), was formed to consider strictly algorithmic considerations. The chair of the AEG was Charles Brookson, then employed by British Telecom (BT).

15.8.1 Position of Ciphering in the Protocol Stack

Ciphering is performed using algorithm A5 (the original A5 came to be known as A5/1, as described below). Ciphering operates at the physical layer unlike the use of SSL, for instance, which operates just above the transport layer, or ciphering in GPRS, which operates at the link layer, layer 2.

SEG decided that ciphering would only exist between the mobile phone and the base station, as it was assumed that most other links afterwards would be along fixed lines, and therefore ciphering would not be required, as GSM had only to be as secure as existing fixed line phone systems. Ciphering therefore had to be somewhere within the physical layer – if ciphering were any further up the protocol stack, this would require the base station to be able to process frames at this layer, whereas it was intended that it be possible for frames above the physical layer to pass transparently through the base station.

SEG, with assistance from radio interface experts in SMG2 decided that ciphering would take place towards the "bottom" of layer 1.

Ciphering is therefore one of the last things done to the data bits in a frame to be transmitted across the radio interface. After encryption, the data is built into "bursts" by the addition of synchronisation and training bits and modulation then takes place. The decision to put encryption so low down had the following consequences:

- The maximum amount of data, both user data and signalling data, is encrypted.
- Ciphering takes place after error correction, and more importantly, deciphering takes place before error correction. There will therefore be errors in the received ciphertext and a stream cipher must be used (see a note on this in the next sub-section).
- The layer 1 frame counter, used for synchronisation at layer 1, can be used as an input to the key stream generator. However, this means that the layer frame counter must be of a greater length than is required for the non-ciphering layer 1 purposes, or the frame counter will repeat during a call and cause considerable weakness in the operation of ciphering. For this reason, there is a "hyperframe" in GSM, which is 1024 times longer than the

superframe, the longest frame aggregation required for non-ciphering purposes. The hyperframe number is input to the cipher and not any smaller frame counter.

- Ciphering (on the uplink) takes place after interleaving. The block of bits that is ciphered is therefore drawn from eight frames of original user data. This makes certain "known plaintext" attacks more difficult as the variation within a block of data to be ciphered (the "plaintext") is greater than if the plaintext were drawn from a single frame of user data.

A major consequence of the position of the ciphering being so low in the stack, along with the design of the ciphering algorithm chosen, is that the ciphering algorithm can be implemented in hardware. Moreover, ciphering can be integrated into the same piece of hardware that contains other low level functions, such as convolutional coding, interleaving and burst building. This means that the ciphering algorithm does not exist in a form where it can be easily extracted from the phone or base station and then used for other purposes.

15.8.2 Stream Ciphers and Block Ciphers

Before plunging into the operation of A5, we must distinguish between block ciphers and stream ciphers.

A block cipher operates by taking a block of text of a certain length (for example, 64 bits as used in DES, or 128 bits as used in the Advanced Encryption Standard (AES)) and encrypting the block as a whole. That is, the block of plaintext is taken and fed as one block into an algorithm along with the cipher key. A block of encrypted text, ciphertext, is the output. Ideally, for security purposes, every bit of ciphertext depends in some way on all bits in the plaintext (and on every bit of the encryption key).

A stream cipher works on a bit by bit basis and not on blocks. Operation is shown in the diagram for A5 (see Figure 15.3).

A "keystream" generator produces a string of pseudo-random bits as a function of the encryption key and frame counter (the frame counter makes sure that the mobile and base station produce the same keystream). This string of bits is XORed with the plaintext to produce the resulting ciphertext. At the decrypting end, the decryptor produces the same keystream and XORs the ciphertext with it to produce the original plaintext.

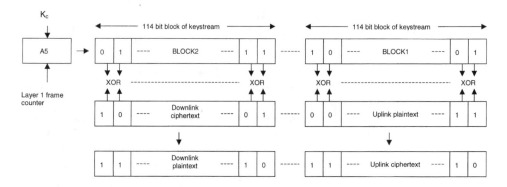

Figure 15.3   Operation of A5 at the mobile station

Block ciphers could not be used for GSM because of the relatively high error rate in wireless environments (there is an uncorrected error rate of about 10^{-3}). As every bit in the ciphertext depends on every bit of the plaintext, and the reverse in decryption, if there is an uncorrected single bit error in the ciphertext received, this error will have a knock-on effect across the whole block in which there was an error. Stream ciphers however operate on a bit by bit basis, so an error in the received ciphertext will only result in the corresponding plaintext bit being in error. Block ciphers are more widely used than stream ciphers, and the cryptographic analysis ("cryptanalysis") of block ciphers is, in the commercial world and in academic circles, certainly better understood than the cryptanalysis of stream ciphers. A block cipher can be used as a key stream generator for a stream cipher application, and it is interesting to note that this approach was taken for 3GPP security where the block cipher KASUMI (based on block cipher MISTY1) was used as a basis for the stream cipher "f8" used in 3GPP.

15.8.3 Operation of the Cipher

The use of A5 in GSM follows traditional stream cipher principles, as shown in Figure 15.3.

The cipher key K_c and the layer 1 frame counter are input to A5. A5 runs and produces two 114 bit blocks of keystream. GSM is fully duplex, i.e. both sides can transmit simultaneously, so within a frame, a mobile or base station both transmits and receives a frame. The first 114 bit block, BLOCK 1, is therefore used to encrypt the plaintext data being transmitted (the uplink plaintext), and the second 114 bit block, BLOCK 2, is used to decrypt the data received in that frame (the downlink ciphertext). At the other end of the communications path, BLOCK 1 is used to decrypt the received ciphertext and BLOCK 2 is used to encrypt the plaintext to be transmitted.

15.8.4 Selection of the Cipher

There were originally four proposals for the ciphering algorithm, from Holland, France, Sweden and the UK. The UK candidate algorithm was changed once following the discovery of weaknesses in the original proposal. At a meeting in June 1988, the AEG, which comprised members from Holland, France, Sweden, the UK and the Federal Republic of Germany, decided to adopt the French algorithm which had been entered into the selection process by the Centre National d'Etudes des Telecommunications (CNET), the research wing of France Telecom. The group decided on the basis of a simple vote. The French proposal was at this stage just for a keystream generator. The exact details of its operation in a GSM context, such as the loading of keys and message dependant counters, were not there and had to be developed within the group.

A major advantage of the French proposal over the others was that it was very amenable to implementation in hardware, indeed it had been designed with hardware implementation specifically in mind. This fact proved fortunate (and of course, not at all coincidental), as implementation in hardware, at the time, eased export restrictions, compared to implementation in software. The French proposal also possessed all the characteristics of a well designed algorithm – it was composed of simple, well-understood components that leant themselves to cryptographic analysis. This fact in particular distinguished it from some of the other candidate algorithms.

15.8.5 Length of the Cipher Key

The length of K_c is 64 bits but the effective length of the key input to the ciphering algorithm is 54 bits, as A3/8s must produce cipher keys where the top 10 bits are set to zero. This condition does not exist within the GSM specifications, but was imposed via the GSM MoU, now called the GSM Association.

In specifying the ciphering algorithm and its effective key length, the AEG had to take note of the restrictions concerning cryptography of the time. These restrictions were considerably more severe than those existing at the time of writing. COCOM regulations, the precursors to the well known Wassenaar agreement, were in force. There were severe controls of the export of cryptography and many countries had controls on the import of cryptography. Some countries, such as France, had controls on the use of cryptography within that country.

The AEG agreed that the strength of the ciphering would be determined by the effective length of the cipher key. The choice of length was pre-conditioned by the fact that the algorithm was designed for a 64 bit key to reduce complexity, and 64 bit keys were regarded as the state of the art at that time. At that time, the export of DES, a cipher with an effective key length of 56 bits, was restricted to use in financial applications. As it was the belief of the AEG that A5 was a full strength cipher and therefore potentially as strong as DES, it was clear that GSM could potentially be put under the same export control as that of DES. The compromise value of 54 bits was recognised as a major achievement by the AEG and SEG at a time when encryption was not in consumer use and where many exportable algorithms could be broken by government agencies.

In the cryptographically liberal times in which this chapter is written (2001), specification committees have the luxury of choosing a cipher length which is appropriate to the level of security that is required. 3GPP SA3, the successor committee to ETSI SMG10 (itself a successor in some senses to the SEG) for instance, chose to have ciphers that took 128 bits without too much discussion at all. That luxury was not open to the AEG and SEG in 1990.

These liberal times are also being taken advantage of with regard to GSM. K_c is 64 bits long, and all places where it is used should be able to handle K_c having an effective length of 64 bits instead of 54. The GSMA SG is therefore conducting the work that will lead to the use of full 64 bit length K_cs (the work in question is verifying that all handsets and base stations will actually take a 64 bit key).

15.8.6 A5/1 and A5/2

In 1992, the MoU SG decided to produce a "more exportable" version of the GSM ciphering algorithm. It was not said that a weaker algorithm was required, though it would seem logical to assume that the more exportable algorithm was weaker. This new algorithm was required because some governments were preventing export of GSM equipment outside Europe, and in some cases, even from outside western Europe. The existing "A5" would be called "A5/1" and the new algorithm would be called "A5/2".

In having two algorithms, there is the potential for inter-operability problems – the base station may have A5/2 only but the mobile phone has A5/1 only. This was avoided by requiring (for GSM phase 2 mobiles onwards) that the mobile supports both A5/1 and A5/2. The inter-operability problem was solved in this way and not by requiring the base station to support both because the export controls were applied to base stations and not handsets.

Handsets are possessed by individuals and move around with individuals – it does not therefore make sense to control their export as they can be "exported" by individuals in any case.

The GSM radio interface allows the mobile phone to state which ciphering algorithm(s) it supports (there is space for up to seven different ciphering algorithms in the relevant message) with the network taking the decision on which algorithm will be used.

Following relaxation of encryption export laws towards the end of the 1990s, the need for A5/2 disappeared, and SMG10 and later the MoU SG began to encourage the general use of A5/1.

15.8.7 Secrecy of the Algorithm

Neither A5/1 nor A5/2 have been officially published (even after it was admitted that the "alleged A5/1" in [3] was in fact A5/1). Again the context in which GSM was developed must be recalled. GSM was the first ever consumer product (it preceded SSL in PCs, which first appeared in Netscape Navigator in 1995, and provided only 40 bit ciphering) to contain cryptography.

Non-publication was a concession to gain the right to use the algorithm. It is difficult to understand the logic of the governments involved in making this requirement. Opening the floodgates to wider use of cryptography may be one. Trying to keep a semblance of an agreement between the US and European governments on how to control cryptography could be another. It has also been alleged that some government agencies knew of weaknesses in A5/1 and wanted the algorithm kept secret so these weaknesses would not be discovered. At the end of the day, if non-publication of the algorithms was the price that had to be paid to get an exportable GSM system with meaningful encryption, it was a price that was worth paying.

Non-disclosure of an element of a specification was not in CEPT or ETSI rules. A5/1 and A5/2 were not therefore part of the ETSI specifications and their specifications were controlled by the MoU SG.

15.9 How Has it Stood Up – Attacks on GSM Security

The most significant attacks on GSM are given below; [7] is also a good reference on this subject.

15.9.1 Microwave Links

A clear gap in GSM security now is the fact that the Base Transceiver Station (BTS, the "base station") to Base Station Controller (BSC, a node controlling a number of base stations, co-ordinating handovers and performing base station co-ordination not related to switching) link is in many cases a point to point microwave link which can be eavesdropped upon as data is at this point un-encrypted. At the time of design, it was expected that the BTS to BSC link would be across fixed links and therefore that ciphering would not be required. In 3GPP, the ciphering extends as far as the Radio Network Controller (RNC), the 3GPP equivalent of the BSC – microwave links - are therefore protected.

15.9.2 SIM-ME Interface

The SIM is functionally required at the start of a call only – once the phone has been authenticated and the SIM has delivered the cipher key to the terminal, the SIM is no longer used. However, the SEG specified that a call must close if the SIM is removed from the terminal during a call. If this requirement is not made, a stolen SIM or a SIM obtained by a false subscription can be used to set up a call in one terminal, then removed and used to set up a call in another terminal. This way, a large number of parallel calls can be set up using one SIM.

Initially it was specified that the terminal have some physical means to detect the presence of the SIM. However the physical detection mechanism implemented by some manufacturers could be fooled in many cases by the use of a bus ticket! The mechanism now specified is that of a frequent polling of the SIM by the ME with a particular command.

It was a design goal of GSM that it be possible for the SIM to be removed from one terminal and put in another when a call was not in progress. As the SIM can therefore be put in a terminal with which it has no previous association, unless some form of public key cryptography is used, there is no way for the SIM to authenticate the terminal or derive keys to protect the SIM to terminal interface. Some analysis will show that the public key architecture required to provide terminal authentication is not at all trivial. The SIM-ME interface is therefore unprotected and it is possible for SIMs to be connected to terminal emulators instead of "genuine" terminals, and if a phone can be physically compromised, for messages on the SIM-ME interface to be tapped. However, unless the algorithms on the SIM are sub-standard, there is no advantage to being able to compromise the SIM-ME interface in this way.

15.9.3 Attack on COMP128

In April 1998, David Wagner and Ian Goldberg of the University of California, Berkeley, announced that they had cracked "COMP128", an algorithm taking the function of A3/8 in the SIMs of many operators [5]. Wagner and Goldberg claimed that they had spotted a flaw in COMP128 within hours of receiving the algorithm specification. Recall that A3/8 is used to derive SRES (the response from the mobile phone that is used to authenticate the mobile phone) and K_c, the ciphering key. Recall also that one of the requirements of A3/8 is that an attacker cannot derive any information about K_i by collecting RAND and SRES pairs, which are sent in cleartext across the radio interface. Wagner and Goldberg correctly claimed that this assumption could not be applied to COMP128. The algorithm had a weakness which would allow complete knowledge of K_i if around 160 000 chosen RAND-SRES pairs could be collected. Some operators knew about the weaknesses in COMP128 at the time of the announcement by Wagner and Goldberg, in fact at least two discovered the same weakness when the algorithm was first shown to them.

How significant is this attack? To assess this we need to examine the difficulty of the attack and what it gains for the attacker – 160 000 pairs is a large number. This is a "chosen plaintext" attack and 160 000 randomly selected pairs is not sufficient. If an attacker were forced to tail a target mobile phone user and obtain the pairs by passively recording the target's radio interface communications, they cannot choose the RAND-SRES pairs. It would take years to collect such a number of pairs let alone the very specific pairs required for the cryptanalysis. The passive attack is therefore not considered feasible in any realistic time frame. However, there are active attacks that can be used to obtain these pairs. The quickest would be to steal the user's phone, remove the SIM and connect it to a phone emulator. As the commands to the SIM are not authenticated in any way and the SIM does not have any knowledge of call protocols, and merely responds to a RAND and a request for the SRES, the emulator can be used to send 160 000 chosen RANDs to the SIM and receive the SRESs. SIMs tend to have relatively slow clock speeds and it can therefore take up to 10 hours to obtain the pairs. For some SIMs, with fast implementations of COMP128 and faster than average clock speeds, the pairs could be obtained in about two and a half hours, but of course, there is no way for an attacker to know how fast a SIM runs before the phone is stolen. Stealing the mobile phone and keeping it overnight, then returning it to the user without her knowledge, therefore seems a plausible way to obtain the K_i. Another method is to use a "false base station" (see below) to send the mobile RANDs over the air. The rate at which pairs can be collected is slower than with the phone emulator and would take a number of days. However, the attacker does not need physical possession of the SIM. One easy way to counteract these attacks is to implement an "authentication counter" in the SIM which counts the number of times the SIM is asked for an SRES. Upon reaching the threshold, which could even be set on a SIM specific basis, the operating system of the SIM will close down and refuse any further operation.

After these efforts, the attacker has the mobile phone's K_i and can therefore masquerade as the user and run up calls on her bill, and also determine the K_c for the user's calls and therefore eavesdrop upon them. As it is anticipated that the legitimate user would notice after some time that calls were being made at her expense, the first fraud could not continue indefinitely. There does not seem to be enough profit for the attacker to take the steps to obtain the K_i, though perhaps with a slick operation the sums might add up. With regard to the latter, knowledge of a user's calls is a more difficult thing to put a value on. Certainly, some sensitive data would be worth the attack. The fact that such sensitive data should not be transmitted over a GSM mobile phone, where it will be in plaintext for much of the journey, is irrelevant, such information is transmitted over GSM, and may be of sufficient value.

In conclusion it can be said that the attack is significant and COMP128 should certainly not be used as an A3/8. A replacement for it was quickly obtained and it is presumed that operators who used COMP128 (not all operators did) replaced SIMs with the new algorithms as soon as reasonably possible (clearly as most users are not sending highly sensitive information over their phones, most users are not likely targets and do not need an immediate SIM replacement). Wagner and Goldberg argued that the fact that COMP128 was not published probably led to the use of a flawed algorithm for a considerable length of time. They were probably right in this regard.

15.9.4 Attacks on A5/1

A5/1 has proved to be a big target for those parts of the cryptographic profession that wish to break algorithms. There have been a number of claimed attacks, but only one legitimate and significant attack.

In 1997 Golic published a paper [4], claiming an attack on A5/1. The algorithm described claimed to have been based on specifications of A5/1 that were obtained by an unspecified method but it was not however A5/1.

The only attack on an algorithm that has been confirmed to be A5/1 was that by Shamir and Biryukov [3], later improved by Wagner [6]. This is the only attack that has made any advance on what was known about A5/1 within the GSM design community.

15.9.4.1 Shamir/Biryukov/Wagner Attack on A5/1

The technique used to attack A5/1 is known as a "time-memory trade-off" [5]. In a pre-processing phase, a large database of algorithm states and corresponding keystream sequences is created. In the attack phase, if a substantial amount of known keystream is obtained, the database is searched for a match with sub-sequences of the known keystream. If a match is found, then with high probability the database gives the correct algorithm state. It is then fairly straightforward to compute the cipher key, and hence to decipher all the rest of that call.

Adi Shamir and Alex Biryukov introduced some new optimisations to this technique that made an attack feasible in practice, if 2 minutes' worth of known keystream could be obtained. Now, without physical possession of the mobile phone or base station, this keystream could in practice only be obtained if 2 minutes of ciphertext and its corresponding plaintext could be obtained precisely and then XORed to return the keystream. Obtaining 2 minutes of plaintext ciphered with a particular key is a tall order; Shamir and Biryukov stressed that their attack was on the A5/1 algorithm in isolation, and made no claims about its impact on the security of GSM encryption as used in practice.

David Wagner spotted a further optimisation that would allow the cipher key to be obtained with only 2 seconds of known plaintext (and a correspondingly larger amount of processing). Obtaining 2 seconds of known plaintext (from both uplink and downlink) is still not trivial - the *precise sequence of bits* that is encrypted must be obtained, and not just the information content (for example, the words spoken) of the bit sequence. However, it is not inconceivable that this 2 seconds could be obtained: for instance, the mobile could somehow be forced into a data call where the attacker knows what data is sent on the downlink and what the mobile phone will return in response.

[6] is a fine piece of cryptanalytic work. However, in practice it would probably not be used, as the false base station attack represents an easier method of eavesdropping.

15.9.5 False Base Station

One of the key assumptions when GSM security was designed was that the system would not be subject to "active" attacks, where the attacker could interfere with the operation of the system or impersonate one or more entities in the system. This assumption was made because it was believed such attacks, which would require the attacker to effectively have their own base station, would be too expensive compared to other methods of attacking GSM as a whole (e.g. wiretapping the fixed links or even just bugging the target).

Use of ciphering on a call does not happen "automatically" on a call – the call begins unciphered (as it must as a shared key cannot be established or re-used until the mobile phone has identified itself to the serving network) and the phone is instructed to begin (or not begin) ciphering at a particular point in call set-up. Ciphering was implemented in this way to give countries the option not to allow mobiles to use ciphering within their borders and some operators took this decision in any case. The instruction to begin ciphering given by the serving network is sent un-encrypted. This means that it can be subject to tampering in transit – the network instructs the mobile to begin ciphering but this is manipulated within transit to be an instruction not to cipher. However, such tampering would not allow eavesdropping as the network would begin ciphering, the mobile would not, and the mobile would then not be able to decode any received data and would drop the call. The more significant feature of the sending of the instruction to begin ciphering in the clear is that there is no "origin authentication" of the command – the mobile cannot know for sure that the command to begin or not begin ciphering has come from the genuine serving network. It is this weakness that is exploited in the false base station attack. The false base station sends the command to not begin ciphering to the mobile, and not the genuine network. The problem of the genuine network expecting ciphering and the mobile not is met by the false base station acting as a mobile and setting up a call to the genuine network itself. The call from the mobile to the false base station is connected to the call from the false base station to the network, so it seems to the caller that they have the call they requested. However, the call from the mobile to the false base station is not encrypted, so the communications can be eavesdropped upon at this point. The call between the false base station and the network is encrypted, so the network does not see that anything is awry. One consequence of this attack is that the call is made on the false base station's subscription and not that of the user's. The attack can therefore be detected afterwards if an itemised bill is available and checked and can be detected at the time if the called party is expecting a particular CLI.

There is a secondary false base station attack which is often termed "IMSI catching" and compromises the identity confidentiality mechanism in GSM. There is a network command that requires all subscribers receiving the command to transmit their IMSI to their base station. This command would be used if a VLR had crashed and lost details of the subscribers registered on it. A false base station can be used to transmit this command and then receive the IMSIs of mobile phones in its vicinity.

The primary false base station attack is a sophisticated attack. It involves a number of tasks in the vicinity of the target, which can be detected by the (very) careful user and are visible to operators carefully monitoring the operation of their base stations. The false base station is a reasonably sophisticated piece of equipment, and has been seen retailing for around 150 000 euros on the Internet. Nonetheless, it is a significant attack and one that the designers of third generation security set out specifically to prevent. This prevention is accomplished by integrity protection of the instruction to begin ciphering, and cannot be achieved solely by "mutual authentication" as is often claimed.

15.10 SMG10 and its Working Parties

The SEG was disbanded in 1988 once it had completed its specification work. It was not felt that a specific security committee was permanently required. Security work continued within the Security Group (SG, initially called "the Security Rapporteur Group") of the MoU. It was the MoU SG that persuaded SMG that a permanent security group was required and in 1995 SMG10 was set up. -The first and only chair in SMG10's history was Professor Mike Walker of Vodafone. Mike had been involved in GSM security from the start. He was a founder member of the SEG, AEG and the AEG's successor, the ETSI Special Algorithm Group of Experts (SAGE). SAGE was set up to design cryptographic algorithms for use in ETSI standards. SAGE did not design A5/1 as has been claimed.

SMG10 began with three sub-committees. Working Party A (WPA) dealt with crypto-graphic issues and was the link between SMG10 and SAGE. Working Party B (WPB) dealt with fraud issues, and looked at security lapses that were caused by poor procedures rather than by defects in cryptography. Working Party C was set up to study security for UMTS, the ETSI third generation project. Its work will be looked at below.

With the establishment of 3GPP in 1999, a committee to specify security for third genera-tion was set up. This was SA3, the third committee within the Services and Architecture (SA) technical sub-group of 3GPP. Mike Walker was elected as chair of SA3. SA3 was initially composed mainly of former members of WPC, but its membership soon expanded to include many other members of SMG10 and members of the other 3GPP partners aside from ETSI. SMG10 co-existed with SA3 until 2000 when it was formally dissolved.

15.11 Other Aspects of GSM Security

As well as monitoring the implementation and effectiveness of the security specified in GSM, SMG10 also specified security for new features introduced into GSM. The major such new features are given below.

15.11.1 GPRS

GPRS security is very similar to that for circuit switched GSM. The SIM is authenticated in the same way as is done for circuit switched GSM. There is the same general method of identity confidentiality (perhaps "identity privacy" would be more accurate), i.e. the use of temporary identities on the radio interface with permanent identities only sent over the radio interface in semi-exceptional circumstances.

It is in the ciphering that GPRS shows most difference. Ciphering for GPRS takes place at the link layer, layer 2, and not at layer 1 as for GSM. This means that security extends from the mobile to the SGSN, the GPRS equivalent of the VLR, and not just to the base station. Ciphering is performed at layer 2 mainly because GPRS "handover" allows for the mobile to receive packets from more than one base station at once – if ciphering terminated at the base station, the mobile would need to have two cipher keys and to know which packet came from which base station – which would complicate matters considerably. Other reasons are so that the layer 1 error correction coding can be changed without impact on ciphering and so that RLC packets can be sent out of sequence. Two advantages of the cipher being at layer 2 are that microwave BTS-SGSN links are ciphered and that the ciphering algorithm is located in fewer locations on the network side so can be updated more easily.

A5/1 is not used for ciphering in GPRS. Algorithm GPRS Encryption Algorithm (GEA1) was designed for GPRS ciphering and this appears in the first versions of the GPRS specifica-tions. SAGE took advantage of relaxations in export restrictions to specify algorithm GEA2, which is in later versions of the GPRS specifications. GEA2 is considerably stronger than GEA1 and should be used in preference to GEA1 wherever possible.

15.11.2 FIGS

The Fraud Information Gathering System (FIGS), was specified within SMG10 WPB with assistance from SMG3 between 1995 and 1997. At this time, pre-paid phones did not exist and billing was conducted after the event. Most subscribers were billed monthly, and billing cycles, especially for roaming subscribers, were long. This gave fraudsters that had obtained subscriptions fraudulently plenty of time to sell (international or premium rate) calls on a subscription before a bill, which would not be paid, arrived.

FIGS was designed to provide call information to operators in a faster timescale that could be provided by most billing systems, so that abuse could be detected more quickly. "Call selling" as described above, generally has a call pattern different from the average user, and so can be detected by reasonably straightforward analysis.

After the requirements for FIGS were written, CAMEL was selected to provide the carrier protocol for FIGS. CAMEL procedures which could be used to monitor call behaviour, almost in real time, were specified.

FIGS as such, has not been used much. However, fraud detection systems based on CAMEL have been developed, and some operators use systems based on CAMEL to achieve the real time billing they need for pre-paid services.

15.11.3 SIM Toolkit

Within GSM, the terminal-SIM relationship is that of a "master-slave". The SIM responds to commands from the ME, and cannot initiate any actions itself. SIM toolkit (or SIM Application Toolkit, the official name from the specification, GSM 11.14) is a method of allowing the SIM to take control in certain situations.

Within SIM toolkit there is a specification for the sending of secure SMS to and from the SIM in such a way that they are not and cannot be read by the terminal, but only by an authorised party, e.g. the operator or a bank, within the network. SMG10 worked in co-operation with SMG9 on the specification of this. Using "SIM toolkit secure messaging", the ciphering capabilities of GSM can be extended for low bandwidth services, such as banking. Commercial banking applications based on SIM toolkit secure messaging have been developed.

15.11.4 MExE

The specification of the Mobile Execution Environment (MExE) began in 1999. MExE was designed to take advantage of the increasing processing power of terminals in order to have a standardised execution environment on a terminal to which applications could be downloaded. In essence, MExE was designed to enable applications to be downloaded to terminals in the same way that they could be downloaded to PCs and PDAs.

Clearly the downloading of executable code to terminals is a potentially dangerous thing. The code could be viruses that would destroy user data on the phone, initiate calls the user had not requested or many other things. SMG10 and then SA3 were therefore called upon by the MExE group to assist in the development of security mechanisms for MExE.

These security mechanisms are based on the generation of two types of executable applications, "untrusted" and "trusted". Untrusted applications are given only limited capabilities, and often rely on user permission to carry out actions. Trusted applications are given much more capability but must be delivered to the mobile with a digital signature indicating the origin of the application and also protecting the integrity of the application as it is transmitted to the mobile. It is hoped that the fact that the origin of the application can be discovered from the signature will prevent or deter the writing of MExE viruses. At the time of writing, there were no MExE products on the market so the effectiveness of the security, or that of MExE as a product, cannot be given. MExE is the first specified use of public key cryptography within GSM and 3GPP.

15.11.5 CTS

The Cordless Telephony System (CTS) was begun as a project within GSM in 1997. CTS was designed to allow a user to have a mini-base station connected to the PSTN within their house. When the user arrived at home, they would camp onto their home base station and make calls via the PSTN, instead of camping onto a cellular base station and making calls via the GSM network.

This proposal had some revenue implications for operators, but also had a number of possibilities for abuse. A fraudster could increase the power of the mini-base station so as to match that of cellular base stations and so provide a cellular like service in a certain area, using spectrum that belonged to a legitimate operator. A fraudster might also camp onto a base station that did not belong to him/her and so make calls at someone else's expense.

SMG10 was therefore called upon to assist in the development of security for CTS. This it did and the result is in an annex of [2]. Security is based on secret key mechanisms. The home base station must possess a SIM and is authenticated by the network through the fixed network. The mobile phone and home base station initially authenticate each other through simple input of a PIN to both and thereafter by a randomly generated secret key.

15.12 Third Generation Security – 3GPP

SMG10 WPC (UMTS) was formed in 1995. It worked for a couple of years and produced an unpublished document GSM 33.20 giving the likely services that would be provided by a third generation system and the security that would be required. It also contained a threat analysis.

The work of WPC was greatly assisted by input from companies involved in an EC funded research project called USECA, short for UMST Security Architecture.

SMG10 WPC, and then 3GPP SA3 analysed the false base station problem and other weaknesses in some detail. The resulting security architecture was based on GSM but had important differences.

Critical signalling commands, such as the start ciphering mode command, would have mandatory integrity protection in order to prevent the false base station attack.

In GSM, a triplet generated for a particular mobile can be re-used any number of times for that mobile. The false base station attack can therefore be used to force use of a particular triplet and the attacker will know what cipher key will be used (some handsets have an indicator showing if ciphering is in use or not – this particular attack overcomes this protection). To prevent re-use of triplets, a sequence number was attached to "triplets" – the sequence number attached to a triplet must exceed the most recently previously received triplet. A Message Authentication Code (MAC) showing that the "quintet" came from the user's operator and no one else and to integrity protect the sequence number is also attached.

Ciphering extends between the mobile and RNC, the 3GPP equivalent of the GSM BSC. Links that may potentially be over microwave are therefore ciphered.

Ciphering is to be performed using a publicly available algorithm taking a 128 bit cipher key. The algorithm is called KASUMI and it is based on the Japanese algorithm, MISTY1. KASUMI is also used for the integrity protection of commands. It has always been intended by ETSI SAGE and 3GPP SA3 that KASUMI be publicly available and it has, at various times, appeared on the 3GPP website, though some bureaucratic difficulties have prevented it being permanently publicly available.

SA3 have also begun specification of protection of network signalling data as it is transmitted between and within networks. This signalling data contains sensitive information such as the "quintets" which can be used to eavesdrop upon user traffic or to masquerade as legitimate subscribers. Further, if false messages can be transmitted along the network, its operation can be interfered with or stopped altogether. It is therefore important that the origin of such commands can be authenticated so that only authorised parties can send them.

The work of SA3 demonstrates a determination by the mobile community to have the best security for 3GPP that is possible and to make sure that a lack of security is not a block to the rich range of value added services that are envisaged for third generation systems.

15.13 Conclusions

GSM security is simple and elegant and, unlike many security protocols in consumer equipment, has not had inter-operability problems. Though weaknesses have been found in GSM security, it must be recalled that GSM was designed as a piece of consumer equipment and these weaknesses are sophisticated and will not cause ordinary "consumers" to be at risk. Further there were severe restrictions on the use of cryptography at the time GSM was designed. GSM has not been subject to any technical fraud, unlike the analogue networks it succeeded, where millions, possibly billions of euros have been lost.

Overall then it can be said that GSM security has been a success. Though for virtually all of its users, its still supplies perfectly adequate security, in principle, it is time for an update of GSM security. This update will come in the form of 3GPP security which has taken full advantage of the present openness concerning cryptography.

Acknowledgements

Thomas Haug, Per Christofferson, Peter Howard, Steve Babbage, Klaus Vedder, Michel Mouly and Charles Brookson are all thanked for either reviewing drafts of this chapter, supplying SEG and AEG documents or answering queries.

References

[1] Digital cellular telecommunications system (phase 2); security aspects (GSM 02.09 version 4.5.1), ETS 300 506.
[2] Digital cellular telecommunications system (phase 2); security related network functions (GSM 03.20 version 4.4.1), ETS 300 534.
[3] Biryukov A., Shamir A. Real time cryptanalysis of the alleged A5/1 on a PC, preliminary draft, December 1999.
[4] Golic J.Dj. Cryptanalysis of alleged A5 stream cipher. Eurocrypt 97, published as lecture notes in Computer science #1233, Springer Verlag, Berlin, 1997.

[5] Babbage S.H. Improved exhaustive search¤ attacks on stream ciphers. ECOS 95 (European Convention on Security and Detection), IEE Conference Publication No. 408, May 1995.

[6] Biryukov A., Shamir A., WagnerD. Real time cryptanalysis of A5/1 on a PC. FSE 2000, published as lecture notes in Computer science #1978, Springer Verlag, Berlin, 2000.

[7] Piper F.C., Walker M. Cryptographic solutions for voice telephony and GSM. In: Proc. COMPSEC 98. London: Elsevier, 1998.

Chapter 16: Short Message and Data Services

Section 1: The Early Years from mid-1982 up to the Completion of the First Set of Specifications for Tendering in March 1988

Friedhelm Hillebrand[1]

16.1.1 The Mandate by CEPT and the First Action Plan of 1982

The mandate given to the Technical Committee GSM by CEPT in mid-1982 requested the "harmonisation of a public mobile communication system in the 900 MHz band".[2] This decision took place during the very hot promotion phase for ISDN. Therefore it is remarkable to note that GSM was not defined as a "mobile ISDN". Instead the decision leaves the nature of the GSM open, but requests the study of the interconnection with ISDN.

The first action plan for the group GSM elaborated by the Nordic and Dutch PTTs was approved at the meeting GSM#1 in December 1982. It mentioned basic requirements for data services[3]

> It is expected that in addition to normal telephone traffic, other types of traffic (non-speech) will be required in the system. However since such predictions concerning the user requirements.... will contain a great amount of uncertainty, a modular system structure allowing for a maximum of flexibility will be necessary.

> The services offered in the public switched telephone networks and in the public data networks at the relevant period of time should be available in the mobile system... The system may also offer additional facilities.

These basic requirements regarding data services were very far-sighted. They were nearly forgotten for long periods due to priorities for telephony.

---

[1] The views expressed in this section are those of the author and do not necessarily reflect the views of his affiliation entity.

[2] GSM Doc 1/82.

[3] GSM Doc 2/82.

16.1.2 Discussions on Data Services from the Beginning of 1983 to the End of 1984

The focus of the GSM work in the period from the end of 1982 to the end of 1984 was on strategic questions and the requirements for the basic telephony oriented system, e.g. the relationship between GSM and emerging analogue 900 MHz interim systems, radio aspects, speech coding and hand-held viability. There was also a lack of data communication expertise in GSM. Therefore the progress was slow in this area during this period. Results were not saved in permanent documents and were often forgotten after some time.

During GSM#2 meeting in February/March 1983 a general discussion on ISDN and OSI and their applicability took place without firm conclusions.

During the GSM#4 meeting November 1983 regarding ISDN it was clarified, that it was not possible to provide the full capacity of B- and D-channels but only the functions.

GSM#6 in November 1984 received a report from a working party on network aspects which had met during the GSM meeting.[4] It proposed to use the ISDN concepts of terminal adaptors and interworking units for data services in the GSM system.

16.1.3 The First Concept for Data Services Agreed in February/March 1985 (GSM#7)

During GSM#7 the Working Party 1 "Services" (WP1) elaborated a document on "services and facilities of the GSM system" which was endorsed by GSM.[5] This was based in the data services part mainly on an input submitted by Germany and France.[6] The output document contained a reference model for data services, definitions of tele-, bearer and supplementary services, network connections and types of mobile stations. The reference model introduced terminal adaptor functions at the mobile station and an interworking unit between the mobile and the fixed network (the diagram is shown in Chapter 10, Section 1, Fig. 10.1.1).

The annexes contained lists of teleservices including the Short Message Service (SMS). SMS had three services: mobile originated, mobile terminated and point to multipoint. It foresaw a maximum message length of, e.g. 128 octets, and an interworking with a message handling system. Other non-voice teleservices mentioned were, e.g. access to message handling systems and to videotex, facsimile, and transmission of still pictures.

A comprehensive range of circuit switched bearer services was listed:

- asynchronous duplex up to 9600 bit/s end-to-end
- synchronous duplex up to 9600 bit/s end-to-end
- asynchronous PAD access up to 9600 bit/s

The necessary connection types to support the mentioned services were defined by a set of attributes.

Another important prerequisite for the work on data services was the emerging consensus on network functions and architecture of the basic GSM system as needed for telephony[7]. This allowed to start work on the reference configuration for data services[8].

---

[4] GSM Doc 90/84 Annex 3.
[5] GSM Doc 28/85 rev. 2.
[6] GSM Doc 19/85, for more details see Chapter 10, Section 1, paragraph 10.1.4.1.
[7] GSM Doc 43/85.
[8] GSM Doc 106/85.

16.1.4 Work in the Period from March 1985 to February 1987, when the Radio and Speech Coding Technologies Were Chosen

In the following meetings GSM and its working parties concentrated again on basic telephony aspects. GSM#12 (September/October 1986 in Madrid) discussed, whether the speech codec should be transparent to DTMF signals. It was qualified as desirable, but not as an essential requirement, since other solutions could be found.

The priority for telephony left no work resources for data. The situation was very critical in WP3 "Network Aspects", which was very busy with system architecture, mobility management, etc. Also WP2 "Radio Aspects" was fully loaded with the evaluation of the different radio techniques and had no time or capacity to work, e.g. on channel coding for data services.

16.1.5 Fundamental Decisions on Data Transmission in GSM#13 (February 1987 in Funchal)

The catalyst for some fundamental decisions on data services was the discussion of requirements on the speech codecs with respect to transparency for DTMF and voice band data. During the selection decision of the voice codec at GSM#13 it became clear, that a transparency requirement for DTMF and voice band data would lead to additional complexity and a deterioration of the speech quality. On the other side it was realised, that data would need additional protection, since the system would provide at cell boundaries only a bit error rate of 1 error bit per 100 transmitted bits (compared to less than 1 in 1 000 000 in ISDN).

GSM#13 decided that the speech codecs should be optimised for speech only, since GSM was intended to be primarily a mobile telephony system. It was further confirmed that the capability of the system to carry DTMF and voice-band data was an essential requirement. Based on the proposal of an ad-hoc group[9] GSM decided that DTMF should be "transmitted as a signalling message over the Dm channel" and that the DTMF tones should be "injected into the audio path at the receive end". "Data services should be supported by the GSM system... The terminal equipment is connected to the mobile station via a terminal adaptor and the transmission is fully digital. Terminal adaptors for the V-series are to be specified."[10]

As a consequence of this decision GSM agreed on the necessity of putting extra resources on this task, since WP3 dealing with network aspects was not able to cope with it due to overload and lack of specialised expertise.

To start the work, a sub-working party under WP3 was created. I was appointed as an interim chairman, since I was one of the very few GSM members with data communication expertise.[11] Furthermore GSM decided, that interim terms of reference should be agreed between the WP3 chairman and myself and that the group should meet independently of WP3 and should not make use of resources presently used by WP3.

[9] GSM Doc 39/87.
[10] GSM Doc 41/87.
[11] I had been responsible for the setting up of the German national packet switching network DATEX-P.

16.1.6 The Agreement on Strategic Issues and the First Set of Specifications for Tendering from May 1987 and to March 1988

16.1.6.1 Setting Up of the New Group IDEG and Initial Discussions

The new group was called the Implementation of Data and Telematic Services Expert Group (IDEG). The first IDEG meeting on 20–22 May 1987 in Bonn was attended by 18 delegates.[12] Draft terms of reference and a set of working assumptions for the work as well as an action plan were elaborated and agreed (see annexes to the meeting report).

The creation of IDEG coincided with the time when the rules for participation in CEPT were relaxed. GSM had asked the superior bodies to allow manufacturer's participation in the detailed technical work after the GSM radio decision. Prior to that time, CEPT participation was limited to representatives of the post and telecommunications authorities only. The new rules allowed industry experts to participate in meetings by the invitation of, and as an advisor to, a CEPT member organisation. They were admitted as experts assisting the CEPT members, not as representatives of their companies. They were also members of the delegation of the CEPT member they were assisting. They could not submit documents in the name of their company, but had to hand them to a CEPT member. Industry participation was limited to two experts per committee per country. This restrictive situation was not sustainable. It lead later to the creation of the European Telecommunication Standards Institute (ETSI) and the transfer of all standardisation work from CEPT to ETSI. In the case of IDEG, there was a majority of delegates coming from industry from the beginning.

The key tool to start the work was a set of *working assumptions*. This concept had been used in the decision on the basic parameters of GSM at GSM#13 in February 1987 in Funchal. The working assumptions were agreed as preliminary conclusions. They were seen as open to change, but the proponent of a change would need to prove their case. This is essentially a process of "slowly drying cement" which allows the chairman of a group to find an agreement on a soft consensus conclusion and to start the process which leads to a firm consensus conclusion.

Key working assumptions with regard to strategic issues were elaborated in the first meeting:

- Confirm telephony as the prime system application.
- Co-use the telephony optimised system for data services to the maximum possible extent.
- Ensure a high quality of service.
- Avoid changes to the system architecture and support data services by add-on modules/functions, which can be implemented as options and can be dimensioned to meet the data traffic needs.
- The concept of "mobile office" lead to the requirement to have identical coverage and velocity requirements from the mobile station as the telephony service.
- Limit the additional complexity, since data was assumed as a small share of the system traffic.

GSM#14 (June 1987 in Brussels) approved the proposed terms of reference for IDEG and confirmed me as chairman.

---

[12] Report in GSM Doc 70/87.

16.1.6.2 The Agreement on Fundamental Aspects

In the second IDEG meeting (6–8 July 1987 in Heckfield, UK) a number of *fundamental aspects* were discussed and brought to conclusion for the specification work.

An important concept was to use existing data terminals and provide the normal interfaces to them by a terminal adaptor function in the GSM mobile station. This meant that the terminals needed to see a quality of service (bit error rate, delay) which is comparable to the quality of service in fixed networks. Quality of service is a critical issue for data services in a radio network. There are short interruptions of a connection during hand-overs or radio fades. At cell boundaries 1 out of 100 bits is corrupted compared to 1 out of 1 000 000 in modern digital fixed networks. Therefore measures have to be implemented in a GSM network to enhance the basic quality of service.

The first measure is a powerful channel coding on the radio interface tailored to data which adds checksums to detect and to correct errors (FEC = Forward Error Correction). This reduces the bit error rate by several orders of magnitude.

The short interruptions during handover and radio fades cannot be corrected by forward error correction, since they last too long if normal transmission delay values are requested. This type of problem can be tackled by protocols providing an automatic re-transmission of lost information. For efficiency reasons such a mechanism can be applied only on blocks of data. The protocol provides an additional checksum per block, which allows the detection of whether the received block is erroneous. A sequence number allows the detection of lost blocks. The receiving side confirms blocks received without detected errors and requests re-transmissions of blocks with errors or lost blocks.

Such protocols are known in the fixed networks as the High Level Data Link Control (HDLC) family of protocols. They cannot be used in a mobile network due to the bad basic quality of service. Due to the great block size (e.g. 1024 octets in fixed packet switching networks) many blocks would have errors both during the first transmission and the re-transmission. Therefore the throughput would fall to zero.

Therefore a special protocol had to be designed which was robust to cope with the difficult radio network environment. This protocol was called Radio Link Protocol (RLP). It was matched to the GSM transmission time slots, had a block length of 240 bits (= 30 bytes) and other means to secure the highest possible throughput.

Services using only FEC were called transparent services. Services using FEC and RLP were called non-transparent services. Both types of services provide a low bit error rate. Transparent services provide a constant throughput and a constant transmission delay. They have however interruptions of service caused by handover and fading. Non-transparent service secure the transmission of all blocks. They re-transmit blocks lost or corrupted by the interruptions caused by handover and radio fades. During such activities the throughput is reduced and the transmission delay time is increased.

IDEG refined, added details and agreed the set of *working assumptions*.[13] They covered all requirements without adding complexity to the basic telephony oriented system.[14] Key detailed working assumptions were:

- Data for connection-less bearer capabilities (e.g. for SMS) are transmitted on a control channel.

[13] GSM Doc 83/87.
[14] GSM#15 Report, Section 7d.

- DTMF information is carried on a control channel on the radio interface.
- Connection mode bearers are carried on one traffic channel, no multiplexing of low speed bearers on one traffic channel.
- A reference configuration for the mobile station was agreed.
- Interworking requirements and architecture were agreed.
- A rate adaptation mechanism was selected.
- One-way transmission delay less or equal 200 ms.
- Transparent services use rate adaptation and forward error correction, no ARQ.
- Non-transparent services use rate adaptation, FEC and RLP
- Bearer services must be capable of transmitting up to 9.6 kbit/s.

I recall a very lively discussion on the maximum bit rate on full-rate channels. Several delegates pleaded to limit the bit rate at 4.8.kbit/s in order to have a very good protection by a "heavy channel coding". They argued also that such a bit rate is more than sufficient for a single user. It was through the enduring efforts of Alan Cox and Ian Harris (both of Vodafone), who succeeded in convincing the meeting that 9.6 kbit/s on a full-rate channel was required in order to be future proof.

In addition the *action plan* was revised. The target was to complete the specification for the essential services (E2) which would be needed for a tendering purposes in early 1988 until the end of January 1988.[15] A list of the planned specifications was elaborated. Major specifications under IDEG's prime responsibility were:

- GSM PLMN connection types
- Technical realisation of the SMS
- Transcoding for data and telematic services
- Rate adaptation (on several interfaces)
- Radio link protocol (on several interfaces)
- Terminal adapters (several specs)
- Interworking with circuit and packet switching data networks
- Service interworking
- In addition contributions to the work of several other WPs were needed:
- WP1 specifications on teleservices, bearer services and charging
- WP3 specifications on network architecture, interworking with PSTN/ISDN, numbering/routing/identification
- WP2 specifications on channel coding

GSM#15 in October 1987 confirmed the working assumptions and the action plan and "promoted" IDEG from a WP3 sub-group to a working party reporting directly to GSM. IDEG was renamed Working Party 4 "Data and Telematic Services" (WP4).

16.1.6.3 The Production of Specifications Needed for Tendering

In the following period WP4 concentrated its effort on producing draft specifications in accordance with the action plan. Two WP4 meetings were held on 26–30 October and on 23–27 November 1987 in Bonn.[16] These meetings elaborated the first draft specifications.

---

[15] GSM Doc 84/87.
[16] Report in GSM Doc 144/87.

Each meeting week had short opening and closing plenaries on Monday Friday morning, respectively. From Monday afternoon to Thursday evening six drafting groups worked on

- SMS
- architecture and connection types
- terminal adaptor functions
- radio link protocol and coding
- interworking
- numbering and routing

Four specifications were completed for examination by GSM#16 in December 1987. These were the most urgent ones, identified by GSM as particularly important for the tendering activities: 03.10, 03.41 (deleted later), 04.21, 07.02. Twelve specifications reached the status "preliminary".

There was another WP4 meeting in Paris on 22–25 February to complete the first draft specifications.[17]

GSM#17 in February 1988 approved the first set of specifications needed for the tendering on 29 February 1988[18]:

- 03.10 GSM PLMN connection types
- 04.21 rate adaptation at the MS/BS interface
- 07.01 principles on terminal adapters
- 07.02 terminal adapters for asynchronous bearer services
- 08.20 rate adaptation on the A interface

16.1.6.4 The Specification of the SMS

WP1 had produced a service description for the three SMSs:[19]

- Mobile originated/point to point
- Mobile terminated/point to point
- Cell broadcast

An additional input with a concept proposal for the technical realisation of the "mobile terminated/point to point" service came from France.[20] It contained a proposal for the functional architecture. It proposed a new entity, the service centre in charge of:

- Dialogue with the user for message submission and status requests
- Handling of messages: storage, status, transmission
- Operation and maintenance functions

It proposed a layered function split between a message application subsystem and a message transmission subsystem as well as several protocols.WP4 created a Drafting Group Message Handling (DGMH) under the leadership of Finn Trosby (Telenor) in July

---

[17] GSM Doc 93/88.
[18] GSM Doc 31/88 rev. 1
[19] See Chapter 10, Section 1, Paragraph 10.1.5.1.3.
[20] IDEG 16/87.

1987.[21] This group was given the responsibility to deal with message handling access services (MHS)[22] and SMS and also to look for a common architecture for both groups of services.

Great efforts and priority were put in the beginning to the MHS access services. But over time they eroded. They were downgraded from teleservices to bearer services and completely deleted in spring 1988, since normal bearer services were seen as providing sufficient functions to support this application. In the background was also a development in the market, that MHS according to CCITT X.400 were not a tremendous success.

But the group produced an initial draft specification GSM 03.40 "Technical realisation of the short message service" in November 1987.[23] It contained a first description of

- Service elements
- Network architecture
- Service centre functions and service centre network
- Functions in other network elements
- Routing principles for the message transfer between the mobile station and the service centre
- Protocols and protocol architecture

The SMS was defined between mobile stations roaming in a GSM network or roaming internationally and a service centre, which had store and forward functions. Both Point to point services and the cell broadcast service were treated within this scope.

I tried to interest the ISDN community to work with us on a compatible SMS service in the ISDN. This would have provided a standardised access to and from ISDN users. But the initiatives did not fall on fertile ground. Therefore the SMS did not provide a standard for the access from and to fixed subscribers.

A close co-operation with the Layer 3 Expert Group responsible for the layer 3 protocols on the radio interface between the mobile and the base station was necessary, since the SMS related messages could be transmitted on normal signalling channels of the GSM system with lower priority than the signalling messages. This would ensure, that short messages could be transmitted to mobile stations in idle mode or involved in a call. The message length was given a ceiling of 180 octets. Later the final value was fixed at 160.

The detailed technical work lead also to a proposal to refine the service descriptions[24] in February 1988. This document covered also the international operation of the point to point services.

The initial technical specification and the revised services description contained the initial results achieved in the first year. They provided a firm basis and framework for the later detailed work,[25] which is the basis of the tremendous success in the market.

16.1.6.5 Achievements of the Data Group in its First Year from May 1987 to March 1988

IDEG/WP4 had despite the very late establishment after GSM#13 in February 87 met the first important target: to complete the work needed for the planned tendering of ten operators on

---

[21] Minutes of second IDEG meeting on 6–8 July 1987 in Heckfield, UK, IDEG Doc. 58/87.
[22] See Chapter 10, Section 1, Fig. 10.1.2.
[23] WP4 Doc. 152/87.
[24] WP4 Doc. 85/88 rev. 1.
[25] See Chapter 16, Section 2.

29 February 1988, contributions to WP1, 2 and 3 as well as the specifications under WP4 prime responsibility. These documents cover the architecture and all functions needed for the asynchronous bearer services. A broad stream of work in other areas was started and initial results were reached. This included SMS.[26]

WP4 had become a committed community of data communication experts who were new to the GSM group. WP4 had grown from 18 participants in the first meeting to more than 50 participants in this period. A very efficient and effective working and co-operation spirit has been developed by the group.

There were many valuable contributions to the work in this early period. Key contributors were, e.g.: Christian Bénard-Dendé (France Telecom), Alan Clapton (BT), Alan Cox (Vodafone), Graham Crisp (GPT), Alfons Eizenhöfer (Philips), Ian Harris (Vodafone), Michael Krumpe (Siemens), Thomas Schröder (GMD), Paul Simmons (Nortel), Finn Trosby (Telenor) and Hans Wozny (Alcatel SEL).

16.1.6.6 Continuation of the Work

Since I only took on the chairmanship of IDEG/WP4 on a temporary basis and had become responsible for the implementation of the GSM network for Deutsche Bundespost Telekom in Germany (D1), I had to find a replacement chairman. During the creation of IDEG and its conversion to WP4, Graham Crisp had been particularly active in the development of the GSM architecture to support the wide variety of data and telematic services and interworking scenarios that had been identified in the requirements. As a result, I proposed Graham Crisp as my successor. However, the idea of an industry representative chairing a CEPT body was not welcomed with open arms by the GSM membership[27]. I, therefore, had a significant task in convincing the members that an industry representative could be entrusted with the chairmanship of a CEPT working party. As a result, Graham was elected Chairman of WP4 by GSM#17bis in March 1988. He was the first colleague employed by a manufacturer to become a working party chairman in CEPT.

The first WP4 meeting to be chaired by Graham was held in Florence in Italy on 5–8 April 1988. Those early years of GSM data standardisation saw some significant changes in the standardisation process, i.e. from CEPT as an organisation open only to PTTs, via industry participation, to ETSI. In 1989 the GSM work was transferred to ETSI and WP4 became GSM4. The GSM phase 1 standards for service opening in 1991 were "frozen" in 1990. The

---

[26] An interesting snapshot of the state of the GSM standardisation in 1988 is contained in the proceedings of the "Third Nordic Seminar on Digital Land Mobile Radio Communication", Copenhagen, September 12–15 1988. During that seminar, a GSM day was held. The GSM day and the rest of the seminar included a number of papers and presentations on the data and telematic aspects of GSM. These included: Implementation of telematic and data services in a GSM PLMN, F. Hillebrand; Architectural aspects of data and telematic services in a GSM PLMN, G. Crisp, A. Eizenhöfer; The Radio Link Protocol (RLP) – a recommendation for the transmission of data in the CEPT GSM Public land mobile network, T. Schroeder, I. Harris, H. Madadi; Rate adaptation and interworking functions for the support of data communication services by a GSM PLMN, A. Clapton, C. Gentile, S. Thomas, G. Ponte, P. Simmons; Support of data transmission services in the European digital cellular 900 MHz mobile communication system, J.C. Benard-Dende; Message communication within the GSM system, B. Kvarnstrom, J. Reidar Rornes, F. Trosby.

[27] Based on a proposal of GSM CEPT had allowed the participation of colleagues from industry in technical working groups. But they were seen as part of a delegation of an administration (see guidelines in GSM Doc 3/87 not available on the CD ROM). This showed clearly the need to open a path for full participation, which was achieved in ETSI.

last GSM4 meeting chaired by Graham Crisp was GSM4#22 held in Vienna, Austria on 13–17 May 1991.

All quoted GSM Plenary documents can be found on the attached CD ROM. The quoted GSM WP4 documents are not copied on the CD ROM. They can be retrieved from the ETSI archive.

Chapter 16: Short Message and Data Services

Section 2: The Development from Mid-1988 to 2000

Kevin Holley[1]

16.2.1 Short Message and Data Services in 1988

In 1988, Graham Crisp of GPT took over the role of GSM WP4 chairman from Friedhelm Hillebrand and continued the good work progressing towards the completion of phase 1. I started to attend GSM WP4 at the same time. It was a small but focused group, about 25 people splitting into four or five groups to do the detailed development in groups of 4–6 people, but joining together at the end of the meeting week to agree results.

In those days, the major focus of the group was in trying to get ISDN terminal equipment to work together with the mobile devices. So the focus was more on what signalling needed to be carried to and from the mobile across the network than trying to squash modem tones into a carrier designed for speech. There were some proposals for "modem codecs" which would convert modem tones, but it was felt that an all-digital system would give a much better performance and that ISDN would anyhow be the way of the 1990s by the time the system was installed.

16.2.2 Radio Link Protocol

In order to carry data across the network efficiently, it was decided that two types of service would be offered, called "transparent" and "non-transparent". The "transparent" services would be characterised by low delay but would suffer from data loss when the signal quality was low. The "non-transparent" services would be characterised by high data integrity but higher delay when the signal quality was low. The "transparent" services were thus protected only by "forward error correction", by sending the important bits more than once, and also using predictive Viterbi techniques just as for the speech encoding.

For the "non-transparent" services it was necessary to detect errors in the transmission path and retransmit packets which were received in error.

---

[1] The views expressed in this section are those of the author and do not necessarily reflect the views of his affiliation entity.

In 1988 two separate simulation studies were undertaken to show how well the retransmission protocol, called Radio Link Protocol (RLP), performed. RLP numbers each data packet and makes sure that each is received intact, and in the correct order. When something goes wrong, the receiver has to tell the transmitter to go back a few data packets and retransmit from there. These studies, reviewed by WP4, showed that the extra complexity required to pick out one data packet and retransmit that one only, contributed significantly to the overall throughput. Whilst of course this made the RLP implementation more complex, it was felt to be beneficial to make the most of the limited data capacity available.

16.2.3 Cell Broadcast

During 1988 interest also increased in the cell broadcast service. The basic concept was to make text available to all phones in a particular area when they are idle (i.e. not in any kind of call). The text could be general information or "teasers" which encourage users to make revenue-generating phone calls. The basic parameters for this service were set this year. An operator could remove one signalling channel (SDCCH) and use this for Cell Broadcasts (CBCH) instead. Each message would be transmitted with a header showing the intended purpose of the message, a serial number and a page number (to allow longer messages to be transmitted). Due to the capacity of an SDCCH it was only possible to send messages of 93 characters of 7 bits, and one of these messages approximately every 1.88 seconds. With this technical capability, it was envisaged that a whole range of information could be made available to users, however as the service proceeded to market it hit problems with handset implementation and investment by network operators.

16.2.4 Facsimile

In 1989 the attention of WP4 was drawn to the development of a reliable way to transmit facsimile images over a mobile network. The existing CCITT recommendation T.30 was the world standard for facsimile machines, but this was designed for consistent data channels, which performed at the same level throughout the duration of the call. With a mobile data connection the channel quality can vary dramatically within a few seconds. It was thus necessary to find a way to re-code the facsimile transmission so that the mobile channel inconsistencies were hidden from standard fax machines. One of the major problems with this approach was the so-called Non-Standard Facilities (NSF) capability. Fax manufacturers had been able to detect the presence of a remote machine of the same manufacturer, and through the NSF negotiation in T.30 decide to use a completely different protocol, different modem transmission rates or a variety of other techniques to make "like" fax machines exchange documents more effectively. Whilst this offered the user some advantages on fixed lines, unfortunately it was impossible to support these features over a mobile channel.

Even without NSF, fax machines use two speeds during transmission, one (higher speed) for the actual picture data and another (lower speed) for negotiation between each page. All of this meant a detailed picking apart of the way fax machines work in order to re-create this reliably at either end of the mobile channel.

In 1990 it was discovered that, whilst the decomposition of the fax protocols for carrying over GSM was sound, the transport mechanism chosen to carry this over the radio was flawed.

This meant that a number of meetings had to be called quickly to almost completely redesign the fax transmission capabilities of GSM in order to meet the deadlines for phase 1 completion. Several fax and data experts were called together to provide reliable framing and frame detection at either side. This was completed in a very short time, thanks mainly to the efforts of Trevor Gill, Ian Harris and Chris Fenton. The "second version" of the fax design was proven to work and is still used today on GSM networks.

In the meantime there was a debate raging over the data transport concept used to carry facsimile. One group of people felt that it was important to provide good quality images and that because fax quality would be visibly affected by errors in the transport, GSM should use the non-transparent data transport for fax. Another (larger) group of people pointed out that fax machines have no capability for flow control (slowing or stopping transmission whilst the data channel ensures that all parts of the image are reliably transmitted), and that any significant radio fades would result in the fax call being prematurely cleared. This argument boiled down to whether it was better for a poor radio channel to result in some visible line errors in the fax, or a premature end to the fax call. Whilst this debate was raging, it was decided to develop two separate specifications for facsimile. GSM 03.45 was the specification for transparent fax and GSM 03.46 for non-transparent fax.

In the end, the proponents of having a few line errors won much of the debate. It was agreed that the default, mandatory part would be based on the transparent mechanism, and that operators could, as an option, also implement GSM 03.46 non-transparent fax. This was one of the key basic ways of working in the early days of GSM, everything was designed with roaming in mind. It was not sensible to leave this completely as an operator option, because people wanted to be able to send faxes when they were roaming as well as at home. If their home network only had transparent fax and the roamed-to network only had non-transparent fax then they would have no service at all whilst roaming.

In 1991, Graham Crisp resigned as chairman and for a while there was no volunteer to carry on the work. This was reported to the main GSM committee and the report shows that there was even a risk of closing the committee at that point unless a volunteer could be found. Fortunately, Michael Krumpe from Siemens stepped forward and we were able to continue the work into GSM phase 2.

16.2.5 Cell Broadcast Enhancements

The basic radio interface design for cell broadcast had been completed in phase 1, but there was little information about how to deliver dynamic information across the network to the BTS. During this time, there was more detailed discussion about standardising the higher layers so that a "Cell Broadcast Centre" could be developed for high-level scheduling and message transmission across the network to the BSC. The BSC-BTS interface was also enhanced. Concern about battery life was always at the forefront of everyone's mind when considering something which happens in "idle mode". Michel Mouly came up with the idea of providing a "Cell Broadcast Schedule", a packet which explained when in time the different cell broadcast messages would be transmitted. Thus a mobile would only need to listen to the CBCH for the cell broadcast schedule and also during the slots when "wanted" messages could be expected. This was added to the phase 2 standard to give battery life improvements.

16.2.6 Call Waiting For Data Calls

There was a long discussion about call waiting on data calls. On analogue systems, call waiting tones are injected into the audio by the switches. However, on GSM call waiting tones are generated locally by the mobile in response to additional signalling on the radio interface. It would thus be conceivable for someone on a data call to receive call waiting information if a voice call comes in whilst a data call is in progress. When used in conjunction with calling line identification, it would be possible to tell, for example, that an important voice call was coming in during a data call. Whilst it would not be possible to put the data call on hold to accept the voice call, it would be possible to clear the data call and take the voice call, or divert the voice call by the procedure for "User determined user busy". As a result of this discussion a request was sent to the other GSM committees requesting that call waiting be allowed on data calls.

In 1992, Michael Krumpe stepped down from the GSM4 leadership to be replaced by Wolfgang Roth from T-Mobil.

16.2.7 Control of SMS by an External Terminal

During the early years of GSM WP4 it was already identified that there was a need to control mobile data functions in general, and SMS in particular from an external terminal. A PC could allow the typing of messages on a full keyboard and display of all the message at once. So a specification was developed for full control of all SMS functions. This was based on the coding used for signalling on the radio interface in GSM 04.08, and included a checksum in case of errors on the serial wire interface. The specification was developed in a very short space of time by Ian Harris, Arthur Gidlow and me and was approved as part of GSM phase 2. It was implemented in early PCMCIA cards for mobiles such as the Nokia 2110, and directly in phones such as the Orbitel 905 which were both around in 1994–1995.

16.2.8 SMS Extras

A number of advanced enhancements to SMS were considered and accepted during 1992–1993. These seemed minor enhancements at the time but are now widely used around the world. The first enhancement was "immediate display" messages, which would be shown on the screen straight away, rather than waiting for the user to select "read messages". These have recently caught the imagination of the more technically aware teenagers, who call them "flash text". They are useful when there is an urgent message which is time critical and needs no storage in the phone or SIM. Some operators are now using these messages to alert users that their SMS storage is full.

Another enhancement was "type 0" messages, which are acknowledged by the phone but not displayed to the user or stored. These messages are useful to detect the presence of a mobile, for example in order for a voicemail system to make a call to push voice messages to the user.

Replace short messages allowed automatic deletion of a message already stored, and replacement with a different message. This was envisaged for voicemail alerts which said "you have one message", "you have two messages" etc. so that the storage wasn't filled up with this type of message.

During this time it was also noted that there would be different storage for SMS, in the mobile memory, on the SIM and in an external data device. So provision was added to enable an SMS originator to direct the messages specifically to one of these storage areas.

16.2.9 How to Extend the SMS Coding Space?

By 1994 so many additions had been made to the SMS header that the 24 byte header was no longer large enough to accommodate new requirements. An extension mechanism was needed for future requirements, one of which was already identified: to allow concatenation of several messages to make a "long message". Several of the key experts who had been involved for many years put their heads together and observed that the coding for the SMS header should have been made more flexible for phase 1, however it was now too late to change this. It was therefore agreed to allow part of the space for the text of the message to be used for additional "header" information. This so-called "user data header" was identified in the main header. To avoid older mobiles displaying the "user data header" as garbage text, we recommended that the end of the user data header included a ⟨CR⟩ character which was defined in phase 1 as meaning "start displaying again at the beginning of the current line", effectively overwriting the user data header characters. The first use of user data header was for SMS concatenation. SMS could now be up to 255 segments of 150 or so characters, and whilst the maximum length of messages would take a long time and be expensive, it was thought that concatenation of three, five or even ten messages would be very useful. This proved to be the case when Nokia introduced its picture messaging based on SMS concatenation in 1998

16.2.10 Message Alert Icons

By 1994 some of the ideas implemented into the operator-specific "GSM1800" specifications were seen to be useful additions to the GSM standard. One of these was the ability to display an icon showing that voice messages were waiting to be read in the network. This concept was extended in 1994 and four indications were defined: voicemail, e-mail, fax messages and "other" messages. It was not decided what "other" meant at the time but there was coding space available so it was recommended that manufacturers implement the capability and it would be decided later what the appropriate use would be. However in the 7 years since the definition of this capability there has not been a need identified for "other"! There were two implementations of the coding for these indications – either by taking part of the text of the message as "user data header" or by using one of the header bytes. If the "user data header" was used then the number of voice, fax, e-mail or "other" messages waiting could be indicated. Whichever mechanism was used, the associated text could either be "discarded" or "stored". If set to discard this would mean that older mobiles would display the text and newer mobiles would just display the icon.

16.2.11 SMS Character Set Enhancements

Many times during 1993–1995 the subject of extending the SMS character set was raised. The original SMS character set was based on the set proposed for the European paging system called ERMES, which was supposed to provide for the majority of European characters, including uppercase Greek characters, however it didn't include several characters needed in Eastern Europe, and even omitted some Northern European characters. Coupling that with the expansion of GSM into the middle east and far east, it was clear that some major enhancement was needed. One option would be to add several code pages to the GSM system, but this would mean that GSM4 would have to define all the characters. Code page switching would result in large tables in the mobile memory, and of course switching code pages means additional characters taken up for switching instructions. So this subject was discussed many times over a couple of years until finally we settled on the use of a standard 16 bit character set used for UNICODE and also known as UCS2. This character set allows transmission of the vast majority of characters in the world but of course uses more than twice the space of the original GSM set. Up to 70 UCS2 characters can be sent in one message, and this can be extended with concatenation.

16.2.12 Compression of Data

In the fixed world, modem evolution was able to occur independently of the transmission medium. But a radio environment is very different and GSM has always provided a proper managed solution to smooth over the roughness of radio technology. When data compression was introduced based on recognising repeated patterns and transmitting smaller symbols instead of the actual data, this potentially made quite a difference to the fixed environment, with gains of around a factor of 4. In order to re-apply this to the mobile environment it was necessary to augment the existing data specification to allow the compression technology, essentially the same as used on fixed lines, over mobile. The main difference between fixed and mobile is the error rate. On a fixed line you can expect very low error rates, so it is very unlikely that something could go wrong when you send a symbol instead of the real data. On a mobile connection things are very different and by corrupting the symbol you corrupt a lot of the data. So the protection of data has to be improved compared with the fixed situation. All of this work resulted in a very useful addition to GSM allowing faster perceived data transmission speeds.

16.2.13 High Speed Circuit Switched Data

During the early years of GSM development, modems at speeds of 300–1200 bits per second were the norm, and 2400 (2.4 K) modems were only just becoming available. So GSM's speed of 9600 (9.6 K) seemed high enough and even a little future proof at that time. Of course with the growth in demand for data over fixed lines, modems became faster and faster and soon 9.6 K was only on the border of "acceptable" and at risk of becoming "poor". In order to compete with other radio technologies, especially in the US, the GSM standard had to be improved. The first improvement was to increase the potential speed from 9.6 to 14.4 K using the same slice of radio spectrum but a different coding technology. The drawback was that the coverage available at 14.4 K would not be as great as that afforded by 9.6 K, although it was found to be quite acceptable. To this we added High Speed Circuit Switched Data (HSCSD). HSCSD enabled operators to devote more than one TDMA timeslot to a single user, allowing them to multiply the speed by a factor between 2 and 8. With 2×14.4 K channels this began to look like fixed modem speeds of 28.8 K, and devoting two channels to one user was not a big overhead (certainly compared with devoting eight channels to one user). HSCSD has now been implemented by several operators and is in a few mobile handsets.

16.2.14 AT Commands

When modems first started to appear, the ITU developed a standard for control of modems from a terminal called V.25bis. Since this was the only standard available when GSM data was first developed, GSM4 extended this standard to be applicable to control of a mobile modem, with all of the extended facilities. However, due to the popularity of the Hayes AT command set, V.25bis was never widely implemented and everyone in the industry became used to AT commands as the way of controlling modems. In 1995, Nokia, Ericsson and Hewlett Packard developed their own specification for AT commands for controlling GSM equipment. In discussions with these companies inside GSM4, they were persuaded that it would be better to have a GSM standard for this, to ensure that it would be compatible with all GSM features. After a couple of intensive ad-hoc meetings and largely due to the very hard work of Petri Heinonen, the first specification was complete. This specification has led to an explosion of utility software, which works with most of the GSM handsets with data facilities, software for PDAs, PCs, etc. to set up the phones (e.g. phonebook), send and receive SMS and of course make data calls.

16.2.15 SMS Compression

By 1996, the small number of characters in SMS was becoming noticed as an issue. Several operators thought that in addition to concatenation, and especially for use together with UCS2, it would be useful to allow SMS messages to be compressed. Compression of a short length character stream is not as easy as modem data compression, but this can be offset to some extent by the fact that the entire message is known before transmission. A number of techniques were investigated and we finally agreed a standard compression technology which could be used with GSM. In addition to the basic compression, "standard dictionaries" for English and German were also defined. The use of these dictionaries enabled improved compression figures.

16.2.16 Mobile Application Execution Environment

During 1997, SMG4 started to talk about enabling third party applications to run on a mobile. The industry was already talking about bringing Internet content to the mobile world, however there were different proprietary schemes being proposed by a number of players. In May 1997 the proponents presented their schemes to SMG4, all of which were received with interest. Along with the interest, SMG4 gave a strong message that the industry could not make a large market through having these disparate schemes and that the proponents should work together to create a single proposal, with the intention to bring the proposal into SMG sooner or later. This indirectly led to the creation of the WAP forum, which brought together the early proposals shown to SMG4 in 1997. Since this time, SMG, 3GPP and the

WAP forum have worked together, often taking ideas from one sphere and making use of them in another.

With the work starting on mobile browsers it was clear that there would be a capability for downloading content from the Internet, but there was concern that the downloading of applications which would run directly on mobiles should be specified in a co-ordinated way. One specific worry was that the always-on capability of mobile networks coupled with mobility and downloaded applications could result in an insecure environment for the user. So the Mobile Application Execution Environment (MExE) work started to develop a secure framework for externally-specified APIs so that operators and manufacturers could be confident of adding value to the terminals in a flexible but secure way. This has resulted in the Release 98, Release 99 and Release 4 issues of the MExE specification.

16.2.17 SMS Icons and Tones

During 1998, the subject of downloaded images and sounds caught the interest of the industry, and it was recognised that it would be useful to have a standard way to convey these using an SMS message. After some hard discussions and particular effort on the part of Lars Novak and Magnus Svensson, finally the specification for enhanced SMS or "E-SMS" was agreed early in 2000.

16.2.18 The End of SMG4

The globalisation of 3G mobile networks was recognised as a necessary step for some time, but it was only towards the end of 1998 when 3GPP was established, that it was clear that SMG4 could not continue in its present guise. SMG4 then comprised several different aspects – the control of real-time data in the form of transparent or non-transparent bearers, or real-time fax at the mobile side and the network interworking side, plus terminal external interface specifications (e.g. AT commands) and applications such as SMS, MExE, etc. The right way to divide the 3GPP work was considered for some time, and in the end it was decided that SMG4 would continue without the bearer control aspects which would be moved into SMG3. This left SMG4 in a good position to be merged into the new 3GPP organisation as T2 "Terminal Services and Capabilities", a working group under TSG terminals. The combined SMG4/T2 group continued to report into TSG terminals and SMG until the close of SMG in July 2000.

Chapter 16: Short Message and Data Services

Section 3: The General Packet Radio Service (GPRS)

Wolfgang Roth and Jürgen Baumann[1]

16.3.1 First Ideas

Even before the work on circuit switched data services was finalised by the end of 1994, the first steps to define a new packet switched data service were taken. It was found, that for short transactions, a better utilisation of resources was needed in order to offer services at lower costs.

The first work item description sheet[2] (SMG 200/93) was discussed in 1993. The arguments were as follows:

- The growing number of dedicated packet radio networks evolving was considered as a need for a competitive service in GSM taking advantage of GSMs Pan-European availability.
- The access to packet networks was possible with the already defined services but it was felt that the provision of a circuit switched channel was inefficient for applications which are bursty in nature. In addition to that the Short Message Service (SMS) provided already a connectionless packet mode service but it was clear that the capacity was not sufficient for new applications. Services in mind were fleet management, logistics, telematics and mobile offices.

The need was clearly seen by network operators. However, manufactures, explicitly some of those in the business of dedicated packet radio networks, had initially some doubts and a controversial discussions took place.

16.3.2 Start of Standardisation (Risks And Delays)

Finally the need for the specification of a General Packet Radio Service (GPRS) was accepted. Different aspects like service requirements, structure and usage of channels, signal-

---

[1] The views expressed in this section are those of the authors and do not necessarily reflect the views of their affiliation entity.

[2] Temporary document SMG 200/93.

ling, new network elements and interworking with the fixed networks had to be considered. Therefore, the work did not start as usual in a specific STC, but a GPRS task group was established in April 1994 under the chairmanship of Gunnar Sandegren.

It was obvious that the subject was very complex and had a major impact on the whole system. The possibility to upgrade existing BTSs was seen as very important. Besides that, the discussion on responsibilities of different aspects of standardisation and the need to evolve the system for all new features in a consistent way which led to the establishment of the system architecture group had a negative impact on the speed of the standardisation process. It was clear by mid-1994 that the work could not be completed by the end of 1994 as originally planned.

In parallel, the need for higher data rates led to the new work item High Speed Circuit Switched Data Service (HSCSD). This had to be considered in conjunction with GPRS as some service requirements could be met with either service and HSCSD was more in line with the existing GSM system. As GPRS was delayed and some very urgent service requirements like road pricing led to the rapid standardisation of the Packet Data on Signalling channel (PDS) which was agreed on as a study item in October 1994. This service was standardised but never implemented as GSM road pricing was not realised. Certainly, this caused an additional delay for GPRS.

The GPRS task Group completed its work at SMG#14 in April 1995. The responsibility was given back to the regular groups (STCs) and a funded GPRS project team was established to speed up the standardisation process. The completion of GPRS was expected by February 1996.

16.3.3 Basic Concept

16.3.3.1 Air Interface

It was soon clear that the basic channel structure on the radio interface should not be changed. A method should be used to share an existing traffic channel between several users. The requirement was then to allocate this channel for a short time to a specific connection and to reuse it afterwards for another connection. This would allow much better utilisation of the capacity as the channel would not be idle in the case of no data to transmit. The frequency of resources reallocation to any user should be high enough to avoid major delays for any user.

Ideas to use the idle periods on speech channels were not seen as an acceptable solution as the impact on data throughput at a given speech quality was very high. Only a best effort data service would have been possible.

Four main proposals were discussed and analysed very intensively. Common simulation requirements, e.g. a set of traffic models were defined and each vendor produced simulation results according to the agreed methods. These results in addition to the level of hardware impact on the BSS were the main criteria for the decision. Finally one main concept was agreed as the basic concept of the channel access principles for GPRS. Due to the high interests of each vendor to push its own proposal as the solution for the standard, it was a huge overall success to have at the end a common decision for one proposal.

Shortly before finalising the concept of the air interface a vendor came up with a proposal called half duplex support. It defined a new channel allocation strategy. In contradiction to the original one (dynamic assignment), the data blocks are reserved in advance. This fixed assignment strategy enables simple MS to support more simultaneous timeslots.

This activity brought an additional complexity into the standard in a very late phase and caused an additional overall delay in the standardisation process. Although no other company supported this feature, there was no possibility to stop this activity. As a compromise this functionality was defined as optional for the network and as mandatory for the MS.

This is one of the rare examples where no common agreement could be reached on a basic conceptual question.

16.3.3.2 Backbone Network

On the interface to the fixed networks it was early accepted that only the interworking with the Internet/intranet was important. Packet networks based on X.25 did not play a major roll anymore. The whole concept was therefore based on Internet Protocols (IPs) and addresses. This implies using internet addresses for all connections.

International roaming was seen as an essential requirement. Different possibilities were required as there are direct links between networks, specific roaming networks and secure connections via the public Internet (tunnelling).

A major task was to define the switching system for the packet transfer. New elements (packet routers) were introduced in order not to impact the MSCs. The concept of an IP backbone was accepted. The definition and the location of the separating function between circuit channels (allocated for a very short period) and the packet backbone had to be done. New protocols were required for the transfer of packets from the radio channel to the Internet gateway.

Another major and extensively discussed architectural question was the transport protocol on the Gb interface which is the interface between the BSS and new packet network. Frame relay and ATM (as connection oriented protocols) and the IP (as connectionless protocol) had been the main discussed choices. Frame relay was chosen which was at that time an already proven and widely used network transport protocol.

16.3.3.3 Mobile Station

On the terminal side, three classes of mobiles were needed: GPRS only mobiles, mobiles for speech and packet data alternatively and mobiles for speech and packet data simultaneously. A major concern was the complexity (i.e. prise) of mobiles. A second receiver was seen as not acceptable.

16.3.3.4 Quality of Service

Support of Quality of Service (QoS) within GPRS was introduced in a very late standardisation phase. At that time it was not possible to define a future proof overall proposal. Therefore, the main idea was at least to add the signalling possibilities between the MS and the network to offer a hook for later implementations and standardisation activities.

16.3.4 Main Standardisation Work

The first progress was noted on the service description in July 1995 and the stage 1 specification was presented for information. But in February 1996 it was obvious that the completion date could not be met and the whole GPRS issue was questioned again. Service requirements should be revisited in conjunction with those for HSCSD and PDS. The SMG#17 (February 1996) report says: "It was clarified that HSCSD will allow the increase of the transfer rate as done in the fixed network, traffic of bursty nature will be covered by GPRS. It seems that first priority shall be allocated to HSCSD and PDS and second priority to the GPRS based on the fact that services have to be mature enough and PDS can be a way to train the users for packet services."

In April 1996 it was requested not to have a further delay of final working assumptions and in October 1996, it was reported that the stage 1 service description (GSM 02.60) was now reasonably stable and should be available at SMG#21 (February 1997) for approval. It was again requested to speed up the standardisation process. To achieve that, it was even proposed to remove the interface to the MSC which would have prevented a common paging.

The progress on the radio part of GPRS was also not as expected. It was reported that it could not be finalised in SMG#21(February 1997) and should now be planned for SMG#22 (November 1997). It was proposed, as a quicker path, to define as much as is essential for the mobile station hardware.

The necessary MAP extensions needed for other services as well were also not ready. All operators and manufacturers stressed the need to finalise the last pieces of Release 96 at SMG#21(February 1997) in order to satisfy market needs.

Security aspects of GPRS were started rather late. It was recognised that a set of new security features was needed, especially on the radio interface protection.

16.3.5 First Release GPRS Phase 1 (GSM Release 97, Completed in March 1998)

Major progress was reported in February 1997. The service description (stage 1, GSM 02.60) was approved. Some functions like point to multipoint were postponed to a second phase. The radio specification (GSM 03.64) was presented for information. There were still some changes needed but finalisation was expected by SMG#22 (June 1997). The general architecture (stage 2, GSM 03.60) was presented for information. Finally in November 1997 the general principles and part of the basic work was completed. On the remaining important parts work continued with priority. In particular progress on security was reported and the ciphering requirements were defined. Type approval and clarification on types of mobiles were other items to be completed. A proposal to shift GPRS into Release 98 was rejected. The workload was so high, that it took until SMG#25 (March 1998) to complete GPRS phase 1 for Release 97.

16.3.6 Major Corrections, Start of Service

The first version of GPRS phase 1 available in March 1998 was far from being ready for implementation. However, network operators urged manufacturers to start development.

With that, the holes and errors became clear very soon. Both, the network part and the radio part needed corrections. There was still major work to be done in the following areas:

- Review of BCCH and corrections on RLC/MAC were envisaged
- Legal interception
- O&M
- Combined mobility management procedures
- PLMN selection
- New ciphering algorithm
- Clarification to timing advance
- Roaming with GSM1900
- Interaction with cell broadcast

In 1999 several concerns about the progress were expressed. It was stated that operational networks were expected by the end of 1999/start of 2000 and that it was too late to change major functions like encryption algorithm. Another statement was that the somehow never-ending GPRS story might have the potential to sooner or later damage the credibility of the entire community; that this should not be misunderstood as a criticism of the GPRS implementers and testers; but that 6 years of announcing GPRS without finally arriving at a common stable specification and, at the same time, proposing an all-IP based network as an achievable Release 2000 goal wouldn't go together.

It took then until SMG#29 (June 1999) for a solution to be implemented. And even then it proved that still more incompatible changes on the radio part and the interface to the core network were necessary so that commercial mobiles could only be based on specifications including SMG#31 Change Requests (CRs). So implementations almost had to be changed again and the first networks could not start with an experimental service before mid-2000.

A further unsolved problem was testing of mobiles.

T-Mobil started as the first operator the GPRS service on 26 June 2000, but corrections and optimisations were necessary until an acceptable service could be offered to the customers.

Figure 16.3.1 shows the number of CRs for the radio system over time. The number of CRs is clearly an indication of the status of specifications regarding the readiness for implementation. This should be considered for similar complex systems in the future.

16.3.7 GPRS Phase 2 (Completed in February 2000)

During the work on the first release of phase 1 (which was rather incomplete) a list of items for phase 2 was already established. All open items which did not need to be fixed immediately were included. Work started in various groups on

- GPRS for PCS1900
- Mobile IP interworking
- Access to ISPs and intranets
- Charging and billing (hot billing, prepaid)
- Point to multipoint
- QoS enhancements
- Support of supplementary services
- Additional external interfaces (other packet networks, modem, ISDN interworking)

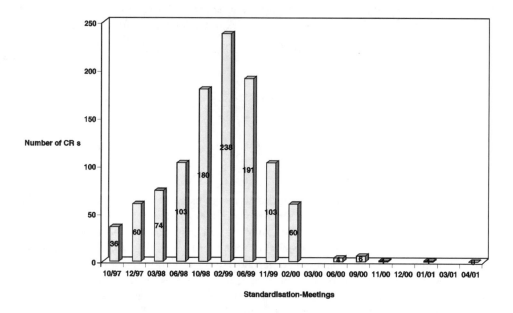

Figure 16.3.1   Number of R97 change requests for GPRS from 10/97

Work was planned to be part of Release 98. However, a possible postponement into Release 99 and with that into the release which contained the first UMTS specifications was seen as not attractive, since the key features were already covered by Release 97, phase 1. Some parts were already available at SMG#28 (January 1999) but as there was so much correction needed for phase 1 the work item slipped into Release 99 and was merged with 3GPP activities. It is therefore not very likely that a GPRS phase 2 will be developed as a GSM functionality.

Features like point to multipoint and additional external interfaces were not seen anymore as necessary. Therefore, it cannot be expected that they will appear in Release 99 or later.

16.3.8 Outlook

The whole standardisation of GPRS showed that it was a hard way to come to a solution. First general concerns, then clarification of requirements and afterwards real hard fruitful work led finally to a complete set of specifications. Now in May 2001 several networks are in operation and all having contributed to this success can be very proud.

It is also a great success that the principles of the network part are the basis for the packet switched part of the UMTS system. This shows the excellent visionary work when defining the basic principles. Certainly signalling and QoS issues have to be improved on the path to an all-IP network.

Chapter 17: Testing and Type Approval of the Mobile Stations

Rémi Thomas and David Barnes[1]

17.1 Introduction

In this chapter we aim to explain the principles and evolutions of the testing and type approval of the GSM Mobile Equipment (ME). After having recalled the context at the end of the 1980s (paragraph 17.2), we outline the technical background (paragraph 17.3) and we explain the first step of the type approval, i.e. the "interim type approval"(paragraph 17.4). Then we describe the TBR regime (paragraph 17.5) and how it evolved to cope with phase 2 (paragraph 17.6) and phase 2+ (paragraph 17.7). We explain the work of two committees, namely ETSI/SMG7 and TAAB (paragraph 17.8), we give some hints regarding the test tools (paragraph 17.9). Finally in the conclusion (paragraph 17.10) we consider the achievements of the GSM type approval and we rapidly explain how the principles of the GSM conformity testing are presently reused by 3GPP for the conformity testing of the UMTS terminals.

17.2 The First Steps in 1988–1990

To understand the ideas underlying the first actions for GSM Mobile Stations (GSM MS) testing, it is necessary to recall some elements which were agreed by the GSM community already at the end of the 1980s:

- ME and SIM together form the GSM MS; the Subscriber Identity Module (SIM) is a chip card provided by the network operator to the mobile subscriber, this card contains in particular all the subscription related data;
- Unlike to the SIM card, the ME is not under the control of the network operator, therefore the network operators have to ensure through some testing process that the MEs will be able to provide service and will not cause any damage to the network.

There were therefore two overriding criteria. Firstly the MSs from different manufacturers should interwork with all networks. Secondly the GSM community required that roaming should work from the beginning of the commercial service. Today this is seen as a straightforward feature but this was not the case in those years and this resulted in some strict conformity requirements on the MEs.

---

[1] The views expressed in this chapter are those of the authors and do not necessarily reflect the views of their affiliation entities.

To cope with these two requirements the GSM community set up two processes:

- A group was established under the leadership of the "Permanent Nucleus" (PN) of the CEPT/GSM committee. The mandate of this group, the "Eleven Series Drafting Group" (ESDG), was to specify conformity tests of the ME. The tests were produced on a voluntary basis by experts provided by companies contributing to the GSM standardisation; these tests were included in a GSM specification, namely GSM 11.10;
- A process for the development, purchase and deployment of a test tool implementing the ME tests defined by the ESDG, this testing tool was called the "*System Simulator*" as the purpose of this tool was to completely simulate a GSM network.

These two tasks were in fact quite ambitious. In order to understand the situation, it is useful to consider a general description of GSM 11.10 and the complexities of developing a system simulator.

17.3 The Technical Context

17.3.1 Some Insight into GSM 11.10

We give some insights in tests domains which are defined in TS GSM 11.10 the purpose of which is quite ambitious as it aims to cover all the aspects of the ME conformity, both at the air interface and at the SIM/ME interface.

17.3.1.1 Radio Tests

This part of the testing checks that the ME is compliant with the radio performance defined in the 05 series of the GSM recommendations. Thus a compliant unit will behave in a foreseeable manner. This is particularly important in order to enable a GSM network operator to make a radio design providing all the advantages of the GSM radio performance.

The main domains of testing are listed hereafter: transceiver, transmitter, receiver, reference sensitivity, usable receiver input level range, co-channel rejection, adjacent channel rejection, intermodulation rejection, blocking and spurious response, timing advance and absolute delay, access times during handover, temporary reception gaps, channel release after unrecoverable errors.

17.3.1.2 Radio Link Management Tests

This comprises the cell selection and reselection tests and the received signal measurements tests.

17.3.1.3 Signalling Tests

All the signalling protocols of the radio interface are tested. First, this comprises the tests of the layer 2 signalling functions which verify the conformity with the LAPDm protocol defined in GSM 04.06.

This comprises as well the layer 3 testing which verifies the conformity with TS GSM 04.08, we indicate rapidly the main domains of testing:

- Handling of unknown, unforeseen, and erroneous protocol data: the purpose of such tests is to cope with future evolutions of the protocol;
- Test of the elementary procedures for radio resource management: Immediate assignment, normal paging, measurement report, channel assignment procedure, handover, ciphering mode setting, channel release;
- Test of the elementary procedures for mobility management: TMSI reallocation, authentication, identification, MM connection;
- Test of the procedures for call control;
- - Finally testing of structured procedures is performed in order to verify that the ME under test is able to perform call set-ups.

17.3.1.4 Testing of the SIM/ME Interface

The purpose of these tests is to check the conformity of the SIM/ME interface of the ME as defined in TS GSM 11.11, testing of the SIM itself is not part of TS GSM 11.10.

It is tested that the ME is able to retrieve information from the SIM and to update them, for example this comprises the following tests: forbidden PLMNs, location updating and undefined cipher key; MS updating forbidden PLMNs, MS updating the PLMN selector list.

Some physical characteristics of the SIM/ME interface are further verified: mechanical tests, and electrical tests.

The test of some services (Short Message Service (SMS), some supplementary services) and of some MS features can be found as well in GSM 11.10.

Last but not least, GSM 11.10 comprises tests of speech teleservices and tests of speech transcoding functions.

17.3.2 Phased Development of GSM – the 1800 MHz Band

In October 1989 the GSM community chose to develop the GSM standard in *phases* and to produce a complete set of standards for each phase.

Phase 1 of GSM was functionally frozen in January 1990 and it is worth recalling that it only encompassed the 900 MHz band thus excluding the operations in the 1800 MHz band which were called "DCS 1800". In order to specify the DCS 1800, ETSI/SMG created delta-specifications which were to be understood together with the corresponding phase 1 specifications. For instance there was a specification named "GSM 11.10-DCS" the first version of which was approved by ETSI/SMG in June 1992.

The basic phase 2 of GSM was frozen in October 1995; among other decisions it was decided to include DCS 1800 in basic GSM phase 2. This implied that 900/1800 operations (mainly 900/1800 handover and 900/1800 selection/reselection processes) were part of basic phase 2.

Then the concept of GSM phase 2+ was introduced in order to add continuously new features to the GSM standard. It is the choice of each network operator, each network infrastructure manufacturer and each MS manufacturer to implement or not to implement any of the GSM phase 2+ features.

To achieve this, new sets of functions are added without changing the protocols for the existing ones and without causing incompatibility problems with already existing equipment.

In other words, there is an unchanged base which is basic phase 2 as approved in October 1995 and then the phase 2+ features are build on it. Based of these rules, new releases of the GSM standard were produced: Releases 96, 97, and 98.

As a consequence a new version of GSM 11.10 was written for each phase of the GSM standard: a limited number of tests were modified to take into account evolutions of already existing features, and new tests were added in order to test new features.

17.3.3 Availability of System Simulators

The key to the testing of GSM MEs for type approval purposes was the availability of a suitable System Simulator (SS). As already explained the operators required a test specification and an SS that was sufficiently comprehensive to ensure interworking between different manufacturers' equipment and enable international roaming. It was apparent that such a development would be extremely expensive and the burning question was, who should pay?

There was no precedent for regulators to fund such system simulators where the requirements to ensure interworking were above and beyond those normally required for regulatory purposes. A similar situation had existed prior to this in the UK with the introduction of the TACS system and in this case a precedent had been set where the operators concerned had funded the development and purchase of a system simulator.

In the UK situation the operators (there were two in the UK at that time, Cellnet and Vodafone) funded the purchase and development of an SS. There was then a factor built into the type approval fee for the operators to recover their outlay over a set number of type approvals. At that stage of course no-one was sure just how many different terminals would be produced and eventually type approved. Needless to say all estimates were considerably less than the actual number and the financing arrangements worked successfully.

A similar approach was therefore considered for GSM and a small number of operators who were members of the GSM MoU undertook to underwrite the development of the SS for GSM type approval. This group of operators was known as the "*buyers club*" and was a vital initiative to get the development and procurement work underway for the delivery of an SS.

After the delivery of the initial SS it became apparent that it would be unfair for the original buyers club to continue to take the full financial burden of the continued development of the SS and the necessary test cases. The GSM MoU however provided the necessary finance and resources by establishing a separtae company within the MoU called GSM Facilities Limited (GSMFL). It was the purpose of GSMFL to see the development of the SS through to the complete implementation of GSM phase 2.

Throughout the lifetime of the buyers club and the GSMFL, Peter Zollman (Vodafone) worked as the project manager for the development of the SS and the associated test cases. In this respect his efforts were particularly appreciated in maintaining the project on track and keeping the GSM community informed of the current status.

The GSM 1800 (then known as DCS 1800) MEs were not catered for in the GSM SS. This was because at that time DCS 1800 was still a separate part of the GSM family. However, the DCS 1800 operators faced exactly the same problem as the GSM 900 operators in terms of their requirements for a suitable SS. At that time the only networks approaching commercial service were in the UK and One-2-One and Orange jointly funded the initial development and procurement of an SS. In actual fact they purchased two simulators.

At a later stage it was seen that there was a continuing process of validation of test cases for the SSs and a DCS validation group was set up under the PCN TAEG. This group was chaired by David Nelson (Orange).

17.4 The Interim Type Approval Procedure

17.4.1 The Technical Process

In 1987, the operators who signed the GSM MoU committed themselves to start the GSM service in 1990. Unfortunately the delays in the different developments resulted in delaying the fulfilment of this demanding commitment. In fact the phase 1 of the GSM standard was functionally frozen in January 1990. As a straightforward consequence, it was not conceivable to have GSM network infrastructure on the field in 1990.

Nevertheless the situation for the GSM terminals was even more worrying. As a matter of fact it appeared in 1990 that it was impossible to foresee when the test of the first GSM ME would be completed. Type approval and testing are inevitably blamed as a source of delay in the availability of MSs for a system such as GSM. One of the reasons for this is that by the very nature of the standards development process the test specifications can only be produced and finalised once the contents of the core specifications is clearly defined. Furthermore, one has to be careful to ensure that the test specifications only reflect what is in the core specifications and that no additional requirements are inadvertently added. Finally it is difficult to place a detailed order for test equipment without a comprehensive test specification. However, this is effectively what had to happen with the test equipment manufacturer having to track the changes to the test specifications as they were refined. This was further complicated by the fact that the test specifications were still trying to track corrections that were made to the core specifications.

It was therefore somewhat inevitable that delays were experienced by the extremely ambitious processes described in the previous paragraphs (namely the drafting of GSM 11.10 and the development of the GSM SS):

- significant work was already accomplished for the radio transmission tests but there remained some important work to produce a sufficient amount of signalling tests;
- the delays in the development of the SS were even more worrying, it became obvious that the SS was on the critical path of the GSM commercial opening, in fact it was not possible to foresee precisely when the SS would be able to test a GSM ME.

To cope with this difficult situation two types of actions were undertaken:

- on the standardisation side more efforts were put into the production of the tests, in particular during the first months of 1990 more signalling and protocols experts were involved in ESDG and a "signalling subgroup" was set up;
- regarding the test tools it became obvious that it was necessary to start testing with a tool including less features than the SS.

As a result of this the regulators and the GSM operators decided at the end of 1990 to set up the "Interim Type Approval" (ITA) in order to speed up the start of the GSM terminals testing. At that time there was considerable discussion between the operators on the minimum level of testing that would be acceptable for ITA. One view called for a high level of testing to ensure confidence in the GSM system. However, there was equally a view that said "we won't know until we start" and there was a need to get terminals into the market place. The famous quote from George Schmitt (Mannesmann) at that time was a new meaning for the term GSM – "God Send Mobiles".

The discussions came to a head in a meeting in Frankfurt where David Hendon (UK DTI) and Armin Silberhorn (German Ministry) were particularly influential in promoting a compromise that was acceptable to all parties and allowed ITA to go ahead.

From a test specification point of view the ITA relied on two principles:

1. reduce the set of tests to be performed;
2. simplify the specification of some tests cases in order that these tests can be performed on a tool simpler than the SS.

Item (b) applied particularly to the handover test cases and to the selection/reselection test cases. As a matter of fact GSM 11.10 required (and still requires) that a total of eight base stations be simulated in order to perform some handover and selection/reselection test cases. This implied having a fine co-ordination of both the radio and protocol activities of eight radio transmitters simulating these base stations. It appeared that such a feature was one of the most difficult features to be implemented. As a consequence it was decided that in the ITA the simulation of only four base stations was required. Evidently this reduces the accuracy of these tests but this was the only solution to speed up the start of the GSM MEs testing.

The ITA test tool was developed and produced by Rohde and Schwarz according to the requirements of the GSM operators.

An early version of the ITA test tool was available mid-1991. The first type approval under the ITA regime was given on 26 May 1992.

The ITA specified the type approval of phase 1 GSM MSs. Again this meant that this tackled MSs operating in the 900 MHz band only. As a consequence it was necessary to set up interim procedures for the type approval of DCS 1800 MSs. Technically it was based on GSM 11.10-DCS. From a regulatory point of view, it relied on a MoU for DCS 1800 type approval signed by the DTI (UK) and the German Ministry as UK and Germany were the first countries introducing DCS 1800. The MoU was quite significant in that it effectively introduced mutual recognition of type approval between the UK and Germany for DCS 1800 terminal equipment. This enabled full international roaming between the signatory countries. Although signed initially by Germany and the UK, other countries also signed the MoU as GSM 1800 began to spread. The MoU was eventually terminated when GSM 1800 equipment became covered by a CTR under the TTE Directive.

17.4.2 Additional Procedures

At the time leading up to the introduction of ITA for GSM the type approval was under the basis of what was known as the EC phase 1 Directive. This effectively gave the possibility of the mutual recognition of test reports based on what were known as NETs. These were European norms for telecommunications and in the case of GSM there was a NET 10.

The procedure at that time left a significant number of gaps with regard to the requirements for GSM terminals. At that stage a manufacturer would still have to obtain individual type approval in every country where the ME was to be placed on the market. There was no provision for the "free circulation" of the terminals for roamers and there was no procedure to deal with such issues as the allocation of International Mobile Equipment Identities (IMEIs).

The GSM MoU was again instrumental in solving a number of these issues and in order to progress work set up a group called the Type Approval Procedure (TAP) group which was chaired by Lilian Jeanty (Netherlands). The group included GSM operators and the type approval authorities involved in the early introduction of GSM.

TAP developed the procedures for the allocation of IMEIs so that a unique IMEI could be allocated to each GSM ME. Essentially the first two digits of the code were allocated by the TAP to a particular country granting type approval. The country concerned could then build up a unique set of IMEI codes in conjunction with the manufacturers concerned. The details for this allocation process were laid down in a Permanent Reference Document (PRD) of the GSM MoU.

The TAP group also worked on the issue of free circulation of MSs within Europe. In this respect the Nordic countries already had experience with NMT of allowing a customer to take a terminal from one country to another and use it on the network provided there. However, the general concept of unhindered international roaming throughout Europe was something relatively new. The TAP group carried out the initial work on a draft CEPT recommendation to enable international roaming throughout Europe. This work was then taken up by the CEPT to produce the necessary ERC decision that formed the basis of free circulation within Europe.

The TAP group had no real legal authority in decisions that it made. However, there were only a limited number of SSs available for the testing of GSM MEs and this effectively limited the number of countries that were issueing type approval certificates. All of the regulatory or notified bodies of those countries were represented in the TAP group and agreed to abide by the procedures laid down in the PRDs developed by the group.

The group reached decisions by consensus and with the agreement of the regulatory bodies involved effectively operated a common European scheme for type approval and free circulation of terminals ahead of the TTE Directive.

17.5 The European Terminal Directive – Technical Basis for Regulation (TBR) for Phase 1

The ITA provided a pragmatic solution which ensured the availability of the first GSM MSs. Nevertheless this emergency solution was not compliant with the EU Directive on terminals (Telecommunication Terminal Equipment Directive 91/263/EEC).

Therefore the GSM community decided in mid-1992 to create *Technical Basis for Regulation* for GSM ME, in accordance with the Terminal Directive.

In the following paragraphs we give some insight into the European Directive on terminal equipment and we explain what are the TBRs for GSM MSs, these documents were produced and maintained by the ETSI/SMG which was the ETSI technical committee in charge of the GSM standard.

17.5.1 The European Directive on Terminal Equipment

The main purpose of this Directive was to create a community-wide market for terminal equipment. This will be accelerated by introducing full mutual recognition based on harmonising conditions for the placing on the market of terminal equipment. To achieve that, rules deduced from this Directive define a type approval procedure which is common to all member states.

According to Article 4 of the European Directive, terminal equipment shall satisfy the following essential requirements:

(a) User safety, in so far as this requirement is not covered by Directive 73/23/EEC;
(b) Safety of employees of public telecommunications network operators;
(c) Electromagnetic compatibility requirements in so far as they are specific to terminal equipment;
(d) Protection of the public telecommunications network from harm;
(e) Effective use of the radio frequency spectrum;
(f) Interworking of terminal equipment with public telecommunications network equipment for the purpose of establishing, modifying, charging for, holding and clearing real or virtual connection;
(g) Interworking of terminal equipment via the public telecommunications network, in *justified cases*.

(In particular the *justified cases* comprise the support of services which the Council has decided should be available Community-wide).

Compliance with the essential requirements (a) and (b) shall be presumed for the terminal equipment which conform with the relevant national standards. Compliance with the essential requirement (c) is verified under another European Directive. Therefore in the following we will only consider essential requirements (d), (e), (f) and (g).

The mutual recognition principle means that type approval granted to a GSM ME in one member state is valid for all member states. Once an ME has been type approved it can be placed on the market of all the countries taking part in this agreement. This creates the conditions for an open and unified market. To be valid, a type approval has to be granted by a regulatory authority. Testing itself is in fact performed by a test house which has been designated by the regulatory authority of its country.

The rules which define the type approval for terminal equipment are specified in documents called "Common Technical Regulations" (CTR). For the phase 1 GSM MSs, CTR 5 and CTR 9 applied.

In fact the CTRs are short documents which refer to the TBRs. The TBRs are technical documents which are produced and maintained by ETSI/SMG, they specify the type approval testing which is to be performed by a designated test house.

17.5.2 GSM 11.10 for Phase 1 and TBR for Phase 1

Now we explain the structure of the TBRs and how they rely on GSM 11.10. For each *conformance requirement* included in a TBR, one or more test purposes are given. For each test purpose a single reference is given to the test method in GSM 11.10. As a consequence the verification of the *conformance requirements* outlined in the TBR is based on the tests described in GSM 11.10. The inclusion of each test is justified by referring to one of the *essential requirements* (d)–(g).

In GSM phase 1 two TBRs define the regulatory testing of the GSM phase 1 MSs, namely TBR5 which copes with the basic attachment to the network and TBR9 which verifies that the speech teleservice is correctly supported end to end.

In TBR5 requirements apply at the air interface and at the SIM-ME interface. It is verified that the MS is able to get access to the network and that it does not disturb it, therefore, TBR5 comprises most of the radio tests, radio link management tests, signalling tests and testing of the SIM/ME interface. The tests in TBR5 are justified under *essential requirements* (d), (e) or (f).

In TBR 9 there are requirements to the speech transmission, it is verified that the speech teleservice is correctly provided end to end. The tests are justified under essential requirement (g). This is possible because speech teleservice is considered by the European Commission as a *justified case*.

It shall be noted that other services (e.g. fax, some Supplementary Services, SMS) are not justified cases, therefore the testing of the end to end interworking of these services is not part of the regulatory testing. However, an MS providing these services has to perform correctly the basic attachment to the network, in other words it shall be compliant with TBR5.

As a summary it can be stated that the European type approval relies on two main sets of documents:

- GSM 11.10 which constitutes the full conformance test suite for GSM;
- the TBRs for GSM which define a subset of GSM 11.10 and the mapping between this subset and the *essential requirements*.

TBR 5 and TBR 9, the TBRs for GSM MS phase 1 were approved by ETSI/SMG at their 4bis meeting in Paris in October 1992. The TBR regime was implemented under European Commission Decision 94/11/EC. This implemented TBR approval from 1 January 1994. There was a transition period for ITA until 1 January 1995.

By transition period it shall be understood that type approval was allowed to work until 1 January 1995.

17.6 Type Approval and TBRs for Phase 2

In order to take into account the phased approach of GSM, it was decided to issue CTRs and TBRs for the type approval of phase 2 GSM MSs. In fact the principles outlined in the previous paragraphs still applied, the only difference being that the CTR regime relied on GSM phase 2, in other words CTRs and TBRs for phase 2 refer to a subset of TS GSM 11.10 phase 2.

TBR19 and TBR20, the TBRs for the 900 MHz GSM MEs phase 2 were approved by ETSI/SMG at their 16th meeting in Vienna in October 1995 after the comments of the public enquiry had been treated.

TBR19 has a scope similar to that of TBR5; TBR20 has a scope similar to that of TBR9, they applied to the GSM 900 phase 2 MSs.

At the same time TBRs were created for the GSM MSs operating in the 1800 MHz band (both 1800 MHz only MSs and dual band 900/1800 MSs), they were TBR31 (access, similar to TBR19) and TBR32 (speech teleservice, similar to TBR20). TBR31 and TBR32 were approved by SMG at their 17th meeting in January 1996 in Edinburgh.

17.7 Type Approval and TBRs for Phase 2+

The type approval of phase 2+ GSM MSs was based on the principles of phase 2+ (see paragraph 17.3). These principles yielded the following scheme which was approved by ETSI/SMG of during their 25th meeting in March 1998 in Sophia Antipolis:

1. TBR19 and TBR31 do not evolve because a new phase 2+ feature has to be type approved;
2. for a phase 2+ feature, a TBR module can be created which relies on TBR19 and TBR31 and which adds some new requirements.

The first TBR module which was approved by ETSI/SMG was the one for HSCSD, ETSI/SMG approved it during their 26th meeting in Helsinki in June 1998.

Other examples of TBR modules were TBR module GPRS and TBR module for R-GSM (GSM for railway applications).

17.8 The Role of Two Committees

17.8.1 ETSI/SMG7

From 1988 to 1993, the specification work needed for testing and type approval was performed in different ad-hoc groups. At the beginning of 1994, ETSI/SMG decided to create a sub-technical committee to cope with these matters, this was ETSI/SMG7.

It was chaired from January 1994 to mid-1995 by John Alsoe (TeleDenmark), then by Rémi Thomas (France Télécom) from July 1995 to March 1999.

ETSI/SMG7 was responsible for GSM 11.10 phase 1, GSM 11.10 phase 2, the phase 1 TBRs, the phase 2 TBRs and the TBR modules.

17.8.2 Type Approval Advisory Board

Once it became apparent that an ITA scheme was required it was clear that a body was needed in addition to the TAP group to oversee the general technical issues, procedures and mechanisms of ITA. A group was therefore established called the "Type Approval Advisory Board" (TAAB) which initially reported to the GSM MoU group. The interim type approval scheme introduced for GSM was based on a CEPT recommendation (CEPT REC T/R 21-08) and a resolution by the Technical Regulations Application Committee (TRAC) on the introduction of NET 10.

The TAAB group handled problems encountered during type approval and gave advice to the type approval authorities concerned on actions to be taken to ensure a harmonised and consistent approach to ITA. TAAB acted as an expert body for all technical problems under the authority of TRAC and as a distribution point for technical problems notified, ensuring that the right organisation or committee was informed.

Initially the group was attended by administrations, type approval authorities, test houses, GSM operators and ETSI specialists. It also had the possibility to invite equipment manufacturers to attend for specific problems. Typically the issues dealt with by the group were:

- Ensuring that common test methods and system simulator software were used;
- Developing concessions to resolve system simulator fault reports;
- Issuing advisory notes on test procedures to ensure a common understanding between test houses;
- Resolving technical problems raised by type approval authorities relating to type approval problems;
- Co-ordinating type approval procedures between the type approval authorities concerned;
- Co-ordination of the introduction of new mandatory tests as required.

TAAB went through several different life spans and the role of the group changed to some degree as it and the regulatory environment evolved. With the introduction of the CTR regime TAAB became New TAAB (NTAAB) and reported to TRAC. TRAC effectively offered advice to the European Commission on technical issues relating to the implementation of CTRs. TRAC actually worked through yet another committee called the Approvals Committee for Terminal Equipment (ACTE) which was established by the TTE Directive.

NTAAB actually made the Directive work in practice and was so successful that TRAC established a number of additional TAABs such as DTAAB for DECT terminal equipment and TTAAB for TETRA. As a logical move, and to maintain consistency with the other TAABs, the NTAAB was finally renamed GSM TAAB (GTAAB). As the group evolved its terms of reference changed and manufacturers became members of the group.

In all of its different phases the TAAB for GSM was continuously chaired by David Barnes (DTI).

As already explained in the previous paragraphs, in the beginning the type approval activities regarding DCS 1800 (former name of GSM 1800) were separated from those regarding GSM 900. Hence at the beginning TAAB only handled type approval of GSM 900 MEs. Therefore, in order to handle the type approval of GSM 1800 MEs the PCN Type Approval Experts Group (PCN TAEG) was set up a with a similar remit to the TAAB for GSM 900.

The PCN TAEG operated under the auspices of the MoU for DCS 1800 type approval referred to in paragraph 17.4. The chairmanship of the PCN TAEG rotated between the UK (David Barnes - DTI) and Germany (Ekkehard Valta - BAPT). Horst Mennenga (BAPT) eventually took on the chairmanship prior to the group being incorporated into GTAAB when the type approval activities for both the 900 MHz band and 1800 MHz band were put under the same roof.

17.9 The Test Tools

As already explained, at the beginning of the testing, procedures were closely linked to the SS developed by Rohde and Schwarz. Then, with the TBR procedures, it became up to the approved test houses to choose testing tools fulfilling the TBRs requirements. These choices had to be consistent with the advice of TAAB. As a consequence a range of tools are used. This includes equipment only aimed at performing signalling tests. This includes as well more complex tools having both signalling and radio capabilities like the SS.

17.10 Conclusion

17.10.1 The Achievements of the GSM Type Approval

It was often argued that the GSM type approval process was too complex, too costly, that it necessitated the development of too many tests and that it delayed the availability on the market of new products.

In fact such debates started as early as the introduction of the TBR regime in 1992. To give an interesting example, it may be recalled that some people suggested that the GSM encryption should not be part of the GSM type approval! If it had been accepted this would have resulted in putting on the market GSM ME without encryption, furthermore, the GSM operators would have been forbidden to deny service to such MEs. As a result the encryption would not have been introduced on the GSM networks.Fortunately such suggestions were not followed, on the contrary it was specified by TBR19 that the two ciphering algorithms, namely A5/1 and A5/2, be mandatory for all GSM MEs.

Another interesting example is linked to the introduction of GSM Enhanced Full-Rate (EFR). The GSM standard specified that a ME implementing EFR shall as well implement the Full-Rate (FR) speech codec which is the basic GSM speech codec.

In addition the signalling mechanisms in basic GSM phase 2 ensure the compatibility between an ME implementing EFR and a network not implementing EFR. Nevertheless, in June 1997, at the time when it was expected that some EFR MEs could have been put on the market, it seemed that some networks had not yet implement the phase 2 mechanisms. As a consequence such networks would have denied service (both in EFR mode and FR mode) to the EFR MEs. Such events would have jeopardised the credibility of GSM. The GSM community took the necessary steps: the network operators were given the necessary technical explanations to upgrade their networks and the type approval bodies were informed of these circumstances. As a consequence no compatibility problem was caused by the introduction of EFR.

In both cases the type approval procedures provided guidelines for both the ME manufacturers and the network operators in order that the implementation and evolution of the features were done in a way which preserved the compatibility.

GSM 11.10 played a key role in the process as it provided the conformity specification utilised by the type approval throughout the evolutions of the procedures. And this was one of the keys for the success of the GSM type approval.

Furthermore, the testing and type approval processes implemented in reality the following principle: any SIM in any ME in any network.

Finally the type approval process implemented for GSM developed a model for the working of the TTE Directive which was utilised to facilitate the application of the Directive in many other areas. In addition to this it also formed the basis of type approval of GSM terminals on a global basis. The use of the system simulators developed in Europe were accepted by every other country in the world adopting GSM. This greatly facilitated global roaming and was a major contributor to the success of GSM as a global standard.

17.10.2 From ETSI/SMG7 to 3GPP/TSG TWG1 (T1)

Due to the success of the testing approach implemented by the GSM community, it was decided by the 3GPP to follow similar principles for the testing of the UMTS ME:

- in January 1999 the 3GPP/TSG TWG1 (also called T1) group was set up inside 3GPP with a scope similar to that of ETSI/SMG7, namely the drafting of conformity tests for the UMTS ME;
- the first meeting of the T1 group was convened by the SMG7 chairman (Rémi Thomas - France Télécom), this meeting adopted the general structure of the conformity specification for the UMTS ME, this structure was quite similar to that of GSM 11.10.

It is difficult to mention individual contributors to the work on type approval and testing because there were so many valuable contributions and all of the TAAB groups worked with a tremendous attitude to make things work. However some of the notable contributors were: Peter Zollman (Vodafone), Nick Sheldon (Vodafone), Peter Collins (Vodafone), David Hendon (DTI), Armin Silberhorn (German Ministry), Horst Mennenga (BAPT), Ekkehard Valta (BAPT), Didier Chauveau (ART), Thomas Jäger (Cetecom and 7Layers), Jürgen Bauer (E-Plus), David Nelson (Orange), Ian Marwood (One 2 One), David Freeman (Motorola), John Alsoe (TeleDenmark), Andrew Howell (Motorola), Ansgar Bergmann (T-Mobil), Isabelle Lecomte (France Télécom), Han van Bussel (T-Mobil), Chris Pudney (Vodafone), Athol Berry (Rohde & Schwarz), S. Hu (PT SMG).

There were also of course the contributions made by the two authors of this section Rémi Thomas (France Télécom) and David Barnes (DTI).

Chapter 18: Operation and Maintenance (O&M)

Gisela Hertel[1]

18.1 What is O&M?

Operation and Maintenance (O&M) of telecommunications networks comprises the following functions:

Alarm handling and fault management: Indications of malfunctions or outages of any network component are transferred back to the O&M system. The O&M technician can then remotely interact with the network component in question and try to repair the problem.

Configuration management: Parameters necessary for network configuration such as frequency plans, next-neighbour relationships or handover algorithms are sent from the O&M centre to all network elements. Configuration management also includes downloading of new software into the network and tracking software versions of all network elements.

Performance management: All network elements generate a large variety of statistical and performance data. These data need to be collected by the O&M centre for further processing and analysis.

Security management: This includes the handling of normal network security measures such as access control or system logs, but also the administration of GSM specific security functions like authentication algorithms.

Subscriber and equipment tracing: In the case of stolen equipment or subscribers engaged in criminal activities, tracing of the affected handsets or SIMs has to be feasible throughout the network.

Subscriber and equipment administration: Subscribers have to be activated or deactivated in the HLR and their service profiles have to be downloaded and updated. Similarly, the EIR databases have to be administrated by the O&M centre.

Charging administration: After each call, data such as calling or called number or time stamps are recorded by the Mobile-Services Switching Centre (MSC) and later sent to the billing system. In the case of some services, additional records may be generated for example by SMS centres or data nodes.

---

[1] The views expressed in this chapter are those of the author and do not necessarily reflect the views of his affiliation entity.

18.2 Why Standardise O&M Functions?

All the functions described above involve data transfer between network elements and the O&M centre, or between the billing system and databases in the Home Location Register (HLR), Authentication Centre (AuC) and Equipment Identity Register (EIR). It is not strictly necessary to standardise these interfaces. Mobile stations and networks need to interact with each other in a standardised manner world-wide to allow roaming, but O&M centres and billing systems are "hidden" in each network, and have no connection with foreign network elements. Internal operations within each GSM system can be considered a matter of each individual operator.

The push to standardise O&M functions came from the network operators: they wanted the freedom to mix equipment of different vendors throughout their networks.

In a multi-vendor environment, integrated O&M functionality is absolutely necessary: the operator needs to collect performance data from base stations throughout the network, and the data need to be comparable. Frequency plans need to be downloaded to all base stations, and adjacent base stations need to be configured to hand traffic to each other, even if they are not made by the same manufacturer. And security management of course needs to work seamlessly throughout the system.

Before GSM, incumbent network operators often had long-term supplier relationships with equipment manufacturers, who were very familiar with the operation of the national fixed or mobile networks. Each new network element was duly connected to the existing O&M centre, mostly through proprietary interfaces. It was not conceivable for these interfaces to be made available to other vendors.

Many new GSM operators however started with a clean slate and wanted to get maximum value from their suppliers. This was only possible by buying whatever equipment best fits their needs at any given time without being tied into a permanent relationship with any supplier.

Many of the equipment manufacturers of course did not favour such a development: opening up proprietary O&M interfaces to competitors, or implementing fully standardised interfaces made their key customer relationships vulnerable.

One aspect of O&M standardisation however was never disputed: The need for all MSCs to provide identical parameters in the call detail records was recognised by everybody without argument. The call detail records are used for charging, and tariff plans can only use those parameters contained in the records. An operator cannot charge for a service, if only one of his vendors makes the appropriate charging parameters available, or if charging parameters vary throughout the network.

In addition, the call and event records contained in the O&M specification GSM 12.05 are used as an input for the transferred account records defined by MoU for the charging procedures of roaming calls. If a parameter is not available in GSM 12.05, it cannot be included into the transferred account procedure and can therefore not be used as a basis for roaming tariffs between operators.

There was another reason why O&M gained more attention as GSM started growing. The analogue networks existing before GSM were quite small and not much effort was required for monitoring the infrastructure by the incumbent operators who had large numbers of skilled O&M staff. But the larger GSM networks soon had thousands of base stations, all of which sent out alarms, created performance data and needed configuration.

Efficient O&M systems started playing a much larger role in providing good service quality and optimising operational costs. This trend was amplified by the fact that the new private GSM operators lacked experienced O&M technicians, and therefore needed intelligent O&M systems. In addition, the incumbent operators were coming under increased pressure from the new competitors, and also tried to limit their operational cost.

18.3 The Elaboration of the GSM Phase 1 O&M Specifications

Originally a group called the Operations and Maintenance Expert Group (OMEG) existed within GSM. This group was chaired by Bernd Haarpaintner until March 1991, and by Gisela Hertel after that date. At the second SMG plenary in March 1992, OMEG was elevated to full GSM sub-technical committee status and renamed SMG6.

At the beginning, work in SMG6 was quite difficult. The operators, particularly the new private operators had a clear interest in comprehensive O&M standards to enable multi-vendor networks, but they lacked the extensive resources necessary for completing the technically very challenging O&M specifications. The manufacturers did have significantly better resources, but they were much less enthusiastic about O&M standards to begin with.

One of the key players during this phase was George Schmitt, then chairman of the GSM MoU association and also technical director of one of the largest private operators, Mannesmann Mobilfunk. He wielded his considerable influence to make sure that manufacturers took O&M issues seriously and contributed to SMG6. George's intervention turned out to be somewhat of a double-edged sword. On one hand SMG6 certainly got everybody's attention and better resources, but on the other hand manufacturers active in SMG6 still had a dilemma. As soon as O&M standards would be stable, operators would demand their implementation. But they had already installed legacy O&M centres developed before standardisation in most networks.

But in the end, a BSS oriented set of O&M specifications for phase 1 was approved in 1991/2. Many companies and individuals contributed to this comprehensive set of standards, but Hans Hauser and his colleagues from T-Mobil deserve special recognition: T-Mobil provided a large team of experts to SMG6 and took responsibility for several key O&M specifications.

The finalisation of these phase 1 specifications was a source of considerable pride and satisfaction among all SMG6 members. To understand this, one has to realise the complexity of standardising the O&M interfaces described at the beginning of this chapter: Every time a new feature or service is introduced into GSM, or every time a protocol is updated, the O&M messages need to be modified too. Any new service needs to be activated in the HLR and may need additional parameters in the call data records. Performance data have to be collected to evaluate the usage of the service, and configuration management may be affected. Also, O&M messages may have to be modified to accommodate additional alarms or fault management actions. Security management has to be reviewed, and subscriber and equipment tracing may have to be updated.

To complete these consistency checking exercises, SMG6 needed the support of PT 12, most likely to a much larger extent than any other STC. Fortunately, support was provided willingly and competently throughout the entire lifetime of SMG6 by Jonas Twingler, Elmar Grasser, Marion Hoenicke, Michael Sanders and Ansgar Bergmann.

Another consequence of the tight link between O&M and all other GSM specifications is an inevitable delay of O&M specifications: They could only be finalised about 6–9 months after the core specifications were stable, because this time was necessary to incorporate all core features into the O&M functionality.

Despite the sense of accomplishment felt after the finalisation of phase 1 specifications, the impact of these standards on the actual GSM world was quite minimal: Due to the late timeframe, O&M systems unaffected by the new standards were largely up and running throughout the GSM world.

18.4 The Elaboration of the GSM Phase 2 O&M Specifications

The situation changed for phase 2 which for SMG6 came not too long after phase 1. The complete set of phase 2 O&M specifications was approved in the summer of 1996, 9 months after the freezing of phase 2 core specifications in October 1995 and within the planned schedule. The only exception was the work on subscriber and equipment tracing which had started late in the process and was not completed until the spring of 1997.

In phase 2, work in SMG6 had become much easier. The big battles of the beginning (which had in hindsight been very useful to define a clear road for O&M standardisation) had subsided, efficient working and decision procedures had evolved and again PT 12 provided exemplary technical and administrative support. In addition, phase 1 specifications had been completed not long before and therefore were quite up to date and provided an excellent basis to work on.

The relevance of phase 2 O&M specifications for the industry was considerably higher than in phase 1. First, the standards were available in a much more timely manner relative to the core specifications, and both operators and manufacturers were much more aware of O&M issues than before. Phase 2 O&M standards were probably not implemented to the letter anywhere in the GSM world, but manufacturers took them into consideration for their new O&M releases, and operators had a better framework to stand on in contract negotiations.

Another issue discussed during phase 2 was the extension of the O&M specifications from the radio subsystem to also include the MSC or related components such as HLR, VLR, EIR or AuC. There was considerable interest in such work, particularly from the operators, but in the end a lack of resources resulted in maintaining the existing scope. MSCs differ much more strongly from vendor to vendor than the radio subsystem, and standardised O&M architectures would have been very difficult. In addition, most of the alarms, faults, performance and configuration parameters are generated by the radio subsystem, so it is most efficient to concentrate standardisation efforts there.

18.5 The Elaboration of the GSM Phase 2+ O&M Specifications

The updating of O&M specifications in phase 2+ proceeded in the same smooth manner as described for phase 2, and phase 2+ O&M standards were completed within the targeted 6–9 months after stabilisation of the corresponding core specification release packages.

The most significant new addition during phase 2 was a new specification describing GPRS charging. Call records incorporating the call duration and parameters designating the circuits used during the connection completely lose their meaning for packet radio, and tariff structures are fundamentally different.

The GPRS charging specification was completed in October 1998. This was the first time an O&M specification was available during the implementation of the service and could be fully taken into account by system manufacturers and operators. This was a major acomplishment of the GPRS working group led by Kai Sjoeblom from Nokia.

For the specification of GPRS charging functionality, consultation was necessary with committees outside SMG6, in particular with the experts in charge of services (SMG2 and MoU SERG) and the group in charge of transferred account records to be exchanged between operators for roaming (MoU TADIG).

Awareness of the need to work on GPRS charging was high among all parties involved and the co-operation offered to SMG6 was excellent. Unfortunately but understandably, nobody was willing to commit to definite charging structures in 1998, so SMG6 had to provide parameters for all conceivable charging scenarios such as data volume, bandwidth used or origin and destination of data. If everybody adopts simple tariff structures eventually, a lot of work will have been unnecessary, but there was no way to avoid this at the time.

18.6 The Elaboration of the UMTS O&M Specifications

For UMTS, O&M issues were again considered from the very beginning, and O&M specifications very similar to those of the earlier phases were included in the first set of UMTS standards. These specifications provided only the general O&M framework, because the exact performance or configuration data could not be determined until all UMTS network elements and functionalities were completely stable.

On the other hand, O&M standards change relatively little between GSM and UMTS: Alarms will still arrive at the O&M centre and subscriptions will still have to be activated in the databases, irrespective of the actual functionality of each node. So once a consistent framework of O&M principles has been created, it can easily be updated to follow future technical developments.

Chapter 19: Professional Technical Support and its Evolution

Section 1: The Permanent Nucleus in the CEPT Environment

Bernard Mallinder[1]

During late 1985 the CEPT Groupe Spécial Mobile (GSM) decided that in the interest of maintaining speed and quality in the development of the technical recommendations, a central support function was required. The mandate for this unit was agreed in early 1986 and the "Permanent Nucleus" (PN) of the GSM was created. The approach and mandate of the unit built on the successful experience that the CEPT had had in the rapid development of specifications for both satellite and ISDN services. The unit was established in Paris during the summer of 1986 and consisted of full-time technical managers and programme managers. The PN had expert full-time resources from France, Germany, Sweden, Norway, Finland, Holland, Switzerland and the UK.

The initial tasks involved supporting the working group chairpersons, ensuring technical consistency, documentation release control, consolidation of work-plans and reporting progress to the plenary of GSM. Within a year the role of the PN extended to the creation of the equipment specifications (11 Series) and the formulation of the network management recommendations (12 Series). In addition, the PN was charged with the co-ordination of the evaluation of the different candidate technologies for the radio interface. These tests were undertaken by CNET in Paris and involved the evaluation of TDMA and CDMA offerings.

After the choice for TDMA with slow frequency hopping, GSM entered into a review of the complexity of the complete system. Initial work undertaken by Televerket in Sweden indicated that the layer 3 protocol on the air interface resulted in an extensive acquisition time during the initial "turn-on" period. The "Complexity Review" commenced during late 1987 and continued until the spring of 1988. During this period all working groups contributed and a number of review meetings were co-ordinated by the PN in Paris. The findings and suggested improvements were tested and examined in various centres including CNET in France, FTZ in Germany, CSELT in Italy, BT and Racal Research in the UK. As a result of

---

[1] The views expressed in this chapter are those of the author and do not necessarily reflect the views of his affiliation entity.

this work a catalogue of "simplifications" was formulated and approved by GSM. The first drafts of the complete recommendations (Series 1-12) were made available to participating parties by the summer of 1988.

In parallel with the "Complexity Review" the PN was charged with evaluating the industrial challenges involved in the design and manufacture of future handsets. This work was principally supported by industry experts who were in a position to create "top-down-design" of complete handsets. Fabrication of the logical functions, the RF entities, casing and battery technologies were considered. Assumptions were constructed concerning the availability of sub-micron technologies, yield rates of new chipsets and performance of external and integrated antennae. The input from this study placed additional pressure on the need to reduce complexity to ensure that the price, weight and performance targets for the handsets could be achieved.

During the evaluation of the early radio interface proposals GSM began to consider the possible impact of intellectual property on the development of the system. Ideas concerning "cross" and "pooled" licensing were considered. There was however a need to identify relevant patents and to consolidate a list of "vital" patents. The PN was charged with the task and Patent Review Group consisting of network operators and manufactures was active during 1987. At the end of the process some 22 previous patents were identified as "vital". Some of these belonged to operators and others to vendors. Numerous initiatives followed the identification of the patents. Members of the GSM group then sought to secure commercial, political and/or technical solutions to the patents.

As the definition of the GSM system stabilised interest in the system began to develop. Various conferences emerged as did requests for technical papers and workshops. Members of the PN were invited to numerous technical presentations and discussions. The important points which needed to be communicated were the progress of the technical work, innovative services, interworking with ISDN and data networks, coverage techniques and the functional architecture. Less technical areas were also addressed including specification methodology, consistency management, release control and eventually system performance.

Chapter 19: Professional Technical Support and its Evolution

Section 2: PT12 and PT SMG in the ETSI Environment

Ansgar Bergmann[1]

When, in 1989, ETSI was created, GSM became a technical committee of ETSI, and the Permanent Nucleus (PN) was transformed to become ETSI Project Team 12 (PT12). Some time later, at the end of 1991, the offices were moved from Paris to the ETSI premises in Sophia Antipolis.

Between 1989 and 2000, PT12 and later PT SMG continued the work of the PN to perform the project management for the Technical Committee GSM and later SMG, to maintain overall aspects of the specifications such as quality, consistency and compatibility, to support working parties of GSM/SMG and to perform specific tasks including specification work. In this period, GSM grow up with breathtaking speed, in terms of users, of countries to introduce GSM, of networks, of variety of services offered to the users and of applicable frequency bands, and the GSM core network became the *de facto* standard platform to connect other radio access technologies.

PT12 was financed from the ETSI funded work program, and there was additional specific funding from the EC, from GSM MoU Group/Association (later GSM Association), and from member companies. This additional funding was often related to specific tasks and deliverables. For the sake of bookkeeping, separate project teams were installed for these specific tasks. However, all these project teams were managed in one integrated structure, and "PT12" was used as a generic term for the varying number of tasks.

Until 1995, PT12 was responsible for GSM only, and did not support the UMTS development within SMG. In 1996, ETSI agreed, on proposal of SMG, to assign project team support to UMTS in order to push the development. The new PT83 was integrated with PT12 as well, and to give things better visibility, the integrated team was named PT SMG.

---

[1] The views expressed in this section are those of the author and do not necessarily reflect the views of his affiliation entity.

19.2.1 Consolidation of GSM Phase 1

In the years 1989–1992, the main focus of activities in GSM and PT12 was the consolidation of phase 1 specifications in order to enable the roll-out of the GSM system and the availability of mobile stations.

An important milestone was the edition of Release 90 phase 1 specifications. For monitoring the status of specifications and to trace all changes, PT12 had established a Change Request (CR) procedure. At GSM#25, this procedure was further improved[2] and it was decided to go on with two parallel series, the phase 1 and phase 2 specifications, cf. Section 17 on working methods. The scope of phase 1 and 2 was defined by GSM MoU.

PT12 also led two expert groups:

- PT12-OMEG, the Operation and Maintenance Experts Group, which was transformed in 1992 into the technical subcommittee SMG6 of SMG;
- PT12-ES[3], responsible for the test specifications for the mobile station and the base station sub-system; in 1994, it was transformed into two SMG subcommittees, SMG7 (mobile station testing) and SMG8 (BSS testing).

Free circulation of mobile stations was one of the essential goals of GSM. Mobile stations had to work correctly and with good quality in any GSM network. This explains the importance of the mobile station type approval regime. PT12 took an essential role, in cooperation with other groups including

- Type Approval Advisory Board (TAAB) and its successor organisations;
- GSM MoU;
- ECTEL-TMS;
- EU organisations such as TRAC and ACTE.

The interim type approval, then the full type approval was installed. Later, the transition to the new legal scheme defined by the EC, the CTR/TBR regime, had–be managed, and with a tremendous effort, GSM was the first–implement the CTR/TBR regime. See also Sections 14 and 17.2.

In 1990, GSM became responsible for standardizing DCS1800, at that time also known as Personal Communication Networks (PCN). It was agreed to specify DCS 1800 as a version of GSM in the 1800 band. In phase 1, there were delta specifications for DCS 1800, mainly concerning adaptation of radio transmission parameters, but also necessary consequences of the added frequency band for signalling, testing, the SIM card, cell selection and network selection, and special requirements of the DCS 1800 operators for equipment sharing and national roaming. The DCS 1800 activities were supported by experts working in an extension of PT12, funded by interested parties (the PCN operators).

19.2.2 Introduction of GSM Phase 2

The need to introduce GSM in phases had already been recognized before 1990. Some features had been deferred rather early to phase 2, for example:

---

[2] Cf. TDoc P-482/89 rev2.
[3] ES stands "Eleven Series": the MS and BSS test specifications were 11.10 and 11.20.

- some services, in particular supplementary services;
- radio resource management functions as the pseudo-synchronous handover.

(Some features like facsimile, data services and the half-rate speech codec were moved later to phase 2 due to technical difficulties arising.) Accordingly, the GSM working parties had started already in 1989 to elaborate and agree phase 2 change requests, however the GSM plenary had in principle postponed phase 2 approvals until meeting #30 in 1991. This way of proceeding had the advantage that

- work concentrated on getting phase 1 working; and
- a more complete picture of phase 2 was achieved before detailed technical decisions were taken.

It had the disadvantage, however, that for some considerable time, phase 2 change requests were prepared in the working parties on a moving basis – changing versions of the phase 1 specifications – a fact leading to several clashes and contradictions. Whence, a consolidation phase became necessary to establish the platform for phase 2. A major part of this work was performed by PT12 and the rapporteurs of specifications.

Other activities demanding high support by PT12/PT SMG related to

- the merging of GSM 900 and DCS1800 specifications;
- phase 2 mobile station type approval, including the creation of phase 2 TBRs and the translation of tests into the formal test description language TTCN.

19.2.3 The Evolution of GSM Phase 2+

In 1995, the phase 2 specifications were frozen as version 4.x.y. The further evolution of GSM took place as phase 2+ (cf. Section 20), the basic concept being to add new features as options. Depending on the completion date of specifications, features were assigned to yearly releases R96, R97 and so on.

With a slightly increased number of experts, PT SMG started to provide support to all SMG Sub-Technical Committees (STCs; the number had grown from 4 to 11; earlier, some STCs did not have PT SMG support) and almost all working groups.

There was substantial discussion about the appropriate way to document the evolving GSM system in the standard, and it was then decided to create a parallel new complete set of specifications for each release.

PT SMG expanded the project management, a roadmap was created to reflect the progress of each work item, and this roadmap was derived from a work item database maintained by PT SMG. For bigger work items, work item managers were nominated and project description documents were created in the 10.xy series.

With the increase of participating companies, the introduction of electronic document handling became more and more important. SMG started to provide all meeting documents in electronic form and made them available on CD-ROM; step by step, the meetings went from paper documents to electronically only distributed documents, and today all ETSI and 3GPP specifications and meeting documents are available on the Web (*www.etsi.org* and *www.3GPP.org*), much work is done in e-mail discussion groups, and meetings use LANs for the document handling.

Most meeting documents had been temporary documents only, however, and had been archived by PT SMG. As they give an important insight into the technical background, ETSI, on the initiative of PT SMG, scanned all GSM and SMG plenary documents.

The regulatory basis for mobile station type approval in the European Economic Area was changed again until finally the R&TTE directive was introduced. The necessary standards were created by SMG7 and PT SMG.

PT SMG participated in the co-operation and liaison on dualband DECT/GSM and GSM/ Mobile Satellite Systems (MSS).

When PCS 1900 (GSM in the 1900 band) was introduced in North America, T1P1 established the corresponding standard, and a co-operation with SMG was established. This led later to a fully integrated standard with joint approval by T1P1 and SMG. PT SMG participated in these efforts and the preparation of the co-operation agreements.

PT SMG also managed the assessment of the Adaptive Multi Rate (AMR) codec together with SMG11.

Several other special tasks were performed by PT SMG:

- A taskforce in PT SMG elaborated major parts of the necessary additions to the GSM standard for the GSM railways applications, funded by the Union Internationale des Chemins de Fer, international railways union (UIC) and European Commission.
- When the GPRS taskforce, a working group installed by SMG in spring 1994 to elaborate the basic settings of GPRS, had concluded and handed over its work to the SMG STCs 1 year later, a special taskforce was established in PT SMG (STF 80V).
- PT SMG made some substantial contributions to the efficient use of the GSM broadcast channel.
- There was a special taskforce on number portability in PT SMG.
- PT SMG continued its function as editor and rapporteur for the GSM TTCN test specifications, but participated in outsourcing the actual production of TTCN code to software houses.

19.2.4 The Introduction of UMTS

Since 1996, PT SMG supported the development of UMTS as well – unfortunately, due to lack of resources with the exception of the UMTS radio access. It became obvious that a global co-operation on GSM standardisation was necessary, and for the third generation this was evident. PT SMG participated in the preparation of this co-operation. In order to strengthen the liaison to ITU, an ITU-T co-ordinator and an ITU-R co-ordinator were installed at PT SMG, funded by member companies. In 1999 and 2000, the work on UMTS and GSM was successively handed over to 3GPP and PT SMG disappeared together with SMG.

Chapter 19: Professional Technical Support and its Evolution

Section 3: MCC in the 3GPP and ETSI Environment

Adrian Scrase[1]

19.3.1 The Need for a Mobile Competence Centre

During the preparatory discussions that led to the creation of the Third Generation Partnership Project (3GPP) the need for dedicated project support had been clearly identified. From the first development of mobile telephone standards within CEPT, and later through their development within ETSI, the availability of dedicated support had been attributed as one of the factors that led to the success of the GSM standard. With the birth of 3GPP it was clear that such support needed to be continued.

The previous sections in this chapter describe in detail how professional technical support had been provided by the CEPT Permanent Nucleus (PN) and PT SMG groups. In both cases the support provided had been undeniably excellent but the creation of an international project such as 3GPP provided an opportunity to consider a fresh approach to project support. History had provided one important lesson though in respect of project support, that the demands made by a technical group may be infinite, whilst the resources available to provide such support will always be subject to close scrutiny.

When considering the support to be provided to 3GPP it was evident from the outset that the timescales were pressing and that support needed to be provided without delay. The first technical meetings of 3GPP took place in December 1998 and some emergency measure was required to ensure that from that day forth a level of support was available. When looking at the organisational partners that had created 3GPP, ETSI was the only partner that had an existing comprehensive support structure in place and to some extent it was a natural decision to base the future support on the ETSI model. However, the dimensions of 3GPP were such that considerably more resource would be required to provide the level of support expected.

The ETSI PT SMG, which had provided support to ETSI TC SMG, provided first level support to 3GPP from its creation, assisted by several ETSI staff members provided on a

---

[1] The views expressed in this section are those of the author and do not necessarily reflect the views of his affiliation entity.

temporary basis. This support enabled the project to get off to a good start but with the cost that resources had been diverted from other support tasks within ETSI. A more permanent and equitable solution was required.

19.3.2 Establishing the ETSI Mobile Competence Centre

The ETSI Mobile Competence Centre (MCC) was created in March 1999. It was formed out of the existing PT SMG, supplemented with additional ETSI staff at both the professional and administrative level. This led to an organizational unit of approximately 20 persons, placed directly under the ETSI director general.

The distinct difference between this and earlier support groups was that it had been established as an organizational unit within the permanent structure of the ETSI secretariat. This provided a more stable basis for the future work and moved the problems associated with funding and administration away from the technical groups and into a domain more suited to such tasks.

By May 1999 the magnitude of the support requirements of 3GPP became more apparent and ETSI members responded by agreeing to an additional injection of 1.4 million euros, raising the budget allocation for MCC for 1999 to approximately 3.5 million euros. This additional injection of funding led to an increase in human resources and a total complement within MCC of 25 persons.

Whilst MCC had been created within a different organizational framework, many of the attributes of the former structures had been retained. The use of a high proportion of contracted experts for example, is a practise that still continues within MCC since this brings a number of benefits. The contracted experts remain under the employment of their home companies and are contracted to ETSI for a fixed duration. The home company receives a daily rate of remuneration for the services provided by the expert. This leads to a very flexible workforce where the skill base can be changed on demand, the level of resources may be raised or lowered very easily, and where the most competent and recently trained experts can be obtained from industry on short-term contracts. There is also a considerable benefit to the home company since the experts invariably return enriched by their experience of having worked within the ETSI/3GPP environment. With a project as complex as 3GPP, there is however a need for a level of permanence which is provided by a complement of staff employed within the secretariat. Such staff are particularly suited to longer-term project management tasks and to the management of the specifications being produced. The combination of both types of human resource has led to an efficient organizational unit. The ETSI secretariat also performs many tasks for the wider ETSI membership and has dedicated groups providing IT assistance, document handing expertise, accountancy services, marketing services, etc. MCC has full and easy access to all of these services.

Another feature that had been retained and expanded within MCC was the international variety of the workforce. By the middle of 1999, 14 different nationalities were represented within the team from four different continents. This included one representative from Japan and one from South Korea both of whom had been provided on a voluntary basis by their home organisations.

By the end of 1999, the importance and magnitude of the task being undertaken by MCC led to its designation as a separate department within the ETSI secretariat.

19.3.3 Organising the Work

MCC contains a number of teams each providing dedicated support to a Technical Specification Group (TSG). This means that every TSG and working group has a nominated support officer. Each team has a "Godfather" who is responsible for ensuring that the TSG as a whole receives the necessary level of support required and that the team works efficiently and effectively.

The support officers are considered as project managers for the technical groups that they support and perform the following tasks:

- support meetings, prepare document list and meeting report, implement agreed actions and present results to the parent body as appropriate;
- manage major work items and update/maintain a work plan for the work item indicating the affected specifications, change requests, new specifications, specification groups, meeting schedules and milestones;
- edit specifications under change control;
- support chairpersons by undertaking administrative functions and carrying out delegated tasks;
- progress liaison statements to relevant groups;
- implement change requests approved by the parent body in the specifications under change control;
- provide the resulting new working versions of specifications under change control to the standardization group experts.

The teams are helped in their work by a team of support assistants that take care of the more administrative day to day tasks. The support assistants each have clearly described tasks and have nominated back-up assistants in case of absence. This information, together with that pertaining to the support officers, is recorded on an intranet enabling every person within MCC to have full access to the information that they require.

A task of high importance within MCC is the management of the specifications being produced by 3GPP. A full-time specifications manager is allocated for this task with responsibility for ensuring that the exact status is known of every specification under MCC responsibility. At the time of writing MCC is managing approximately 1300 specifications.

A high proportion of the workload for each support officer involves the implementation of changes to specifications that have been agreed by the TSGs. Each change is formulated in the form of a "Change Request" (CR) and once agreed have to be carefully implemented within the related specification. The changes may vary from a simple editorial correction to many pages of technical data and in each case a check is required to ensure that the specifications as a whole retain their integrity with the implementation of the change. During 2000, MCC processed more than 5700 CRs, which equates to approximately 25 changes implemented per working day.

Another task for which dedicated resources are provided is for the project management of the 3GPP work plan. Following the creation of 3GPP a new approach was considered in terms of structuring the work plan and this has resulted in a functional decomposition of the work into three hierarchical levels; features, building blocks and work tasks. A comprehensive plan has been created which lists all of the work ongoing within the project and categorises that work into one of the three categories given above. Included within this categorization is an

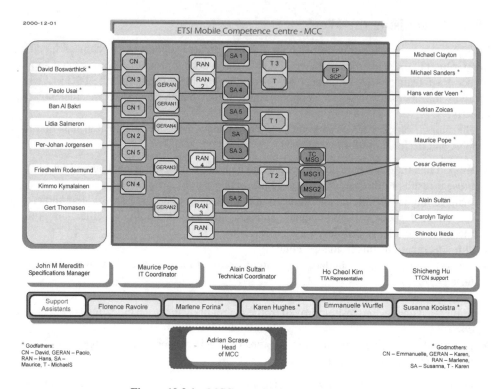

Figure 19.3.1   MCC organisation at end of 2000

understanding of the interdependencies that exist between each activity. The result of this management process is portrayed in Gantt form and enables users to view each feature, to determine what progress has been made, and when the specifications required for that feature are likely to be completed. A technical co-ordinator is appointed within MCC to manage the work plan.

3GPP makes extensive use of modern IT tools and MCC plays a leading role in the development and deployment of such tools. An IT co-ordinator within MCC has the task of evaluating new tools and for developing methods to improve the efficiency of the document editing process.

Since MCC is an integral part of the ETSI secretariat it is required to meet the requirements of the ISO 9002 quality system and regularly undergoes stringent auditing to ensure that the requirements are met.

Figure 19.3.1 gives a snap shot of MCC taken at the end of 2000 and shows the distribution of work at that point in time.

19.3.4 Funding Issues and Financial Control

The provision of comprehensive project support carries with it an associated cost, and with a competence centre employing more than 25 persons it is clear that the cost will be significant. During the first year of operation, the vast majority of the support costs were met by the ETSI budget, but, for the future, a more balanced funding method was required involving contributions from all organisational partners.

Decisions made within 3GPP fall into two categories: political/administrative decisions taken by the partners where each partner has effectively one vote; and technical decisions taken within the TSGs where each individual member has effectively one vote. In determining a funding mechanism for 3GPP it was decided to reflect both of these categories of decision making. This was achieved by deriving half of the required budget by equal proportion among the organisational partners and by deriving the other half of the budget from the organisational partners in proportion to the number of their individual members within the project. This is the basic principle upon which the funding is derived for MCC.

In addition to the support provided by MCC there are other standardization tasks which urgently require completion and for which voluntary resources are not always available. Additional taskforces are created for this work and a budget is set aside each year to cater for this requirement. Since the results of this work will be directly published by the participating organisational partners, they contribute equally to this budget.

The direct costs of running MCC can easily be derived from salary payments, travel claims, etc. but there are many hidden overhead costs which are of a substantial order. The partners had decided from the outset that they should share the true cost of running MCC and this has led to an annual budget for 2001 in excess of 6 million euros. It should be borne in mind however that this cost is very small in comparison to the contribution made by industry through their direct participation in 3GPP.

With the increase in expenditure comes a greater demand for financial transparency and the 3GPP organisational partners have established an ad-hoc group specifically to oversee such matters. An open accountancy policy has been established with access provided to detailed financial records such that each organisational partner can have visibility of MCC expenditure.

19.3.5 Later Developments

By the end of 2000, MCC had become firmly established as a team of 26 persons providing comprehensive support to 3GPP. By this time, the former ETSI TC SMG had closed, having transferred all remaining GSM specification activities to 3GPP, most of which was accommodated in a new TSG entitled "GSM/EDGE Radio Access Network (GERAN). A new entity called ETSI TC Mobile Standards Group (MSG) was created to take care of European regulatory standards for mobile telephone systems. The former ETSI EP UMTS had also closed and the former SMG9 entity had been transformed into the ETSI Project Smart Card Platform (EP SCP). By the end of 2000 MCC was providing support to 3GPP, TC MSG and EP SCP.

Chapter 20: Working Methods

Ansgar Bergmann[1]

SMG[2] has always treated its working methods using a pragmatic approach. Working methods were discussed and agreed when problems arose or in anticipation of problems, and were adapted when practical experiences were made.

The fora to elaborate working methods were:

- Ad-hoc groups of the SMG plenary and of working parties;
- PT SMG;[3]
- SMG co-ordination group;[4]
- A proper subgroup on working methods, WOME, was established in March 1998 (at SMG#25).

Papers on working methods were agreed and maintained since the 1980s, but a permanent document, GSM 01.00, was only approved in 1995. The working methods were agreed in SMG as a complement to the ETSI rules of procedure, which fix for example the rules for membership, elections and voting.

Major areas for working methods relate to:

- Ways to write specifications, in particular the usage of description techniques, for example formal description techniques;
- Handling of documents;
- Phased approach;
- Structure of specifications;
- Change Request (CR) procedures;
- Work item management.

20.1 Who Uses the Specifications?

The main goal of SMG/3GPP is to produce specifications. Working methods are a means to achieve this goal. In order to understand what a good specification is, it must first be understood who uses the specification and for what purpose.

---

[1] The views expressed in this chapter are those of the author and do not necessarily reflect the views of his affiliation entity.

[2] Throughout this chapter, "SMG" is used to denote the standardisation group called GSM before 1992. The working methods of SMG are, with some modifications, also applied by 3GPP.

[3] In this chapter, "PT SMG" is used to denote the project team earlier called PN and PT12.

[4] This group existed under different names.

The GSM specifications, being the definitive description of the GSM system, are used as the basis for:

- Product planning: this is the process to devise new features to be introduced into the GSM system – an activity between manufacturers and operators.
- Work on GSM evolution: this is mainly triggered by the needs of product planning. Other reasons for evolution are corrections and general enhancements, for example in the field of compatibility. The publicly visible part of the work is done in the specification groups; it is based on a publicly invisible part done within the companies, with much higher efforts.
- Product development: for the manufacturer this means the development of the necessary hardware and software components, for the network operator the provision of necessary functions in subscriber management, network planning, but also fields like customer care, and advertising. This area is rather complex, and keeps major parts of companies busy, and many departments are engaged.
- Procurement: the technical parts of contracts typically refer directly to certain versions of the specifications, stating compliance and sometimes non-compliance – and sometimes reference is even made to single CRs. For example, Release 90 of the GSM specifications was called *tender list of specifications,* and was planned mainly as the reference for procurement.[5]
- Testing: the GSM specifications include several test specifications, for the mobile station, base station sub-system, SIM-ME interface and for codecs. For other interfaces, testing is also important, and tests are written based on the GSM specifications. For example, major parts of MAP concern the interfaces between different networks; tests for international roaming have been prepared by the GSM association and the ETSI technical body SPS.
- Network planning and operation: an obvious example is given by the radio parameters, defined in the specifications, which have to be optimised during operation.
- Reference for curriculum, research and development: GSM is today's most successful mobile communications system, and the GSM specifications are a reference for education, a starting point for research and a reference model for development – nobody would today specify a mobile communications system without studying the GSM specifications.

Of course, in some of the fields mentioned above, secondary literature can be used, and an expert working on one area does certainly not have to study all specifications.

20.2 Achieving High Quality Specifications

As explained in the previous section, there is a surprisingly wide readership working directly with the GSM specifications. It is not possible to satisfy the needs of all readers.

For example, it would be nice if specifications were always easy to understand without any prerequisites in education and knowledge. But on the other hand, specifications should not contain unnecessary parts; in particular, they should not repeat or otherwise duplicate information; also, they should be precise, correct and complete. Unfortunately, these qualities are almost opposed to the goals of easy understanding.

Quality of specifications relates to:

- The described system: the system has to really work and to support the intended services with good performance; efforts of implementation must be reasonable. Ideally the system

[5] See TDoc GSM 223/88.

is optimised for its purposes. Methods to achieve these goals are simulations, validations, field trials, and evaluations of the running system. In order to adapt to future requirements, the system must provide means for compatible evolution. This goal is mainly achieved by (abstract) analysis.

- The way in which the system is described: requirements to the description relate to criteria as completeness and absence of contradictions.
- The way in which the description is developed: requirements to the development of the specification mainly relate to traceability of changes.

Quality of specifications is the major goal of working methods. Among the contributions discussing the issue, a document on *Dimensions of Recommendation Review*[6] is a good example, and its results are still up-to-date. It gives definitions of completeness, consistency and non-ambiguity as necessary qualities of GSM specifications.

Completeness: TDoc 103/87 defined completeness as the property that what was targeted is to be covered in the specifications. This is a rather laconic definition, replacing the question "what means completeness of the specifications" by the question "what content of the specifications should we target at". But the area is difficult, and here are some aspects that have been stressed during discussions:

- Specifications should conform to a general formal presentation and should contain definitions of vocabulary and abbreviations, contents list, scope and references; sentences and paragraphs should be complete.
- All indications of open questions and of items for further study in the specifications should be carefully checked: in a phased approach, specifications can well contain such issues, however they must be known and intended, not simply result from oblivion.
- Referenced documents should be publicly available and have the necessary quality.
- For the open interfaces, the actions and reactions necessary to achieve the necessary functionality must be defined. For the sake of compatible evolution, it is often beneficial if reaction to unforeseen events is defined aswell.
- In a competitive environment, a full specification of functions, applications and services is not always wanted. This point of view, however, taken to the extreme, would lead to the position that the GSM specifications should be restricted to the events on open interfaces. This is too restrictive, and problems have been experienced if behaviours were specified in GSM without the corresponding stage 1 and stage 2 descriptions.

Consistency: this means the absence of contradictions and the fitting-together of concepts in the different parts of the GSM specifications. Whereas there are different views on the necessary degree of completeness, it is clear that consistency must be achieved as much as possible. When a possible inconsistency has been found, there is normally no disagreement between experts whether it is a true inconsistency or not. Consistency can be improved by systematic checks of:

- logical aspects
- time conditions
- modularisation and layering and the restriction to defined service primitives
- consistent usage of terms and concepts within a specification and between different specifications

---

[6] See TDoc GSM 103/87.

- consistency with working assumptions and general decisions
- usage of well-understood and proven description models and techniques (see paragraph 20.3).

Non-ambiguity: this means that experts will not interpret specifications in different ways. Similar for consistency, it is clear that a maximum non-ambiguity must be achieved. However, naturally, when a possible ambiguity has been found, even experts may disagree whether it is a real ambiguity. Therefore, an *in dubio contra reum* should be applied: If there is a doubt whether a specification is ambiguous, this normally shows that it is.

It is sometimes difficult to discover ambiguities in a text you have written yourself. Committee work helps to discover ambiguities, because many experts discuss their understanding. Also, during the generation of test specifications, many ambiguities of a text are identified.

TDoc 103/87 of GSM#15 recognizes that completeness, consistency and non-ambiguity are not sufficient to ensure that the system functions, and sketches a verification program based on specification review (walk-through process), simulations, emulations and validation in a real system environment. Validation was an issue in the following years.[7] The validation process was mainly performed within the companies; simulations and emulations were also performed by universities and research laboratories. The role of SMG was mostly restricted to monitoring and reviewing these activities. An area where SMG took a very active role in validation, characterisation and selection is the development of speech codecs.

Specification review meetings were organised many times in SMG. At these meetings, rapporteurs and other interested delegates participated. This was an occasion to check the specifications, and also to elaborate more complex improvements, for example re-structuring. For bigger, complex and central specifications, several meetings of several days could be necessary during a year.

20.3 Rigorous and Formal Descriptions

For achieving high quality specifications, rigorous and formal description techniques are used.

20.3.1 Rigorous Descriptions

Most parts of the GSM specifications are written in "prose" (as opposed to descriptions in a formal language like TTCN).

To write specifications in prose in a clear way is sometimes called a rigorous description technique. A rigorous description applies rules to the language and format, which are sometimes directly opposed to good English style:

- A reduced vocabulary is used. For the same item, the same designation is used whenever possible.

---

[7] GSM#13 asked the PN to elaborate a system verification program, allowing continuous validation of the GSM specifications. But WP2 commented at GSM#14 that this would mean a delay, as several administrations had already started validations. GSM#14 decided that each administration should conduct its own verifications and that a coordinating committee should be established. Cf. also TDoc GSM 51/87, TDoc GSM 107/87, TDoc GSM 96/88, TDoc GSM 222/88 and GSM#21 Report, Section 8.

- The logic is clearly expressed. Care is taken to express clearly which conditions apply in conjunction or as alternatives.
- Optional and mandatory features are clearly distinguished.
- Defined attributes are used.

This is just what should be applied in any scientific and technical document. For ETSI and other standardisation bodies, a set of rules has been agreed fixing the details. These are the Production of Norms in Europe (PNE) rules.

Behaviours are described by communicating processes. These processes are modelled as (extended) finite state machines ("extended" means that each state may have additional parameters), or using a procedural approach or in a combination of both (in fact, they are equivalent). When possible, a layered structure is used, where each layer uses the services provided by the lower layer, and offers services to the higher layer.

Using these techniques, a rigorous specification is established. For such a specification, verifications can be carried out by proving certain "health properties" like pre/post-conditions and invariants: pre/post conditions describe changes of attributes after a transition or the execution of a procedure; invariants define properties that remain stable during a sequence of transitions or for the lifetime of a process. Care is taken to avoid deadlocks and live-locks. Exceptional events like time-outs and lower layer failure are considered.

Such verifications have been performed in GSM, mostly not in meetings but by delegates as homework. Specifications having undergone such a review process typically contain very few errors and ambiguities.

20.3.2 Formal Descriptions

Several specifications make use of formal description techniques. "Formal description technique" means here a technique that

- has a defined syntax;
- has defined semantics; and
- can be compiled by a computer.

The following formal description techniques are used in the GSM specifications:

- SDL is the most popular specification language applied in telecommunications. It is defined in ITU recommendation Z.100. SDL is supported by tools that can check the syntax and simulate the described processes. There are companies using an SDL based development environment where SDL is used from design up to implementation and can be compiled into executable code. SDL is used in many parts of the GSM specifications: in MAP, in most specifications of supplementary services, in most stage 2 descriptions and in several stage 1 descriptions. Specification 04.06 of the signalling layer 2, and specification 04.08 of the layer 3 protocols Radio Resource (RR) management, Mobility Management (MM) and Call Control (CC)[8] do not use SDL descriptions. In fact they earlier contained SDL descriptions, but these have been removed at GSM#25. Instead, state-transition diagrams for MM and CC have been introduced into 04.08. The official reason to delete the SDLs in these specifications was that GSM3 couldn't find the necessary resources for keeping them updated. However, there were also some other reasons. One is that SDL is

---

[8] From Release 99 onwards, 04.08 has been split into 04.18, later 44.018, for RR and 24.008 for MM and CC.

normally applied with modified semantics (and not the one defined in Z.100). Inconsistencies discovered by simulations often just reveal these differences of semantics. Another is that the semantics of SDL are restricted: Whereas concurrent process languages like Lotos or Milner's CCS [1] can simulate SDL, the converse is not true. A third point is that in order to write neat SDL specifications, some auxiliary modelling is necessary which is "artificial" and not relevant for the real system (SDL doesn't have clear notions of abstraction, encapsulation and equivalence of processes). A fourth point is that for many cases, state-transition diagrams are equivalent to SDL descriptions but are much more compact.

- Abstract Syntax Notation One (ASN.1)[9] is used in many GSM specifications to define the syntax of messages and operations. As the name "Abstract Syntax Notation" implies, ASN.1 itself does not specify the encoding; this is done by encoding rules attached to ASN.1. The idea behind this comes from the concept of abstract data types, proposed in the scientific field, where it is required that data structures should be first designed abstractly (as so-called *initial algebras*) and that the coding should be developed independently. ASN.1 however does little to support this algebraic concept, and typical ASN.1 specifications tend to use mainly bit strings as data types. Different sets of encoding rules have been defined for ASN.1. The free availability of an ASN.1 compiler, at least a syntax checker, for SMG experts was discussed on several occasions, e.g. at SMG#2. Instead, rapporteurs took over these checks using tools of their companies. In other specifications, the message contents are defined in tables, showing the position of information elements and defining the meaning of the different bit combinations. The resulting coding is a byte oriented encoding, typically with indication of type, length and value part (however, for optimisation purposes, type and length indicator can be suppressed under certain conditions). In GSM, it became obvious that a byte orientated encoding with type and length indication is not efficient enough for the radio interface, mainly on the common control channels. Therefore, it was proposed to use a context-free syntax description, the Backus–Naur form, at least for the radio interface. The underlying grammar of each message is bit-orientated; it has to have "nice" properties that make en/decoding efficient (typically, a look-ahead 1 grammar). This scheme was then elaborated to become Concrete Syntax Notation One (CSN.1) [2].[10] This approach gives a very powerful mechanism for efficient coding. It is mainly applied in 04.08. It allows economic coding schemes taking into account the *a priori* probability of occurrence of values. Context free grammars are known in computer science for a long time,[11] and play a major role in compiler building, one of the best-understood techniques in computer science. A big number of generic tools are available for the Backus–Naur form; specific tools for CSN.1 are available.

- Tree and Tabular Combined Notation (TTCN)[12] has been applied since phase 2 in the mobile station test specification 11.10 for the protocol related parts. The prose test specifications are developed first and are then translated into TTCN. The dynamic part of TTCN uses a notation for allowed sequences of observable actions that is inspired by Hoare's CSP [3] calculus. For the data part, a tabular notation or ASN.1 can be used. Tool environments are available and are used by experts of the ETSI secretariat for checking

---

[9] Defined in ISO/IEC 8824.
[10] Cf. also GSM 04.07.
[11] Even if they have been studied first in the field of linguistics by Chomsky and others.
[12] Defined in ISO/IEC 9646.

the syntax and static semantics. Also, software companies are developing test implementations based on TTCN, and have taken rapporteurship as the main parts of the TTCN in 11.10. Between phase 1 and phase 2, the prose test descriptions have been transformed into a rigorous format that follows a structure very close to TTCN. These have then been translated into TTCN. For TTCN, PT SMG elaborated a modified CR procedure allowing a fast debugging process. During the translation of GSM tests into TTCN, the need for several improvements of TTCN were discovered, resulting from design problems of the language, relating for example to the handling of parallel processes. Hence, GSM contributed to the development of TTCN.

- In the area of test specifications, ICS and IXIT are applied. Implementation Conformance Statements (ICS), and Implementation Extra Information for Testing (IXIT), were first introduced for protocols. That is why acronyms "PICS" and "PIXIT" ("P" standing for "Protocol"), are more familiar terms, sometimes also used where protocols are not engaged. ICS and IXIT are declared for equipment to be submitted to conformance tests. While PICS declare which options permitted by the standard are implemented, PIXIT declares how features and functions are realised. (In 11.10, the distinction is sometimes not precise.) For example, a PICS could declare that a mobile station does not implement the optional feature "fax transparent". A PIXIT statement could, e.g. explain how a service is initiated at the user interface of the mobile station. This kind of information is essential for test houses. 11.10 specifies a proforma (a kind of questionnaire) for ICS. It also classifies features as mandatory, optional or conditional and gives logical constraints for optional and conditional features. These constraints use a kind of pseudo-formal description. It has been proposed to use instead a formal description, the well-known and proven calculus of Boolean algebra; appropriate tools could be provided easily. Discussion in this area will certainly go on.
- In the area of Telecommunications Management Network (TMN), the object orientated description method GDMO[13] has been applied in the 12.xy series. In 3GPP, there are tendencies to replace GDMO by CORBA, and more recently by other description techniques.

For the usage of formal description techniques, the availability of tools seems essential. They can check syntax and static semantics and perform simulations and thereby validate the description. Also, the implementation cycle can become shorter. A drawback is that the number of experts being able to review the description is reduced. Also, the formal description technique may be insufficient or may require artificial auxiliary modelling.

Maybe it is often the best approach to use a rigorous and a formal description in parallel. A question is often raised when this is done. In the case of contradictions, does the formal or rigorous description prevail? This question is surprising, or, more exactly, it is surprising that the question is often found adequate: Contradictions are not planned. When they are discovered, they must be resolved, and it is not predictable where the error has crept in. This was at least the position taken by SMG, when it was decided that both the prose and TTCN descriptions in 11.10 are normative.

---

[13] GDMO = Guidelines for the definition of managed objects; see ITU X.722 (ISO 10165-4).

20.4 Handling of Documents

Handling of documents, in particular electronic handling, has been a non-trivial issue for a long time. Until 1996, documents were normally not available in electronic form. For electronic documents, the limitations of file names to 8 + 3 characters were cumbersome. The introduction of fully electronic document distributions at meetings required some learning and conviction processes. But with the improvements of network facilities, these problems have gone. Today, 3GPP working groups have their e-mail groups and meeting documents are maintained on the 3GPP server at ETSI.

20.5 Phased Approach

At GSM#20 (October 1988), three phases of GSM activities were distinguished:[14]

- Phase 1: development of the recommendations;
- Phase 2: validation and consolidation of the technical aspects;
- Phase 3: revisions based on pre-operational experience.

The document also remarks that it can be expected that revisions will continue once commercial service commences.

These phases, however, should not be confounded with the phased development and the phased implementation of GSM as expressed at GSM#25, where it was decided to create two parallel series, version 3 (phase 1) and version 4 (phase 2).[15] This phased development meant a first phase, to be implemented and operated, followed by an upgraded phase 2 with additional services and improved functionalities.

In order to concentrate on the introduction of phase 1, specifications were *functionally frozen*.[16] This means that no new functionalities should be added to phase 1, and that phase 1 changes should be restricted to necessary corrections. For example, VLSI companies commented that specifications should be stable 2 years prior to the intended 1991 Ready For Service (RFS) date.

At GSM#25 it was decided that GSM#25bis (spring 1990) should be the last opportunity for changes except ones vital for operation of the system; in other words, the core specifications were frozen at GSM#25bis.

Then, for some time, SMG plenary refused in principle to work on phase 2 specifications until the phase 1 specifications were stable. After the stabilisation of phase 1, option pruning was an important point. This meant to delete options in the specifications, which nobody intended to implement any more. For the introduction of phase 2, compatibility was a requirement: phase 1 mobile stations should be able to work in an evolved network without loss of quality.

The concept of phase 2+ and the phase 2+ program were approved at SMG#7 (June 1993).[17] The basic idea was to introduce new functionalities as options, in a compatible way. CRs should be approved in a complete package so that the specifications would always be complete and consistent.

---

[14] See TDoc GSM 222/88.
[15] See TDoc GSM 482/89 rev2.
[16] Cf. for example TDoc GSM 222/88 and TDoc GSM 227/88 at GSM#20.
[17] See TDoc SMG 475/93. Cf. also TDoc SMG 517/93.

There was a fundamental problem. To elaborate complete packages of CRs required some time. As work went on for several features in parallel, there was a danger of creating contradicting CRs in the piles for the different features. Another danger was to elaborate ad-hoc isolated solutions for different features, and to miss common solutions.

It was proposed to create working versions or *shadow versions* of specifications; these could then show the current status of specifications in preparation. It was however a common feeling that the CR procedures had to be maintained. One proposed method was to elaborate, after the freezing of versions 4, a new version 5, to apply the CRs procedure for the elaboration of version 5, then to agree on a set of new features the specifications of which were stable, and then to re-integrate the changes into version 4. The principle would have been to maintain a version 4 as basis for implementation, but version 4 would change from time to time to incorporate new features. At the same time, a version 5 would be used for the working versions under development.

There were several reasons why this scheme was not adopted. One reason was that the re-integration would have been a difficult task; another one was that mainly for mobile station type approval, the "original" phase 2 had to be kept, because it had to be possible to continue mobile station type approval on its basis.

The pragmatic decision was taken to freeze the version 4 specifications and to create a new version 5; this was intended to allow some time to identify a better solution later, to be applied from version 5 onwards.

In the following years, two principle alternative ways ahead were identified and discussed:

1. To create new releases from time to time, for example every year.
2. To identify new description parts in the version 5 specifications with formal markers. The markers would relate the description parts to features (one or several), and a master table would indicate when a feature was stable.

The issue was complex, and discussions went on for some time. It turned out that scheme (2) was too difficult, and finally solution (1) was chosen, to create a new release, that is a full set of specifications, every year. This scheme was at least simple and easy to understand, but it meant also that the amount of specifications under maintenance increased considerably; also it became soon obvious that the scheme could not be applied as such for testing and type approval.

There is a principal asymmetry in the concepts of telecommunication networks, which has consequences for version management: the terminal is considered to be the slave of the network. The GSM network should know the different types of mobile stations and know how to handle them. Therefore, there should be a clear distinction between a phase 1 mobile station and a phase 2 mobile station whereas the notion of "phase 1 network" or "phase 2 network" does not make much sense.[18] By the way, there are approaches to change this philosophy, and to require operators to specify their access interface, so that terminals can adapt to it. Anyhow, neither networks nor mobile stations normally fully conform to one single GSM release.

---

[18] This philosophy has however not really been applied in strength. For example, new phase 2 features were step-by-step implemented in phase 1 mobile stations, before phase 2 type approval had started.

20.6 Structure of Specifications

There are several grouping schemes that apply to the GSM specifications. The structure is complex, mainly due to

- the complexity of a fully detailed system specification;
- the need of version management; and
- the integration of GSM and UMTS specifications.

20.6.1 Semantic Structure

The principle structure of specifications, at that time called recommendations, was established very early, also the concept of reports versus recommendations (later called specifications), preliminary drafts (version 0.x), first drafts (version 1.x) and final drafts (version 2.x).[19]

The structure until Release 98 was as follows:

- 01 series: requirement specifications; general documents (e.g. 01.04, abbreviations and acronyms); administrative and managerial documents, (e.g. 01.00, working procedures for SMG and 01.01, list of specifications for a given release);
- 02 series: stage 1 descriptions; man-machine interface of the mobile station (02.30);
- 03 series: stage 2 descriptions, network functions and architecture;
- 04 series: protocols of the radio interface (the interface between the mobile station and the network);
- 05 series: layer 1 of the radio interface;
- 06 series: speech codecs;
- 07 series: data services;
- 08 series: protocols of the A and Abis interface;
- 09 series: protocols of the core network;
- 10 series: work item management;
- 11 series: test specifications for the mobile station and base station subsystem; specifications of the SIM and SIM-ME interface;
- 12 series: operation and maintenance;
- 13 series: harmonised standards for mobile station type approval.

From Release 99 onwards, specifications were re-structured due to the integration of the GSM and UMTS core network. This is described in paragraph 20.6.4.

20.6.2 Stages

For complex features, typically a stage 1 specification (in the 02 series), a stage 2 specification (in the 03 series) and several stage 3 descriptions (often in the 04 series, the 08 series, the 09 series) exist in all releases containing the feature. Possibly there is also a performance specification. Other aspects including testing, AT commands, SIM ME interface, charging,

---

[19] See for example TDoc GSM 23/86 rev.2, the GSM action plan, a living document updated periodically; also TDoc GSM 13/86 proposing some revisions to GSM 01.01, the list of GSM recommendations; also TDoc GSM 91/ 85, partially based on earlier GSM 28/84 rev 2 (not available).

billing, event records, TAP records (described in GSM Association specifications), security aspects, operation and maintenance aspects, and tracing have to be covered as well, often in separate specifications, cf. paragraph 20.8 on work item management. For major work items, in the initial phase of the work, a requirement specification (in the 01 series) and a dedicated report on project scheduling and open issues (in the 10.xx series) have been created; these specifications are normally not maintained in the later phases of work.

For the specifications from Release 99 on, see paragraph 20.6.4.

20.6.3 Versions

In this section, the versions of GSM specifications are described.

The GSM specifications are structured in phases and releases. A specification has versions in different releases. For example, the protocol specification of the voice group call service, *Group Call Control (GCC) protocol,* has versions in Release 96 and all subsequent releases (Figure 20.1).

The GSM phase 1 specifications have been the basis for the GSM implementations (network and mobile stations) in the first years, roughly from 1992 to 1995. Important features of phase 1 included

- telephony;
- some supplementary services like call forwarding;
- international roaming.

The GSM phase 2 specifications have been the basis for GSM implementations roughly between 1995 and 1998. GSM phase 2 added mainly

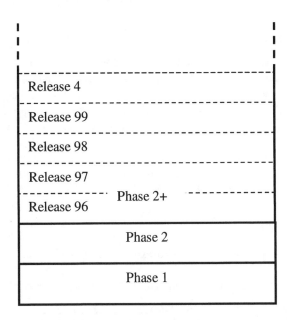

Figure 20.1 Phases and Releases

- telefax;
- data services;
- full range of supplementary services.

Phase 2+ is a mostly compatible evolution of phase 2. The principle is to add features as an option for the network and for the mobile station. Hence, for the implementations, there is neither a clear border between phase 2 and phase 2+ nor between Release 96, Release 97 and so on. The main features introduced in phase 2+ include

- services customisation and service portability: this is realized by functions like

  - Intelligent Network (IN) functions; and
  - SIM Application Toolkit (SAT) which allows the network operator to download intelligence via the radio interface to the smart card in the mobile station;

- circuit data (High Speed Circuit Switched Data (HSCSD)) and packet data (General Packet Radio Service (GPRS)) with data rates up to 64 kbit/s and higher;
- Advanced Speech Call Items (ASCI): Voice Group Call Service (VGCS), Voice Broadcast Service (VBS) and priority, officially called enhanced Multi-Level Precedence and Preemption Service (eMLPP);
- quality and capacity enhancements, in particular the enhanced full-rate codec and the adaptive multi-rate codec;
- localization;
- global roaming between GSM 900, GSM 1800, GSM 1900 and also inter-system roaming, for example with mobile satellite systems.

Between 1996 and 1999, yearly releases have been issued. Later, in 3GPP, this was found too restrictive, and the later releases are now called Release 4, 5, etc.

20.6.4 New Structure of GSM/UMTS Specifications

From Release 99 on, the GSM and UMTS core network specifications have been integrated. For example, the specification GSM 09.02, *Mobile Application Part (MAP) specification,* was turned into a 3GPP specification, common to GSM and UMTS, and became 3GPP 29.002, *Mobile Application Part (MAP).* The transformation is shown in Figure 20.2.

Other specifications were only relevant for GSM but not for UMTS; they kept their numbers in Release 99.

The next release was Release 4, as explained above. From Release 4 on, the remaining GSM-only specifications were transferred from SMG to 3GPP and renumbered as well. Their transformation scheme is shown in Figure 20.3.

20.6.5 Versions and CRs

Evolution of GSM is performed by CRs to existing specifications and by creation of new specifications.

Within a release, specifications are updated in new versions. For example:

- the first version of a specification in Release 96 is version 5.0.0;
- the first version of a specification in Release 97 is version 6.0.0;

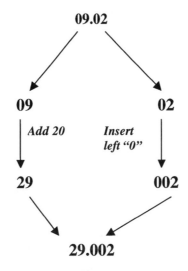

Figure 20.2   Renumbering of Specifications

- the first version of a specification in Release 98 is version 7.0.0;
- the first version of a GSM-only specification in Release 99 is version 8.0.0;
- the first version of a joint GSM/UMTS specification in Release 99 is version 3.0.0;
- the first version of a joint GSM/UMTS specification in Release 4 is version 4.0.0;
- the first version of a GSM-only specification in Release 4 is version 4.0.0.

When one or more CRs to a specification have been accepted by SMG/3GPP, a new version of the specification is created. (Typically, there have been four SMG plenaries per

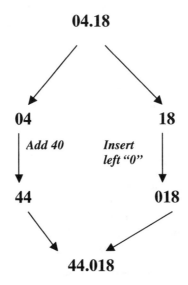

Figure 20.3   Renumbering of GSM-only Specifications

Table 20.1   Versions of a GSM/UMTS Specification

Release	Specification number	Version
R96	03.67	5.1.1
R97	03.67	6.1.0
R98	03.67	7.2.0
R99	23.067	3.2.0
R4	23.067	4.0.0

year.) After approval of a non-editorial CR to version 7.0.0 of a specification, the resulting version is 7.1.0. If at a SMG/3GPP plenary, only editorial changes had been approved, the right digit[20] is increased: for example version 7.0.0 becomes version 7.0.1.

Typically, a GSM specification exists in several, often in all releases, and parallel CRs are approved for several releases of the same specification at the same SMG/3GPP plenary.

20.6.6 Examples

The GSM specification *Enhanced Multi-Level Precedence and Preemption Service (EMLPP); stage 2* has been created in Release 96 as GSM 03.67 version 5.0.0. As the service is supported in all following releases as well, the specification became also part of Release 97 and Release 98. Then it was decided to include the service in UMTS as well. As a consequence, the Release 99 specification got number 23.067 and version 3.0.0; the following release, Release 4, also includes 23.067 as version 4.0.0.

CRs to Releases 96, 97, 98 and 99 have been approved. The most recent versions are shown in Table 20.1.

The GSM specification *Voice Group Call Service (VGCS); stage 2* has been created in Release 96 as GSM 03.68 version 5.0.0. As the service is supported in all following releases as well, the specification became also part of Release 97 and Release 98. Then it was decided *not* to include the service in UMTS. As a consequence, the Release 99 specification is a GSM-only specification, kept number 03.68 and got version 8.0.0; in the following release, Release 4, it was re-numbered as 43.067 version 4.0.0.

CRs to Releases 96, 97, 98 and 99 have been approved. The most recent versions are shown in Table 20.2.

20.7 CR Procedures

As GSM is a living system, specifications have to be improved and enriched. It is important that every change in the specifications is documented and can be traced. A CR procedure was agreed already in the 1980s.[21] The purpose of a CR is the correction of errors and of inconsistency, a necessary clarification or the addition of a new feature.

Later, a CR database was established. The scheme was completed when, after 1995, all

---

[20] For a version x.y.z, for example 4.1.1, the numbers x, y, z were normally called digits; this is not very precise, as for example a version 4.17.1 may well exist.

[21] See TDoc GSM 161/88 and 161/88 rev. 1.

Table 20.2   Versions of a GSM-only Specification

Release	Specification number	Version
R96	03.68	5.5.1
R97	03.68	6.3.0
R98	03.68	7.2.0
R99	03.68	8.2.0
R4	43.068	4.2.0

CRs were made available in electronic form (this included scanning of previous CRs, which had been archived on paper so far).

During the years, the CR regime was perfected, and has, since then, been adopted by several other institutions, in particular by 3GPP. The early discussions concluded to apply the CR procedure as the only way to change specifications:

- GSM#20 agreed that consolidations of the GSM specifications were to be carried out only by means of the CR procedure.
- GSM#21 defined four CR categories.
- GSM#27 clarified that clarification statements have to be presented as CRs.

It was seen as essential not only to record all changes, but also the reasons for changes and further information. Also, working methods were installed to systematically check the quality of CRs (consistency, completeness, etc.).

Final approval was reserved to the SMG plenary, so that companies had time to assess feasibility and consequences. Before presentation to SMG, the CRs had been scrutinised by the responsible STCs in consultation with the rapporteur, with PT SMG and secondary responsible STCs (if any).

Separate CRs to all affected releases had to be produced.

To ensure a consistent presentation and documentation, standardised CR forms were used. The purpose of the CR form itself was to provide the relevant management information of the proposed changes: version number of the affected specification, source of the CR, reason for the proposed change, consequences if the CR was not accepted, category of the CR and cross-phase compatibility aspects.

For functionally frozen specifications, only CRs were allowed for essential corrections.

Impacts on other specifications had to be examined and the necessary corresponding CRs had to be produced.

When CRs were finally presented to SMG, they were normally approved, however two other decisions were possible: A CR was *postponed* if the concept seemed acceptable in principle but further work was necessary: the CR was sent back to the STC for revision and possible re-submission at a later SMG plenary. A CR was *rejected* if it was not found acceptable in any sense.

If at an SMG plenary there was at least one approved CR, a new version number of the specification was allocated (changing the least significant digit if all CRs were editorial, and the second significant digit for everything else), and the new version was issued.

The same procedures are – *mutatis mutandis* – applied in 3GPP.

20.8 Work Item Management

New features are introduced into GSM depending on market demands. This term "feature" is intentionally vague: A feature can be

- directly relevant for the user (as, for example, the presentation of the calling number to the called user);
- indirectly perceivable by the user (as, for example, the enhanced full rate codec, offering a better average speech quality); or
- completely invisible to the user (as, for example, an improved error handling on the A interface, the interface between the MSC and BSS).

A feature may require changes to the GSM specifications or not – in fact, a considerable part of new features can be realized outside of the GSM specifications.

Features may depend on other features, and it is essential for the quality of the system to identify building blocks that can be the basis for several features.

The phase 2 + concept was intended to add new features as options to the GSM phase 2. For the introduction of a feature, a corresponding SMG work item had to be created, see further below.

The intention to introduce new features results from the product planning processes in the companies engaged, an activity between

- the manufacturing companies, which have many operators world-wide as clients;
- operators, often multi-national, normally having several vendors.

As each vendor and operator wants to be the first to place a new feature on the market, intentions are not always shown fully in public. On the other hand, for some features, Pan-GSM availability is important, in particular for the roaming user. Traditionally, most work items were proposed by SERG, the GSM association expert group on services. During the evolution of GSM, the focus moved from specification of services to the specification of generic platforms enabling each operator to introduce specific features without the need of additional specification work, ideally so that they are available even when the user is roaming.

The management of work items in SMG was improved during the years, resulting in the final scheme described here.

The following steps were performed in order to realize a work item in SMG:

- Creation of a work item: a new work item was first proposed to an STC of SMG, typically to SMG1 (responsible for services); this STC then produced a work item description sheet for approval at SMG. The work item description sheet indicated the necessary changes to existing specifications and the necessary new specifications.
- After approval by SMG, requirements and main service aspects were defined, resulting in a requirement specification and a draft stage 1 specification.
- Then, architectural aspects and the main principles of the realization were elaborated; the result was a draft stage 2 specification. A paradigmatic structure for phase 2 specifications was discussed at SMG#11.[22]
- The next step was to work out the detailed technical solution (resulting in stage 3 descrip-

---

[22] See TDoc SMG 438/94.

tions). This was accompanied by a feedback process to refine and improve stage 1 and stage 2.

- During the stage 3 elaboration, work on all consequential areas started: test specifications of the mobile station and base station system, operation and maintenance aspects, changes to event records (necessary for charging), additional AT commands, sometimes a report on performance characteristics (to replace the requirement specification, which became obsolete once the feature was defined), security related issues, additions to the SIM-ME interface, consequences for the user interface (man-machine interface of the mobile station), changes to the organization of subscriber data and others.
- Completion of work item: finally, the complete set of CRs and new specifications was approved by SMG.
- After that, a phase had to be foreseen where experiences from implementation led to clarifications and corrections of the specifications.
- Very often, when the work item was completed, the need for enhancements and additional functionality became obvious leading to new phases of the work item.

Sometimes, the technical development came to a point where a fundamental decision had to be taken, often the choice between several ways forward, where however not all consequences of such a decision were clear. This could have led to a blocked situation. In order to achieve progress, the principle of *working assumptions* was agreed:

- The presumably best solution was selected as the mainstream of future development.
- It was to be accepted at the end of the optimisation and detailed specification phase if an alternative solution was not shown to be better.
- The working groups had to concentrate on the working assumption.
- The elaboration of the "challenging solution" and the proof of its supremacy had to be carried out by the challenging parties outside of the working groups, but the final assessment had to be taken inside of SMG.

This principle was defined by GSM#13 in February 1987,[23] a long time before phase 2 + was invented, when the basic parameters for the GSM radio were defined.

Work items together with their milestones were recorded in the work item database from which a roadmap was generated, indicating the planned completion dates of work items. All work item description sheets, updated when necessary, were available in specification 10.00. For major work items, dedicated reports on project scheduling and open issues were created in the 10.xx series. These reports, containing a project plan for the work item and describing issues under discussion as well as the relevant technical decisions taken during the specification process, provide a valuable source of background information.

In 3GPP, work item management has been further structured, and the distinction of features, work items and building blocks has been introduced. A building block is a technical functionality, devised with the intentions of re-use and generality when possible; a work item can use several building blocks, a building block may be used by several work items. The most general term is "feature"; features are components of work items and building blocks.

---

[23] See TDoc GSM 45/87, an accompanying document for TDoc GSM 41/87, which fixed the package of narrowband TDMA working assumptions at the famous Funchal meeting in February 1987.

References

[1] Milner R. A Calculus of Communicating Systems, LNCS, Vol. 92. Berlin: Springer, 1980.
[2] Mouly M. The CSN.1 Specification. perso.wanadoo.fr/cell.sys/csn1_fr.htm.
[3] Hoare C.A.R. Communicating Sequential Processes. Englewood Cliffs, NJ: Prentice Hall, 1985.

Chapter 21: The Contribution of the GSM Association to the Building of GSM and UMTS

Section 1: Cooperation of the Operators from the Agreement of the GSM MoU to the Opening of Service in 1991/1992

Renzo Failli[1]

21.1.1 The Main Objectives of the MoU Group in the First Period of its Activity and its Organization

The MoU group had its first meeting on 14 October 1987, shortly after the signature of the MoU by the mobile operators of 13 European Countries: Federal Republic of Germany (FRG), France, Belgium, Denmark, Ireland, Finland, Italy, Norway, Netherlands, Portugal, Sweden, UK and Spain.

The numbers of operators was 14 since at that time a single mobile operator was active in each country apart from the UK where two operators (Cellnet and Vodafone) were active.

In the following months the MoU was signed by the mobile operators of another three countries: Luxemburg, Switzerland, Austria. In such a way Western Europe was substantially represented (Greece and Turkey joined later).

The main objectives may be summarized as follows:

- to give evidence to the manufacturing industry of the actual willingness of a large number of European mobile operators to invest significantly in GSM networks in order that the industry might dedicate sufficient resources to the development of the new technology;
- to avoid that holders of potentially essential IPR for the GSM technology might create problems for the economical development of the new system;
- to define all technical, commercial and regulatory aspects necessary to allow a true Pan-European service allowing each GSM client to roam in every country using the same GSM terminal and keeping the possibility to originate and to receive calls to and from all other fixed and mobile telephone terminals;

---

[1] The views expressed in this section are those of the author and do not necessarily reflect the views of his affiliation entity.

- to support the standardization work in GSM and to take decisions by majority voting in cases where no unanimous decision could be reached in GSM;
- to ensure the correct operation and interworking of network equipment of different manufactures and, above all, the compatibility of all GSM terminals with all networks;
- to promote the GSM technology in countries outside Western Europe with the purpose of both reducing network and terminal costs due to economy of scale, and to improve the service by enlarging the roaming area.

In order to reach such objectives in the proper time, the MoU group immediately established seven expert groups with the mandate to study and report on specific aspects, and asked seven signatory countries to offer the support with a rapporteur:

P – group: Procurement (rapporteur from France: Ghillebaert)

- harmonization of procurement policy;
- IPR co-ordination;
- configuration control (freeze technical configurations in order to allow the GSM network to interwork each other: choice of GSM recommendations and their version to be used in contracts).

BARG: Billing and accounting rapporteur group (rapporteur from UK: Maxwell)

- administration of subscribers;
- billing harmonization;
- credit control;
- fraud prevention measures;
- accounting operations;
- statistics;
- definition of harmonized billing and accounting software requirements.

EREG: European roaming expert group (rapporteur from Italy: Failli, followed by Panaioli and subsequently Eynard)

- coordination of all technical and operational principles and plans for the support of European roaming, including:
- mobile numbering plans;
- routing of terminated calls and of signaling messages;
- technical implications of tariff principles on international interworking;
- establishment of international signaling links;
- interworking between PLMNs using different work functions, quality and availability of service.

MP: Marketing and planning (rapporteur from Sweden: Tadne followed by Magnusson and subsequently Zetterström)

- coverage maps;
- GSM name and logo;
- coordination of awareness campaign and public relations;
- commissioning of market survey;
- identification of selling features to guide system developments.

TAP: Type approval administrative procedures (rapporteur from The Netherlands: Hoefsloot followed by Jeanty)

- review of existing or emerging directives and identification of possible difficulties;
- control and issue of International Mobile Equipment Identities (IMEIs).

TC: Technical coordination (rapporteur from Norway: Bliksrud)

- validation of CEPT/GSM recommendations;
- commercial perspectives of MOU signatories on technical GSM options.

N: European network and equipment implementation (rapporteur from FRG: Hillebrand followed by Linz)

- validation of system components;
- network implementation phases;
- technical type approval matters.

The activity of the MoU group and of its rapporteur groups in the first years had no centralized support. Each operator contributing either with the MoU chairman or with one of the seven rapporteurs took care of the secretariat support while operators/administrations inviting the groups had to take care of all support for the meeting.

It was a very light organization, thanks to the reduced number of people involved. Just to give an idea the number of people attending MoU plenary meetings was in the order of 25-30 persons.

The actual work was done within the rapporteur groups while the MoU plenary dealt mainly with the definition of tasks for the rapporteur groups, the approval of the main results, the discussion of the most relevant/controversial issues.

The MoU group chairman remained in power for a period of 6 months, so that many signatories might contribute with their experience to that task. The main task of the MoU chairman, in addition to the co-ordination of the activities, was the relationship with external bodies, including potential new signatories.[2]

In 1989, due to the evolution of activities, it was necessary to review the rapporteur groups' mandates. Consequently it was decided to move to an organization with ten groups. In particular three completely new groups were established with the following mandates:

TADIG: Transfer account data interchange group (rapporteur from FRG: Giessler) with the task to define a MoU standard for the interchange of billing data for international roaming. The precise mandate and the main results obtained by TADIG are reported in Chapter 3, Section 3.

SERG: Service expert rapporteur group (rapporteur from the UK: Barnes followed by Hall and subsequently by Toepfer from FRG)

- maintenance of commercial aspects of GSM 02 series, recommendations following transfer of responsibilities from ETSI/GSM;
- allocation/revision of the status of implementation;
- categories of services and dates of introduction;
- review of compatibility of services for roaming.

---

[2] A list of all MoU plenary meetings can be found in Annex 1, a list of the MoU chairpersons in Annex 3.

The main activities and results of this group are described in Chapter 3, Section 2.
SG: Security group (rapporteur from the UK: Brookson)

- maintenance of algorithms and test sequences;
- administration of non-disclosure undertaking for algorithms;
- monitoring of adequacy of system security

Moreover the N group and the TC group were substituted by the following two groups:
RIC: Radio interface co-ordination group (rapporteur from France: Maloberti)

- coordination of technical aspects of type approval and identification of problems affecting type approval as results of validation and conflicting interpretations of recommendations;
- resolution of technical problems with regard to type approval in different countries;
- organization of compatibility testing of mobiles in different networks to ensure adequacy of type approval;
- review of results of GSM validation activities and assessment of effects on implementation plans;
- review of system simulator activities through a specific sub-group System Simulator Project Group (SSPG).

CONIG: Conformance of network interfaces group (rapporteur from The Netherlands: Vocke followed by Pujol from France) with the task to solve network implementation incompatibilities through:

- list/definition of tests for conformance of "MAP", "A" and "Abis" interfaces;
- harmonization of test activities.

Due to both the enlargement of the MoU group with new signatories, starting from the second mobile operators in the already represented countries, and the larger responsibilities of the MoU following transfer of certain responsibilities from GSM to the MoU, particularly in the fields of services, security and terminal equipment identity, in 1990 the need arose for a Permanent Secretariat. It was established and supported by the assistance of the Dutch Administration. But in September 1991 R. Hagedoorn, the Secretary provided, was no longer available and the GSM MoU Group decided to accept the offer of Telecom Eireann to establish the Permanent Secretariat in Dublin under the co-ordination of John Moran.

In 1990 it was also decided to promote the GSM standard in Eastern European countries and the MoU organized a specific seminar in Budapest under the coordination of Friedhelm Hillebrand. Similar initiatives were subsequently decided in order to promote the GSM in the other continents (for details see Chapter 22).

21.1.2 The Main Subjects Dealt with and the Main Results Got by the MoU Plenary and its Rapporteur Groups

21.1.2.1 The IPR Undertaking

Most probably the IPR issue has been the single subject which took up the most activity time of the procurement group and of the MoU plenary within the first 2 years of activity.

Before entering into the description of problems and facts I wish to mention two simple events which can demonstrate the dimension of the effort produced by the MoU.

During my working experience I spent a large number of days in meetings of international bodies. All the meetings are well organized and their afternoon sessions ended at reasonable times. Only twice I experienced meeting sessions initiated on one day which finished on the subsequent one.

The first one was in Funchal (Madeira, February 1987) on the last day of the most exciting GSM meeting, when the main GSM radio access parameters were decided. The second one was in Albufeira at the end of 1988 during the MoU plenary meeting, under my chairmanship, when we were discussing the IPR issue with the manufacturing industry, and particularly with Motorola. Unfortunately the results of this meeting were much less exciting and fruitful than the one in Funchal.

The other fact that I wish to recall was an extraordinary MoU plenary meeting, a telephone conference, held during the summer of 1988 on the same subject. Friedhelm Hillebrand was on holiday in Italy in South Tirol. To participate in the meeting he had to use the hotel telephone, for which the hotel applied a substantial surcharge. If I remember well he had to pay more for the telephone bill than for the hotel room. That was an experience which today, with the GSM coverage and roaming extension would not happen.

Entering into the specific facts relating to the IPR, the MoU group, following a specific objective defined in the MoU, agreed the IPR clauses to be included in the call for tender that signatories had to send to the manufacturing industry during the first quarter of 1988 for the GSM network implementation.

The text used by operators, entitled *"Tenderers undertaking on intellectual property rights"* can be found on the attached CD-ROM.[3] Such undertaking had to be signed by each tenderer before sending it jointly with his offer to the inviting operators. The text gives clear determination for IPR clauses to be incorporated in the contracts resulting from the tendering action.

The most important and stringent clause of the undertaking relates to essential IPRs which should have been given on fair, reasonable, non-discriminatory and royalty-free terms and conditions (free of any payment of any description).

The main problem experienced was that, while the European manufacturing industry signed such an undertaking, or a similar version, Motorola refused to sign.

The MoU P group and the MoU plenary had been working with this problem, both with normal and extraordinary meetings, for 2 years (1988-1989) without being able to find a solution which might be acceptable for the MoU, the European industry and Motorola.

The practical solution was then found through bilateral agreements of Motorola with other industries involved. Problems have been experienced by some industries, especially relatively small terminal manufacturers, not having their own IPR basket to be exchanged with Motorola.

Reference to this experience and the intervention on the matter of the US government, is mentioned in Chapter 2, Section 4 with a more political approach.

21.1.2.2 Networks Implementation

Probably the main result of the GSM MoU group in its initial activity phase was to convince

---

[3] See "MoU IPR Declaration" on the attached CD-ROM in folder D.

the main telecommunication manufacturers about the importance of the project and the actual willingness of European mobile operators to invest in GSM networks.

In fact, in 1987, after the main decisions on the GSM radio interface had been taken, only very few manufacturers were deeply involved in the development of the GSM system, while a large part of them had not yet decided to concentrate enough resources on the GSM project, and were consequently trying to delay the procurement process. The co-ordinated activity of European mobile operators, through the MoU group, was determinant in convincing the manufacturing industry to dedicate the necessary resources to GSM.

The implementation of GSM networks in signatory countries between mid-1991 (agreed target date) and mid-1992 was reached also thanks to the joint effort of numerous groups within the GSM MoU, namely P, TC, N, RIC and CONIG.

The main activities co-ordinated or carried out by these groups can be summarized as follows:

- publication in January 1988, in the official Journal of the European Community, of the intention to call for tender by most of the signatories in a coordinated way;
- co-ordinated calls for tenders in the first quarter 1988;
- exchange of test results on validation systems in order to ensure compatibility;
- co-ordination of requests to manufactures in terms of GSM specifications to be used and their version for the opening of the service.

21.1.2.3 Terminals

The aspects relating to terminals were followed by N and RIC groups for the technical aspects and from the TAP group for the regulatory ones.

An important activity was carried out for the procurement and the development of the system simulator, specifically designed for type approval test of GSM terminals.

Such a simulator, supplied by Rhodhe & Schwartz, was procured by a few signatories of the MoU (the buyer's club). The buyer's club oversaw the development and several test houses bought the system simulator without having to pay for the development costs.

From the administrative point of view the main activities were to follow the draft EEC Directive on "Mutual recognition of type approval for telecommunication terminal equipments" in order to monitor that it would be applicable to GSM terminals according to the MoU view, and to administrate IMEIs.

Due to late availability of terminals and above all of the system simulator, the RIC group was also involved in the selection of a restricted number of essential tests to be performed on terminals to get interim type approval.

On the other hand the limited number of tests increased the risk of incompatibility between networks and terminals. In order to manage such a risk, an advisory board was established by MoU/TAP and Mou/RIC to study and follow this kind of compatibility problem.

The main principle concerning the responsibility in case incompatibility problems would happen were established as follows:

1. Mobile Station (MS) manufacturers had the full responsibility that equipment met all requirements;
2. interim type approval had a time limited validity;

3. should incompatibility between a network and a mobile station occur, the relevant problems shall be resolved among the MS manufacturers, the network operator and the type approval house.

21.1.2.4 European/International Roaming

The European roaming capability, subsequently extended to international roaming, was followed by the EREG (subsequently IREG) group for the technical aspects and BARG for tariffs/billing and accounting. The contributions given by TADIG and by SERG have been already presented in Chapter 3, Sections 2 and 3.

In terms of charging principles the main decisions related to mobile terminated calls.

Specifically it was decided that a mobile client will not be charged for air time either when he/she is in his/her home country and when he/she is roaming in another country. In the last case the mobile client will have to pay for the international extension from his country to the visited one.

Long discussions took place in relation to optimized routing, that is a routing which reduces to the minimum the connection length and optimizes the quality. EREG proposed technical solutions already in 1989, but the commercial position expressed by BARG was contrary to an extensive use of such technique. Finally, accepting the proposal from BARG, the MoU plenary decided not to introduce optimized routing for international calls.

Years later the optimized routing for calls originated in the same Country where the mobile client is a visiting roamers was recommended, but, as far as I know, it is not yet practically used.

Commercial issues against optimized routing have been:

- the small percentage of interested cases;
- the technical implications;
- the risk that the calling party might receive information giving an idea of the called party position, which might be against privacy rules/laws.

An important tariff principle adopted at that time for mobile originated calls in roaming environments was that the tariff should have been the same as for a local subscriber increased by a surcharge (15%) to cover accounting/administrative costs.

Within the technical aspects of roaming, it is also worth mentioning all the studies performed for identifying adequate signaling relations and routing among the involved networks, which were not so immediate due to the initial phase of development of the signaling system no. 7 at that time, and the definition of all tests to be performed, within bilateral agreements, between MoU signatories, in order to make the roaming between two operators a reality.

21.1.2.5 Marketing and Planning

A set of activities quite important for the development of the GSM system all around the world was carried out by the MoU group.

First was the decision about the name and the logo. In 1988 GSM was just the "Groupe Spécial Mobile" and not many people believed that GSM might be used as a brand. In fact the

MP group was asked to find an appropriate name and during the MoU plenary, John Peet, from Vodafone, offered a case of champagne to the person who suggested the name which at the end was adopted. I forgot to ask John whether he paid or not due to the fact that GSM was an already known name, or whether he considered as valid for the prize the definition of "Global System for Mobile Communication".

Other important tasks performed by MP had been the continuous update and emission of coverage maps, the development of advertising material as well as the organization of the GSM presence in important events, starting from the Geneva 1991 telecommunication exhibition.

21.1.2.6 Co-operation Agreement with the Association of PCN operators

In 1990 three licences were given in the UK for the operation of cellular networks denominated Personal Communication Networks (PCN) using GSM at 1800 MHz. The following merger of two companies reduced the PCN operators to two, One Two One and Orange.

The term PCN derived from the desire to differentiate the new service from the "normal" cellular service. In practice the only actual differentiations were of regulatory nature, while from the technical point of view the PCN simply used the extension of GSM from 900 to 1800 MHz. The main regulatory differences with respects to the rules applied to Cellnet and Vodafone, the two operators using the 900MHz band, were:

- the new operators could commercialize the service directly, so that they could keep direct contact with the client, while the existing ones had to use service providers;
- the fixed operators were obliged to apply a lower transfer price for fixed to mobile calls, so that the new operators might be able to offer lower tariffs for this kind of traffic.

Substantially it was a way to compensate the newcomers for the disadvantage they would have experienced in entering the competitive market with a coverage considerably smaller than that offered by the existing operators. The same objective has been reached initially in Italy, and subsequently in other countries, by obliging incumbent operators to offer national roaming to new entrants for a defined period of time at cost oriented prices.

The UK PCN operators preferred to differentiate themselves also in the international arena, creating a new body, the European Association of PCN Operators. However, part of the GSM recommendations had been already transferred to the GSM MoU group, so that an agreement had to be defined with the new association for an ordinate development of 900 and 1800 MHz networks in order to avoid future and potential incompatibilities for dual band networks and terminals. The main terms of the agreement reached in 1991 were the following:

- the PCN association had to partially compensate the GSM MoU group for the historical costs, and had to contribute to future costs;
- the PCN association had access to the documentation of the GSM MoU subgroups and the right to participate at their meetings as well as the commitment to contribute to their activities.

We can say that PCN operators have participated to the activities of the MoU group, while keeping their differentiation which disappeared some years later when also the regulatory differences disappeared.

21.1.2.7 Modification of the MoU to Allow the Entrance of new Operators and the International Expansion of GSM

In mid-1991 the GSM MoU group was still limited to 17 countries and 29 signatories, four of which were administrations.

The need was identified to amend the MoU with two main objectives:

- to allow new operators of the already represented countries without the obligation of opening the service in 1991;
- to enlarge the GSM presence around the world as much as possible, with advantages in terms of economy of scale and roaming extension.

The agreed amendment to the GSM MoU included the following main concepts:

- the MoU might be signed by any telecommunication administration and by any operator authorized to provide public digital cellular mobile telecommunication service at 900 MHz;
- new network operator signatories had to procure operational networks using GSM recommendations with the objective of providing public commercial service within 2 years from their signature.

The limitation to 900 MHz was subsequently removed both in order to accept 1800 MHz and 900/1800 operators as well as to extend to the 1900 MHz band. The occasion of the MoU amendment was taken also in order to improve the continuity of the MoU group management by increasing the time duration of chairmen from 6 to 12 months.

Chapter 21: The Contribution of the GSM Association to the Building of GSM and UMTS

Section 2: God Send Mobiles, the Experiences of an Operator Pioneer

George Schmitt

21.2.1 Setting the Scene

It was 14 December 1989 and the Mannesmann consortium was awarded the D2 license to build a GSM network in Germany. It was a surprise award as most expected the license to go to one of the other 13 consortia that bid for the license. It was the first competitive license awarded in the world that had to use GSM technology. The terms of the license were simple: Build a network and have it in service within 18 months and have it cover 95% of the people and the area of then West Germany within 5 years.

Pacific Telesis, a major US based telephone company and a major player in the US mobile telephone market with analog cellular networks covering over 60 million Americans, led Mannesmanns' partners in the consortium from a technical point of view. The other consortium partners included DG Bank, Cable & Wireless, Lyonaise de Eaux with two small German unions. Pacific Telesis was to lead the technical development of the network. Just before Christmas of 1989, Telesis announced that I would be their lead representative.... And so the work began.

On arrival in Duesseldorf on 3 January 1990, I made my way to Schanzenstrassse 66 in Oberkassel, our new headquarters, with about 20 other Americans. We soon discovered that we had a lot more work to do than we thought.

21.2.2 Building a Network and Selecting Suppliers

The GSM standard was not yet completed and the version then available, abis21 could not be used to build a commercial system, but the license requirements were set in stone and had to be met. Along with our new colleagues, we prepared a request for proposal and began general planning for the network. In February 1990, the RFP was released. The proposals were due to be returned by mid-March 1990. Among other conditions, Mannesmann required the winning bidder to agree to pay a penalty of up to DM 1 million per day if they failed to deliver the network elements on time and the government levied fines against Mannesmann. They also needed to agree to a penalty of DM 1 billion if the license was revoked because of their non-performance. Bids were received from Motorola and AT&T in the US, Siemens with Philips and Bosch as their partner, Ericsson with Matra as their partner and Nokia with SEL as a partner. German and American teams evaluated the bids looking at switching, O&M, GSM radios, size and weight of equipment, features to be provided, compliance to GSM abis21 and any agreed changes to it, price and warranties and acceptance of the penalty provisions. Any winning bidder had to accept the penalties in full or would be eliminated from the award instantly.

AT&T was the first vendor to be eliminated. With typical arrogance, they bid a non-compliant system with almost total disregard for the GSM standard as it existed. The next consortium eliminated was the SEL/Nokia consortium when the lead partner SEL announced that they could not make the dates and were unwilling to accept the penalties. This infuriated Nokia, who had not been consulted prior to the announcement, which eliminated them from the bidding. The other three consortiums were in heavy contention with Ericsson and Motorola leading the way. Motorola lost credibility quickly when their lawyer kept overruling their business people in the negotiation process. Motorola was totally eliminated when the consortium learned that Craig McCaw was replacing the system in New York City with Ericsson equipment due to continuing technical problems with Motorola.

In the end, Ericsson became the primary supplier due largely to customer references from McCaw in the US, Televerket in Sweden, and several South American companies. Siemens won a 25% portion of the contract with its partners. Both signed up for the penalties but without any liability for the others failure to perform. Looking back, this probably kept both companies from working together to reduce problems for us, and we were later to find out that Deutsche Telekom had explicitly ordered Siemens not to partner with Ericsson over a year earlier. This cost Siemens dearly when they had trouble meeting standards a year later as their equipment was removed from the network and replaced with Ericsson base stations.

21.2.3 The Next Big Hurdles: Interconnection and Type Approval

During this same timeframe, we began to understand two other major issues:

1. The interconnection of the D2 network with Deutsche Telekoms network and the rest of the world.
2. Type approval for mobiles and system integration testing and the difficulty of getting that accomplished on time.

We had no idea how horrible the type approval process was and would become and what the cost of delays in having commercial handsets available would be, we will discuss this issue later.

21.2.3.1 Interconnection

We knew Deutsche Telekom would be a problem for interconnection but were able to secure the "yellow book", a publicly available document to guide us. With the data from that book, we were able to decode the Deutsche Telekom network and determine where we could interconnect with them.

As it ended up, we agreed to interconnect our gateways in Frankfurt and Duesseldorf, which required us to purchase more back-haul to interconnect our network than other locations. But as a trade-off for our cooperation with Deutsche Telekom's bidding on interconnection, the ministry agreed to allow us to use microwave which ended up being a huge cost saving for us and ultimately broke Deutsche Telekom's monopoly on 2 MB circuits. They could no longer stop our network from being connected and in service on time!

21.2.3.2 The Availability of Mobiles and the Type Approval Process

The availability of mobiles became a nightmare. In 1988 ten network operators had issued contracts for infrastructure, This triggered the development of GSM in many infrastructure companies. Nobody had issued contracts for mobiles. Everybody had believed the market would generate a broad supply automatically. This was overoptimistic, since in 1990/1991 there was a reasonable supply of infrastructure, but no mobiles.

The biggest mistake we made was not "jumping on" a 30 day valid offer from Ericsson to guarantee to supply mobiles when we gave them the infrastructure contract. By the time we had managing board approval to sign a supply contract the offer had expired and Ericsson wisely decided not to extend the offer. In the end, a great decision on their part and the start of our biggest problem, getting type approved mobiles for our customers.

Type approval for mobiles was initially put in place to assure that every mobile works in every network and that no harm was done to any network in Europe when mobiles moved across borders. GSM was designed to be the first Pan-European system and the fixed network operators were concerned about "Harms to the network" much as AT&T was in the US in the 1970s.

Type approval was a nightmare from the start. It was originally envisaged as a software system with over 400 tests that had to be "passed" before a mobile could be sold to the public. The problem was we needed a perfect network before the type approval software could be put together and tested. The type approval software manufacturer Rohde & Schwarz also required working mobiles before it could manufacture software. The set of requirements were a classic chicken/egg problem and became painfully apparent in late 1990.

We struggled with the type approval process until mid-1991 and then decided it could not and would not work as designed. We had a series of meeting with the EU, BMPT (Ministry of Post and Telecommunication) in Germany and the various working groups of ETSI and the GSM MoU group.

By 30 June 1991, with over $500 million invested in the D2 network, we had exactly 50 20 W non-type approved mobiles from Motorola, at a cost of $20 000 each, but they worked!

We had a commercial system installed by Ericsson and Siemens, but no manufacturer close to having a commercial mobile of any size or type. But, we had saved our license and gotten through our first tests.

By the end of 1991, we knew that the type approval process had to go away. By then 15 switches and 375 base stations were installed and 750 were being built and installed.

The start of the campaign to change the type approval regime began in early 1992 when the slogan GSM – "God Send Mobiles" was introduced at the GSM world congress in Berlin in early February. By then we had received permission from Minister Christian Schwartz-Shilling to open our network in Germany without type approved mobiles. The administration of BMPT had assigned people to figure out how to do the interconnection without type approved mobiles even if it meant no intentional roaming. We went to the EU and met with the director general and his staff and explained to them that the D2 network would open using non-type approved equipment. They were shocked that the German government would allow this. Mr. Hubert Ungerer of DG13 used his power to help cause an interim type approval regime to come into place without EU interference.

The GSM MoU met in late February 1992 and determined by a close vote that an interim type approval regime would be put in place, with the deciding votes cast by Deutsche Telekom by direct order of the ministry.

By May, Ericsson, Nokia, and Siemens were providing a few commercial handsets with interim type approval and on 16 June 1992 the first GSM mobiles were installed in customers' cars! The worst of the journey to type approval was over, from there the rest is history.

By the end of 1992, D2 had 120 000 paying customers and had purchased and sold 25 000 Ericsson £1 handhelds called "Hot Line", by far the smallest handhelds in Europe at the time. Every GSM mobile manufactured was shipped and sold at a profit as fast as we could get them. Demand far outstripped supply. By mid-1993, nearly 350 000 customers were on the network and by the end of 1993 D2 was approaching breakeven cash flow and 1 million customers. Every step of the way was a fight. Billions of dollars were invested in networks, but there were no mobiles due to the bureaucracies and process.

21.2.4 The Lessons not Learned

One would think that future generations would attack the device problem early and often.... Unfortunately, that was not the case.

In late 1992, work began on General Packet Radio Service (GPRS), by late 1999 software for the network was available but mobiles only started to arrive in 2001. I will predict 3G will suffer even more before it becomes fully commercially available.

The reason is always the same. Operators issue infrastructure contracts early. Then the development and supply begins. Everybody believes mobiles will "fall from heaven".

The real problem is we continue to have the same issue over and over.

21.2.5 And Now for the Good Things

Competition: every country in Europe has at least three GSM operators, and most have four. And in accordance with capitalistic economic theory as competition thrived prices fell and demand increased dramatically. Europe leads the rest of the world by far in the development and use of mobile telephony by its citizens. It is the best example of Europe leading North America in any technology.

Business development: the decision in 1992 to export GSM technology and algorithms beyond Europe has added dramatically to the dominance of GSM technology around the world. The technology is embraced by the vast majority of the countries of the world and has more than five times the number of customers than any other digital wireless technology. This is in spite of the ill-advised interference of the Clinton Administration on behalf of the

Qualcomm narrowband CDMA standard. Unless Europe falters, it should lead the way and dominate the 3G standard development and implementation as well.

Jobs and industrial growth: the development of GSM and its success around the world revitalized European telecom manufacturers to grow and prosper throughout the 1990s. Nokia becoming the dominant worldwide leader in handheld mobiles, LM Ericsson becoming the dominant force in GSM base station development, with Siemens and Alcatal claiming larger shares of the worldwide market. All at the expense of Lucent formally Western Electric and Motorola of the US who had dominated sales of analog cellular technology but failed to understand the worldwide markets.

European pride: In the end, the thousands of people who worked so hard on the standardization and development of the GSM standard have much to be proud of and those of us who engineered, installed and marketed these networks can be equally proud of what was accomplished in a relatively short period of time.

Customers: the miracle of GSM is best demonstrated by the use of the systems around the world, with over 500 million customers by mid-2001. And in many lesser developed parts of the world, people now have communications; modern and up to date and countless people have used GSM mobiles to make their first telephone calls. What a long way in less than 10 years!

It is my hope that we have learned from the past and that our successors will develop networks and mobile devices at the same time so that never again will frustrated operators have to utter the equivalent of the words "God Send Mobiles".

Chapter 21: The Contribution of the GSM Association to the Building of GSM and UMTS

Section 3: The Evolution from the Informal GSM MoU Group to the GSM Association from 1992 to 1998

Arne Foxman[1]

At the start of my chairmanship of the GSM MoU Group in September 1991 the GSM system had been implemented in a number of member countries as described in Chapter 21, Section 1, thus fulfilling the fundamental requirement of operation in 1991, but in a non-commercial role, hampered by the lack of mobile stations. As an example Telecom Denmark had on 1 July 1991 13 operational base stations but only two test mobiles!

In a formal sense the basic requirement in the GSM MoU action plan of having an operational system by 1 July 1991 had been met, but a commercial start was not to happen until mobile stations were available in reasonable numbers, which turned out to be in spring 1992!

A GSM MoU Group Secretariat/support function had been established and supported by the kind assistance of the Dutch administration, but in September 1991 the secretary provided was no longer available due to internal re-allocation of responsibilities and the GSM MoU Group decided to accept the kind offer of Telecom Eireann to make John Moran available as GSM MoU Group co-ordinator in facilities provided in Dublin. This was to become the established headquarters until the present day.

The increased workload made it necessary to have a chairman in office not for 6 months but for a full year and this was implemented in 1992, starting with the chairmanship of Kari Martinnen.

21.3.1 The Initial GSM MoU Group Structure

The commercial success of GSM was dependant on the resolution of a substantial amount of practical and administrative details, what caused the GSM MoU Group to initially establish a number of working groups, which have been outlined in Chapter 21, Section 1.

A few comments on the results.

---

[1] The views expressed in this section are those of the author and do not necessarily reflect the views of his affiliation entity.

21.3.1.1 The IPR Question

This was an effort to assist in the procurement process to take care of the Intellectual Property Rights (IPR) issue as a result of the GSM MoU Group deliberations and its contacts with the manufacturing industry.

Some GSM MoU Group members were very much concerned with securing a supply of mobile stations from independent manufactures not being part of the infrastructure consortia. This was finally achieved in the relevant GSM infrastructure contract appendices as part of the original procurement activity as a kind of "Musketeer Oath" mentioned in Chapter 21, Section 2.1.[2]

The basic idea was two-fold:

1. To prevent a manufacturer from blocking access to essential IPR's to be used by a supplier to another GSM MoU Group member; and
2. To provide access to essential IPR's for manufacturers of GSM mobile stations independently of infrastructure supply contracts.

This approach, based on "Fair, reasonable and non-discriminatory terms and conditions", proved to be a success in the sense that no blocking has been reported and at the same time it has respected the commercial IPR interests

21.3.1.2 European Roaming

It was obvious that economy of scale could only be achieved by having a truly European system, fully integrated and everywhere allowing customers the same basic services and uncomplicated roaming between the GSM MoU country systems. The Nordic NMT system had proven the success of this approach albeit at a smaller and simpler scale.

The GSM MoU Group decided that outgoing calls made by roamers should be charged at the national standard tariff with a surcharge (15%) to cover administrative expenditures. This surcharge represented an agreed ceiling and the surcharge was an option at the discretion of the HPLMN operator.

Mobile terminated calls to roamers would be handled as national calls in the visited network, whereas the originating GSM MoU network was free to apply an international call charge on top of the normal national mobile tariff.

It should also be noted that as a basic principle the roamer, receiving a call from the national GSM network, has to pay for the transfer of the call from the national GSM network to the visited network for two reasons:

1. The caller is – and has by requirement to be – unaware of the location of the mobile subscriber for privacy reasons. He/she is therefore unaware of the cost of the call. The switching systems charges are based on the number called, rather than on the routing of the call.
2. 2. The roaming party is enjoying the service of having the call transferred and is thus the logical party to face the cost.

This solution has worked extremely well in connection with the GSM authentication

---

[2] The IPR statement used in a GSM systems contract between Telecom Denmark and L.M. Ericsson AB illustrates the approach and can be found on the attached CD-ROM in folder D.

process as an uncomplicated approach, which has now migrated to be truly global based on bilateral roaming agreements between operators, outlining commercial and testing aspects based on the activities defined below.

21.3.1.3 Billing and Accounting

This group has faced and solved an enormous task of accommodating all national charging principles and procedures into a manageable, common framework for all GSM MoU signatories on a global scale as a prerequisite for the interchange of billing data.

21.3.1.4 Transfer Account Data Interchange

In the early days of roaming, file transfer was based on exchange of data tapes on a monthly or bi-monthly basis. Soon it was realised that practical issues, as well as fraud, caused by running up large bills in a visited network with no intention of paying, necessitated an electronic data transfer method.At the same time the number of roaming agreements increased steeply making a bilateral accounting procedure unattractive. As a consequence a small number of clearing houses (data clearing houses) were established acting as a hub for transmission of call data, which the clearing house then arranged in one batch for each participating operator at a nominal fee per record.

   The exchange of calling data has been constantly developed and is now almost real-time (so-called "hot billing"). (For a detailed report see Chapter 21, Section 6.)

21.3.1.5 Type Approval

The problem of securing a Pan-European system interworking with all brands of mobile stations (terminals) in all GSM networks was from the outset considered a major problem and a prerequisite for success.

   As part of the initial procurement co-operation the GSM MoU Group had established a common set of specifications for type approval test equipment. Procurement of a simulator from the elected manufacturer (Rohde and Schwartz) was agreed by the individual test houses in order to facilitate the handling and validation of test cases and thus assuring a common test platform.

   This turned out to be a crucial activity, which is partly covered in Chapter 21, Section 2, 21.3.2 and 21.3.3 (below).

21.3.1.6 Marketing Planning

The GSM MoU (MP) group spent considerable effort in collecting and distributing GSM coverage information in a comparable and easily understood manner. This information was crucial in the national networks for customer information on roaming and for securing realistic comparison between national, competitive networks.

   The GSM MoU Group developed a database in Dublin and a "Map-Info" application for the presentation.

   Considerable effort was spent on developing and protecting the GSM logo. France Telecom, on behalf of the GSM MoU Group, initially performed this task and later this task was transferred to Dublin and France Telecom was reimbursed for the considerable expenditure involved.

21.3.1.7 Services Group (MoU-SERG)

This group secured the availability of services in the same form at an agreed timetable to secure that roaming customers could reasonably expect the same service independent of location. (See detailed report in Chapter 21, Section 5.)

Finally – as a most important aspect – the GSM MoU Group had established:

21.3.1.8 The Security Group (MoU-SG)

- Administration of non-disclosure undertaking for algorithms
- Maintenance of algorithms and test sequences
- Monitoring of adequacy of system security and proposals for enhancements as required.

It should be noted that the GSM MoU Group handles the authentication algorithm (A3/A8) for its members based on a sample algorithm, which can be used at the individual GSM MoU Group operator's discretion. As the A3/A8 algorithm is used exclusively inside the individual operator's network, the operators are free to use propriety algorithms to improve security and it should be noted that the authentication registers are specified to handle more than one algorithm. This allows introduction or phasing-in of new algorithms without recalling or re-issuing Subscriber Identity Module (SIM) cards.

The encryption algorithm (A5) used in the radio part of the network must for technical reasons be identical in all networks and the GSM MoU Group SG is responsible for control-ling the restricted distribution of the algorithm specification and safeguarding the system.

It should be noted that the encryption of the GSM system is designed to protect the privacy over the radio part of the system to a reasonable, non-military level. A5 is a 64 bit algorithm, which met some initial protest by relevant authorities, forcing the GSM MoU Group to activate only 54 bit initially and to accept that some member countries would not at all accept encryption over the radio part of GSM.

The security group also plays an important role in spreading knowledge about any attacks on the security. In addition it provides know-how to new members of the GSM MoU Group on all security issues.

21.3.2 The Initial Regulatory Aspects

A European system allowing for uncomplicated roaming and border crossing with mobile stations was not an obvious feature at the creation of GSM.

The GSM MoU Group achieved in close cooperation with CEPT regulatory bodies free circulation (border crossing) of mobiles without requirement for a formal license document via recognition of the type approval marking as a substitute for a license.

At the start of GSM operations, the specification was part of the NET regime and as no validated mobile station type approval test equipment was available from day one, the MoU-RIC group had to define a subset of 266 tests from ETSI's 11.10 (NET 10) to form the basis for an Interim Type Approval (ITA) intended to expedite the availability of mobile stations at an acceptable level, while waiting for the availability of test facilities for the full NET 10 testing.

Problems with the availability of ITA equipment forced the MoU to apply a two step procedure, where the 160 most essential tests formed the basis of the ITA, but still requiring testing to the full 266 test complement as and when these test cases were available.

It should also be noted that a manufacturer submitting equipment for ITA had at the same time to request full type approval testing to the NET 10 level and was obliged to rectify any problems encountered after the ITA that could endanger the GSM system operation.

The MoU had to set up a type approval advisory board (MoU-TAAB) to assist in interpretation of any disputes in the testing process and to secure action to be taken by the responsible parties, which may be either ETSI-TC-GSM, MoU-RIC or, via MoU-TAAB, the regulatory bodies.

This MoU-TAAB was envisaged as an interim measure, which would later be substituted by the general recognition arrangements and type approval procedures as these became available and as the GSM standard achieved stability based on operational experiences. MoU-TAAB has however been solving the practical problems in a very efficient way under several regimes and several names, and has proven its value many times over.

The GSM MoU Group obtained recognition of the International Mobile Equipment Identity (IMEI) as part of the type approval procedure and established the register in the permanent GSM MoU Group secretariat in Dublin – initially as a document – but now as a data base accessible for operators, type approval authorities and mobile station manufacturers.

The Equipment Identity Register (EIR) is functionally specified in the GSM standard, but the fact that the legal relation between the GSM operators and the subscribers is vested in the SIM card, which is independent of the physical equipment and its IMEI, provided several problems, especially in the handling of stolen mobile stations via the "Black List".

Although the EIR is not a mandatory feature in the GSM system it is worth noting that some regulatory bodies have set the condition that only type approved mobile stations must be used in the nationally licensed networks, thus *de facto* making use of the EIR and its "White List" a mandatory requirement in these countries.

The "Black List" is not a mandatory feature, but is to be seen as a service to the mobile customer and a support to the police in the dealing with stolen mobile stations.

21.3.3 The Type Approval Development: Financing, Creation of GSM-Facilities Ltd

The initial procurement of system simulators (from Rohde and Schwartz) was financed mainly by a buyer's club of leading operators. They financed the system development. This allowed national test houses, often owned by the national operators, to buy the system simulator more or less at hardware cost. The operators (mainly Monopoly Operator (PTT)) considered this as part of the investment necessary to implement a Pan-European mobile radio system.

This worked well for the initial phase 1 of the GSM specification, but at the end of 1992 the GSM MoU Group "buyer's club", dealing with the subject, indicated that there was no intention to finance the further activities in Phase 1 testing or the development and validation of the phase 2 test cases.

The main reason was the explosive increase in investments, which the test houses could not possibly recover from type approval testing fees. No subsidies from mother companies were available any longer.

At the same time lack of manpower in ETSI PT 12 delayed the handling and production of test cases and the GSM MoU Group consequently endeavoured to make such resources available.

The GSM MoU Group decided during 1993/1994 to establish an organisation, GSM-Facilities Ltd, which would finance the test activities via contributions from the GSM MoU Group members at an estimated investment of 15 million ECU (about 16 million Euro). The GSM MoU Group had to create this legal entity in order to be able to place contracts and to manage these activities.

The main initial activities were:

1. Financing of upgrading the Rohde and Schwartz simulator to GSM phase 2. (The contract covered the hardware development and software costs, whereas the individual test houses had to procure the physical modification at their own cost.)
2. Financing of the validation of test cases.
3. Financing a SIM simulator development (ORGA)
4. Financing a signalling tester (CRAY)

The GSM MoU Group was in close contact with the European Union, where General Directorate XIII offered financial support for the test cases, provided these were written in the TTCN language, which was finally achieved via ETSI support.

GSM-Facilities Ltd played an important role in achieving type approval, first against the initial net regime and later the CTR regime (CTR 5 and 9). The above-mentioned TAAB group served as an important "problem solver" in its more formal role as NTAAB inside the ACTE framework. The NTAAB chairman maintained close relations with the administrative GSM-Facilities Ltd thus securing "value for money" in the process.

It was a prerequisite for membership of the GSM MoU Group and later the GSM Association that the member operator was also party to GSM-Facilities Ltd and paid the cost involved. Considerable effort was spent on developing a fair contribution algorithm penalising neither the small new operators nor the very big operators. It is my impression that the result achieved by Mrs Gretel Hoffmann in her chairmanship fulfilled the criteria.

GSM-Facilities Ltd served its role until October 1998 when it was finally liquidated. The remaining activities were of a minor maintenance nature and these were transferred to the GSM Association, which was at that time established as a legal entity. It is worth noting that GSM-Facilities Ltd stayed within the budget of 15 million ECU.

21.3.4 The Co-operation Agreement with ETSI

As the GSM MoU Group became more and more international, the ETSI policy of restricting access to documentation to ETSI members and also the policy for non-European GSM MoU Group member participation in the standardisation work became quite an issue.

In close co-operation with Friedhelm Hillebrand a co-operation agreement between ETSI and the GSM MoU Group was signed in 1996 as described in Chapter 5, Section 2, paragraph 5.2.5.1.

In the "considering" section it recognises the role of both organisations and their contributions to GSM i.e. ETSI standards and GSM MoU permanent reference documents on services, charging/accounting, international roaming, security and fraud. It confirms relevant elements of the ETSI reform. In the "agreement" section on information and document exchange it was agreed that the GSM MoU Group would be entitled to send observers who can submit documents and have the right to speak to relevant ETSI technical committees (i.e. mainly SMG). GSM MoU Group members outside of Europe got access to all GSM documents without additional payments. The GSM MoU Group contributed a substantial fixed sum to the SMG project team budget. It was further agreed to "make any effort necessary in order to maintain the integrity of the GSM standards by close liaison with ANSI..."

21.3.5 The Creation of GSM Association and its Organisation

The above sections have described the ground work, which played a very important role in the success of GSM as a system, not only based on technical standards – however important – but also by making available all the practical facilities for operating a network, such as billing and accounting, roaming and type approval.

From 1992 and onwards the GSM MoU Group became more and more interesting in a global perspective, starting with the attendance from the Arab Emirates at the 1991 plenary.

At the same time 1800 MHz (PCN) networks were implemented, starting in the UK and eventually the standards of the two systems were unified and a formal co-operation between the GSM MoU Group and the PCN operators developed into full membership.

When 1900 MHz frequencies were auctioned off in the US, also American operators opted for GSM MoU Group membership and the practical problems of different signalling systems and roaming data interchange were all solved.

The GSM MoU Group then changed its legal status from a Memorandum of Understanding (MoU) between members into a legal institution, the GSM Association, formally based in Switzerland, but with the headquarters in Dublin. The details of the new organisation and its membership categories can be found below.

As the GSM Association grew in membership, the participation at the plenary meetings became very large with attendance of up to 500 persons in 1998. It was obvious that a plenary of this size had to concentrate on policies and principles as well as dealing with the budget. This and the geographical spread made it obvious that working groups had to be the answer.

21.3.5.1 Working Group Activities

Over the years new GSM Association working groups have been established and especially the **Terminals Working Group (TWG)** played an important role as a link between operators, ETSI, type approval authorities and manufacturers with the following tasks:

- Key interface with manufacturers (MS, SIM, test equipment) and standards bodies on all MS technical issues.
- The development of GSM Association positions on MS features and services.
- Regulatory and access to market issues.
- Central Equipment Identity Register (CEIR) and IMEI issues.
- Service and network feature interworking and interoperability.

The European type approval developed into the RTTE Directive without any requirements for formal testing by accredited test houses before marketing the terminal, thus solely relying on manufacturer declarations of conformance. The GSM Association decided to examine a voluntary certification process implying formal testing under the influence of the GSM Association in order to secure terminal functioning in all GSM networks. This requirement was further enhanced by the fact that the termination of European testing, which had been globally recognised, left a vacuum to be considered in order not to compromise the technical possibilities of global roaming. The further development in this area is described in Chapter 21, Section 4.

Electromagnetic compatibility issues were soon major issues and especially interference from GSM TDMA frames (217 Hz) to hearing aids caused much anxiety. The GSM MoU Group co-operated closely with European Hearing Aid Manufactures Association (EHIMA) and made national research results available. The main issue was an argument of permissible field strength, where the GSM standards people had anticipated a 10 V/m value, and some even 17 V/m, whereas the hearing aid standard called for only 3 V/m as a permissible field strength. The awareness and research results resulted in much improved hearing aid designs, overcoming the problem, but creating many problems for the hearing aid wearers in the early days of GSM.

Also the health aspect gave rise to much anxiety and the GSM MoU Group and its members participated in financing studies from operator independent research institutes to shed light on the issue, but of course no firm scientific assurance of absolutely no effects have been achieved.

The GSM MoU Group established the **Environmental Working Group (EWG)** with the following objectives:

> The EWG provides expert advice on the range of environmental issues. Technical advice focuses on potential health effects, interference and other environmental issues related to GSM mobile communications. EWG also monitors the GSM Association sponsored EMF research program.

21.3.5.2 Regional Interest Groups

The ever increasing membership and the global aspect necessitated a stronger commitment to local national and regional interests and nine groups were created with the objectives:

> Each of the nine regional interest groups aims to represent the interests of GSM operators at the highest level. The primary objectives are to:
>
> Promote and facilitate GSM development within the region or country
> Support the evolution of GSM technology and interoperator relationships
> Represent regional issues and concerns at an international level
> Facilitate, enhance, protect and support the investment in network infrastructure within the geographical area
> Co-ordinate contributions to and lobbying of regulatory bodies
> Identify and pursue future directions and opportunities.

The nine groups are:

- GSM Europe
- GSM Asia Pacific

- GSM Arab World GSM
- GSM Russia GSM
- GSM North America
- GSM South America
- GSM Africa
- GSM Central Asia
- GSM India

21.3.5.3 Present Working Group Structure[3]

The working groups mentioned above were reorganised and today, as listed below, there are ten working groups within the GSM Association. These groups meet regularly to develop and set the pace of policy direction and innovation for the GSM industry. These groups have brought about major technical breakthroughs through the pooling of expertise, the sharing of resources, and an efficient flow of communication.

- International Roaming Expert Group (IREG)
- Legal and Regulatory Group (LRG)
- Security Group (SG)
- Fraud Forum (FF)
- Billing and Accounting Roaming Group (BARG)
- Services Expert Rapporteur Group (SERG)
- Smart Card Application Group (SCAG)
- Terminal Working Group (TWG
- Environmental Working Group (EWG)
- Transferred Account Data Interchange Group (TADIG)

21.3.5.4 The GSM Association: Creation, Organisation, Membership Types and Facilities

The GSM Association was formally established on 15 July 1995 as an association registered in Geneva, Switzerland, but still with its headquarters in Dublin. The structure was formalised with an elected chairman and deputy-chairman, but now with a managing director in charge of the daily operation of the Dublin office and all activities decided by the plenary. The chairman had up until then had the support of an informal group of former chairmen (The MoU Chairmens Group (MCG)), but now an executive committee elected by the plenary was established. An electronic GSM MoU infocentre was created to deal with the ever-increasing amount of documents and the maintenance of databases (equipment identity register and type approval register, etc). The infocentre went live in September 1996. In the process of creating the GSM Association great effort went into defining the criteria for membership and the obligation associated with membership. Great emphasis was placed on the requirement of members to operate a full GSM system, i.e. that the potential member had received a licence

---

[3] The detailed description of the GSM Association working groups and their mandates is contained in Annex 2, Section 4.

for the frequencies used, had Switching Centre (MSC), Authentication Centre (AUC) and a full radio network (BSC/BS).

The formal definition today is:

- Licensed mobile network operators committed to building and implementing GSM based systems. This includes GSM 900, GSM 1800, GSM 1900, 3G network operators and satellite operators.
- Government regulators/administrations who issue commercial mobile telecommunication licences.

It is worth noting the satellite operators, who where applying the GSM system, but of course the satellite radio networks were not part of the ETSI GSM specification.

Great interest was shown by many interested parties to have access to the information contained in the GSM Association permanent reference documents and to have access to working groups and to participation in plenaries.

After considerable discussion in the membership the associate membership was created in principle in 1998 and currently associate membership is open to GSM suppliers and manufacturers world-wide, such as:

- Billing systems suppliers
- Data clearing houses
- Financial clearing houses
- Infrastructure suppliers
- Mobile terminal suppliers
- Roaming brokers
- SIM card suppliers
- Security systems suppliers
- Signalling providers
- Simulators suppliers
- GRX carriers
- Application providers

It is worth noting, what the benefits allocated to this membership category currently are:

- Attendance at the GSM Association's twice yearly plenary meetings.
- Attendance at various working group, interest group and regional group meetings.
- Right to participate on issues being discussed in the above meetings including tabling input papers and voicing opinions.
- Access to proprietary and confidential information owned and maintained by the association.
- Active involvement in the ongoing business of the association.
- Associate members may wish to form their own interest group within the GSM Association.
- Right to use the internationally recognised GSM technology logo and trademark.

In the creation process the multiple benefits for all parties involved were stressed:

- The opportunity to work closely with customers.
- Ensuring joint action and co-operation in promoting GSM as the core of 3G technology.

- Allowing network operators and manufacturers/suppliers to work together in a common forum will produce synergy by pooling of expert information, the sharing of resources and an increased and more efficient flow of information.
- Improving and accelerating the definition of technical requirements for new GSM features and services thus improving the lead time in offering such services in the marketplace.
- Promoting co-operation and agreement on issues facing all entities involved in the GSM industry.
- Helping to provide ever-increasing leadership and direction for the GSM industry.

The latest information on the GSM Association can be found at the GSM Association information centre, GSM World at *www.gsmworld.com* and you are encouraged to visit this site for further information.

Chapter 21: The Contribution of the GSM Association to the Building of GSM and UMTS

Section 4: The Life and Role of the GSM Association from 1998

Petter Bliksrud[1]

Since I have been attending all plenaries up to now, it is tempting to linger somewhat on the substantial changes implemented in the organisation to adapt to the changing environment. The MoU, and later the GSM Association, has evolved from a tiny founder putting a new and strong obligation into the concept of a Memorandum of Understanding (MoU), and into a comprehensive organisation. The early days had quite a narrow focus, to convince the manufacturers that the operators were serious and really intended to implement the new system, and maybe also convincing each other. The world was not totally certain of the mobile success, then. The present situation is kind of reversed, one task of the GSM Association is now to dissuade politicians and regulators that the success is not so huge that it can easily bear any cost of licence acquisition and straitjacket sector regulation.

21.4.1 Standards and Intellectual Property Rights (IPR)

The organisation has stepped out of its daring and optimistic adolescence and reached what could be called the pragmatism of maturity. To exemplify: as described earlier in this chapter, the IPR issue was hot and difficult. Today, in contrast, the focus lies more with the Association's own efforts, as formulated in its objectives for IPR regulations, which are:

- to clearly state the rules of the Association in relation to the creation, use and protection of IPR
- to protect the Association's IPR
- to create value for the Association and its members, associate members, rapporteurs and Forum members from the IPR and its use
- to give guidance to the Association members, etc. in relation to IPR issues
- to give guidance to third parties who deal with the Association in relation to IPR matters

[1] The views expressed in this section are those of the author and do not necessarily reflect the views of his affiliation entity.

The fact that a policy is deemed necessary, signifies the increasing importance of the organisation's own contributions to the wireless world.

The Association indeed seeks to have influence in the standardisation bodies, it should in particular be mentioned that the GSM Association participates as a market representative partner in the 3GPP. Accordingly it has been necessary to elaborate a set of guidelines, as a basic opinion of what the Association, as well as individual members, should promote when working in standardising; for example, standards should be:

- open and readily obtainable
- capable of being implemented without payment of royalties or other fees
- derived based on actual need and supported by a representative community of interest
- maintained and revised by the standard organisation which is responsible for integrity, consistency and clarity of the standard
- developed in a manner which makes efficient use of resources

and, maybe even more clarifying, standards should *not* be:

- rubberstamped product specifications of one of the members of the standardisation body
- so vague or conceptual that they lend themselves to different interpretations or implementations
- pursued purely for the economic benefit of a single member of the Association or standards organisation
- intellectual exercises or philosophical debates that drag on and on

One can but wonder who the individuals might be that have inspired the inclusion of the last item.

21.4.2 UMTS

Of course, the GSM system has been, and still is, the basis for the GSM Association activities. However, this does not mean that the third generation system is neglected, and the Association members all share the common view that:

> evolved second generation and true third generation services will take the personal communications user into the information society. They will deliver information comprising voice, raw data, pictures, graphics and multi-media services and applications directly to people. They will provide them with access to the emerging generation of information based services and service offerings.[2]

Maybe not surprising viewpoints, but if taken seriously require attention, and a dedicated interest group (the IMT200 steering group) has been set up to:

- identify, lead and communicate IMT2000 issues for the Association
- develop actions, campaigns and measures to promote the interest of the members in the IMT200 environment
- ensure and facilitate roaming in the IMT200 family in accordance with market needs
- have a coordinating role within the Association on 3G issues
- be a decision-making body for 3G issues in accordance with the GSM Association agreed policy and its overall strategy.

[2] Permanent Reference Document (PRD) TG.01.

The need for focus on data communications in the next generation has been reflected in the admission of application providers as associate member and, contacts with the WAP Forum and the membership is constantly reminded of this important realm by Mike Short, the ardent Association data crusader.

21.4.3 Co-operation

Attention has been paid to the benefits for the GSM Association of co-operating with other, and sometimes competing, interest groups towards common goals. Formal agreements have been elaborated, e.g. with the UMTS Forum and the Universal Wireless Communications Consortium (UWCC).

Two of these working areas, however, deserve mentioning in more detail, namely voluntary terminal certification and interstandard roaming.

Since the European regulators decreased the contents of type approval procedures, the operators have been increasingly aware that reduction of formalities might create unwanted burdens to the network if terminals were not to be as good as they should. A voluntary certification scheme has been contemplated for quite some time, and at last negotiations with the manufacturers entailed a common agreed framework in August 1999.

The GSM Certification Forum has been established to administer this certification scheme to ensure the interoperability of GSM terminals globally. This forum provides for the verification of devices against the Forum's technical requirements, with global recognition and acceptance of results, the avoidance of multiple testing and cost efficiency. The program includes all GSM frequency bands and multimode devices. It shall be complementary to all regulatory regimes with participation on a voluntary basis.

The Forum and testing shall not cover commercial or quality aspects of a terminal and any such testing shall remain a part of the commercial agreement between the customer and the manufacturer. New requirements and tests may be incorporated into the Certification Forum when they have become part of the GSM core specifications, tests have been published by a standard development organisation and validated test equipment is commercially available.

It is intended that the Certification Forum will evolve as the GSM technology evolves and will be extended and modified to encompass future technologies such as the third generation.

The GSM has indeed proved a very successful system, but there are still areas around the world that are served by other cellular technologies only. For the benefit of mobile users in general the GSM Association has therefore initiated work to facilitate roaming over technology barriers through the establishment of the Global Roaming Forum to elaborate on issues related to interstandard roaming. This forum shall provide a working place for network operators and suppliers to address problems and propose solutions in relation to interstandard roaming.

The ordinary GSM standards should be adjusted to the realisation of interstandard roaming between GSM and other wireless solutions (e.g. ANSI95-CDMA, iDen, ANSI136-TDMA). At the moment the focus lies with 2G systems, but the Forum offers as well means for building up convergence with 3G systems.

21.4.4 Organisational Adjustments

The early GSM MoU group's attendance consisted of people with about the same level of competence and operator experience; information flow went accordingly. Today's membership includes operators at very advanced stages of mobile operations and beginners at the start of their operational careers. This implies that the Association tasks comprise a learning function, informing the new delegates both of the Association and the business as such. It is therefore customary now that the plenaries include workshops on particular topics, offering a kind of midwife function to novices, thus easing the introduction of the GSM in less evolved parts of the world.

Not less important is the increased efforts of making the Association website easily accessible and the issuing of various news and information letters.

The first phase of the MoU life could manage with an ad-hoc secretariat function; then a small secretariat was set up in Dublin and now the present workload requires the assistance of a professional staff of around 40. As mentioned earlier in the chapter, John Moran was the first secretary, building up the Association Secretariat through searching out engaged and competent people, as well as assisting individual members when required and the community by taking meeting minutes. It is proper to say that an era came to an end when John left the Dublin headquarters for retirement after the 42nd plenary.

The working means have changed, from hand-written agendas in the early days to colourful slide-shows for presentations and use TV-screens during plenaries. In itself, the sheer number of delegates to a plenary necessitates means to maintain innovation and alertness in the organisation. To further avoid bureaucratic tendencies, the introduction of the CEO round table discussions seems to be beneficial, where invited high ranking officials from some of the members, different from time to time, highlight issues they find particularly important at the moment for the Association to act on.

The enlarging of the membership, or associate membership base, has led to called some voices of warning. The few early players had a pointed focus, hence clear-cut opinions were obtained. The differentiation in membership, both with regard to evolvement and business basis, e.g. not operators only, may reduce the basis for possible concerted action and opinions.

An example of this difficulty surfaced when it was debated whether regulators could retain membership eligibility in the Association. Few applications have been presented in the last few years, but the already established did not favour the idea of expulsion. Neither did the plenary.

21.4.5 Future Vision

Although formally only a teenager, the GSM Association has nevertheless come a long way. Gone are the days when it was not necessary to use all digits in the A5 ciphering key, and gone is the time when old folks just did not have enough Association days to reminisce about.

Even the Association itself has become retrospective and created a GSM roll of honour where history is reflected through a list of names of individuals assessed to have had the strongest influence on the development of the GSM system. This list, which can be found on *http://www.gsmworld.com/events/awards/cat5/honour.html* is not to be regarded as a token of complacent lethargy.

The GSM Association is very forward-looking with the vision of leading the way into the future, worded as follows:

To grow and prosper, the wireless industry must adapt and change to meet the ever-changing wireless consumer needs. We at the GSM Association are leading the way towards universal, ubiquitous, wireless services.

Our position is a direct result of our planning. GSM was originally introduced as a voice technology, but its data capabilities were built-in from the start. Adaptable to local needs but meeting a global requirement, the GSM family has built-in flexibility.

Ours is therefore a robust and proven wireless platform, which the GSM Association has nurtured, encouraged, and developed through a series of initiatives, innovations and partnerships that have produced a platform for today and for tomorrow.

Chapter 21: The Contribution of the GSM Association to the Building of GSM and UMTS

Section 5: The Work on Services Aspects – from the Early Days to Current Affairs

Armin Toepfer[1]

The GSM MoU was founded at a time when the regulatory framework for the provision of cellular services was determined by national administrations and, to some degree, co-ordinated by CEPT. Deregulation and liberalisation was not an issue at that time. In this regulatory environment, the early scope of the Services Experts Rapporteur Group (SERG) which was entrusted to work on services, was very wide: Of course striving for common service requirements, which is still the role of the group – but also looking after types of subscriptions, agreements on mandatory service implementations and even pure regulatory issues. Over time the role of SERG changed considerably.

In the beginning, i.e. from 1985 to the end of 1987 all services work was carried out in GSM WP1. GSM WP1 had agreed on a complete set of services specifications, including operational guidelines, which was elaborated in due course.[2] The first set of specs for tendering were approved in 1 Q 88.

A first split-off of responsibilities for work on GSM services happened after the formation of ETSI and its technical committee GSM. SMG1, the services group under SMG, took over responsibility for the technical specifications. SMG1 was supposed to focus on basic service requirements, at that time called "stage 1 description". Consequently, during 1993, work on documents not addressing technical standards matters was transferred to SERG. This was the time I took office as chairman of SERG, after Bruno Massiet du Biest of SFR, Ken Hall, Cellnet and David Barnes, DTI who acted as the first chairmen in 1988.

Although this is not the right place to discuss technical details it might be helpful to address a couple of former hot topics to understand the role of SERG for the success of GSM:

---

[1] The views expressed in this section are those of the author and do not necessarily reflect the views of his affiliation entity.

[2] See also Chapter 10 on services aspects.

Regulatory framework

- Prior to the advent of GSM, the European cellular market was fragmented and characterised by diverse national systems. When starting GSM development this was identified as a blocking stone for success and wide penetration of cellular systems in Europe. In the framework of regulatory issues GSM WP1 established the guidelines for the licensing process and the framework for Europe-wide free circulation of terminals through harmonised type approval. After the foundation of ETSI and the creation of TC GSM under ETSI those documents on licensing and free circulation were transferred to SERG.
- The idea to introduce GSM with similar licence conditions in all CEPT countries and the goal to get rid of import/export procedures for terminals was presumably born outside SERG – at least the group under GSM MoU ensured that official requirement documents were created. After the general regulatory framework was set by the European Commission, these guideline documents[3] became obsolete and were subsequently dropped.

Operational requirements and support

- In order to stimulate subscriptions, the GSM MoU was keen to generate commonly available subscription profiles and guidelines for the use of services.[4] Whereas such guidelines may have eased the market entry for new operators, the concept of making these guidelines mandatory discouraged operators from inventing different ideas. It was not very suitable for a liberalised environment later on. Today it could be even considered as anti-competitive.
- GSM 02.20/SE.10 started with the following phrase in its scope:"Since a homogeneous, compatible service spectrum is the objective for the services of GSM PLMNs, standardisation of the charging mechanisms for collection charges is the minimum requirement. This is especially important as mobile subscribers can move in different GSM PLMNs." GSM 02.20 served quite well as requirements specification for the development of the technical systems, it also set the framework for roaming conditions. With this changed scope, focussing on requirements for roaming, MoU Billing and Accounting Rapporteurs Group (BARG) took over responsibility from MoU#24 onwards. Currently BARG is looking after charging requirements in a very generic way. Those requirements are applicable to the international roaming case. The means for publishing "wholesale" tariffs among the operators are set up via their secure website (IOT = Inter Operator Tariffs).
- Most of the agreed text in this area was withdrawn during 1993, some useful text was kept as a recommendation. It is obvious, looking at this attempts to regulate everything, that people's mind in the late 1980s were still biased by their history. During those days nobody anticipated the choice of subscription and service options we encounter today. However, the technical system GSM was able to cope with the changed commercial environment.
- SERG also performed some roles in assisting newcomers and as administrator/registrar: how and to whom to apply for SIM IC card identifier (SE.14) and how operators should select the network name for appearance on handsets displays (SE.13: GSM mobile

---

[3] GSM 02.12/SE.06: Licensing, deleted MoU#24. GSM 02.15/SE.09: Circulation of mobile stations, deleted MoU#24.

[4] GSM 02.05/SE.04: Simultaneous and alternate us of services, deleted MoU#24. GSM 2.10/SE.05: Provision of telecommunication services, deleted MoU#24. GSM 02.13/SE.07: Subscription to the services of a GSM PLMN, kept. GSM 02.14/SE.08: Service directory, deleted MoU#24.

network codes and names). Meanwhile the administrative work has been absorbed by the GSMA headquarters in Dublin. During the early days, discussion on the permissible length of network names was an issue to entertain the whole working group for some time as display space on handsets was very limited. It also happened that operators, in the rush for starting commercial operations thought of everything but of registration of the IC card identifier at the ITU via national administrations. As ITU administrations are not really business focussed, time became an issue in some cases.

- Security elements of the GSM standards are a strong sales argument. Two elements are best known: the SIM and the International Mobile Equipment Identity (IMEI), the latter allowing tracing of faulty handsets and prevention of network access of stolen equipment. A couple of recommendations were created in this area, starting with general instructions or suggestions[5] on "IMEI checking strategy". This was very useful as we all know that standards normally contain the minimum compromise. Options often reflect that no single compromise was reached, but also contain various operational conditions. The operators active in SERG had a vital interest in harmonising the security level, e.g. checking the IMEI at similar time intervals. In addition operational specifications were created to run the Central Equipment Identity Register (CEIR) and provide on-line interfaces with EIR operators (the local register at operators premises, SE.16 and SE.17). The CEIR was operated by GSMA HQ for some years. There was also a "mandatory" requirement to members to implement the EIR (originally 1 January 1995). Again, this was an issue for long disputes as the community of operators had grown a lot with a collection of diverging business ideas. After all, the requirement was compromised, permitting also off-line solutions. The number of operators connected to the CEIR is still very limited. As handset value decreases rapidly due to frequent change of models meanwhile, one can question the benefit of this security measure. However, the IMEI concept has proven to be very useful in particular when operators expand and upgrade network capabilities frequently and need to know how handsets can cope with these new functions.

Service requirements and implementation

- SE:03 classification of services – this simple document which tries to gather high level service implementation requirements is the most debated document in SERG, in particular the GSM Association plenary loved to discuss the content at length. Why so? There is a truth, people familiar with committee dynamics know it: the simpler the issue the more comments one can expect and the longer it takes for resolution. But this does not fit well for the debate we had for a couple of years. If we analyse the history of this document and its content one can imagine the spirit of the early adopters of GSM which is in contrast with the rather different approach of today's operators. GSM was born, simply speaking, as the wireless complement to ISDN. ISDN services were just adopted by the GSM standard bodies. The documents as such contained a list of (ISDN) bearer services, tele-services and supplementary services. All were marked with a mandatory or optional implementation date by which the operator, being a member of the GSM Association should have deployed the service. Most were mandatory. As the community of operators was limited in the early days (just Europe) and the choice of services was limited too, there was little to argue. The content of the document just described more or less the feature being part of the basic network roll-out. In addition this early approach was beneficial for

---

[5] SE.11: Security aspects.

market creation, avoiding market segmentation in the early phase. The market was just crying out for mobiles (George Schmitt, CTO Mannesmann Mobilfunk in those days: God Send Mobiles → GSM). In 1994, the standards bodies did progress the ideas on service enabling technologies and respective means to implement these (the first was CAMEL), SERG also started to look at this option with the view to integrate this into their implementation requirement document. This move characterised the change of attitude. Operators were forced by market competition to think of differentiation through operator specific services. Complex and proprietary services are a challenge for network integrators and implementers because of various vendors involved. The concept was to standardise the platforms and tools which allow the creation of services and provision even while roaming. The obligation for a new licensee, probably operating in a third world country, to implement such sophisticated network approach from the very beginning cannot really be justified. This operator might be proud to offer telephony service to the people. All these "nice to have" features may come later when the market has matured. During those days the plenary of the GSM Association (MoU) started to argue on the meaning of mandatory in a highly competitive environment. The conclusion reached was obvious: The assembly of operators cannot interfere in individual business cases of their members by implementation obligations. So far so good. Then we had the option to drop the document or to change the objective. The latter was decided. The document on service implementation requirements was converted into a recommendation document, which collects the services, features and service enablers in wide extent. In addition there was an exercise to identify priorities from a commercial perspective, also in terms of availability of certain functions at certain dates. This was used to push activities in ETSI GSM and to set focus for standardisation (with limited success). In addition the content also was used as a collective wish list in view of vendors development schedules. The document is still in place, again alive, while dormant for a while.

The above illustrates the operational support SERG developed during the early days by elaborating various guideline documents. It also shows the benefit of harmonised service ideas during the start-up phase. Over the years the group objectives evolved substantially.

Beyond SERG's role as described, SERG maintained a close relationship with ETSI SMG1, which was now called 3GPP SA1. This close relationship was supported even by means of cross-group chairmanship, i.e. Alan Cox, chairman of SMG1 and SA1 for many years served as vice-chairman whereas I served him as vice-chairman in SMG1. SERG took initiative for the SMG work programme after GSM Phase 2. We all now that Phase 2 was created just because the original GSM work programme was not finalised in 1990 and operators were pressing for completion which led to the split: Phase 1 comprised everything that was ready, Phase 2 the rest. Operators soon noticed that the world was moving on, plenty of new ideas came up – but there was nothing to cater for those new proposals, there was no appropriate mechanism for the release of technical specifications. SERG devoted some meetings to collecting and classifying new service ideas. Those activities were performed under the chairmanship of Bruno Massiet du Biest, whereby I took care of these ad-hoc meetings. The draft work programme for GSM Phase 3 was submitted to SMG and turned out later as Phase 2+ work programme, emphasising the evolutionary aspect of GSM.

In 1994 SERG had a close look at UMTS and the respective operator requirements as specification work in ETSI SMG5 was carrying on notably without the attention of the operators.[6] As SERG was still GSM-only focussed an ad-hoc group was formed, chaired by me which later was promoted to the 3. Generation Interest Group (3GIG) in GSMA.[7]

I resigned after 5 years chairing the group and Philippe Lucas was elected as new chairman and took over responsibility with the first meeting in 1999. With the advent of Internet-type services the group adapted their scope to look at WAP services, provision of applications and contents, but still struggled to establish an operator view.

[6] Gerd Grotelüschen, who observed the activities for Mannesmann Mobilfunk classified the group as a committee of observers.

[7] See Chapter 21, Section 6 on 3GIG.

Chapter 21: The Contribution of the GSM Association to the Building of GSM and UMTS

Section 6: The GSM MoU Work on Data Interchange for International Roaming

Michael Gießler[1]

21.6.1 Billing and Inter-Operator Accounting in the International Roaming Field

The eighth plenary of the *GSM MoU Group*[2] established a working group – named *Transferred Account Data Interchange Group (TADIG)* – which should specify the procedures on how to exchange roaming related charging information between GSM operators/networks. This specification should allow for a common and harmonised technical implementation of the rules for billing and inter-operator accounting in the case of roaming. These procedures are also referred to as *Transferred Account Procedure (TAP)*. Due to the foreseen high volume of data to be transferred between the networks, the *GSM MoU Group* decided not to use the signalling capabilities between the networks also for the transfer of charging information but to implement additional interface(s). Later on, it became obvious that this basic decision made it impossible to implement *real time data interchange* between GSM networks using the additional TAP interface. The reason is that the involvement of a visited GSM operator's billing system (to analyse visitors' traffic tickets and to calculate the corresponding roaming charges) "per se" adds a certain time delay before charging information is transmitted to the home GSM operator.

Based on the rules on how access in the case of roaming is provided to roaming subscribers:

- Subscriptions valid for usage in foreign networks are issued by the home GSM operator;
- Only services are accessible (a) for which the subscriber has a valid subscription with the

[1] The views expressed in this module are those of the author and do not necessarily reflect the views of his affiliation entity.

[2] Was then called GSM Association.

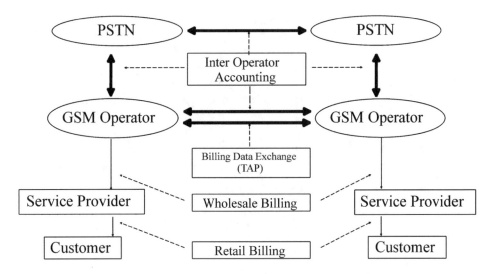

Figure 21.6.1   Accounting and billing interfaces in the international roaming field

home GSM operator and (b) provided these services are supported by the visited GSM network;

.. the principles of the *Transferred Account Procedure* are as follows:

- The home GSM operator is responsible for all roaming charges for GSM services used in a visited GSM network by his roaming subscribers;
- The home GSM operator is responsible for the collection of charges from his subscribers in the case of roaming;
- Billing records sent by the visited GSM network via TAP are used by the home GSM operator as a basis for charging the roaming subscribers.

The corresponding billing and inter-operator accounting processes in the international roaming field form a complex accounting and billing chain. The interfaces in this chain are outlined in Figure 21.6.1.

The charging principles for international roaming have been specified by the "*Billing and Accounting Rapporteurs Group*" (BARG), recently renamed "*Billing, Accounting and Roaming Group*" of the GSM Association.

According to the originally defined charging principles for international roaming, the operator of the visited network charges visitors for their roaming calls on the basis of the "*Normal Network Tariff*" (*NNT*) that is applicable in that particular network plus a possible supplementary charge of up to 15% ("*VPLMN mark-up*"). The home network operator pays in advance on behalf of all his roaming subscribers and then bills the service providers or subscribers for the roaming costs incurred abroad, the "*HPLMN mark-up*" on those costs and his own costs for forwarding incoming calls to roaming subscribers ("*roaming leg*").

New competition-oriented charging principles for international roaming have been introduced in 1998. These new principles have replaced the originally applicable rules providing operators a transition period of at least 1 year for implementation.

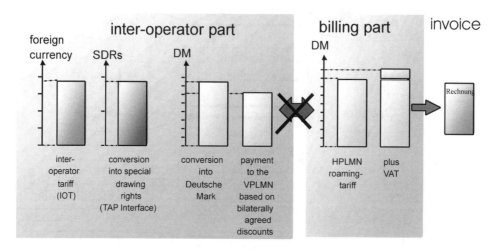

Figure 21.6.2 Diagram of the new charging principles from a German operator's point of view (Assumption: discounts are not processed via the TAP interface)

The fundamental feature of the new charging principles – known as "*Inter-Operator Tariff*" (*IOT*) – is that the prices from the visited network plus the HPLMN mark-up are no longer passed on to the particular network's own subscribers by means of a "cascading method." Instead, the *home* network operator can design roaming prices for his subscribers as if he were a manufacturing plant that purchases supplies from roaming partners (uses an outside network) at wholesale conditions. This splits the functions, namely into a part that covers the relationship of GSM operator ⇔ GSM operator (the inter-operator part) and a *logically separate billing part* that covers the relationship of GSM operator → customer (Figure 21.6.2).

In addition, the new charging principles make it possible to design and organise tariffs for international roaming in entirely new ways. This is because the introduction of competition-oriented charging principles marks the end of an era of limited freedom in tariff arrangements between network operators. The decoupling of inter-operator tariffs from customer tariffs creates greater freedom in designing tariffs vis-à-vis end customers and/or service providers. The flexibility this allows in offering transparent, attractively priced roaming tariffs ensures competitiveness in the roaming area. Limited offers involving special tariffs for specific target groups are also possible. However, the requisite groundwork must be implemented into an operator's billing system.

Summary of the essential IOT elements:

- Services are offered by a VPLMN[3] for roaming partners on the basis of a generally applicable wholesale tariff – the *IOT* – which the *visited* network operator determines.
- "Tariff sovereignty" – setting roaming tariffs for subscribers – is now in the hands of the *home* network operator in all call scenarios.
- The new charging principles do not contain any stipulations whatsoever regarding possible bilateral pricing agreements (discounts, also known in more neutral terminology as incentives).

[3] Recently the abbreviations "HPMN" and "VPMN" have been introduced.

- The IOT should apply for at least 6 months. Exceptions that would allow changes to be made at an earlier point in time are however permissible under defined circumstances.
- The level of IOT charges can be increased by an operator only on the first day of a month after a minimum period of 6 months with a minimum of 60 days advance notice.
- IOT charges can be decreased at any time.
- There are no more restrictions on the pricing of incoming visitor calls.

21.6.2 The TAP Specification Process and the Key Milestones

The first meeting of TADIG – consisting of nine delegates representing seven GSM operators – was held at the end of May 1989 under the chairmanship of Michael Gieβler, Deutsche Telekom/DeTeMobil. Michael Gieβler chaired the group for more than 6 1/2 years until December 1995.

TADIG's *Terms of reference* as specified by the eighth plenary of the *GSM MoU Group* were as follows:

1. To specify the detailed file interchange mechanisms to include tape and data transfer between billing entities, to facilitate the transfer account procedures as defined by GSM Rec. 02.21 and related recommendations, taking into account the necessary security and quality of service requirements.
2. To specify the format of data records to be exchanged either by tape or by data transfer, based on GSM Rec. 12.20.
3. To specify [a] standard set of protocols for such data transfer and [to] ensure compatibility between current MoU signatories.
4. To satisfy additional requirements (such as timely data interchange) as identified by MoU subgroups, especially MoU BARG and MoU EREG.

During one of the first meetings the group decided to go for a tape procedure as well as an EDI[4] procedure, the latter allowing the file transfer in a multi-vendor environment based on OSI standards using FTAM profile A/111. The standard data interchange mechanism was defined to be direct machine-to-machine file transfer with magnetic tape as a back-up medium in case of failure or problems. As such, both interchange methods needed to be implemented into an operator's billing system. It should be noted that, at that time, FTAM was unknown to most of the GSM operators and was also implemented in different ways by the billing systems' suppliers – if at all available at that time!

Based on the specifications and other groups' requirements referred to within the terms of reference TADIG defined the data record format of the TAP and the respective procedures. For example, at a later stage of the specification process, PRD BA.12 defined daily transfer by EDI as the standard mechanism for the exchange of billing records between GSM networks and PRD BA.08 defined 72 hours after call completion as the standard time scale for the exchange of billing records. As GSM specifications and PRDs were not stable at that time various modifications to the TAP standard had to be made throughout the whole specification process.

---

[4] EDI: Electronic Data Interchange.

It became obvious that all details on the implementation of the TAP necessary for network interoperability as well as for commercial implementation had to be fixed within the annexes to the roaming agreements. This method is valid today.

Also during 1989 TADIG proposed a set of rules on how the GSM community should deal with *data clearing houses*. These rules are summarised within two documents: *"Considera-tion as to the use of the clearing house in the interchange of data between signatories to the GSM system"* and *"Specification relating to interchange of data through a data clearing house"*.

The *role* of a data clearing house (as described by one of them) includes:

1. One point of contact for communication and administration issues
2. Data media conversion
3. Customised services
4. Value added services (TAP validation, reports, scheduling)
5. Financial settlement and clearing
6. Full roaming administration services (invoicing, bilateral netting, VAT reclaim)

At least at the very beginning of having talks with representatives from clearing houses there were some strong reservations from the operators' side to have clearing houses becom-ing important players in the field of international roaming.

The formal approval of the first set of TAP related PRDs defining the data record format as well as the tape procedure for the exchange of charging information related to roaming subscribers between GSM networks by plenary #11 of the *GSM MoU Group* early 1990 was the kick-off to many (European) GSM operators to start with their TAP implementations. TADIG decided to change frozen specifications "only if there is a catastrophic error".

With respect to the EDI procedure at that time no security features have been incorporated into the FTAM standard with the exception of user ID and user and file password protection. Therefore there was a need for TADIG to come up with additional security requirements. During 1990 the *GSM MoU Group* approved the basic set of EDI related specifications. According to BARG EDI should become the preferred transmission method and should be in operation on 1 January 1993.

In order to implement international roaming successfully the GSM community had to follow a commonly accepted TAP testing approach. The respective TADIG PRD TD.06 is concerned with the specification of tests related to the TAP to be carried out prior to a GSM network offering a roaming service. The objectives of testing the TAP can be summarised as follows:

- To prove that a network can both send and receive files by tape and EDI according to the standards;
- To prove that the correct data is transferred when *all* types of services are utilised both in mobile originated and mobile terminated capacities;
- To ensure the interworking capabilities of the billing centres of different networks;
- To limit the costs of participating networks when introducing new networks to the TAP.

As the basic test strategy TADIG developed the concept of *"Testing against an accredited PLMN"*. An *accredited PLMN* is one authorised by TADIG to validate another network's ability to operate the TAP. All other networks shall accept the decision of an *accredited PLMN* that another network is capable of operating the TAP. Additionally EDI testing on an individual basis is always necessary – not least because FTAM implementations are vendor specific and (normally) outside the control of the operators. Serious testing became necessary to "survive" as some operators detected major problems of interoperability with implementations of FTAM even if they had successfully passed conformance tests in accredited test labs! In 1991 the *GSM MoU group* formally approved TADIG's test concept.

First experiences with TAP implementations as well as successful TAP testing were reported by some Nordic operators in April 1992: *Telecom Finland* successfully transferred files of live data on tape to and from *Televerket Radio* (Sweden) and *Telecom Denmark*.

However during this phase of collecting first experiences with TAP implementations, many problems in the day-to-day business were reported to the TADIG plenaries, mainly due to non-compliance with the TAP standards. From the moment of having real TAP implementations in operation, it became one of the main tasks for TADIG to deal with all identified TAP problems. Beside the already mentioned FTAM problems the most critical areas of potential problems were:

- At that time GSM operators were forced to upgrade their networks more or less continuously to meet updates of the technical specifications as well as modified "own" technical and commercial specifications. This process resulted in changing the output into other networks possibly not in accordance with the TAP standard;
- Non-compliance with BARG's requirements (a) to implement daily transfer by EDI as the standard mechanism for the exchange of billing records between GSM networks and (b) to cope with the 72 hours after call completion as the standard time scale for the exchange of billing records.

During 1992 TADIG started to work on the specification for the next TAP version ("*Data record format– version 2*"), also known as "*TAP2*". This work was done by TADIG's "*TAP specification subgroup*".

Due to the problems identified in the day-to-day business TADIG specified test scenarios for how the data should be structured and how the data interchange should be handled for all cases applicable while roaming internationally. The main goal of this exercise was to develop a *common understanding* on how to create call related data in the case of roaming. PRD TD.10 detailed the "*General scenarios to test the transferred account procedure*". One challenge for operators was to comply with the respective interface requirements also in the case of not having implemented all services (or traffic cases) into their live networks at the time of performing TAP testing.

In 1992 and sufficiently in advance to the commercial launch of international roaming throughout Europe TADIG had performed their major tasks. The set of specifications related to TAP (version 1) or TAP1 is:

PRD TD.01   Data record format (version 1)
PRD TD.02   Transferred account tape procedure
PRD TD.03   Transferred account EDI procedure
PRD TD.04   The use of FTAM in the transferred account EDI procedure
PRD TD.05   Security model for the transferred account EDI procedure
PRD TD.06   Testing the transferred account procedure
PRD TD.10   General scenarios for the transferred account procedure
PRD TD.13   TAP PLMN naming conventions

The vision to allow for the usage of a GSM terminal also in the case of travelling outside the home country became a reality by, beside other things, the implementation of commercially relevant interfaces into numerous different billing systems of the GSM networks concerned.

In December 1992 the *GSM MoU Group* accepted TADIG's proposal that *Vodafone* should become an *accredited PLMN*. The *GSM MoU Group* clarified that a GSM operator cannot be forced to undertake testing against an *accredited PLMN*.

In July 1993 the *GSM MoU Group* approved PRD TD.17, data record format (version 2) or TAP2. All GSM networks should have implemented TAP2 by 1 January 1996. Effectively the implementation date was delayed until 1 September 1996. The set of specifications related to TAP2 consists of PRDs TD.17, TD.13, TD.37, TD.24, TD.20, TD.40, (TD.27, TD.31/NAIG).

In reality most of the GSM network operators used, at least for a certain period of time, clearing house facilities to convert TAP files from version 1 to version 2 (and vice versa).

By end of 1995 the basic strategy for testing the TAP– testing against an *accredited PLMN* – was dropped mainly due to the fact that this method has never been translated into action as GSM operators went for bilateral testing while introducing new roaming relations.

21.6.3 Further Improvements: TAP3 Specification and the TAP Testing Toolkit

Following an "interim" TAP standard – known as "TAP2 + " – TADIG came up with a new version: "TAP3". The official effective date for TAP3 Release 1 (TAP3.1) was 4 June 2000; TAP3 Release 2 (TAP3.2) came into effect on 4 December 2000.[5]

The improvements provided by TAP3 can be summarised as follows:

- Support of GSM phase 2 + services (GPRS, HSCSD, CAMEL phase2; MSP; SPNP,...);
- Interest group requirements embedded;
- Tariff information for validation and HPLMN re-pricing;
- "Reject and returns process";
- Binary (ASN.1), variable length format.

The set of specifications related to TAP3 consisted of PRDs TD.57, TD.60, TD.32.

The *TAP Testing Toolkit* ("TTT") shall simplify and automate TAP testing. The first release of the TTT, which caters for TAP1, TAP2, TAP2 + , NA TAP2 and TAP3.1, has been completed and distributed to the members of the GSM Association. The TTT is provided for free. The major features are:

- Validation;
- Implementation of test scenarios;
- Cross referencing from TAP file to scenario;
- TAP file "viewer".

The preparation for a major release in spring 2001 is on its way...

---

[5] TAP releases are managed by the TAP Release Management Taskforce. The taskforce is a joint BARG/TADIG effort to produce high-level guidelines for management of TAP and RAP releases.

21.6.4 Summary on Critical TAP Problems

Beside the "old" problem of non-compliance with BARG's requirements (to implement daily transfer by EDI as the standard mechanism for the exchange of billing records between GSM networks and to cope with the standard time scale for the exchange of billing records), there are numerous TAP file problems such as:

- Incorrect service codes used;
- Records on conditional call forwarding missing;
- Taxes not applied correctly;
- Exchange rate incorrect and consequently incorrect charges;
- Incorrect roaming tariff set-up within the billing system.

The latter one is the most critical as this normally results in incorrect billing towards the roaming subscriber by the home GSM operator.

Also the general non-availability of TAP3 is a serious bottleneck for the implementation of new services such as prepaid-roaming based on CAMEL. Only a couple of operators had commercial implementations in place at the end of 2000.

21.6.5 TAP and Inter-Standard Roaming

Based on the successful implementation of TAP throughout Europe it became obvious to use the TAP standard also in the case of *inter-standard roaming*. The very first implementations of inter-standard roaming – better known as "GlobalRoam®" – using TAP as the method for the exchange of charging information related to roaming into different standards (GSM, analogue AMPS and PDC) have been realised by DeTeMobil (Germany), GTE (USA) and NTT DoCoMo (Japan).

21.6.6 Other TADIG Activities

In April 1992 TADIG started to look into the technical solutions for the interconnection of *Equipment Identity Registers (EIRs)* and the *Central Equipment Identity Registers (CEIRs)*. The "*CEIR Subgroup*" was established to elaborate the respective specifications PRDs TD.18 and TD.19.

End of 1994 a TADIG Subgroup started to look into the technical solutions on how to handle all types of documentation from the GSM MoU Group. Additionally, consulting support was added by the GSM MoU Group. Thus TADIG became the leading working group during the process of implementing the wellknown and nowadays quite frequently visited *GSM Association*'s *InfoCentre*. The *InfoCentre* is a web based application operated by the GSM Association's permanent secretariat (address: *infocentre.gsm.org*). Access is provided via Internet, X.25, ISDN or telephone modem. The access is free of charge for members. The *InfoCentre* contains information such as:

- Permanent Reference Documents (PRDs) of the GSM Association;
- Marketing information (coverage maps, etc.);
- Newsletter;
- Meeting organisation information.

Also secured information is part of the *InfoCentre* – such as individual annexes of GSM Association members' roaming agreements or working groups' bulletin boards – and restricted access is guaranteed.

Both TADIG and BARG have seen the need for the GSM Association to keep track of the application and content billing that is going on in the Internet world and to influence the work according to the requirements of the GSM Association's members. Therefore they announced to enter into co-operation with bodies such as the *IPDR Billing Initiative* and the *WAP Forum Billing Experts Group*.

One of the key questions is: will the TAP standard be able to "survive" within this fast moving industry?

Chapter 21: The Contribution of the GSM Association to the Building of GSM and UMTS

Section 7: The Third Generation Interest Group (3GIG)

Neil Lilly[1]

21.7.1 The Formative Years (1994–May 1996)

At the beginning of 1994 "third generation" was a little known term. Activities were concentrated in technical groups such as the ITU Radio Communication Taskgroup 8/1, the European Commission led RACE projects such as CODIT and SMG's working group SMG5 (see Chapter 7, Sections 1 and 3, Chapter 8, Section 1).

At this time, the prime focus of the GSM MoU was the development and globalisation of GSM, incorporating not only GSM 900 but also DCS 1800 and PCS 1900 operation. GSM was still in its operational infancy, so why did the GSM operators become involved then in 3G?

The answer to this question revolves around events at GSM MoU#28 in Cairns in July 1994. In an ad-hoc meeting chaired by Fred Hillebrand a number of leading GSM operators expressed their concerns regarding the existing 3G activities. Firstly, SMG and ITU TG8/F activities were progressing to define radio access across technologies for spectrum allocated at WARC 92, but these activities were dominated by manufacturers – now a familiar story! The operators present at the Cairns ad-hoc meeting commented that because of this situation the entire 3G focus was technical, there being virtually no market or business input.

Secondly and of more concern, was the fact that both ITU and SMG5 were then defining 3G solutions that were incompatible with GSM – radio and network designs were based upon very different technologies to GSM. Those operators investing heavily in GSM roll-out did not want to hear this and neither did their investors – another familiar story today with the 3G–4G situation!

Thirdly, many US companies were already seeking to position IS95 technology as *the* way

---

[1] The views expressed in this section are those of the author and do not necessarily reflect the views of his affiliation entity.

to the future; this "clash of technologies" was already becoming evident in the tripartite governmental FAMOUS discussions (US, Europe, Japan). At Cairns documents were tabled summarising the European Commission's Mobile Green Paper, the fourth FAMOUS meeting results and there was one from Fred Hillebrand entitled "From GSM to UMTS/FPLMTS/ IMT2000".

The result of these discussions was a small ad-hoc group being established to formulate a recommendation to the next MoU plenary in Jersey regarding what the MoU should do to redress their concerns. This group was chaired by Armin Toefer, chair of GSM MoU SERG.

Unfortunately MoU#29 was preoccupied with more urgent matters such as the supply and type approval of terminals for GSM, multi-band GSM operation, and health and safety concerns. A document entitled "Migration towards UMTS" was merely noted and not discussed. Several operators left the meeting with the impression that the GSM MoU really wasn't interested in UMTS and this was one of a number of factors which led to the establishment of the UMTS Forum.

It was to be another four months, at plenary #30 in Cape Town in February 1995 before the GSM MoU plenary had any substantial discussion of 3G. However, in the meantime the SERG ad-hoc group had been doing its homework! It had met a number of times since Cairns and had already defined a set of business driven "guiding principles" for UMTS and formulated a number of technical 3G requirements. At a small meeting in Dusseldorf on 3 January 1995 a number of key requirements coalesced. This had absolutely nothing to do with the gathering in the bar of the Ramada the evening before, the preparatory discussions over copious quantities of Weizenbier and the local Altbier. Among these requirements were virtual home environment, software downloadable terminals and a system design very much based upon open-standard interfaces and application programming interfaces. These requirements today are very familiar and well accepted, but in 1995 they were considered to be very much leading edge and they were hotly debated. Contributions had been submitted to SMG5 but had met with a limited success because of a lack of active operator support and resistance from manufacturers opposed to any change or proposal they didn't initiate.

The 3G guiding principles were tabled in a number of documents at Cape Town[2] and key points contained therein were noted in the plenary minutes:

- With respect to the GSM platform the development of UMTS/FPLMTS should be evolutionary and not revolutionary.
- Network operators should provide more input to SMG5. At present SMG5 mainly reflects the interests of manufacturers.
- The need for a formal ad-hoc group was identified, together with the need for a chairman who could devote time to it.

The SERG ad-hoc group continued its work and a proposal was submitted to MoU#31 (Budapest) in May 1995 to establish a " 3rd. Generation Interest Group (3GIG)". A co-chairmanship arrangement was proposed, Michael Davies (Bell South, New Zealand) and myself (Orange PCS Ltd, UK). This was approved for a trial period of one year. Within that period 3GIG was to address all aspects of 3G/UMTS, establish itself along the lines of a working group and represent the GSM MoU externally on 3G matters.

During the following twelve months 3GIG built up a dedicated membership of 20–30

---

[2] SERG input the same "top-level" document to both GSM MoU Plenary #30 (MoU doc 25/95) and to ETSI TC SMG #13 (Tdoc SMG 110/95) – see CDROM.

participants, drawn mainly from the big, early adopter GSM operators. Contributions were input to ETSI SMG, SMG5 and ITU-R TG8/1 covering many aspects of 3G including service requirements, system architecture, adaptive terminals and spectrum allocation. Reports were made to GSM MoU#32 (September 1995) and #33 (January 1996). 3GIG was charged with preparing a seminar for GSM MoU#34 (Atlanta, May 1996) to give the membership a comprehensive briefing on all aspects of 3G.

This was a significant period for the formulation of 3G, both from the UMTS and the broader IMT2000 context. The UMTS Forum had been established, SMG was being "infiltrated" by more GSM operator members from 3GIG, TG8/1 was becoming more business led and in the inter-governmental FAMOUS discussions NTT DoCoMo was beginning to align itself with the GSM world.

The pre-plenary seminar in Atlanta was a big success and many members who had previously been unaware of the opportunities and risks of 3G were now well briefed. A large number of documents were tabled in the plenary meeting addressing 3GIG progress, MoU 3G policy, preparation for ITU WRC 97, interaction between 3GIG and other GSM MoU working groups, the outcome of the March FAMOUS meeting in Dallas and a review of W-CDMA technology.

The plenary ratified the continued existence of 3GIG under my chairmanship, with Michael Davies as deputy chairman – other business commitments meant that Michael had a limited amount of time available for 3GIG. 3GIG was then charged with producing a set of 3G Permanent Reference Documents (PRDs) (the "TG.xx" series) which would capture all the GSM MoU's 3G requirements and working assumptions. 3GIG was now moving into top gear!

21.7.2 The Definitive Years (June 1996–Dec 1998)

The next two years were the "adolescence" years of 3G. This was characterised in 3GIG, in the UMTS Forum and ETSI SMG by unbridled enthusiasm, trying out new technologies and new techniques. VHE, open services architecture and "softer terminals" were beginning to be assimilated into the predominantly "2G" consciousness and accepted not only as keystones of 3G, but also to be included in a preliminary form in "GSM 2.5G". There was close co-operation between the UMTS Forum and the GSM Association (as the GSM MoU was now called) each addressing topics well suited to their own specific set of skills and participating in each other's meetings. Both input their market/business requirements into ETSI SMG with key members from both organisations participating in SMG to ensure a "smooth passage" for their contributions. That is not to say that everything was accepted without debate – there was many a heated exchange of views on issues such as a new network architectures, the choice of radio interface, terminal equipment capabilities and OA&M, necessitating many late night sessions in hotel bars with manufacturers!

At the next GSM MoU meeting (Hong Kong, September 1996), 3GIG tabled documents formally defining 3G principles. Among these was the need for an "UMTS MoU", a draft 3GIG work plan and a list of the TG.xx series of PRDs to be produced. Also tabled was a market vision for 3G which was to prove to be remarkably clairvoyant in its predictions. This vision was produced at a very productive 3GIG meeting in Munster and it contained a number of market scenarios. Perhaps the most striking scenario is the one which predicted substantial industry consolidation such that within 7 to 8 years there would be only six global players/

mega-alliances. Each alliance would have its own proprietary features and enhancements to the generic 3G standards and its own preferred set of suppliers. In hindsight, writing just as 3G enters the "grown up" era with the launch of the first 3G networks saddled with large mortgages, this was a remarkable piece of crystal ball gazing! The only inaccuracy was that the consolidation occurred in less than 5 years rather than 7 to 8 years.

The deliverables from 3GIG amounted to nearly 40 PRDs, which were structured as follows:

PR.0x series	administrative PRDs, statement of commitment, etc.
PR.1x series	market vision and evolution
PR.2x series	service requirements
PR.3x series	system/network requirements
PR.4x series	security requirements
PR.5x series	OA&M requirements
PR.6x series	evolution from GSM to UMTS
PR.7x series	terminal equipment testing and type approval requirements

Remarkably, there was no shortage of editors for these PRDs. The regular 3GIG participants soon "got wise" to the idea that whenever a delegate proposed the need for a new document then the chairman automatically assigned them as editor! This "modus operandi" did not seem to inhibit enthusiasm or good ideas.

An enormous amount of hard work was done and a lot was accomplished during this period. The 3G regulatory and market environments were fast taking form with the prospect of spectrum auctions, in which operators might have to part with a few hundred million dollars to use the ether! This eventuality and the "great SMG 3G radio interface battle" in Madrid in December 1998 and in Paris in January 1999 ensured that there was a sustained and keen level of interest and participation in 3GIG activities. This high level of activity is noted in the GSMA plenary reports of that period. However, there was some compensation for all the hard work. To "globalise" 3GIG and involve the regions, more 3GIG meetings were in "exotic" and diverse locations – Atlanta, Auckland, Boston, Durban, Riga, Slough - to name just a few !

During this period there was also a significant increase in external GSMA activities and publicity. Besides the close, ongoing links with ETSI SMG and the UMTS Forum discussions were being held with NTT DoCoMo to bring them into the GSM Association as 3G members. Initial discussions were also taking place among global operators who had inherited or acquired networks with diverse 2G technologies (GSM, IS95 and IS136) with the intention of converging 3G technologies as far as was practical. This led to the establishment of the Operators' Harmonisation Group. Conferences were being targeted for 3G presentations and at the GSM World Congress in Cannes in 1998, 3G was firmly on the agenda.

GSM MoU#38 in Cyprus in September 1997 was the 10th anniversary of "the GSM MoU". At a pre-plenary seminar day 3G stole the limelight, a number of organisations giving 3G presentations. The 3GIG presentation gave an overview of GSMA activities and the operators' 3G requirements. To reinforce some of the key points such as system modularity, open standard interfaces, competition, continual evolution and enhancement, the presentation compared the mobile cellular industry with the automobile industry. For those members who had not been so aware (or concerned!) of the potential opportunities – and threats – of 3G, this was their "wake-up call !" In the plenary meeting itself, nine 3G PRDs were elevated to version 3.0.0, thus bearing the hallmark of plenary endorsement.

The "wake-up call" also galvanised a number of GSMA working group chairs to realise that they needed to become involved in the 3G aspects of their remits. At the next GSM plenary meeting (Warsaw, April 1998) a proposal was accepted identifying which working groups should have final responsibility for the many 3GIG PRDs. Another significant decision at this plenary was to make the 3G PRDs publicly and easily accessible by posting them on the GSM world website. Seven PRDs were updated to version 3.0.0. or 3.1.0 and most of the remaining ones were presented for information and feedback, being version 1.0.0 or later.

At plenary #40 in New Delhi (October 1998) all the remaining 3G PRDs were given plenary approval. This was a record achievement – 30 plus PRDs being approved within a two year period. Plenary #40 was also the "swan song" of 3GIG. It was agreed that the working groups should become the "prime movers" for the associations' 3G interests and activities and thus there was little need for 3GIG to continue. This, together with the fact that I was unable to continue chairing 3GIG (like many other 3GIG veterans I was " recalled to base" in preparation for the forthcoming 3G license bids) led to the decision to wind-up 3GIG on this high note. However, a number of delegations had reservations about closing down 3GIG and with hindsight, maybe they were correct.

21.7.3 3G Reality

The beginning of 1999 to June 2001 was the "coming of age" of 3G. This also corresponded with a period of phenomenal success for GSM, establishing it beyond doubt as the 2G system of choice, providing not only "plain, ordinary cellular service" but via GRPS a mobile means of accessing the Internet. Many of the original 3G concepts and requirements were being tested in the GSM phase 2 + environment – and lessons were learnt! During this period more than fifty 3G licenses were awarded to "proud parents" – but in many cases at costs that surpassed even the wildest 3G predictions and fears!

The closing down of 3GIG at the end of 1998 left a vacuum. Although the GSMA working groups were beginning to address 3G matters, GPRS and other 2.5G specification work had become more expansive, needing more time and effort. Aware of the need for continuity and focus, especially towards external organisations such as the newly-created 3GPP and the ITU (with WRC 2000 on the horizon), Plenary #41 (Helsinki, April 1999) appointed Jonas Twingler as 3G co-ordinator. He presented a document entitled "An approach to affirm and utilise 3G policy" which besides proposing a way forward addressed "areas of immediate concern'. Many other 3G documents were submitted encompassing such topics as 2G/3G mixed environment impacts, roaming and billing in a 2G/3G environment, "cross-mode" operation and inter-standard roaming. A "3G spectrum briefing session" was given by David Court, regulatory and strategic advisor to the Association. Because of the number of significant issues raised, it was decided to create an ad-hoc 3G taskforce to assist the 3G co-ordinator. Many of the experts who had participated in 3GIG were now full-time committed to 3G license applications and had little time available for "extra-mural activities". At Plenary #42 in Montreal in October 1999 it was agreed to create a "IMT2000 steering group" to consolidate and focus the Association's 3G policy and activities. Of particular relevance to its remit were the preparation and participation needed for in WRC 2000 (Istanbul, May/June 2000), deployment and operational issues of 3G (e.g. cross-border co-ordination and roaming), 2G/3G roaming and 3G inter-standard roaming issues.

Concurrent with WRC 2000 the GSMA was lobbying extensively for a "GSM + UMTS solution" for Brazil. There were many discussions in Brazil, in parallel with WRC 2000 and at GSMA Plenary #43 (May 2000, Santiago, Chile). All of these preparations paid huge dividends, as Brazil announced in July its intentions to open up its markets to GSM and UMTS technologies. The rest of South America is now following suit, but the mixed technology environment had added urgency to the work of the global GSM roaming forum and the need to define 3G inter-system roaming.

During this period there had also been a proliferation of forums addressing various aspects of 2G/3G business, applications and operation. Many GSM members expressed a desire to have again a single group to act as a focus for not just 3G matters but all future needs – the term "4G" had already made its public debut! Accordingly, at Plenary #45 (Seattle, March 2001) it was agreed to create a "operators' 3G forum" which would assimilate the IMT 2000 steering group and once again give the GSM Association a "future focused" group.

To conclude, since the beginning of 1994, the GSM Association has taken a lead in defining the design and operational requirements for 3G. Many participants have made major contributions to its success in converting the original 3G technology-led focus into a market and business-led phenomenon. Much has been achieved, in particular the 3GIG PRDs. To date most of the concepts and requirements contained in these documents have been assimilated into 3G specifications and standards, with the remainder now being addressed in the evolution of 3G.

Acknowledgements

During the four year lifespan of 3GIG many delegates made their mark on the 3G debate and the TG.xx series of PRDs. Some of these made a sustained contribution throughout those four, hectic years and among these special thanks are due to the following delegates.

- Michael Clayton, for keeping the meetings on course, trying to keep me an honest chairman whilst injecting humour and still drafting excellent meeting minutes.
- Michael Davies, for co-chairing 3GIG with me in the first two years and for driving much of the visionary work
- Alan Cox, for acting as "king of service requirements" and for unwittingly acting as a proof-reader of the meeting minutes.
- The many editors, including Chris Friel, Ian Goetz, Massimo Mascoli, Christian Casenave, Jussi Hattula, Jorg Kramer, David Court, Bernhard Diem, Mario Polosa and Philippe Lucas for taking on demanding editorships in the face of very able critics.
- The key advisors and "keepers of our conscience", among them David Barnes, Horst Mennenga, Thomas Beijer, Bernd Eylert, Petter Bliksrud.
- Adriana Nugter for persistent "behind the scenes" encouragement and for broadening the "3G sphere of operations" to address convergence.
- Finally, and by no means least, all those organisations who hosted our many meetings, permitting us to sample their kind hospitality and large quantities of wine!

Chapter 22: GSM and UMTS Acceptance in the World

Section 1: Global Acceptance of GSM

Friedhelm Hillebrand[1]

22.1.1 Introduction

Nobody amongst the early key players expected that GSM would become a global success. There was even deep concern whether it would become a success in all Western European countries. The GSM MoU signed in September 1987[2] differentiated between commitments for CEPT countries and the rest of the world:

- The signatories shall...make efforts...to extend a 900 MHz Pan-European digital cellular...service to cover the territories of all CEPT administrations (GSM MoU Article 14)
- ..signatories shall... provide...advice and other appropriate support to administrations outside CEPT considering the introduction of a...system (according) to the CEPT/GSM standard" (GSM MoU Article 16)

When the name of the standardisation group GSM (Groupe Spécial Mobile) was used as system name and interpreted as "Global System for Mobile Communication" in 1991/1992, many people – including myself – expressed concern about this wording. We felt this wording as over-ambitious and even arrogant. But there are also self-fulfilling prophecies.

Interest in mobile communication grew considerably during the late 1980s and early 1990s in many countries around the world. The situation varied considerably from country to country. The main difference for a decision on which mobile system to use, was, whether the interest of a country was driven primarily by the need of a mobile infrastructure or whether there was a strong industry policy interest in system development and manufacturing.

The acceptance of GSM needed promoters. The promotion within Western Europe was done mainly by leading network operators. Outside Europe generally the lead was with

---

[1] The views expressed in this section are those of the author and do not necessarily reflect the views of his affiliation entity.

[2] The text can be found in the attached CD-ROM Folder D.

Table 22.1.1 GSM promotion seminars under the auspices of the GSM MoU Group

Date	Location	Purpose	Co-ordination
12–14 October 1988	Hagen (Germany)	To present the first GSM specifications prepared for tendering of infrastructure to a wide audience of operators, manufacturers and academia	A. Silberhorn, P. Dupuis, L. Kittel (University of Hagen)
16–18 October 1990	Budapest (Hungary)	To "familiarise Eastern European network operators with the GSM standard and its inherent benefits", "...study the possibility of implementation of...GSM...in Eastern Europe"	F. Hillebrand
June/July 1993	Prague (Czech Republic)	To present the benefits of accepting GSM and how to use GSM to several Middle/Eastern European ministries	G. Schmitt
27–28 September 1994	Beijing (China)	The provision of information to the Chinese authorities: GSM services, system, features and their evolution, GSM acceptance in the world, comparison between second generation systems	F. Hillebrand
17–19 August 1994	New Delhi (India)	Th provision of information to the Indian authorities: SM standard, security issues, regulation, licensing of operators, planning and operation	F. Hillebrand
4–5 December 1995	Buenos Aires (Argentina)	Promotion of GSM in Latin America	F. Hillebrand

globally active manufacturers who saw the chance of creating a world market for GSM, e.g. Alcatel, Ericsson, Lucent, Motorola, Nokia, Nortel, Siemens. There were also efforts to promote GSM by network operators with international ambitions, e.g. Deutsche Telekom, France Telecom, Vodafone.

Seminars played a key role, organised under the auspices of the GSM MoU Group later with participation of manufacturers' associations. They provided a neutral discussion place where interested parties in "new" countries could meet key players and discuss with them. These events often played a catalytic role in the GSM acceptance process. The most important events are mentioned in Table 22.1.1.

The GSM MoU Group opened up in 1991 for non-European members by a revision of the original MoU in an Addendum[3]. This opened up a basically European club for non-European members.

With the greater number of non-European members a "globalisation" was needed. To achieve this, the GSM Association undertook during 1995 a complete review. This lead to the

---

[3] The text can be found in the attached CD-ROM Folder D.

removing of terminology like Pan-European System etc. from the GSM MoU. The membership fee structure was reviewed, arising from both the growing globalisation of GSM and ongoing interest in GSM as a wireless local loop candidate technology in addition to

- cellular from lower GDP/ head countries. A balance needed to be struck
- between the interests of these newer potential members (including in some
- cases their higher relative scepticism to whether the roaming benefits could
- be seen and the need for some of advanced services) and the more
- traditional GSM operators who may have been involved for some time.

An appropriate set of compromises were reached and the original membership

- fee structure was changed formally in 1996 to one based on a weighted GDP
- per head basis, including an assessment of the number of competing
- operators, so the burden could be shared within a given country, and
- regional or local licensed footprints could also be taken into consideration
- as well. This membership fee structure is still in place today and has
- helped with the constructive development of membership throughout the globe.

The adoption of roaming and advanced services may not yet be everywhere, but

- by holding the GSM ASSOCIATION together in this way the benefits are greater
- overall for GSM customers and the industry as a whole.

22.1.2 Europe

22.1.2.1 Western Europe: 15 States Now in the European Union, Switzerland and Norway

The GSM Memorandum of Understanding (MoU) was signed in September 1987 by 14 network operators from 13 countries. In November 1988 there were already 19 signatories from 18 countries. These network operators committed themselves to open service to their capital city and its airport in 1991. This agreement was a good starting point. But the implementation was not easy.

The biggest problem was that manufacturers needed to be convinced that GSM was meant seriously. This was achieved finally in autumn 1998 by the award of ten contracts to a number of manufacturers for the supply of infrastructure. But this meant also that nine network operators who had signed the GSM MoU had not ordered infrastructure in the autumn of 1988. In addition several of the ten contracts ordered only a validation system. There were only two contracts with substantial quantities of infrastructure equipment, the contracts of Deutsche Telekom and France Telecom. Both operators had decided not to implement a large capacity analogue "interim" system. These volumes were the first orders of large amounts of real serial equipment.

GSM had a difficult competitive situation in the case of operators who had implemented large capacity analogue "interim" systems[4] GSM had an inferior coverage in the beginning. Analogue hand-helds were available since the late 1980s. GSM hand-helds became available in quantities in 1993 and were in the beginning bigger than advanced analogue hand-helds.

[4] E.g. NMT900 in Scandinavia, the Netherlands, Switzerland and TACS in the UK, Italy and Austria.

The second wave of volume orders for serial equipment came from the new private competitors who got GSM licences in the 1989/1990 time frame (e.g. Mannesmann Mobilfunk in Germany, SFR in France, Omnitel in Italy) and the GSM 1800 operators in the UK and Germany.

At this stage GSM created a much larger demand than the expansion of existing high capacity analogue systems could cause. The strong development efforts of mobile station manufacturers lead to an unexpectedly early appearance of several hand-helds already in 1993. Finally the superior services portfolio (including international roaming) and the higher security lead to an energetic construction and expansion of GSM networks. Every GSM MoU signatory had an operational GSM network in mid-1994. In Western Europe the number of GSM subscribers passed the number of analogue subscribers in early May 1996 with 13.8 million each. Most analogue networks became irrelevant during the 1990s. Many were switched off in 2000. At the end of 2000 only 2.3% of all European mobile subscribers were still served by the analogue systems NMT and TACS.

22.1.2.2 Central and Eastern Europe

Strong promotion efforts of manufacturers and two seminars organised under the umbrella of the GSM MoU Group (see Table 22.1.1) prepared the ground. All countries of the region had an underdeveloped fixed network. Most countries rapidly issued two GSM licences. Large network were developed in Hungary, Czech Republic, Poland, Romania and Slovakia (Table 22.1.2). Smaller networks exist in Albania, Belarus, Bulgaria, Bosnia, Croatia, Macedonia, Moldavia, Montenegro, Serbia, Ukraine and other countries.

Table 22.1.2   Large GSM networks in Middle/Eastern Europe

Country	Start of 2 GSM operators	Users at the end of 2000 in millions
Hungary	1994	3.8
Czech Republic	1996	4.0
Poland	1996	5.2
Romania	1997	2.3
Slovakia	1997	1.3

These networks had a growth which was much higher than forecasted. They played a critical role to the economic recovery of these countries, since they rapidly provided a powerful infrastructure for economic development. International roaming with Western Europe was very attractive. The network operators attracted many talented people since they offered future-proof new jobs. These licences attracted Western investors, who provided capital and know-how.

Russia was in a more difficult situation due to the revolutionary changes of the economy and the sheer size of the country. They issued only regional licences and small coverage islands appeared. They achieved 2.2 million GSM users by the end of 2000.

The three Baltic states Estonia, Latvia and Lithuania found investors and know-how from their Scandinavian neighbours and developed from 1993 to seven GSM networks with 1.5 million users at the end of 2000.

22.1.3 The Arab World

The Arab countries (about 40) share one language and they communicate intensively amongst themselves. Therefore GSM was very attractive, since it was the only system offering fully developed proven international roaming. Another attractive feature for the country leaders was the high level of protection against eavesdropping and unauthorised use. During the 1990s telecommunication was a monopoly in all Arab countries. This also included mobile communications.

The acceptance of GSM started by an agreement between the Gulf states in the Gulf Co-operation Council to adopt GSM in the early 1990s. Etisalat (United Arab Emirates) was the first operator from the region who applied for membership in the GSM MoU Group, already in May 1991 (MoU#16). Their manager Hatim Lutfi was the first overseas participant in a MoU meeting at MoU#17 in September 1991 in Brighton. The Gulf states planned to become a services and trading centre for the region in the period after the exhaustion of the oil. Therefore they invested heavily in telecommunication infrastructure and were interested in becoming leaders in GSM in the Arab World.

The second phase was the agreement of the Arab League to adopt GSM in 1992. Within the framework of an ITU project called MODARABTEL, the Arab League had established a Telecom office in Tunis. In 1993 Nina Danielsen, a GSM expert from Norway was posted in this office. One of her achievements was the preparation of a workshop open to the mobile communication experts of all Arab countries. This workshop took place in Amman on the 22-25 November 1993 and was extremely successful. This contributed to a blossoming of GSM networks in many Arab countries, e.g. Saudi Arabia, Egypt, Jordan, Syria, Algeria, Tunisia, Lebanon, Morocco. Today all the Arab countries have a GSM network.

GSM provides to mobile users a seamless infrastructure for the Arab World. In total 10 million people used GSM at the end of 2000.

22.1.4 Asia Pacific

22.1.4.1 Australia, New Zealand and Pacific Islands

Australia's regulator Austel was the first non-European organisation which applied for membership in the GSM MoU Group in February 1991 (MoU#15) since it had decided to implement GSM. The Australian regulatory authority had thoroughly reviewed all available analogue and digital alternatives. The Australian incumbent operator had an analogue AMPS network. The regulator wanted modernisation and competition of at least two operators for the incumbent. Therefore it was decided to go for a digital system. GSM was seen as superior to other alternatives. A serious disadvantage for the analogue and digital systems developed in the US was, that they allowed only two operators to compete in a given area.

New Zealand successfully implemented a GSM network very early.

Vodafone acquired a licence in *Fiji*. The French operators covered the *French overseas territories* in the Pacific.

22.1.4.2 The Asian Tiger States

South Korea was driven by a strong industry policy interest in developing and manufacturing their own system. They acquired licence rights to CDMA ANSI 95 radio transmission technology and became a leader in that technology. GSM was not admitted. Korea achieved a penetration of the home market of 41%. But they are isolated from the world in the second generation. They have drawn conclusions from this experience for their third generation plans and are active members of 3GPP, at least for the terminal and radio part.

Singapore and *Hong Kong* had a liberal regulation regime. In both states networks with different technologies were implemented by several operators. In both cases GSM was the clear winner of the competition. This is especially remarkable in Hong Kong with its difficult radio coverage situation (skyscrapers, sea, mountains). Here GSM took 84% of the market in 1999 (Table 22.1.3).

Table 22.1.3 Distribution of mobile users between competing standards in Hong Kong

System	Subscribers in millions (end of 1999)	Share in %
GSM	3.4	86
ANSI54	0.05	1
ANSI95	0.51	13
Total	3.96	100

In *Taiwan* the incumbent operator implemented GSM 900 very early. Then competitors and GSM 1900 operators were licensed. Taiwan reached 17.6 million users by the end of 2000.

Also *Malaysia* successfully licensed several GSM 900 and 1800 operators.

22.1.4.3 China

The GSM promotion activities were mainly done by many major manufacturers who were already active in China in fixed or even mobile communications. The first success was the GSM network in the southern province of Guangdong, an industrialised special zone. They became indeed the first Chinese signatory of the GSM MoU. Guangdong was to my mind seen by the Chinese Ministry of Post and Telecommunications (MPT) as a test case.

There were intensive promotional activities of the two main groupings in 1993/1994:

- "CDMA": ANSI95, lead promoter Qualcomm
- "GSM": promoted mainly by Alcatel, Ericsson, Nokia and Siemens

The MPT invited several field trials in early 1994. Qualcomm declared victory for ANSI 95 CDMA in a very aggressive seminar in spring 1994.

The GSM "camp" provided a lot of information, but it became clear, that a bigger effort was needed. So the idea of a GSM promotion seminar under the auspices of the GSM MoU Group was born (see Table 22.1.1). I worked at that time for the GSM MoU Group and was charged with arranging such a seminar. Richard Midgett and his colleagues from Hong Kong

Telecom looked after the local organisation. They had Chinese colleagues who travelled every week to Beijing. They interested the MPT in supporting this seminar. In addition they provided invaluable support in cultural questions.

The program provided comprehensive information about GSM: standards, operator's co-operation, system evolution and acceptance of GSM in the different regions of the world. We did not avoid the comparison between GSM and CDMA. Wilhelm Heger of Siemens held a University lecture of one and a half hours on that subject, which provided facts after facts and made a deep impact.

There were about 15 speakers, one-quarter of them were Chinese born and about 40 participants. More than one-third of the time was allocated to discussions. Some took place in plenary, many in little groups. Hong Kong Telecom and MPT provided interpreters.

After this seminar all "camps" intensified their lobby efforts. In late 1994 MPT adopted GSM as a Chinese national standard. The decision regarding CDMA was postponed.

Then China Telecom Mobile and China Unicom began to construct GSM networks in all provinces. Manufacturers formed joint ventures with Chinese companies. There were only very limited possibilities for foreigners to participate in the network operators: some financing in the beginning, and recently a small share of Vodafone in China Telecom Mobile.

The result is that at the end of 2000 all Chinese provinces are covered and China Telecom operates the largest GSM network with 52.8 million users. They enjoy international roaming with the whole GSM world. In total there were 71.6 million GSM users in China at the end of 2000.

22.1.4.4 India

India has licensed GSM operators in two phases. At first eight operators were licensed in the most densely populated four regions. These licence decisions were appealed by parties not considered in the Supreme Court. These cases were resolved. The operators had a difficult start in finding finance, by high license fees and by the fact that they had only a regional license. This was a barrier to nation-wide roaming and also to international roaming, since so many small operators were not attractive roaming partners. This could be overcome by an intensive co-operation of the operators.

Before the Ministry of Post and Telecommunication initiated the license competition for the 48 telecom circles, covering the rest of the country, they asked the GSM MoU Group to provide information about the GSM standard, security, spectrum matters, licensing of operators, etc. A seminar was organised in autumn 1994 (see Table 22.1.1). The subsequent licensing process went smoothly without problems.

Indian operators achieved 3.1 million GSM users by the end of 2000 despite the difficulties and the late start.

22.1.4.5 Indonesia

Indonesia is with its extent of several thousand kilometres east to west and several ten of thousands islands a very large country. Several manufacturers (e.g. Alcatel and Siemens) were successful in promoting GSM. The networks cover only densely populated areas and support the economic development of Indonesia.

22.1.4.6 Philippines

The Philippines were successful in implementing GSM networks.

22.1.4.7 Japan

The Japanese observed the GSM standardisation work in CEPT, but were – to my mind – not convinced that the Europeans would succeed in agreeing on one standard and implement it successfully. Japan is of course like South Korea very much interested in development and manufacturing. On the world market for analogue systems they had been fairly successful in the terminal business.

After the very limited success of the analogue mobile system in Japan they developed a standard for a second generation digital system called Personal Digital Communication System (PDC.)[5] The radio solution uses 30 kHz channels and applies a TDMA with three channels on each radio carrier. This basic TDMA system is similar to the ANSI 136 standard developed in the US. The Japanese also standardised a fully digital core network based on ISDN with extensions for mobile communications which are similar to GSM.

The first system standards were completed about a year after the first GSM standards. PDC was very successful in Japan. It achieved a penetration of the market of about 40%, but it was not accepted abroad.

In order to break the roaming isolation, work started on inter-standard roaming between GSM and PDC in 1996/1997, but it achieved little success in the market.

Finally for the third generation the leading Japanese players and the GSM community agreed on a common path to the future (see Chapter 8, Section 2, paragraph 2.6)

22.1.5 North America

Analogue cellular mobile systems had been invented in the US by Bell Laboratories in the early 1970s. A single system was standardised: the Advanced Mobile Phone Service (AMPS). This was very successful in the US. This early success influenced almost all countries of the Americas, which adopted AMPS 800 MHz and later its succeeding solutions TDMA and CDMA in the same frequency band. Total Access Communication System (TACS) is a variant modified for European spectrum plans. It was very successful in Italy, UK and Austria. US operators and manufacturers were very successful in the analogue era. AMPS had annual terminal productions of more than 1 million units per year in the 1980s. Such volumes were not reachable in the fragmented European analogue market. But such numbers were the dreams for GSM.

The key event for the long-term development was a decision of the FCC immediately after President Reagan came into office. The FCC decided not take action to encourage the agreement on one national standard for the emerging digital cellular technologies. Instead the market should decide which technology would be used.

Two digital radio solutions were standardised and implemented to provide additional capacity in AMPS networks:

- ANSI 136: TDMA in 30 kHz bandwidth
- ANSI 95: CDMA in 1.25 MHz bandwidth

[5] PHS (Personal Handyphone System) is a Japanese standard for a "cordless" system like Telepoint systems.

An opportunity for GSM emerged, when new spectrum became available in the regulatory framework of the Personal Communication Service (PCS) in 1993/1994 (see Chapter 6, Section 1). All licences were issued on a regional basis. The rich features of GSM, including fraud minimisation, and the advanced proven technology were attractive. An adaptation of GSM for the US was needed (see Chapter 6).

The competitive situation for GSM was very difficult due to the different industry policy interests of several players. The first North American PCS network was opened by American Personal Communications in the Washington DC area in November 1995.

Today GSM is operated in the US by Cingular, Voicestrem and Powertel. In Canada the operator Microcell opened service in 1996 and reached 922 000 users at the end of 2000. At the end of 2000 there were 8.9 million GSM users in North America.

The US cellular operators have seen a strong consolidation process. Most smaller GSM operators merged into Voicestream. The CDMA ANSI-95 operators formed two country-wide operators: Verizon Wireless and Sprint PCS. The domestic US wireless properties of BellSouth and SBC were joined to form Cingular Wireless in late 2000. Cingular operates AMPS (both analogue and ANSI 136 TDMA) and GSM networks in the US. ATT Wireless has AMPS (both analogue and ANSI 136 TDMA) networks. ATT Wireless has declared their plans to join GSM for generation 2.5 (i.e. GPRS and EDGE) and third generation (i.e. UMTS). Instrumental in this development was the very rich set of features and functions and the strong momentum for their evolution. International roaming and the high level of security were also very attractive. This sets the scene for a very promising future for GSM in North America.

22.1.6 Africa, South of the Sahara

The leading country is South Africa with more than 7.5 million users at the end of 2000 (42 million population). Two GSM 900 licences were issued in the 1993–1994 timeframe. At that time the total long-term forecasts had been about 500 000 users. The PSTN monopoly network, Telkom, attended GSM MoU Association Plenaries from 1992. Membership was transferred to its 50% subsidiary Vodacom in 1993 when the latter, together with MTN, were awarded national GSM licences.

For me two elements are remarkable in South Africa. The regulator had stipulated as licence conditions contribution to the development of the under-serviced areas (i.e. a community service obligation). The GSM operators had to install about 30 000 GSM pay phones. The most common *modus operandi* was the leasing of community phones and their booths (or phone shops) to small entrepreneurs in the townships. These entrepreneurs created many new jobs and provided telephone services in previously under-serviced areas.

The second remarkable element is that the pre-paid service with its low entrance barrier was a great opportunity for people with a lower income to get access to a personal telephone service. South Africa was one of the very first countries to introduce a fully GSM integrated Intelligent Network (IN) based pre-paid service, this being in 1996, when many networks were still planning non-IN service node solutions, which the GSMA subsequently advised against because of inherently greater risks of fraud.

In the *rest of Africa South of the Sahara* the economic situation is difficult. But most countries implemented small GSM networks which covered the capital city and densely populated areas. Both Vodacom and MTN have established other GSM networks in sub-Saharan Africa.

Africa, South of Sahara, had 10.3 million users at the end of 2000. And GSM was the only digital mobile communication system in commercial operation.

22.1.7 South America

With the adoption of AMPS 800 MHz in the late 1980s throughout the continent, the North American evolution model was set in the region. Not only ANSI 139 TDMA and ANSI 95 CDMA were introduced to expand and to replace 800 MHz AMPS networks, but the deployment of US-PCS 1.9 GHz was decided by some countries of that region. Dominated mostly by North American suppliers, the continent was entirely aligned with the US and condemned to live without GSM. Chile became an exception, it started with GSM 1900 MHz networks in 1997. Other minor GSM networks have been introduced later as it happened in Venezuela and El Salvador, deploying fragments of the regionally scarce 900 MHz spectrum.

In total there were 1.6 million GSM users in South America at the end of 2000.

Looking back into history, a first seminar organised under the umbrella of the GSM MoU Group supported by manufacturers associations was held in Buenos Aires in December 1995 (see Table 22.1.1). In several conferences presentations were made. These efforts provided much information, but the situation remained difficult, as it was not possible to change the deep-rooted model the region had embraced previously.

The turning point happened in 2000 in Brazil, where Alcatel, Nokia and Siemens joined forces to change the model, from regional to global. These efforts resulted in a public enquiry released by the Brazilian regulator, ANATEL, which called society to decide between the use of North American PCS 1.9 GHz or the GSM 1.8 GHz and global IMT-2000 core-band (1.9–2.2 GHz). Innumerable debates, seminars and lobbying activities took place during a 6 month period with the aid of the GSM Association, UMTS Forum and the a.m. suppliers. They worked very closely together to face the fierce dispute against dominant suppliers and organisations defending PCS 1.9 GHz. Perhaps never before was a question so deeply discussed, involving so many experts and authorities even from outside of the telecommunications branch. At the end, the forces defending the global GSM/UMTS model finally won.

The decision of Brazil, the largest market in the region. will most likely influence neighbouring countries in the next years, contributing to the their inclusion in the world of global GSM/UMTS communications. Venezuela, with GSM in 900 MHz, is considering the possibility of deploying GSM1.8 GHz in combination with the IMT2000 core-band/UMTS.

So the wide adoption of GSM and its evolution to third generation UMTS in the continent is not considered a question of "if" anymore, but only a question of "when".

22.1.8 East Central Asia

The lead country in this region is Iran. Together with neighbouring countries they achieved 2 million users by the end of 2000.

22.1.9 Israel

Israel opened a GSM service in the late 1990s. At the end of 2000 they had 834 000 users. This is 20% of the total number of mobile users in the country.

22.1.10 Conclusion

During the short period from 1992 to 2000 GSM succeeded in becoming the only digital system used in Western and Central/Eastern Europe, Australia, the Arab World and Africa South of the Sahara. It became the dominant system all over Asia. It covers the US and Canada and several South American countries.

GSM already reached 500 million users in May 2001. GSM networks operate in 168 countries. GSM systems serve 69% of all digital mobile users. Therefore GSM can bear its name "Global System for Mobile Communication" by right and with pride.

Acknowledgements

Nearly all subscriber figures were provided by EMC World Cellular Data Base. The text on the US has been reviewed and augmented by Don Zelmer (Cingular) and Quent Cassens (Connexant). The text about the Americas has been reviewed and augmented by Mario Baumgarten (Siemens, Brazil). The whole text and especially the section on Africa has been reviewed and augmented by Barry Vlok (Vodacom). Mike Short (BT Wireless) reviewed the text and contributed to the description of the development in the GSM Association.

Chapter 22: GSM and UMTS Acceptance in the World

Section 2: Global Acceptance of UMTS

Bernd Eylert[1]

22.2.1 3G: a Service-oriented Picture

In forecasting the global acceptance of 3G, it is important to regard third generation mobile communications as essentially a *service* environment, rather than as a set of technology choices. While the major ongoing work programme by the Third Generation Partnership Project (3GPP) deals in detail with the technical and standardisation aspects of delivering UMTS/3G family systems, the end-user is most interested not in the air interface linking their personal terminal to the network, but rather the choice, value and utility of services that they are offered. That said, it is interesting to note that at the time of writing W-CDMA represents the preferred evolutionary choice of nearly 90% of the global 3G market place – including all European 3G licensees. Furthermore, with GSM as the *de facto* 2G standard choice with almost 70% of the current global cellular base, the clear possibility exists for roaming between different systems using multi-mode terminals. This likelihood of interoperability – driven as it must be by market demand – will offer mobile users over the next few years the appearance of a "unified standard", with all the benefits in terms of service choice and accessibility that it brings.

Unlike GSM – that was launched by operators with only the experiences of their analogue user base to learn from – UMTS will be introduced predominantly by players with experience in offering feature-rich services in a competitive marketplace to customer bases that in some instances number several million individuals. This deep understanding of "big number" customer behaviour will be even more crucial with the introduction of UMTS. Here, operators – many of them GSM incumbents – will have to build new relationships with subscribers who are presented with a more complex array of network, service, terminal and pricing options than ever before.

Ultimately, only 3G can deliver the all-important global roaming capability, capacity and economies of scale necessary to satisfy the expectations of users. But what confidence does the industry have that UMTS will succeed in a marketplace where it will co-exist – in the

---

[1] The views expressed in this section are those of the author and do not necessarily reflect the views of his affiliation entity.

early years at least – alongside other technologies from GSM to evolved 2.5G systems like GPRS and EDGE? Popular speculation among some observers is that operators have over-committed to UMTS without a clear roadmap to profitability. But while such cynicism may be fashionable, a growing voice from many quarters of the industry suggests that this view-point is an incorrect one.

22.2.2 Assessing the Revenue Opportunity for Operators

Mindful of the possible long-term impact of this major commitment by operators on the global economic picture as a whole, the UMTS Forum has conducted a major market study that seeks to identify the revenue opportunities for operators worldwide, and to identify which services will drive uptake of third generation systems on a regional basis. Based on extensive qualitative and quantitative data from a broad cross-section of industry organisations as well as original research, this conservative forecast takes a service-based approach to predicting revenues, rather than a more usual subscriber-based approach. Instead of trying to predict the pricing strategies of individual operators, the crucial question in gauging 3G revenue streams is *"what will people be using their phones for, and how much will they be prepared to pay for the privilege?"*

In the UMTS Forum's recent report titled *The UMTS Third Generation Market– Phase II: Structuring the Service Revenue Opportunities*[2] regional and worldwide forecasts for opera-tor-retained revenues are evaluated for service categories including mobile Internet access, multimedia messaging service for business, location-based services, rich voice and simple voice. Other key service categories, including so-called "customised infotainment" and mobile intranet/extranet access for business users, have already been studied in previous Forum reports. By considering the relative size of revenues from both consumer and business market segments for all these service categories, the study presents a compelling overall picture of revenue growth between now and 2010 (Figure 22.2.1). In particular, it is clear that 3G will arrest and eventually reverse the pressures on monthly revenues that mobile operators are experiencing now due to ever-tighter market competition and the tendency for end-users to increasingly view mobile communications as a low-cost commodity.

- By 2010, total 3G services provider-retained worldwide revenues from all 3G services are forecast to reach $322 billion. This represents cumulative revenues of over 1 *trillion* dollars from now to 2010.
- By 2010, the average 3G subscriber will spend around $30 per month on 3G data services.
- By 2010, with 3G only representing 28% penetration of the total worldwide mobile base, additional revenues from 3G data services will add $9 per month to total worldwide cellular Average Revenue Per User (ARPU).
- Revenue streams to 3G services providers are highly dependent upon the business models adopted by themselves and their partners. The industry structure is still evolving, so no clear business models are yet established. It is clear, however, that the role of the 3G services providers will change from a simple voice-only, direct relationship with the user to one that involves multiple partners, revenue sharing, with third parties also targeting the end-user.

---

[2] Report No. 13 from The UMTS Forum, The UMTS Third Generation Market – Phase II: Structuring the Service Revenue Opportunities, is available for free download at www.umts-forum.org/reports.html

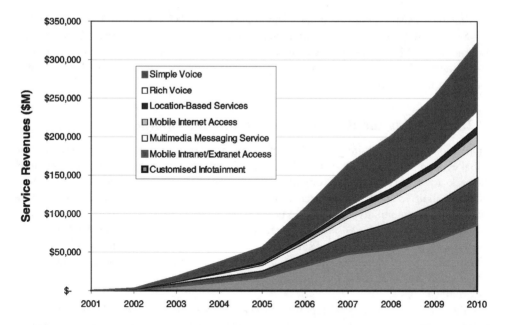

Figure 22.2.1   Worldwide Revenues - All Services. Source: Telecompetition., February 2001

- Both business and consumer market segments are forecast to have significant revenue potential, with the consumer segment contributing about 65% of the revenue on a world-wide basis.
- Throughout the forecast period, customised infotainment is the earliest and single largest revenue opportunity among the forecast services, contributing $86 billion annually in 2010.
- Non-voice service revenues will overtake voice revenues in the 3G environment by 2004 and comprise 66% of 3G service revenues in 2010.

While these predictions paint an optimistic picture for 3G operators, it is equally important to appreciate that UMTS will not – and indeed cannot – be a route to overnight profitability for operators. The study cautions that service revenues will exhibit a modest linear increase from first service launches to around 2005, followed by a sudden acceleration in growth until 2010. This characteristic profile of slow initial uptake followed by rapid growth is of course a common feature of countless other industries. For example, now-ubiquitous hardware formats like the compact disc standard took many years to penetrate the mass market, long after major investments had been made by manufacturers in standards, technology, software industry support and marketing. In the case of 3G, subscriber and revenue growth will be governed by factors ranging from the spread in service launch dates over the next few years, availability of compelling services and content, and widespread availability of affordable terminals.

22.2.3 Taking the Long-term View

Worldwide investment in UMTS and 3G is indeed huge, representing an aggregated spend on licenses, network infrastructure and terminal equipment worth hundreds of billions of dollars. But when putting these sums into perspective, it must be acknowledged that a UMTS license necessarily represents a long-term opportunity. With the typical lifespan of a 3G license of between 15 and 20 years – and longer still in some territories such as Russia – there is ample scope for operators to recoup their investment. The world-wide success for UMTS, however, will be driven by services, and research by the UMTS Forum and others suggests that there are in excess of 20 000 companies world-wide developing third generation services and applications. This expression of confidence from the IP and content communities underlines the long-term importance of UMTS as a conduit for business and personal communications of all kinds.

UMTS demands a major change in end-user's attitudes to their "mobile phone". Primarily a tool for point-to-point communications dominated by voice and SMS, today's GSM user experience effectively mirrors that of using the traditional fixed-line phone – with the added benefits of mobility and value-added services like voicemail. With UMTS, however, the relationship between the customer, network operator and service provider is far more complex. Whereas today's GSM terminals exhibit obvious similarities in terms of form factor, functionality and user interface, terminal equipment for UMTS will offer far greater diversity. While "simple voice" is likely to remain a component of most operators' service offerings, its importance will be displaced by other services. As an access point to IP based services – from Internet and intranet access to media-rich information, education and entertainment – terminal design will fragment as niche products emerge to serve personalised service needs. A business user may need a large screen and keyboard to work on documents and spreadsheets, while a teenage consumer demands extreme portability and high-quality audio reproduction quality to listen to streamed music and video clips. Non-IT literates, on the other hand, may prefer keyboard-free input/output devices featuring speech recognition or pen-based user interfaces.

The relationship between end-user and network operator becomes similarly more complex. In return for their monthly subscription, users will access an array of information and services, often personalised to suit their own needs. This "mobile portal" approach to delivery of 3G services also points to a very significant revenue opportunity for new players such as wireless application service providers (W-ASPs) and content providers as well as for the operators themselves. Establishing and maintaining customer profiles will be an important activity for 3G operators, and the winners on the world stage will undoubtedly be those capable of building and branding an end-user experience that customers stay loyal to. Ultimately, UMTS is not about technology and standards: it is instead about delivering services that make a difference to people's professional and personal lives. While the financial obstacles to success may today appear daunting, 20 years from now we may reflect on the billions of dollars spent by some operators on their UMTS licenses, and conclude that it was one of the shrewdest business decisions that our industry ever made.

Chapter 23: GSM's Success Factors

Friedhelm Hillebrand[1]

23.1 Acceptance in Europe, the First Step

The first big step was the success and acceptance in all European countries. GSM was certainly the right product at the right time. There were countries like France and Germany who needed capacity in the early 1990s. Other countries like the UK, Italy, the Nordic countries had high capacity analogue systems. They were attracted by superior GSM features like small hand-held terminals as well as superior services like international roaming, SMS and data services. Also the high security, capacity and quality played an important role. In addition the newly licensed operators put forward demanding requirements. These demanding market requirements led to an advanced future proof system design and specification and evolution.

The way all players in Europe worked together towards a common goal and the energy this created was essential. In this process a capability was developed to reach decisions even on the most controversial questions, which had not been anticipated.

A key activity to build momentum was the co-ordinated infrastructure procurement by the operators in 1988 which lead to ten contracts which convinced manufacturers to take GSM seriously.[2]

23.2 Globalisation by World-wide Opening

For the globalisation again the principle "the right product at the right time" was the key. In the early 1990s GSM existed in many networks. Advanced terminals and infrastructure were available. International roaming was in operation. This worked together due to the comprehensive GSM system standard. There was no real competition in digital systems by other standards. Therefore the advantages over existing analogue systems became the key for the countries outside Europe. And GSM was future-proof.

The globalisation of GSM was driven by the fact that the major manufacturers understood the opportunity to play a leading world market role in mobile communication promoting GSM. This was spearheaded by Ericsson and Nokia, but all other major manufacturers like

---

[1] The views expressed in this chapter are those of the author and do not necessarily reflect the views of his affiliation entity.

[2] For terminals no co-ordinated procurement had taken place. There was a lack of terminals at service opening.

Alcatel, Lucent, Motorola, Nortel and Siemens were active as well. Amongst the operators Vodafone exploited this potential best. But also France Telecom, Deutsche Telekom, BT, Sonera and Telenor participated. The promotion was supported by actions of the GSM Association and associations of manufacturers.

The open welcome and participation offered to the non-European operators was a key element. The globalisation of the European GSM MoU Group to the world-wide open GSM Association and the transformation of the European standardisation groups in CEPT via ETSI to the Third Generation Partnership Project (3GPP) were other key reasons for the global acceptance of GSM. The very fast evolution of GSM even towards third generation (UMTS) made GSM/UMTS the most future proof solution world-wide leading to a market share of 69%.

23.3 GSM's Superior Services and Systems Features, the Basis of the Success

The "right product" encompasses in the rapidly evolving markets a very advanced services and systems concept. GSM offers the most comprehensive range of services: telephony, short message, data and a very comprehensive range of supplementary services. Above this a superior set of toolkits for service creation was defined. GSM is the only system allowing a global roaming to all continents.

GSM offers an unmatched voice quality due to the basic high quality radio transmission and a range of voice codecs, even with an adaptive system solution offering the optimum combination of capacity and quality in each cell.

GSM systems offer a very high capacity by employing a very high frequency efficiency by using an advanced TDMA technology with adaptive power control, strong channel coding and equalisation, slow frequency hopping and discontinuous transmission, microcells under the umbrella of macrocells. GSM is very robust due to the mobile assisted handover.

GSM is optimised for hand-helds. Digital technology allows the use of very largescale integration. The technology is less complex than some competitors. A sleep mode allows battery saving. Slow frequency hopping ensures a high quality for slowly moving hand-helds.

GSM's superior security and anti-fraud measures protect the users privacy and the operator's revenue. GSM is an open standard, that is not dominated by the IPR of a single manufacturer. GSM provides the only complete system standard defining services, system architecture and selected interfaces: SIM/ME, terminal/base station, two intra-network interfaces and an inter-network interface enabling international roaming. GSM terminals, infrastructure and test systems have the highest market volume and the widest number of manufacturers leading to the widest choice and lowest cost.

Annex 1: Plenary Meetings of GSM, SMG, 3GPP TSG SA and GSMA

A1.1 GSM Plenary Meetings

GSM#	Date	Location	Comments
1	7–9 December 1982	Stockholm	First meeting chaired by Thomas Haug, agreement on the GSM action plan including basic requirements
2	23–26 March 1983	The Hague	
3	11–14 October 1983	Gothenburg	
4	28 February–2 March 1984	Rome	First working party meetings during plenary
5	26–29 June 1984	Berne	
6	12–16 November 1984	London	
7	25 February–1 March 1985	Oslo	Discussion with EEC, first discussion about permanent nucleus
8	10–14 June 1985	Paris	Revision of the strategic requirements (doc. 78/85)
9	30 September–4 October 1985	Berlin	
10	17–21 February 1986	Athens	Decision on mandate and location of the permanent nucleus
11	9–13 June 1986	Copenhagen	
12	29 September–3 October 1986	Madrid	
13	16–20 February 1987	Funchal	Decision on the basic parameters of GSM
14	9–12 June 1987	Brussels	Complete agreement on GSM parameters
15	12–16 October 1987	London	"Candlelight meeting"[a]
16	14–18 December 1987	The Hague	Examination of the specifications for tendering of infrastructure, first approvals (first time 150 000 copies)
17	1–5 February 1988	Florence	Approval of further specifications for tendering

GSM#	Date	Location	Comments
17 bis	15–16 March 1988	London	Completion of the approval of specifications for tendering
18	25–29 April 1988	Vienna	
19	20–23 June 1988	Helsinki	
20	24–28 October 1988	Paris	
21	30 Jan–3 February 1989	Munich	
22	6–10 March 1989	Madrid	Last meeting of CEPT/CCH/GSM
23	5–9 June 1989	Ronneby	First meeting of ETSI/TC GSM
24	2–6 October 1989	Fribourg	
25	11–15 December 1989	Rome	
25 bis	23–25 January 1990	The Hague	Freezing of GSM phase 1 specifications
26	12–16 March 1990	Sophia-Antipolis	
27	11–15 June 1990	Stavanger	
28	1–5 October 1990	Corfu	
29	14–18 January 1991	Saarbrücken	
30	11–15 March 1991	Bristol	
31	10–14 June 1991	Kiruna	
32	30 September–4 October 1991	Nice	Last meeting of ETSI TC GSM. UMTS group set up in GSM

[a] During the meeting, Southern England was hit by a severe storm. The railway traffic was interrupted on many lines and power lines were broken by falling trees so that most of London was without electricity for many hours. Most of the foreign visitors were staying in nearby hotels and could come to the meeting venue (The Cavendish Hotel) while the British delegation did not show up most of the day. In the Cavendish Hotel, we had no electricity so those of us who were present, went on with the meeting illuminated by candles, supplied by the hotel. That gave rise to the name of the meeting.

A1.2. SMG Plenary Meetings

SMG#	Date	Location	Comments
1	20–24 January 1992	Lisbon	First meeting of ETSI/TC SMG
2	30 March–3 April 1992	Ostende	Last meeting chaired by Thomas Haug
3	22–26 June 1992	Copenhagen	First meeting chaired by Philippe Dupuis
4	28 September–2 October 1992	Madrid	
4 bis	30 October 1992	Paris	Approval of the phase 1 TBRS
5	18–22 January 1993	Amsterdam	Review of the work on the half-rate speech coding
6	29 March–2 April 1993	Reading	First proposal to extend the scope of GSM evolution, invention of the term phase 2+

SMG#	Date	Location	Comments
6 bis	28 May 1993	Paris	Consideration of the comments on the phase 1 TBRS
7	21–25 June 1993	Eindhoven	Peter Hamelberg, chairman of the technical assembly of ETSI, takes part in the meeting
8	27 September–1 October 1993	Berlin	Discussion on the use of TTCN
9	17–21 January 1994	Nice	Selection of the half-rate speech codec
10	11–15 April 1994	Regensdorf	
11	27 June–1 July 1994	Dusseldorf	Completion of the core specifications for GSM phase 2
11 bis	13 September 1994	Copenhagen	Adoption of the half-rate speech coding algorithm
12	3–7 October 1994	Helsinki	Five UMTS reports approved
13	23–27 January 1995	Sophia Antipolis	Approval of the half-rate speech coding specifications
14	3–7 April 1995	Rome	First version of the phase 2 TBRs approved
15	3–7 July 1995	Heraklion	Specifications for multiband operation approved
15bis	28 August 1995	Paris	Methodology for phase 2+
16	16–20 October 1995	Vienna	GSM phase 2 specifications frozen Establishment of a permanent relationship with T1P1
17	29 January–2 February 1996	Edinburgh	EFR specifications approved
18	15–19 April 1996	Bonn	50th GSM/SMG plenary meeting Last meeting chaired by Philippe Dupuis
19	24–28 June 1996	Kista, Sweden	First meeting chaired by Friedhelm Hillebrand
20	7–11 October 1996	Sophia Antipolis	
21	10–14 February 1997	Paris	GSM Release 96 approved, first release of GSM phase 2+ UMTS strategy consensus reached
22	9–13 June 1997	Kristiansand	
22 bis	18 August 1997	London	Approval of IMT-2000 contributions to the ITU
23	13–17 October 1997	Budapest	First Chinese participation
24	15–19 December 1997	Madrid	Approval of GSM Release 97 Endorsement to create a global ETSI partnership project
24bis	28–29 January 1998	Paris	Selection of one UTRA concept
25	16–20 March 1998	Sophia Antipolis	Completion of UMTS basic concepts and parameters including services, radio and network aspects
26	22–26 June 1998	Helsinki	

SMG#	Date	Location	Comments
27	12–16 October 1998	Prague	
28	8–12 February 1999	Milan	Approval of GSM Release 98
			Approval of UMTS reports and raw specifications
			Transfer UMTS work to 3GPP
28 bis	12 March 1999	Frankfurt	
29	21–25 June 1999	Miami	
30	9–11 November 1999	Brighton	
30 bis	6 December 1999	Frankfurt	
31	14–16 February 2000	Brussels	Approval of GSM Release 99
31 bis	17 April 2000	Frankfurt	
32	19–20 June 2000	Düsseldorf	Last meeting of ETSI TC SMG
			Transfer of the remaining GSM work to 3GPP
			Last meeting chaired by Friedhelm Hillebrand

A.1.3 TSG SA Plenary Meetings

TSG SA#	Date	Location	Comments
1	7–8 December 1998	Sophia Antipolis	Inauguration meeting
2	2–4 March 1999	Fort Lauderdale	Real start of the technical work, election of officials
3	26–28 April 1999	Yokohama	
4	23–26 June 1999	Miami	Outlook beyond Release 99
			Birth of the all-IP idea
5	11–13 October 1999	Kyongju	
6	15–17 December 1999	Nice	Freezing of Release 99
7	13–17 March 2000	Madrid	Conclusions on the all-IP concept based on the workshop in February 1999
8	26–28 June 2000	Dusseldorf	
9	25–28 September 2000	Hawaii	Agreement on the revision of the release concept
			Definition of the content for Release 4 and 5
10	11–14 December 2000	Bangkok	
11	19–22 March 2001	Palm Springs	Freezing of Release 4

A.1.4 GSM MoU and GSM Association Plenary Meetings

Number	Meeting date	Meeting venue	Comments
MoU 1	14 October 1987	London, UK	Agreement on action plan
MoU 2	17 November 1987	Bonn, Germany	
MoU 3	19 January 1988	Paris, France	
MoU 4	17 March 1988	London, UK	
MoU 5	8–9 June 1988	Rome, Italy	
MoU 6	15–16 September 1988	Copenhagen, Denmark	
MoU 7	29–30 November 1988	Albufeira, Portugal	Decision on cipher key length
MoU 8	22–23 February 1989	Madrid Spain	
MoU 9	23–24 May 1989	The Hague, Netherlands	
MoU 10	25–27 September 1989	Stockholm, Sweden	
MoU 11	9–10 January 1990	Taormina, Italy	
MoU 11 bis	23 January 1990	The Hague, Netherlands	
MoU 12	2–4 April 1990	Berlin, Germany	
MoU 13	5–7 June 1990	Rovaniemi, Finland	
MoU 13 bis	14 June 1990	Stavanger, Norway	
MoU 14	22–24 October 1990	Porto, Portugal	
MoU 15	25–27 February 1991	Berne, Switzerland	First non-European membership application (Austel)
MoU 16	7–8 May 1991	Madrid, Spain	
MoU 16 bis	21 June 1991	London, UK	
MoU 17	23–24 September 1991	Brighton, UK	First non-European participant (Etisalat)
MoU 17 bis	30 October 1991	Taastrup, Denmark	
MoU 18	17–19 December 1991	Taastrup, Denmark	
MoU 19	17–18 March 1992	Helsinki, Finland	
MoU 19 bis	27 April 1992	Frankfurt, Germany	
MoU 20	10–12 June 1992	Harstad, Norway	
MoU 21	16–18 September 1992	Luxembourg	
MoU 22	9–11 December 1992	Madrid, Spain	
MoU 23	16–18 March 1993	Düsseldorf, Germany	
MoU 24	8–10 June 1993	Killarney, Ireland	
MoU 25	20–22 September 1993	Tallinn, Estonia	
MoU 26	7–9 December 1993	Abu Dhabi, UAE	
MoU 26 bis	24 January 1994	Frankfurt, Germany	
MoU 27	6–8 April 1994	Paris, France	
MoU 28	11–13 July 1994	Cairns, Australia	
MoU 29	18–19 October 1994	Jersey, Channel Islands	
MoU 30	22–24 February 1995	Cape Town, South Africa	

Number	Meeting date	Meeting venue	Comments
MoU 31	31 May–2 June 1995	Budapest, Hungary	
MoU 32	27–29 September 1995	Rhodes, Greece	Tenth anniversary of the GSM MoU
MoU 33	17–19 January 1996	Rome, Italy	
MoU 34	29–31 May 1996	Atlanta, USA	
MoU 35	25–27 September 1996	Hong Kong	
MoU 36	26–28 February 1997	Salzburg, Austria	
MoU 37	28–30 May 1997	Fiji	
MoU 38	24–26 September 1997	Nicosia, Cyprus	
MoU 39	27–30 April 1998	Warsaw, Poland	
MoU 40	20–23 October 1998	New Delhi, India	
PL 41	20–23 April 1999	Helsinki, Finland	
PL 42	19–22 October 1999	Montreal, Canada	
PL 43	4–7 April 2000	Santiago, Chile	
PL 44	10–13 October, 2000	Montreux, Switzerland	
PL 45	27–29 March 2001	Seattle, USA	

Annex 2: Organisation Evolution of the Technical Groups

Section 1: Organisation of the GSM Work in the CEPT Era

Thomas Haug[1]

The idea of CEPT COM-T seems to have been that the new baby, Groupe Spécial Mobile (GSM), should only harmonise the mobile systems that existed. This was not a realistic idea, however, since it would lead to a perfectly conventional system, and it was quite clear from an early date that GSM could not only rely upon the output from other groups if new development was needed. One reason for this was that the development of a totally new system with very close interaction between the various technologies involved, would require strong co-ordination by the group responsible for the work. This was in line with the work plan,[2] accepted by CEPT/CCH in November, 1982. In other words, GSM wanted to be in charge on the essential points, but other CEPT groups were utilised when it was found desirable.

For the first three meetings, GSM worked in the plenary only. The number of delegates was quite small, so it was easy to deal with the group as one body. Furthermore, the goal of the group was not very clearly defined besides presenting an outline specification of a Pan-European system in 1986, so it was seen preferable to have all delegates taking part in the discussions.

After three meetings, however, some problems had been identified in sufficient degree to justify a discussion in greater depth than we had had in the beginning. Many of the delegates were specialists in a particular technical area and thus able to contribute more to the progress than if they all had to attend the plenary all the time. Starting at meeting no. 4 (Rome, 28 February–2 March 1984) we therefore split into three working parties for a part of the meeting, an arrangement which was then thought to be for that meeting only. As is usually the case when an organisation sets up subgroups, the arrangement became permanent. Splitting into subgroups was perfectly normal in CEPT, and we did not have to ask any of the superior bodies for permission to organise our own work in the meetings. Contacts with CEPT/CCH resulted in a reminder that we did not have permission to set up subgroups

---

[1] The views expressed in this section are those of the author and do not necessarily reflect the views of his affiliation entity.

[2] Doc 2/82.

which met independently of the main group meetings, however, since that committee was well aware of the risk of proliferation of committees once that development is started.

The Working Parties (WPs) were dealing with the following areas:

WP1   Services and facilities
WP2   Radio questions
WP3   Network questions

This arrangement worked well for a while, and in meeting no. 7 (Oslo, February 1985), the ad-hoc WP4 for system requirements was set up. In addition, we had several times occasion to set up temporary working parties of the heads of delegations, usually chaired by myself, when there were policy questions involved. By October 1985, the work of GSM had reached the point where it was felt that the WPs had to be permanent (which in fact they had been for a while) and also had to be able to meet independently of the main group since there was much more work to do than could be done during the meetings of the main group alone. This was accepted by CEPT/CCH without much debate, since it was felt that the work of GSM was progressing well and that the group needed increased resources in time and manpower.

As a general remark it must be said that through the 10 years of my tenure, GSM had a great degree of autonomy, since CCH never tried to interfere in the technical matters of the group. Since the membership in the group was quite stable and the meeting frequency was high, the delegates came to know each other very well and probably felt an allegiance to the group and its decisions, a fact which was very helpful for the creation of an atmosphere of co-operation.

Basically the same structure was kept when the Permanent Nucleus (PN) was set up in the spring of 1986, initially led by Bernard Mallinder and later by Eike Haase. By mid-1985, it had become obvious that a PN was needed to cope with the increasing amount of work, but it was unclear what role the PN should have. Therefore, the preparation of the mandate for the PN caused considerable discussion in GSM, in particular at a number of meetings of the delegation heads during meeting no. 9 of GSM in September–October 1985. Two quite different approaches were considered. One was to follow the approach sometimes seen in large international organisations, i.e. to leave it to the PN to work out the detailed specifications on the basis of broad agreements in the main group only, perhaps even to the point of discontinuing the WPs. The other approach was to stay with the well functioning mode of operation that we had established, according to which we would keep the present structure in which most of the detailed work would be performed by the WPs, and use the PN to provide support to the WPs in the production and management of the documentation. We chose the latter approach, and I don't think that anyone among us ever regretted it.

One huge benefit of the PN was that we now had a body with program managers for each WP, able to assist the WPs in their work as well as keep an overall view of the system work such as identifying the effects in one area of the system when changes were made to another. It also had the resources to keep track of the various documents and the changes to them, a monumental task as it turned out to be when the number of Change Requests (CRs)rose to several hundred in each plenary meeting. In addition, the PN also took on several other tasks which were of a general nature and thus not necessarily belonging in any particular WP. One example of this was the initiative concerning the IPR issue, another one was the work done in the very important field of operation and maintenance, which after some time led to a regular working party.

Gradually the organisation grew. For instance, a new expert group (later WP4) was established for data services, chaired by Fred Hillebrand, and an expert group for security issues under the chairmanship of myself was established.

The period from 1987 brought very intensive work. The workload on all members of the group was very high, which again generated an enormous amount of documents on all sorts of details. However, the major decisions were already taken, and it is safe to say that the period was characterised by hard, detailed work rather than dramatic events. The fact that the group left CEPT and joined ETSI did not lead to any immediate changes in the organisation. According to an unwritten agreement with the technical assembly chairman, GSM was accepted as a working entity and went on more or less as before with a high degree of autonomy. Thus, the move to ETSI had no deleterious effects on the efficiency of the work, but the nomenclature was brought into line with that of ETSI, i.e. GSM now became a Technical Committee (TC/GSM, after October 1991 TC/SMG), the working parties became sub-technical committees and the PN became a project team, until January 1992 still located in Paris. As I mentioned in an earlier chapter, I retired from TC/SMG in April 1992.

Annex 2: Organisation Evolution of the Technical Groups

Section 2: ETSI GSM and SMG

Ansgar Bergmann[1]

In 1990, TC GSM had four subgroups:[2]

- GSM 1, working on service aspects; a subgroup SIMEG, the SIM expert group, was formed in summer 1991;
- GSM 2, working on radio aspects;
- GSM 3, working on network aspects, with subgroups WPA (radio interface protocols), WPB (supplementary services protocols) and WPC (network protocols);
- GSM 4, working on data aspects.

Further groups reported to GSM:

- the Operation and Maintenance Experts Group (OMEG);
- ES, responsible for the test specifications for the mobile station 11.10 and the base station sub-system 11.20;
- the half-rate codec group, which later became Traffic Channel Half rate Speech (TCH-HS, a subgroup of ETSI/TM/TM5). It existed until the selection of the half-rate codec at SMG#11bis in September 1994;
- specification 09.02, the central core network specification on MAP, was at that time under prime responsibility of ETSI/SPS 2. The responsibility of SMG and SPS 2 for MAP was further discussed and re-arranged in the following years, until it came under full responsibility of SMG and later 3GPP. Areas of responsibility were maintenance of phase 1, evolution of phase 2 and MAP test specifications. Aspects of the discussion included the fact that GSM was initially the only user of MAP; that there were some ideas to make MAP a CCITT/ITU standard and the usage of MAP for third generation (UMTS).

At GSM#32, TC GSM became responsible for the specification of UMTS, and an STC on UMTS was formed. This was taken as a reason to rename TC GSM to TC SMG. Consequentially, the next meeting was SMG plenary no. 1, and the STCs became SMG1–SMG5.

In the following years:

---

[1] The views expressed in this section are those of the author and do not necessarily reflect the views of his affiliation entity.

[2] Officially called Sub-Technical Committees (STCs).

- More and more companies became involved in GSM for which it was essential to participate in SMG. In 1990, around 100 delegates participated in a plenary, in 1995 around 150, and later the number increased up to 316. The amount of direct contribution to SMG groups attained an exceeded 500 man years per year.
- Around 20–40 work items were completed every year. In an evolving system, additional features have dependencies and interactions with existing features, and the work necessary for integration grows exponentially. The workload was augmented additionally by the necessity to maintain different releases in parallel.
- A joint development of a single GSM standard had to be ensured with a co-operation on equal basis between SMG and T1.P1.
- UMTS development got into the hot phase. Sufficient time had to be made available during meetings.
- Decisions became more and more momentous, due to the huge and fast increasing investments in GSM.

Structure and working methods of SMG were to be reviewed and optimised for the changing demands:

- The principle of one plenary was maintained. This was seen essential in order to allow companies to get prepared for decision taking considering the interdependencies in the various areas.
- The frequency and duration of SMG plenaries was adapted. The basic scheme was four meetings of 5 days a year. Extraordinary meetings of 1 or 2 days were added when necessary, typically with a focused agenda when very essential decisions had to be made.
- Transparency of planning and decisions was improved. Decisions had to be based on written documents, which could be taken home by the delegates.
- Sufficient time was allocated for UMTS.
- Debates were tightened by focusing on one subject, concentrating to reach a decision and using indicative voting; for controversial subjects, after an initial discussion in SMG plenary, an ad-hoc group was installed to resolve the problem; it reported back and a decision was taken on the last day of the plenary week, which was reserved for postponed items.
- Meeting reports were made available some days after the meetings.

Meetings have to be prepared. Critical items have to be recognised and highlighted; sufficient discussion time must be planned, and alternatives have to be prepared. For SMG, this preparation was done in the SMG steering group, consisting of SMG officials. In order to make this process very transparent, this group was later replaced by three new advisory groups with clear mandates approved by SMG plenary:

- Co-ordination Group (CG), a management team consisting of the SMG chairman, vice-chairmen, all STC chairmen, liaison officers from GSMA and ECTEL TMS, PT SMG leader, chaired by the SMG chairman.
- Advisory Group (AG), discussing strategic questions and making recommendations to the plenary; the participation was open to all SMG members. AG elected a convenor per meeting.
- Working Methods Group (WOME), discussing and recommending issues related to working methods. It had a permanent chairman.

The SMG working methods were constantly improved, see Chapter 20. Technical support was given by an integrated project team, see Chapter 19.

The hierarchical structure was flattened and the decision paths were reduced. New STCs were installed when appropriate and closed when their work had been completed:

- In spring 1992, OMEG was transformed into SMG 6.
- SMG7 (mobile station testing) and SMG8 (base station testing) were installed in January 1994, SMG9 (SIM aspects) in April 1994.
- SMG Security Group (SMG-SG) was installed in April 1995. It was transformed into SMG10 in June 1996.
- SMG11 on speech coding aspects was created in October 1996, continuing the work of the Speech Experts Group (SEG) and Speech Quality Strategy Group (SQSG).
- SMG12 on system architecture was created in March 1998.
- In June 1997, the "New UMTS" concept had been agreed and all UMTS work had been successfully distributed to the GSM STCs in order to benefit from their competence and experience, SMG5 was closed.
- SMG8 was closed after the completion of the main part of the work; the remaining issues were transferred to SMG2.
- Several taskforces were installed with limited lifetime, for example the GPRS task force (between April 1994 and April 1995) and the joint ERM/SMG taskforce for the production of European harmonised UMTS/IMT-2000 standards needed for regulatory purposes.

Annex 2: Organisation Evolution of the Technical Groups

Section 3: 3GPP

Adrian Scrase[1]

A2.3.1 December 1998 to Mid-1999

During the preparatory talks that led to the creation of 3GPP, many discussions took place to find the optimum organizational structure. The ETSI TC SMG model had worked well for many years and it was very tempting to adopt a similar structure and just widen the sphere of participation. However, some voices called for a more radical approach in order to streamline the structure and to reduce the time taken for specifications production. As a result of these discussions, the following key principles were established on which 3GPP was structured:

- Minimum number of hierarchical levels;
- Large degree of distributed autonomy;
- Clear separation of technical activities from political and administrative activities.

When 3GPP was created, four Technical Specification Groups (TSGs) were formed to undertake the preparation of technical specifications. The four TSGs were as follows:

- TSG CN – core network
- TSG RAN – radio access network
- TSG SA – services and system aspects
- TSG T – terminals

Each of the TSGs was authorized to develop and approve specifications and reports within its terms of reference. This represented a departure from the more traditional approach where a single entity (i.e. a plenary) within a project has the authority to approve a project's output. It was believed that by distributing the approval authority, the time taken to produce specifications would be reduced since this effectively removes one level of hierarchy from the approval procedure. However, it was apparent from the outset that distributing the approval of specifications would lead to a greater requirement for technical co-ordination and thus TSG SA was tasked to perform a co-ordination role across all TSGs. This co-ordination role

---

[1] The views expressed in this section are those of the author and do not necessarily reflect the views of his affiliation entity.

has been aided by the collocation of the TSG meetings and by concerted efforts from the industrial members within 3GPP.

On the creation of 3GPP, a large amount of the work previously undertaken by ETSI TC SMG was transferred to the four TSGs. It was important for all involved to track the transfer of work carefully and meticulous care was taken to map the work from its old home in SMG to its new home in the 3GPP TSGs. This mapping information was made openly available on the 3GPP and ETSI websites to ensure that the telecommunications community could, as a whole, follow the work. This transfer of work was a form of "soft handover", with groups existing in parallel within SMG and within 3GPP for a period of time and items of work being transferred at the most appropriate point. The complete transfer of work was achieved within a period of 6 months.

The scope of 3GPP had been a subject of much debate and at the time of creation the scope covered the 3G system incorporating the UTRA radio access technology. This implied that not all of the work that existed within ETSI TC SMG was to be transferred to 3GPP. There remained a lot of work to be done for the evolving GSM radio interface (i.e. GPRS and EDGE) and this work would remain within SMG for the time being. In addition, the generic work relating to IC cards did not belong in 3GPP either and this too remained within ETSI TC SMG. SMG also retained the responsibility for European issues relating to both 2G and 3G, particularly for regulatory matters, and was also responsible for the transposition of 3GPP specifications into ETSI deliverables.

3GPP had no responsibility for the long-term evolution of the 3G system nor any responsibility for the fixed access component of UMTS. An ETSI project was therefore created (EP UMTS) to take care of these aspects.

A2.3.2 Mid-1999 to Mid-2000

3GPP was an entirely new concept and the first few months of operation were, in effect, experimental. However, in a very short time the project proved to be successful, and the industrial members gained confidence in the new method of working. The preparation of the first release of specifications proceeded at an alarming speed with more than 300 specifications being completed within the first year of operation. At the same time, the development of GPRS and EDGE continued within the ETSI TC SMG environment with active participation from North America. It was not long before serious consideration was to be given to the transfer of all remaining work and the closure of ETSI TC SMG.

An ad-hoc group was created within 3GPP in January 2000 to give full consideration to the widening of the 3GPP scope, particularly to include GPRS and EDGE. It was clear that not all 3GPP partners had a commercial interest in GPRS and EDGE and assurances were required that the ongoing UTRA based activities would not be unduly delayed by such a change in the 3GPP scope. By July 2000 the necessary agreements had been obtained by each 3GPP partner and the scope of 3GPP was formally changed to include the development and maintenance of GSM specifications, including the GSM evolved radio access technologies (such as the General Packet Radio Service (GPRS) and Enhanced Data Rates for GSM Evolution (EDGE)). This was achieved by the creation of a new TSG called TSG GERAN – GSM/EDGE Radio Access Network.

The scope of 3GPP was also modified to make clear that the responsibility for the long-term evolution of the 3G system was vested there. This enabled the ETSI group EP UMTS to be closed, thus focusing efforts firmly within 3GPP.

With the transfer of GSM into 3GPP it was a natural progression for ETSI TC SMG to be closed. However, ETSI still had the important task of transposing the 3GPP results into ETSI deliverables and the preparation of harmonised standards required to meet European regulations. This activity was not expected to be particularly onerous but it was nevertheless of high important for the European industry. To accommodate this work, the ETSI TC Mobile Standards Group (TC MSG) was created.

The only remaining activity to be accommodated was the generic activity pertaining to IC cards. ETSI had earned a high reputation for this work and since it was not specific to mobile telecommunications systems it was not appropriate for this to be placed within 3GPP. This led to the creation of an ETSI project later to be called Smart Card Platform (EP SCP).

A2.3.3 Mid-2000 Onwards

By mid-2000 the focus of attention was now clearly on 3GPP where all UTRA based and GSM radio based activities were now taking place within five TSGs. The European regulatory interests were being taken care of by ETSI TC MSG, and the generic IC card activities by EP SCP. (The former TC SMG and EP UMTS had been closed by this time).

Within Europe, interest had been shown by the railway community to adapt the GSM system and to use it as the basis for a European railway telecommunications system. This work had progressed well within the former ETSI TC SMG with much of the work having been completed before its closure. The systems were now close to deployment and it was desirable to have a permanent home for these activities. This led to the creation of a new ETSI Project called Railway Telecommunications (EP RT).

By late 2000, 3GPP had grown used to having five TSGs and had gained some experience of operating with its expanded scope. Part of the agreement reached for the expansion of the scope was that a review should be held after 6 months of operation to ensure that the best organizational structure had been found. At the time of writing that review had just begun.

Annex 2: Organisation Evolution of the Technical Groups

Section 4: GSM Association Working Groups[1]

Arne Foxman[2]

A2.4.1 Working Groups

A2.4.1.1 International Roaming Expert Group (IREG) – Chairman: Xavier Palacios, Servei De Tele D'Andorra

The International Roaming Expert Group (IREG) specifies technical, operational and performance issues supporting international roaming. IREG focuses on the study, from compatibility and interoperability perspectives, of the signalling and interworking of roaming issues between Public Land Mobile Networks (PLMNs), Public Switched Telephone Networks (PSTNs), Integrated Services Digital Networks (ISDNs) and Public Packet Switched Networks (PPDNs) modes, to define the end-to-end functional tests of bearer services, teleservices, and supplementary services.

E-mail: ireg@gsm.org

A2.4.1.2 Legal and Regulatory Group (LRG) – Chairman: Joost Batelaan, Dutchtone N.V.

LRG provides advice and support to the GSM Association and its members on policy issues relating to legal and regulatory matters. LRG's work involves careful consideration of competition law issues, both globally and in discussion with regional competition authorities. LRG is supported in its work by the Association's corporate affairs and external relations department.

E-mail: lrg@gsm.org

---

[1] Information current as of 15 April 2001.

[2] The views expressed in this section are those of the author and do not necessarily reflect the views of his affiliation entity.

A2.4.1.3 Security Group (SG) – Chairman: Charles Brookson, DTI, UK

SG was established to maintain and develop GSM Association algorithms and protocols, technical security aspects of customer apparatus and to examine and recommend infrastructure solutions to combat fraud. The Group consists of technical representatives from Association members who study the security threats to GSM, its interfacing with third generation and converging technologies, and advises members of possible security issues, or required countermeasures. James Moran, Fraud and Security Director supports SG in its work at headquarters.

E-mail: sg@gsm.org

A2.4.1.4 Fraud Forum (FF) – Chairman: Joao Pedro Do Rosario, Telecom Italia Mobile, Italy

The main focus of the GSM Association's FF is to identify and analyse the various techniques that are used throughout the world to perpetrate fraud against member networks and to recommend practical, cost effective solutions. In conjunction with the SG, the FF monitors and reports on all types of cellular fraud and security breaches throughout the world.

E-mail: ff@gsm.org

A2.4.1.5 Billing and Accounting Roaming Group (BARG) – Chairman: Luc van den Bogaert, Belgacom Mobile, Belgium

The GSM Association's Billing and Accounting Roaming Group (BARG) supports international roaming through the on-going evaluation and assessment of the specification, focusing on financial, administrative and procedural issues.

This work includes the definition and implementation of charging principles for international roaming, together with the related inter-operator procedures, billing harmonisation, credit control and liaison with other groups regarding fraud control.

E-mail: barg@gsm.org

A2.4.1.6 Services Expert Rapporteur Group (SERG) – Chairman: Philippe Lucas, Microcell Telecommunications Inc., Canada

The main focus of the Service Experts Rapporteur Group (SERG) is to develop service requirements from GSM operators on GSM and 3rd Generation, taking into account billing and customer care issues. SERG also provides a connection between operator marketing requirements and technical realisation, involving standardisation.

E-mail: serg@gsm.org

A2.4.1.7 Smart Card Application Group (SCAG) – Chairman: Paul Aebi, Swisscom, Switzerland

The Smart Card Applications Group (SCAG) is responsible for all commercial aspects of 'smart card' technology within the context of the GSM Association's membership. Terms of reference are to monitor and report on smart card evolutionary potential, to indicate to the relevant standardisation bodies the commercial prioritisation and to act as a primary liaison point between the Association and smart card manufacturers.

E-mail: scag@gsm.org

A2.4.1.8 Terminal Working Group (TWG) – Chairman: David Nelson, Orange PCS Ltd, UK

The Terminal Working Group (TWG) is the GSM Association's primary body dealing with GSM Mobile Station (MS) issues. Core issues of focus for TWG include:

- Key interface with manufacturers (MS, SIM, test equipment) and standards bodies on all MS technical issues.
- The development of Association positions on MS features and services.
- Regulatory and access to market issues.
- Central Equipment Identity Register (CEIR) and International Mobile Equipment Identity (IMEI) issues.
- Service and network feature interworking and interoperability.

E-mail: twg@gsm.org

A2.4.1.9 Environmental Working Group (EWG) – Chairman: Brent Gerstle, Optus Communications

The EWG provides expert advice on a range of environmental issues. Technical advice focuses on potential health effects, interference and other environmental issues related to GSM mobile communications. EWG also monitors the GSM Associations sponsored EMF research program.

E-mail: ewg@gsm.org

A2.4.1.10. Transferred Account Data Interchange Group (TADIG) – Chairman: Christer Gullstrand, VoiceStream Wireless Corporation, USA

TADIG is responsible for defining data interchange procedures. Its main focus is billing, including TAP, the rejects and returns process, TAP testing and the requirements for the Tap Testing Toolkit (TTT). TADIG is also responsible for other technical interfaces, e.g. the interface between operators' Equipment Identity Registers (EIRs) and Central EIR (CEIR).

E-mail: tadig@gsm.org

A2.4.2 Technology Interest Groups

A2.4.2.1 Satellite Interest Group (SATIG) – Chairman: Erwis Sinisuka, ACeS International Limited

SATIG co-ordinates all matters relevant to the co-operation between the GSM Association and all satellite network operators participating in the Association. It provides operators of mobile satellite networks based on the GSM platform with a collaborative forum.

E-mail: satig@gsm.org

A2.4.2.2 IMT2000 Steering Group (ISG) – Chairman: Mario Polosa, BLU, Italy

The IMT2000 steering group will provide a focus for 3G operators, incumbent operators as well as new entrants. Specifically tasked to review the high level aspect of the migration from a 2G to a 3G operational environment, the group will cover a wide range of issues – including policies and strategies for standardisation, regulation and operation – as well as liasing with other 3G bodies around the world. ISG is responsible for identifying, analysing, leading and communicating 3G issues for the GSM Association, and for co-ordinating the overall 3G work programme distributed across the experts working groups. It focuses primarily on high level and strategic aspects and relies on the working groups and their subject matter experts for detailed undertakings.

E-mail: isg@gsm.org

Annex 3: List of Chairpersons

A3.1 List of Chairpersons in GSM and SMG[1]

Group (plenary and permanent groups reporting to the plenary[a])	Name	Terms of office start	Terms of office end
GSM/SMG chairmen	Thomas Haug	June 1982	April 1992
	Philippe Dupuis	April 1992	April 1996
	Friedhelm Hillebrand	April 1996	July 2000
GSM/SMG vice-chairmen	Bernard Mallinder	June 1989	June 1994
	Bernard Ghillebaert	June 1989	October 1993
	Alain Maloberti	July 1995	January 1997
	Gunnar Sandegren	October 1993	July 2000
	Alan Cox	October 1997	July 2000
WP1/SMG1	J.F. Wallingford	February 1984	June 1984[b]
	Ganesh Nilakantan	June 1984	February 1985[c]
	Martine Alvernhe	February 1985	March 1991
	Gunnar Sandegren	March 1991	April 1994
	Alan Cox	April 1994	July 2000
WP2/SMG2	Didier Verhulst	February 1984	November 1984
	Alain Maloberti	November 1984	June 1995
	Niels P.S. Andersen	July 1995	July 2000
WP3/SMG3	Jan Audestad	February 1984	January 1989
	Panaioli	January 1989	October 1989
	Per Bjpörndahl	October 1989	January 1993
	Michel Mouly	January 1993	June 1998
	Harald Dettner	March 1998	February 2000
	Ian Park	February 2000	July 2000
IDEG/WP4/SMG4	Friedhelm Hillebrand	March 1987	April 1988
	Graham Crisp	April 1988	October 1991
	Michael Krumpe	October 1991	June 1992
	Wolfgang Roth	June 1992	February 1997
	Kevin Holley	February 1997	July 2000

[1] Editor: Friedhelm Hillebrand.

Group (plenary and permanent groups reporting to the plenary[a])	Name	Terms of office start	Terms of office end
SMG5	Stein Hansen	January 1992	April 1993
	Juha Rapeli	January 1994	June 1997
OMEG/GSM6/SMG6	Bernd Haarpaintner		March 1991
	Gisela Hertel	March 1991	1999
	Michael Sanders	1999	July 2000
11.10 group/GSM7/SMG7	David Freeman	January 1992	January 1994
	John Alsoe	January 1994	July 1995
	Remi Thomas	July 1995	March 1999
	Jean-Marc Recouvreux	March 1999	July 2000
11.20 group/SMG8	William Jones	January 1994	January 1995
	Simon Pike	January 1995	February 1999
SIMEG/SMG9	Gerald Mazziotto	January 1988	April 1994
	Klaus Vedder	April 1994	March 2000
SEG/SMG10	Thomas Haug	May 1985	December 1988[d]
	Mike Walker	July 1995	July 2000
SQSG/SMG11	Phil Gaskell	October 1995	June 1998
	Kari Järvinen	June 1998	July 2000
SMG3 SA/SMG12	Michel Mouly	June 1997	June 1998
	Francois Courau	June 1998	July 2000
PN/PT SMG/MCC	Bernard Mallinder	January 1986	January 1989
	Eike Haase	January 1989	Autumn 1992
	Jonas Twingler	Autumn 1992	Spring 1995?
	Ansgar Bergmann	Spring 1995	March 1999
	Adrian Scrase	March 1999	July 2000

[a] The working parties were permanent and met also outside the plenaries only after the GSM#9 plenary in Berlin in September/October 1985. However in the time from the Rome meeting (GSM#4 in February/March 1984) they met regularly as ad-hoc groups during the plenary.

[b] During this period general aspects of services and facilities were treated.

[c] During this period there was a focus on hand-held requirements.

[d] The Security Experts Group was put "on ice" after the completion of the key recommendations GSM 02.09 and 03.20. It was re-activated later, when the need arose.

A3.2 Secretaries to the GSM/SMG Plenary

Name	Terms of office start	Terms of office end
Thomas Beijer	December 1982	March 1991
Bo Olsson	June 1991	April 1992
Francois Courau	April 1992	April 1996
Ansgar Bergmann	June 1996	June 2000

A3.3 List of the Chairpersons in T1P1 and JTC[2]

Group	Name	Terms of Office Start	Terms of Office End
T1P1			
T1P1 chairs	Mel Woinsky	February 1994	February 1998
	Asok Chatterjee	February 1998	Expires February 2002
T1P1 vice-chairs	Mel Woinsky	February 1991	February 1994
	Jim Papadouplis	February 1994	February 1996
	Stephen Hayes	February 1996	June 1996
	Asok Chatterjee	June 1996	February 1998
	Mark Younge	February 1998	Expires February 2002
T1P1 working groups			
T1P1.4 chair	Ed Ehrlich	April 1995	July 1996
T1P1.5 chair	Ed Ehrlich	February 1996	February 2000
T1P1.5 vice-chair	Quent Cassen	February 1996	February 1998
	Don Zelmer	February 1998	February 2000
JTC			
Co-chairs	Gary Jones	February 1993	July 1996
	Charles Cook	February 1993	April 1995
	Ed Ehrlich	April 1995	July 1996

A3.4 Officials of 3GPP[3]

TSG/WG	Position	Name	Start date	End date
CN	Convenor	Stephen Hayes	1998-12-07	1999-03-03
	Chairman	Dettner Harald	1999-03-03	2000-03-15
	Chairman	Stephen Hayes	2000-03-15	
CN 1	Chairman	Hannu Hietalahti	1998-12-07	
CN 2	Convenor	Keiijo Palviainen	2000-03-17	2000-05-26
	Convenor	Masami Yabasaki	1998-12-07	1999-03-01
	Chairman	Ian David Chalmer Park	1998-12-07	2000-03-17
	Chairman	Keijo Palviainen	2000-05-26	
CN 3	Chairman	Norbert Klehn	1999-03-01	
	Chairman	Oscar Lopez-Torres	1998-12-07	1999-03-01
CN 4	Convenor	Yun Chao Hu	2000-05-26	
CN 5	Chairman	Yun Chao Hu	1999-11-04	2000-03-17
	Convenor	Lucas Klostermann	2000-03-17	2000-05-25
	Chairman	Lucas Klostermann	2000-05-26	

[2] Editor: Don Zelmer.
[3] Editor: Adrian Scrase.

TSG/WG	Position	Name	Start date	End date
CN ITU-T	Chairman	Masami Yabusaki	2000-03-17	
GERAN	Convenor	Niels Peter Skov Andersen	2000-07-31	2001-04-02
	Chairman	Niels Peter Skov Andersen	2001-04-02	
GERAN 1	Convenor	Niels Peter Skov Andersen	2000-08-28	2001-04-03
	Chairman	Niels Peter Skov Andersen	2001-04-03	
GERAN 2	Convenor	Jean-Francois Minet	2000-09-04	2000-11-10
	Chairman	Bruno Landais	2000-11-10	
GERAN 3	Chairman	Ake Busin	2000-08-28	
GERAN 4	Convenor	Jean Marc Recouvrex	2000-08-28	2000-11-22
	Chairman	Jean-Marc Recouvreux	2000-11-23	
RAN	Convenor	Akio Sasaki	1998-12-07	1999-03-01
	Chairman	Yukitsuna Furuya	1999-03-01	2001-03-13
	Chairman	Francois Courau	2001-03-13	
RAN 1	Convenor	Yukitsuna Furuya	1998-12-07	1999-02-22
	Chairman	Mr. Antti Toskala	1999-02-22	
RAN 2	Convenor	Denis Fauconnier	1998-12-07	1999-03-08
	Chairman	Denis Fauconnier	1999-03-09	
RAN 3	Chairman	Per Willars	1998-12-07	2001-02-26
	Chairman	Martin Israelsson	2001-02-26	
RAN 4	Convenor	Howard Benn	1998-12-07	1999-02-15
	Chairman	Howard Benn	1999-02-15	
SA	Convenor	Fred Harrison	1998-12-07	1999-03-01
	Chairman	Niels Peter Skov Andersen	1999-03-01	
SA 1	Convenor	Alan Cox	1998-12-07	1999-03-10
	Chairman	Alan Cox	1999-03-10	2001-02-08
	Chairman	Kevin Holley	2001-02-08	
SA 2	Convenor	Yukio Hiramatsu	1998-12-07	1999-03-01
	Chairman	Teuvo Jarvela	1999-03-01	2001-02-26
	Chairman	Mikko Puuskari	2001-02-26	
SA 3	Chairman	Michael Walker	1998-12-07	
SA 4	Convenor	Kari Järvinen	1998-12-07	1999-03-01
	Chairman	Alain Ohana	1999-03-01	2000-06-28
	Chairman	Kari Järvinen	2000-06-28	
SA 5	Convenor	Inaki Cabrera	1998-12-07	1999-03-01
	Chairman	Albert Yuhan	1999-03-01	
T	Convenor	Sang Keun Park	1998-12-07	1999-03-01
	Chairman	Sang Keun Park	1999-03-01	
T1	Convenor	Remi Thomas	1998-12-07	1999-03-01
	Chairman	Bjarke Nielsen	1999-03-01	
T 2	Convenor	Kevin Holley	1998-12-07	1999-03-01
	Chairman	Kevin Holley	1999-03-01	
T 3	Convenor	Klaus Vedder	1998-12-07	1999-03-01

TSG/WG	Position	Name	Start date	End date
	Chairman	Klaus Vedder	1999-03-01	
PCG	Chairman	Karl Heinz Rosenbrock	1999-03-04	2000-12-31
	Chairman	Akio Sasaki	2001-01-01	

A3.5 List of the Chairpersons in the GSM MoU Group/Association and GSM Association[4]

Name	Start date	End Date
Armin Silberhorn	September 1987	March 1988
Philippe Dupuis	March 1988	September 1988
Renzo Failli	September 1988	March 1989
Ted Beddoes	March 1989	September 1989
Gunnar Fremin	September 1989	March 1990
Dick Hoefsloot	March 1990	September 1990
Petter Bliksrud	September 1990	March 1991
Miguel Menchen	March 1991	September 1991
Arne Foxman	September 1991	March 1992
Kari Marttinen	March 1992	March 1993
George Schmitt	March 1993	March 1994
Bruno Massiet du Biest	March 1994	March 1995
Mike Short	March 1995	March 1996
Gretel Holcomb Hoffman	March 1996	March 1997
Adriana Nugter	March 1997	April 1998
Richard Midgett	May 1998	April 1999
Michael Stocks	May 1999	April 2000
Jim Healy	May 2000	April 2001
Scott Fox	May 2001	April 2002

[4] Editor: Friedhelm Hillebrand.

Annex 4: Key Abbreviations

3G	Third Generation (mobile communication system)
3GPP	Third Generation Partnership Project (based on GSM and UTRA)
3GPP2	Third Generation Partnership Project 2 (based on ANSI 95 and 41)
ACTS	Advanced Communication Technologies and Services (research program of the European Union)
AMPS	Advanced Mobile Phone Service (leading analogue cellular system)
AMR	Adaptive Multi-Rate (Codec System)
ANSI	American National Standards Institute
ARIB	Association of Radio Industry Businesses
CAMEL	Customised Applications for Mobile Enhanced Logic
CDMA	Code Division Multiple Access (either general meaning or ANSI 95 CDMA)
CEPT	Conference Europénne des Postes et Télécommuniçations
CN	Core Network
CR	Change Request (to existing approved specifications)
CWTS	China Wireless Telecommunication Standards (Institute)
DCS1800	Digital Communication System 1800 (term replaced by GSM1800)
DECT	Digital Enhanced Cordless Telecommunication
Doc	Document
EDGE	Enhanced Data Rates for the GSM Evolution
EGPRS	GPRS with EDGE
ETSI	European Telecommunications Standards Institute
FPLMTS	Future Public Land Mobile Telecommunication System (term replaced by IMT-2000)
GERAN	GSM EDGE Radio Access Network
GHz	Giga Hertz
GPRS	General Packet Radio Service
GSM	Global System for Mobile Communication Groupe Spécial Mobile
GSM#1, 2, 3, etc.	GSM Plenary Meeting Number 1,2, 3, etc.
GSM1, 2, 3, etc.	(TC GSM) Working Group 1, 2, 3, etc.
GSM400, 900, 1800, 1900	GSM in 400, 900, 1800 or 1900 MHz

GSMA	GSM Association
HSCSD	High Speed Circuit Switched Data
IMT-2000	International Telecommunication 2000
IN	Intelligent Network
IPR	Intellectual Property Right
ISDN	Integrated Services Digital Network
ITU	International Telecommunication Union
MAP	Mobile Application Part
ME	Mobile Equipment (= Mobile Station without a SIM)
MexE	Mobile Application Execution Environment
MHz	Mega Hertz
MoU	Memorandum of Understanding
MS	Mobile Station (= ME with a SIM)
MSG	(ETSI TC) Mobile Standards Group
NA	ETSI TC Network Aspects
NMT	Nordic Mobile Telephone
PCG	Project Co-ordination Group (in 3GPP)
PCS	Personal Communication Services (US regulatory term)
PN	Permanent Nucleus
PRD	Permanent Reference Documents (of the GSM Association)
PT	Project Team
PTT	Post Telegraph and Telephone (Administration)
RACE	Research in Advanced Communication technologies in Europe (research program of the European Union)
RAN	Radio Access Network
SIM	Subscriber Identity Module
SMG	Special Mobile Group
SMG#1, 2, 3, etc.	SMG Plenary Meeting Number 1,2, 3, etc.
SMG1, 2, 3, etc.	SMG Sub-Technical Committee 1, 2, 3, etc.
SMS	Short Message Service
SPS	ETSI TC Signalling, Switching and Protocols
STC	Sub-Technical Committee
TACS	Total Access Communication System (European adaptation of AMPS)
TAP	Transferred account Procedure
TC	Technical Committee
TDMA	Time Division Multiple Access (either general meaning or ANSI 54/136 TDMA)
Tdoc	Temporary Document (in GSM or SMG)
TSG	Technical Specification Group (in 3GPP)
TTC	Telecommunications Technology Committee
UMTS	Universal Mobile Telecommunication System
UTRA	UMTS Terrestrial Radio Access
UTRAN	UMTS Terrestrial Radio Access Network
UWCC	Universal Wireless Communication Consortium
WAP	Wireless Application Protocol

WCDMA   Wideband CDMA
WP1, 2, 3, 4   Working Party 1, 2, 3, 4
WRC   World Radio Conference (organised by the ITU, formerly
                         called WARC)

Index

CHECK FOR ___1___ PARTS

(1 CD)